# Universitext

T0202880

Roger Godement

# Analysis II

## Differential and Integral Calculus, Fourier Series, Holomorphic Functions

 Springer

Roger Godement
Département de Mathématiques
Université Paris VII
2, place Jussieu
75251 Paris Cedex 05
France
E-mail: rgodement@aol.com


*Translator*

Philip Spain

Mathematics Department
University of Glasgow
Glasgow G12 8QW
Scotland
E-mail: pgs@maths.gla.ac.uk

Mathematics Subject Classification 2000: 26-01, 26A03, 26A05, 26A09, 26A12, 26A15, 26A24, 26A42, 26B05, 28-XX, 30-XX, 30-01, 31-XX, 32-XX, 41-XX, 42-XX, 41-01, 43-XX, 44-XX, 54-XX

Library of Congress Control Number: 2003066673

ISBN-10  3-540-20921-2 Springer Berlin Heidelberg New York
ISBN-13  978-3-540-20921-8 Springer Berlin Heidelberg New York

Springer is a part of Springer Science+Business Media
springeronline.com
© Springer-Verlag Berlin Heidelberg 2005
Printed in The Netherlands

Typesetting: by the authors and TechBooks using a Springer LaTeX macro package

Cover design: *design & production* GmbH, Heidelberg

Printed on acid-free paper      SPIN: 10866733     41/TechBooks     5 4 3 2 1 0

# Contents

# V – Differential and Integral Calculus

## § 1. The Riemann Integral

The theory of integration expounded in this Chapter dates from the XIX[th] century; it was, and remains, of great use in classical mathematics, and its simplicity has rewarded all who have written for beginners in the subject. For professional mathematicians it has been dethroned by the much more powerful, and in some respects simpler, theory invented by Henri Lebesgue around 1900, and perfected in the course of the first half of the XX[th] century by dozens of others; we present a small part of it in the Appendix to this Chapter. The "Riemann" theory expounded in this Chapter therefore has only a pedagogic interest.

### 1 – Upper and lower integrals of a bounded function

Let us first recall the definitions of Chap. II, n° 11.

A scalar (i.e. complex-valued) function $\varphi$ defined on a compact, or more generally, bounded, interval $I$ is said to be a *step function* if one can find a partition (Chap. I) of $I$ into a finite number of intervals $I_k$ such that $\varphi$ is constant on each $I_k$; no conditions are imposed on the $I_k$. Such a partition will be said to be *adapted to* $\varphi$.

When $I = (a, b)$ this is the same as requiring the existence of a finite sequence of points of $I$ satisfying

$$(1.1) \qquad a = x_1 \leq x_2 \leq \ldots \leq x_{n+1} = b$$

and such that $\varphi$ is constant on each *open* interval $]x_k, x_{k+1}[$, because the values it takes at a point $x_k$ have no connection with those it takes to the right or left of this point, and are irrelevant to the calculation of traditional integrals[1].

A sequence of points satisfying (1) is called a *subdivision* of the interval $I$. A subdivision by the points $y_h$ is said to be *finer* than the subdivision (1) when the $x_k$ appear among the $y_h$, in other words when the second subdivision is obtained by subdividing each of the component intervals in (1). The definition is similar for two partitions $(I_k)$ and $(J_h)$ of $I$: the second is said to be finer than the first if every $J_h$ is contained in one of the $I_k$, in other words if the second partition of $I$ is obtained by partitioning each of the $I_k$ themselves into intervals (namely, those $J_h$ contained in $I_k$).

If $\varphi(x) = a_k$ for every $x \in I_k$ one calls the number

$$(1.2) \qquad m(\varphi) = \sum a_k m(I_k) = \sum \varphi(\xi_k)m(I_k)$$

the *integral of $\varphi$ over $I$*, where, for every interval $J = (u,v)$, the number $m(J) = v - u$ denotes the length or *measure* of $J$, and where $\xi_k$ is any point of $I_k$. Since the $I_k$ of zero measure do not matter in (2) one can replace the partition by a subdivision (1) and write

$$(1.3) \qquad m(\varphi) = \sum \varphi(\xi_k)(x_{k+1} - x_k) \qquad \text{with } x_k < \xi_k < x_{k+1}$$

since $\varphi$ is constant, so equal to $\varphi(\xi_k)$, on $]x_k, x_{k+1}[$.

Since there are infinitely many ways of choosing the $I_k$ – every finer partition, for example, will equally be adapted to calculating the integral –, we have to show that the sum (2) does not depend on the choice of the $I_k$. So let $(J_h)$ be another partition of $I$ into intervals such that $\varphi(x) = b_h$ for every $x \in J_h$. Since each $I_k$ is the union of the pairwise disjoint intervals $I_k \cap J_h$, as is shown by the relation

$$X = X \cap I = X \cap \bigcup J_h = \bigcup X \cap J_h,$$

valid for every subset $X$ of $I$, we have

$$m(I_k) = \sum_h m(I_k \cap J_h)$$

and similarly

$$m(J_h) = \sum_k m(I_k \cap J_h)$$

where, by convention, $m(\emptyset) = 0$. Thus

$$(1.4) \qquad \sum a_k m(I_k) = \sum a_k m(I_k \cap J_h),$$
$$(1.5) \qquad \sum b_h m(J_h) = \sum b_h m(I_k \cap J_h),$$

---

[1] This is not the same in generalisations of the classical theory. See n° 30.

where, on the right hand sides, we sum over all the pairs $(k, h)$. We thus have only to prove that

$$m(I_k \cap J_h) \neq 0 \quad \text{implies} \quad a_k = b_h,$$

which is clear: on $I_k \cap J_h$, which is nonempty since its length is not zero, the function $\varphi$ is equal simultaneously to $a_k$ and to $b_h$.

This argument shows immediately that

(1.6) $$m(\lambda\varphi + \mu\psi) = \lambda m(\varphi) + \mu m(\psi)$$

for any step functions $\varphi$ and $\psi$ and constants $\lambda$ and $\mu$: consider partitions $(I_k)$ and $(J_h)$ of $I$ adapted to $\varphi$ and $\psi$, write $a_k$ for the value of $\varphi$ on $I_k$ and $b_h$ for that of $\psi$ on $J_h$, and calculate the integrals of $\varphi$, $\psi$ and $\lambda\varphi + \mu\psi$ using the intervals $I_k \cap J_h$ on which $\varphi$, $\psi$ and $\lambda\varphi + \mu\psi$ are equal respectively to $a_k$, $b_h$ and $\lambda a_k + \mu b_h$; in effect we are adding the relations (4) and (5), multiplied respectively by $\lambda$ and $\mu$, term-by-term.

Since it is clear that the integral of a positive function (i.e. one whose values are all positive) is positive, we see that

(1.7) $$\varphi \leq \psi \qquad \text{implies} \qquad m(\varphi) \leq m(\psi)$$

for real-valued $\varphi$ and $\psi$, since $m(\psi) - m(\varphi) = m(\psi - \varphi) \geq 0$ by (6) and $\psi - \varphi$ is positive.

Finally, the triangle inequality applied to (2) shows that

$$|m(\varphi)| \leq \sum |\varphi(\xi_k)| m(I_k) = m(|\varphi|) \leq \sum \|\varphi\|_I m(I_k)$$

always, where, as in Chap. III, n° 7, we write in a general way that

$$\|f\|_I = \sup_{x \in I} |f(x)|.$$

Since $\sum m(I_k) = m(I)$ we finally obtain the inequality

(1.8) $$|m(\varphi)| \leq m(|\varphi|) \leq m(I)\|\varphi\|_I.$$

This completes the "theory" of integration as it applies to step functions. It rests on two properties of lengths which are the starting point for all later generalisations:

(M 1): the measure of an interval is positive;
(M 2): measure is additive, i.e. if an interval $J$ is the union of a finite number of pairwise disjoint intervals $J_k$ then $m(J) = \sum m(J_k)$.

There are many other interval-functions which have these properties. One can, for example, choose a continuous function $\mu(x)$ which is increasing in the wide sense on $I$ and put[2]

---

[2] For an arbitrary increasing function one has to take account of its discontinuities and modify the formula to obtain a reasonable theory. See n° 32 on Stieltjes measures.

$$\mu(J) = \mu(v) - \mu(u) \quad \text{if } J = (u, v).$$

One can also take a finite or countable set $D \subset I$ and assign to each $\xi \in D$ a "mass" $c(\xi) > 0$, with $\sum c(\xi) < +\infty$, and then put

$$\mu(J) = \sum_{\xi \in J} c(\xi)$$

for every interval $J$, so that the measure of a singleton interval can very well be $> 0$; in this example property (M 2) reduces to the associativity formula for absolutely convergent series. We obtain *discrete measures* in this way.

For a "measure" $\mu$ satisfying (M 1) and (M 2) the integral of a step function is, by definition, the number $\mu(\varphi)$ given by the formula (2), replacing the letter $m$ by the letter $\mu$. For a discrete measure, one clearly finds that $\mu(\varphi) = \sum c(\xi)\varphi(\xi)$, summing over all the $\xi \in D$. These generalisations will be studied at the end of this chapter, but the reader may be interested to observe, every time we use the traditional integral, those results which depend only on the properties (M 1) and (M 2) of "Euclidean" or "Archimedean" measure, or, as one now calls it, of "Lebesgue measure" (since it was for this that Lebesgue constructed his grand integration theory) because these properties extend to the general case. Certain results which, on the contrary, use the explicit construction starting from the usual measure, mainly concern the relations between integrals and derivatives, Fourier series and integrals, partial differential equations, almost all applications to physical sciences, etc. They rest on an obvious though fundamental property of the usual measure: it is invariant under translation; see below, (2.20).

Now let us pass on to arbitrary *bounded* real functions on a bounded interval $I$ (in general compact).

Given a bounded real-valued function $f$ on $I$ there exist step functions, even constant functions, $\varphi$ and $\psi$, such that $\varphi \leq f \leq \psi$, i.e. $\varphi(x) \leq f(x) \leq \psi(x)$ for every $x \in I$. By (7) we must have $m(\varphi) \leq m(\psi)$, and every reasonable definition of $m(f)$ must satisfy $m(\varphi) \leq m(f) \leq m(\psi)$. We therefore examine the *lower* and *upper* integrals of $f$ over $I$ defined by the formulae

$$(1.9) \qquad m_*(f) = \sup_{\varphi \leq f} m(\varphi), \qquad m^*(f) = \inf_{\psi \geq f} m(\psi)$$

where $\varphi$ and $\psi$ range over the sets of step functions such that $\varphi \leq f \leq \psi$.

As we have seen in Chap. II, n° 11, we have $m_*(f) \leq m^*(f)$ since every number $m(\varphi)$ is less than the $m(\psi)$, so is less than their lower bound $m^*(f)$, which, larger than all the $m(\varphi)$, is also larger than their upper bound $m_*(f)$. Since the constant functions equal to $-\|f\|_I$ and $+\|f\|_I$ feature among the functions $\varphi$ and $\psi$ respectively, we even have

$$(1.10) \qquad -m(I)\|f\|_I \leq m_*(f) \leq m^*(f) \leq m(I)\|f\|_I.$$

Relation (6) does not extend to the lower and upper integrals of arbitrary functions; if it did, the theory of integration would finish with n° 2 of this chapter. However, we always have the inequalities

$$(1.11) \quad m_*(f+g) \geq m_*(f) + m_*(g), \qquad m^*(f+g) \leq m^*(f) + m^*(g).$$

Among the step functions less than $f+g$ are the sums $\varphi + \psi$, where $\varphi$ is less than $f$ and where $\psi$ is less than $g$; consequently, $m_*(f+g)$ is greater than all the numbers of the form $m(\varphi + \psi) = m(\varphi) + m(\psi)$. It remains to note that if $A$ and $B$ are two sets of real numbers, and if one writes $A+B$ for the set of numbers $x+y$ where $x \in A$ and $y \in B$, then

$$\sup(A+B) = \sup(A) + \sup(B)$$

with a similar relation for the lower bounds (exercise!), so that every number larger than the $x+y$ is larger than $\sup(A) + \sup(B)$. Whence the first relation (11). The second is proved in the same way, reversing the inequalities.

It is easier to show that

$$(1.12) \quad m_*(cf) = cm_*(f), \qquad m^*(cf) = cm^*(f) \qquad \text{for every } c \geq 0$$

and

$$(1.13) \qquad\qquad m_*(-f) = -m^*(f);$$

it is enough to note that multiplication by $-1$ transforms the step functions below $f$ into those above $-f$.

## 2 – Elementary properties of integrals

The most natural definition of integrable functions with real values is that they should satisfy the condition

$$m^*(f) = m_*(f),$$

the common value of the two sides then being the value of the integral $m(f)$ of $f$; one extends the definition to functions $f = g + ih$ with complex values by requiring both $g$ and $h$ to be integrable and putting

$$m(f) = m(g) + im(h).$$

This definition, adopted in the First French Edition for reasons of simplicity, has several drawbacks; in particular, it is not obvious — although, of course, true — that the absolute value $|f| = [\mathrm{Re}(f)^2 + \mathrm{Im}(f)^2]^{\frac{1}{2}}$ of a complex-valued integrable function is again integrable, as Michel Ollitrault, a reader of the First Edition, has justly remarked to me. We shall therefore abandon this definition temporarily, to recover it later, and we shall adopt a method used in the modern theory too. We shall develop it for complex-valued functions, but it will also apply to functions with values in a finite dimensional vector space, or even a Banach space, which is not the case for the first simplistic definition.

We shall say that a function $f$ is *integrable* if, for any $r > 0$, there is a step function $\varphi$ (with values in the same space as $f$ if one is integrating vector-valued functions) such that

$$(2.1) \qquad\qquad m^*(|f - \varphi|) < r.$$

If $f$ has real values this means, intuitively, that the *numerical* (and not algebraic) measure of the area in the plane included between the graphs of $f$ and $\varphi$ is $< r$; there is no point in assuming $\varphi$ "above" or "below" $f$. It comes to the same to require the existence of a sequence of step functions $\varphi_n$ such that

$$(2.1') \qquad\qquad \lim m^*(|f - \varphi_n|) = 0$$

or, as one says, which *converges in mean* to $f$. One says "in mean" because the fact that the upper integral of a positive function $h(x)$ is very small does not prevent $h$ from taking very large values on very small intervals: $10^{100} 10^{-200} = 10^{-100}$.

To define the integral of an integrable function $f$ one uses the relation (1'). By the triangle inequality we have

$$|\varphi_p - \varphi_q| \le |\varphi_p - f| + |f - \varphi_q|$$

and so

$$|m(\varphi_p) - m(\varphi_q)| = |m(\varphi_p - \varphi_q)| \le m^*(|\varphi_p - f|) + m^*(|f - \varphi_q|),$$

by (1.11). The sequence with general term $m(\varphi_n)$ therefore satisfies Cauchy's convergence criterion (Chap. III, n° 10, Theorem 13). Its limit depends only on $f$. For if $\psi_n$ is another sequence of step functions satisfying (1') the relation

$$|\varphi_n - \psi_n| \le |f - \varphi_n| + |f - \psi_n|$$

shows, in a similar way, that $m(\varphi_n) - m(\psi_n)$ tends to 0.

It is natural to call the limit of the $m(\varphi_n)$ (common to all sequences of step functions converging to $f$ in mean) the *integral* of $f$, and to denote it by $m(f)$. This kind of argument, used in many other places, is similar to the one we used to define $a^x$ for $a > 0$ and $x \in \mathbb{R}$, by approximating $x$ by a sequence of rational numbers $x_n$ and showing that the sequence $a^{x_n}$ converges to a limit independent of $x$ (Chapter IV, § 1, end of n° 2).

*If an integrable function $f$ has real (resp. positive) values then its integral is real (resp. positive).* If $f$ is real, and if in (1') one replaces $\varphi_n$ by Re $\varphi_n$ one decreases the function $|f - \varphi_n|$ and so its upper integral, so that the sequence of real functions $\mathrm{Re}(\varphi_n)$ again converges to $f$ in mean, whence the first result. If, moreover, $f$ is positive, in which case one may assume the $\varphi_n$ real, one argues in the same way, replacing the $\varphi_n(x)$ by 0 on the intervals where $\varphi_n < 0$: this can only decrease the value of $|f(x) - \varphi_n(x)|$, and so of the upper integral.

*If $f$ and $g$ are integrable then $f + g$ is integrable and*

$$m(f + g) = m(f) + m(g).$$

Take step functions $\varphi_n$ and $\psi_n$ converging in mean to $f$ and $g$, write

$$|(f + g) - (\varphi_n + \psi_n)| \le |f - \varphi_n| + |g - \psi_n|$$

to show that $\varphi_n + \psi_n$ converges to $f + g$ in mean, and use (1.6).

*If $f$ is integrable then so is $\alpha f$ for any $\alpha \in \mathbb{C}$, and $m(\alpha f) = \alpha m(f)$.* Obvious: multiply $f$ and $\varphi$ by $\alpha$ in (1) and apply (1.12).

These first results already show, for real integrable $f$ and $g$, that

$$f \le g \text{ implies } m(f) \le m(g),$$

since $0 \le m(g - f) = m(g) + m(-f) = m(g) - m(f)$.

*If $f$ is integrable then so is $|f|$, and*

(2.2) $$|m(f)| \le m(|f|) \le m(I)\, \|f\|_I$$

where, we recall, $\|f\|_I = \sup |f(x)|$ is the norm of uniform convergence on $I$ (Chap. III, n° 7). For any complex numbers $\alpha$ and $\beta$ we have $\bigl||\alpha| - |\beta|\bigr| \le |\alpha - \beta|$, whence, in the notation of (1'),

$$\bigl||f(x)| - |\varphi_n(x)|\bigr| \le |f(x) - \varphi_n(x)| \qquad \text{for all } x \in I$$

and so $m^*(|f| - |\varphi_n|) \le m^*(|f - \varphi_n|)$; this proves that $|f|$ is integrable like $f$, since the $|\varphi_n|$ are also step functions. Since the integrals of $\varphi_n$ and $|\varphi_n|$ converge to those of $f$ and $|f|$, by definition of the latter, and since (2) applies to the $\varphi_n$, one obtains the first inequality (2) in the limit. The second follows from the fact that $|f(x)| \le \|f\|_I$ everywhere on $I$, so that $m(|f|)$ is less than the integral of the constant function $x \mapsto \|f\|_I$.

*The complex-valued function $f$ is integrable if and only if the functions $\mathrm{Re}(f)$ and $\mathrm{Im}(f)$ are. If so,*

$$m(f) = m[\mathrm{Re}(f)] + i.m[\mathrm{Im}(f)].$$

Since $|\mathrm{Re}(f) - \mathrm{Re}(\varphi_n)| \le |f - \varphi_n|$, with a similar relation for the imaginary parts, it is clear that $\mathrm{Re}(f)$ and $\mathrm{Im}(f)$ are integrable if $f$ is; the relation to be shown then follows from the linearity properties already obtained; these show no less trivially that $f$ is integrable if $\mathrm{Re}(f)$ and $\mathrm{Im}(f)$ are.

*A real function f is integrable if and only if* $m^*(f) = m_*(f)$.

Suppose first that $m_*(f) = m^*(f)$. Then, for every $r > 0$ there are step functions $\varphi$ and $\psi$ framing $f$ whose integrals are equal to within $r$. Since $|f - \psi| = f - \psi \leq \varphi - \psi$ it follows that $m^*(|f - \psi|) \leq m(\varphi - \psi) < r$, whence the integrability of $f$.

Suppose conversely that $f$ is integrable and consider a step function $\varphi$ such that

$$m^*(|f - \varphi|) < r;$$

one may assume $\varphi$ real as above. Since $m^*(|f - \varphi|)$ is, by definition, the lower bound of the numbers $m(\psi)$ over all step functions $\psi \geq |f - \varphi|$, the strict inequality proves the existence of a step function $\psi$ such that

$$|f - \varphi| < \psi \qquad \& \qquad m(\psi) < r.$$

Since $\varphi - \psi \leq f \leq \varphi + \psi$ we have thus framed $f$ between two step functions whose difference has integral $\leq 2r$; so $m^*(f) = m_*(f)$. Moreover,

$$m(\varphi - \psi) \leq m^*(f) \leq m(\varphi + \psi);$$

since $f$ is integrable we already know that this relation is preserved if one replaces $m^*(f)$ by $m(f)$, whence $m(f) = m^*(f)$, since the extreme terms in the preceding relation are equal to within $2r$.

To sum up:

**Theorem 1.** *Let I be a bounded interval. (i) If the bounded functions f and g are integrable on I, then so likewise is* $\alpha f + \beta g$ *for any constants* $\alpha$ *and* $\beta$, *and*

$$(2.3) \qquad m(\alpha f + \beta g) = \alpha m(f) + \beta m(g).$$

*(ii) If f is defined, bounded and integrable on I, then the function* $|f|$ *is integrable, and*

$$(2.4) \qquad |m(f)| \leq m(|f|) \leq m(I)\|f\|_I = m(I). \sup |f(x)|.$$

*(iii) The integral of a positive function is positive.*

The standard notation

$$m(f) = \int_I f(x)dx$$

will be explained later (n° 3).

The definition of integrable functions shows immediately that, on a *compact* interval, *every regulated function is integrable*; for every $r > 0$ there exists, by the definition (Chap. III, n° 12) a step function $\varphi$ such that $|f(x) - \varphi(x)| < r$ for every $x$; then, by (1.10), $m^*(|f - \varphi|) < m(I)r$, whence the result. We shall prove later (n° 7) that, on a compact interval, every continuous function is regulated, so integrable. One hardly needs more subtle results in elementary analysis.

It is not difficult to construct non-integrable functions: it is enough to take the Dirichlet function $f(x)$ on $I$, equal to 0 if $x \in \mathbb{Q}$ and to 1 if $x \notin \mathbb{Q}$. Now, if a step function $\varphi \leq f$ is constant on the intervals $I_k$ of a partition of $I$, it must be $\leq 0$ on every nonsingleton $I_k$ since such an interval contains rational numbers where $f(x) = 0$; likewise, every step function $\psi \geq f$ must be "almost" everywhere $\geq 1$. Thus $m_*(f) = 0$ and $m^*(f) = m(I)$. The Lebesgue theory allows one to integrate the function $f$, with the same result as if one had $f(x) = 1$ everywhere, and this because $\mathbb{Q}$ is countable. It may appear bizarre to consider such functions – Newton would have said that one does not meet them in Nature[3] –, but it is one of those which led Cantor towards his great set theory, not to be confused with the trivialities of Chap. I. Even though the function in question is strange, one cannot deny it the merit of simplicity; if analysis is incapable of integrating such functions, one might begin to suspect that this is the fault of analysis and not of the function ...

We said above that the integral of a positive function is positive; could it perhaps be zero? This is one of the fundamental questions which the complete Lebesgue theory allows one to resolve. For the moment we make just two elementary remarks.

*If the integral of a* continuous positive *function f is zero, then f = 0.* For if we have $f(a) = r > 0$ for some $a \in I$, then the continuity of $f$ shows that $f(x) > r/2$ on an interval $J \subset I$ of length $> 0$; if $\varphi$ is the step function equal to $r/2$ on $J$ and to 0 elsewhere then $m(f) \geq m(\varphi) = r\, m(J)/2 > 0$.

This result (which presupposes the integrability of the continuous functions and uses the fact that, in the traditional theory, the measure of a nonempty open interval is $> 0$) does not extend to discontinuous functions. For a positive step function for example, it is clear that the integral vanishes if and only if the points where the function does not vanish are finite in number. In the much more general case of a regulated function, the apposite condition is that the set defined by the relation $f(x) \neq 0$ should be *countable* (n° 7).

Before stating the next theorem let us note that if we have real functions $f$ and $g$ defined on any set $X$ we can construct the functions

$$\sup(f, g) \ : \ x \mapsto \max[f(x), g(x)],$$
$$\inf(f, g) \ : \ x \mapsto \min[f(x), g(x)];$$

these definitions generalise in the obvious way to a finite number of functions (and even to an infinite number on replacing max and min by sup and inf) and

---

[3] We will meet them in computer science when there exist machines capable of distinguishing the rational numbers automatically from the others.

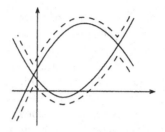

**Fig. 1.**

lead us to the *upper* and *lower envelopes* of the given functions. In particular, for every real function $f$ we can define the functions

$$
\begin{aligned}
f^+ &= \sup(f, 0) &:& \quad x \mapsto f(x)^+, \\
f^- &= \sup(-f, 0) &:& \quad x \mapsto f(x)^-, \\
|f| &&:& \quad x \mapsto |f(x)|
\end{aligned}
$$

where, for every real number, we put (Chap. II, n° 14)

$$ x^+ = \max(x, 0), \qquad x^- = \max(-x, 0); $$

it is trivial to show that, for every $x \in \mathbb{R}$,

$$ x = x^+ - x^-, \qquad |x| = x^+ + x^- $$

with similar relations for real-valued functions. An elementary argument, which Figure 1 makes obvious, shows that

$$ \sup(f, g) = f + (g - f)^+, \qquad \inf(f, g) = g - (g - f)^+; $$

these operations are defined pointwise, using only the values taken at each $x \in X$ by $f$ and $g$, so these relations follow from the same relations for real numbers. See Chap. II, n° 14, where this notation has already been used.

**Theorem 2.** *If the real functions $f$ and $g$ are integrable on $I$, so are the functions $\sup(f, g)$ and $\inf(f, g)$.*

By Theorem 1 and the formula above it is enough to show that if $f$ is integrable then so is $f^+$. This follows immediately from the definition, (1) or (1'), and from the inequality $|f^+ - \varphi^+| \le |f - \varphi|$.

The preceding "theorem" shows more generally that the upper and lower envelopes of a *finite* number of integrable real functions are again integrable. When we try to extend this result to a countable family of functions we embark on integration theory proper; see Appendix (L 16).

**Theorem 3.** *Let $f$ and $g$ be two bounded integrable functions on a compact interval $I$. Then the function $f\bar{g}$ is integrable and (Cauchy-Schwarz inequality[4])*

---

[4] Hermann Amadeus Schwarz, German mathematician of the end of the XIX[th] century. The Soviet mathematicians remarked several decades ago that one ought to

(2.5)                          $|m(f\bar{g})|^2 \leq m(|f|^2)m(|g|^2).$

In checking that $f\bar{g}$ is integrable we may assume $f$ and $g$ real, and even positive, since every integrable real function $f$ is the difference of the integrable functions $f^+$ and $f^-$. Given $r > 0$ we may choose positive step functions $\varphi'$ and $\varphi''$ framing $f$, and $\psi'$ and $\psi''$ framing $g$, both less than a fixed constant $M$ which simultaneously majorises $f$ and $g$. The product $fg$ is framed by $\varphi'\psi'$ and $\varphi''\psi''$, so we need only evaluate the integral of the difference

$$\begin{aligned} \varphi''\psi'' - \varphi'\psi' &= \psi''(\varphi'' - \varphi') + \varphi'(\psi'' - \psi') \\ &\leq M(\varphi'' - \varphi') + M(\psi'' - \psi'). \end{aligned}$$

The integrals of $\varphi'' - \varphi'$ and $\psi'' - \psi'$ can be chosen to be $< r/2M$, making that of $\varphi''\psi'' - \varphi'\psi' < r$ by a suitable choice of these functions, whence the integrability of the product.

An immediate consequence of this result is that if $f$ is integrable on $I$ and if $J \subset I$ is an interval, then the function

(2.6)            $\chi_J(x)f(x) = f(x)$  on $J$,     $= 0$  on $I - J$,

is again integrable. On multiplying step functions $\varphi_n$ converging in mean to $f$ on $I$ by the characteristic function $\chi_J$ of $J$ (Chap. I) one finds step functions converging in mean to $\chi_J f$. Since it is clear that

(2.7)                          $$\int_J f(x)dx = \int_I \chi_J(x)f(x)dx$$

is true for the $\varphi_n$ we get the same result for $f$. From this we deduce that

(2.8)                          $$\int_J f(x)dx = \sum \int_{J_p} f(x)dx$$

if the intervals $J_p$ form a partition[5] of $J$: the function $\chi_J$ is actually the sum of the characteristic functions of the $J_p$. This is the additivity (it would be

---

speak of the Cauchy-Buniakowsky-Schwarz inequality, but their ancestor being less known, even unknown, compared to the other two, the "Matthew effect" to which we have alluded in Chap. III, n° 10, has applied in his case. Moreover, in my youth, we spoke simply of the Schwarz inequality, despite the fact that Cauchy already had quite a reputation …

[5] This hypothesis is not needed in the case of the usual measure – it is enough that the intersections $J_p \cap J_q$ contain at most one point – for the integral over an interval $J \subset I$ is clearly unchanged if one adjoins the end-points to $J$. But it is essential in the case of a measure which includes discrete masses. This explains the need to integrate over bounded rather than compact intervals: it is impossible to construct a non-trivial finite partition of an interval into *compact* intervals.

better to say: the associativity) of the integral considered as a function of the interval of integration, and not of the function being integrated. This confirms in passing the existence of many interval functions that enjoy the properties (M 1) and (M 2) of n° 1: choose a positive integrable function $\rho$ and put

$$\mu(J) = \int_J \rho(x)dx;$$

physically, this is a "distribution of mass" having a "density" $\rho(x)$ at each point $x \in I$; we write $\mu(J)$ for the total mass of the interval $J$; the traditional integral is obtained when $\rho(x) = 1$, a "homogeneous" distribution of mass.

The proof of (5) is an exercise in algebra (Appendix to Chap. III) not specifically to do with integration theory; more exactly, it follows from the formal properties (i) and (iii) of Theorem 1 alone, and not from the explicit definition of an integral. We call the number

(2.9) $$(f \,|\, g) = m(f\bar{g})$$

the *scalar product* of the functions $f$ and $g$ on the interval $I$. The inequality to be established is then

(2.10) $$|(f \,|\, g)|^2 \leq (f \,|\, f)(g \,|\, g).$$

It is clear that $(f \,|\, g)$ is a linear function of $f$ for $g$ given, that $(f \,|\, g) = \overline{(g, f)}$, and that $(f \,|\, f) \geq 0$ for any $f$. For every constant $z \in \mathbb{C}$ we then have

(2.11) $\quad (f + zg, f + zg) \quad = \quad (f \,|\, f) + (f, zg) + (zg, f) + (zg, zg) =$
$$= \quad c + b\bar{z} + \bar{b}z + az\bar{z} > 0 \quad \text{for every } z \in \mathbb{C},$$

with $c = (f \,|\, f)$, $b = (f \,|\, g)$ and $a = (g \,|\, g)$, notation chosen to evoke the well-known second degree trinomials, although here the variable is complex; we know in advance that $a$ and $c$ are $\geq 0$.

If $a \neq 0$ we can put $z = -b/a$, a value for which the right hand side of (11) can be written $c - b\bar{b}/a - b\bar{b}/a + ab\bar{b}/a^2 = (ac - b\bar{b})/a$; since the left hand side of (11) is $\geq 0$ like $a$, the numerator of the result is $\geq 0$, whence (10) in this case.

If $a = 0$, the expression (11) cannot be $\geq 0$ for every $z$ unless $b = 0$, in which case (10) does not require proof. Indeed, if we replace $z$ by $tz$ with $t \in \mathbb{R}$, we must then have $(b\bar{z} + \bar{b}z)t \geq -c$ for any $t$, which forces $b\bar{z} + \bar{b}z = 0$, whence $b = 0$ since $z \in \mathbb{C}$ is arbitrary.

The Cauchy-Schwarz inequality shows that

$$(f + g \,|\, f + g) = (f \,|\, f) + (f \,|\, g) + (g, f) + (g \,|\, g) =$$
$$= (f \,|\, f) + 2\mathrm{Re}(f \,|\, g) + (g \,|\, g) \leq (f \,|\, f) + 2|(f \,|\, g)| + (g \,|\, g) \leq$$
$$\leq (f \,|\, f) + 2(f \,|\, f)^{1/2}(g \,|\, g)^{1/2} + (g \,|\, g),$$

whence, on taking the square roots of the two sides,

$$(2.12) \qquad (f + g \mid f + g)^{1/2} \leq (f \mid f)^{1/2} + (g \mid g)^{1/2}.$$

The expression

$$(2.13) \qquad \|f\|_2 = (f \mid f)^{1/2} = m\left(|f|^2\right)^{1/2} = \left(\int_I |f(x)|^2 dx\right)^{1/2}$$

is called the $L^2$ *norm* of the function $f$ on $I$; the inequality (12) shows that

$$(2.14) \qquad \|f + g\|_2 \leq \|f\|_2 + \|g\|_2$$

and clearly $\|\lambda f\|_2 = |\lambda| \cdot \|f\|_2$ for every constant $\lambda \in \mathbb{C}$. This justifies the word "norm", apart from the fact that the norm can be zero for functions which are not identically zero. The calculation preceding formula (12) also shows that

$$(2.15) \qquad (f \mid g) = 0 \Longrightarrow \|f + g\|_2^2 = \|f\|_2^2 + \|g\|_2^2,$$

the integral version of Pythagoras' Theorem; one says then that $f$ and $g$ are *orthogonal*.

We define the $L^1$ *norm* also, by

$$(2.16) \qquad \|f\|_1 = m(|f|);$$

and we again have (14) in this case, and much more easily, since $|f + g| \leq |f| + |g|$.

For every real number $p > 1$, one defines more generally the $L^p$ *norm* by

$$(2.17) \qquad N_p(f) = m\left(|f|^p\right)^{1/p} = \|f\|_p;$$

n° 14 on convex functions will show that again in this case

$$(2.18) \qquad \|f + g\|_p \leq \|f\|_p + \|g\|_p$$

and that

$$(2.19) \qquad |(f \mid g)| \leq \|f\|_p \cdot \|g\|_q \qquad \text{if } 1/p + 1/q = 1;$$

these are the famous (but, at our level, largely useless) Minkowski and Hölder inequalities.

As for the notation $L^2$, or $L^1$ or $L^p$, these allude to the "grand" integration theory. These calculations play a fundamental rôle in the theory of Fourier series, as we shall see a little below.

On several occasions we have remarked that the explicit construction of the integral does not feature in establishing Theorems 2 and 3, nor, as we shall see, in many other cases. It occurs elsewhere because of the *translation invariance* of the Euclidean measure of length. To translate this into the language of integration one writes the formula

$$(2.20) \qquad \int_{a+c}^{b+c} f(x)dx = \int_a^b f(x+c)dx,$$

to be interpreted as follows: if $x \mapsto f(x)$ is integrable on $[a+c, b+c]$, then $x \mapsto f(x+c)$ is integrable on $[a, b]$ and (20) holds. In other words, if one has an integrable function $f$ on an interval $I$ and if one submits both $I$ and the graph of $f$ to the same horizontal translation, then nothing changes. This is quite clear for step functions, and we leave the epsilontics for the reader to check.

This result may appear (and is) trivial. Yet not only is it of constant use, it characterises Euclidean measure up to a constant factor among all those measures which satisfy conditions (M 1) and (M 2) of n° 1. This is also the key to the generalisations of Fourier analysis to group theory, a boom topic for more than fifty years.

To give an application we shall use in n° 5, let us consider a function $f(x)$ of period 1 on $\mathbb{R}$ and show that

$$(2.21) \qquad \int_a^{a+1} f(x)dx = \int_0^1 f(x)dx,$$

in other words that the left hand side is independent of $a$. To do this we consider the integer $n$ such that $a \le n < a+1$. By the additivity formula (8), the integral over $[n, n+1]$ is the sum of the integrals over $[n, a+1]$ and $[a+1, n+1]$. By (20), the second is also the integral over $[a, n]$ of the function $x \mapsto f(x+1) = f(x)$. The integral over $[n, n+1]$, equal for the same reason of periodicity to the integral over $[0, 1]$, is thus the sum of the integrals of $f$ over $[a, n]$ and $[n, a+1]$, which is the integral on $[a, a+1]$, qed.

## 3 – Riemann sums. The integral notation

The relation (1.2) or (1.3) allows one to show how to calculate the integral of a complex-valued function $f$ approximately from the *Riemann sums* (or Cauchy, not to go back to Fermat or even to Archimedes ...). Assume $f$ *regulated*, enough for elementary use, and, given a number $r > 0$, let $(I_k)$ be a partition of $I$ into intervals on each of which $f$ is constant to within $r$. Choose a $\xi_k \in I_k$ at random in each $I_k$ and consider the step function $\varphi$ which on each $I_k$ takes the value $c_k = f(\xi_k)$; now $|f(x) - \varphi(x)| < r$ for each $x \in I$, so $\|f - \varphi\|_I < r$, whence, by (2.4),

(3.1)
$$\left| m(f) - \sum m(I_k)f(\xi_k) \right| \leq m(I)r.$$

On replacing this partition by a subdivision

(3.2)
$$a = x_1 \leq x_2 \leq \ldots \leq x_{n+1} = b$$

of $I$ as in n° 1, and choosing a point $\xi_k$ at random in the *open* interval $]x_k, x_{k+1}[$, we obtain

(3.3)
$$\left| m(f) - \sum f(\xi_k)(x_{k+1} - x_k) \right| < m(I)r;$$

the fact that a singleton interval is of zero measure, which does not feature in deriving (1), justifies (3) in the case of the usual measure. We may note that this argument applies *verbatim* to vector-valued functions.

What is more, these inequalities remain valid for every partition finer than $(I_k)$; for they rely only on the hypothesis that $f$ is constant to within $r$ on each of these intervals, a hypothesis true also for every partition finer than $(I_k)$.

Relation (3) explains the notation

$$m(f) = \int_I f(x)dx = \int_a^b f(x)dx$$

used to denote an integral. In this notation,

$$(f \mid g) = \int_I f(x)\overline{g(x)}dx, \quad \|f\|_2 = \left( \int_I |f(x)|^2 dx \right)^{1/2}, \quad \|f\|_1 = \int_I |f(x)|dx.$$

The analogy with the notation for series would be complete if one wrote

$$\int_{x=a}^{x=b} f(x)dx \text{ or } \int_{a \leq x \leq b} f(x)dx \text{ or } \int_{x \in I} f(x)dx.$$

It seems quite curious that the sign $\int$, invented by Leibniz in 1675, appeared fully 150 years before the sign $\sum$ of which one finds no trace in Fourier nor in Cauchy's *Cours d'analyse* of 1821. On the other hand, Leibniz and his XVIII$^{\text{th}}$ century successors never wrote the limits of integration explicitly, which can be rather a nuisance; the modern notation appeared in Fourier's *Théorie analytique de la chaleur* of 1822; but in 1807, when he was composing his fundamental memoir, refused by the *Académie des sciences*, Fourier still wrote, for example, $S(\sin .x\varphi x dx)$ for what we now write as

$$\int_0^{2\pi} \varphi(x) \sin x \, dx.$$

Leibniz' notation is explained by his conception of the integral, inherited from certain of his predecessors and notably from the Italian Cavalieri. For them it was to calculate the area bounded by the axis $Ox$, the graph of $f$, and the verticals through the end-points of $I$. They imagined $I$ to be composed of "infinitely small" or "indivisible" intervals, which Leibniz denoted by $(x, x + dx)$ and, consequently, that the area to be calculated is composed of infinitely thin vertical slices having these intervals for bases and the numbers $f(x)$ as their heights. The area of such a slice is "clearly" $f(x)dx$, so that the area to be calculated is the "continuous sum" (in contrast to the "discrete sum", i.e. to the series) of these infinitesimal areas; whence the notation, in which the sign $\int$ is an abbreviation of the word "sum" or of its Latin equivalent. All this is metaphysics. But since, three centuries after Leibniz, Mankind has not felt the need to change his notation, whether dealing with integrals for neophytes or with their most abstract generalisations, it looks as though no one knows how to do better.

Before Leibniz, Cavalieri used the word "omnia", all, or "omn.", instead of the sign $\int$; after reading Cavalieri, Leibniz wrote in 1675 in a Latin that one can understand untaught, "Utile erit scribi $\int$ pro omn. ut $\int l$ pro omn. $l$ id est summa ipsorum $l$" (Cantor, III, p. 166; *chez* Cavalieri, one adds the lengths, denoted $l$). Others, like Wallis and Newton, wrote a square before the integrand[6], as in the formula

$$\square x^2 = b^3/3 - a^3/3,$$

the square evoking the word "quadrature" which, at the time, meant precisely: to construct a square whose area is equal to the area bounded by a curve, as in the problem of the "quadrature of the circle". Here again we see to what extent the choice of good notation can contribute to the advancement and to the comprehension of mathematics.

Further, Leibniz' notation led directly to the definition of the integral given by Cauchy. Instead of considering the infinitesimal expressions $f(x)dx$ Cauchy used a subdivision of $I$ as above, and considered the sum

$$\sum f(x_k)(x_{k+1} - x_k),$$

traditionally denoted $\sum f(x_k)\Delta x_k$ because the letter $\Delta$ is the initial of the word "difference". The integral of $f$ is, for him, the limit of these sums as the subdivision becomes finer – which is indeed the case, as we shall see, for continuous functions.

## 4 – Uniform limits of integrable functions

The relation

---

[6] During his controversies with Leibniz at the start of the XVIII[th] century, Newton claimed to be the first to have invented a symbol to denote an integral. Quite possible, but his was perfectly unusable, principally because of its typographical clumsiness. Leibniz' notation is furthermore neatly adapted to the change of variable formula, to multiple integrals, etc.

(4.1)
$$|m(f)| \leq m(|f|) \leq m(I)\|f\|_I,$$

valid for every bounded integrable function $f$ on a compact (or more generally bounded) interval $I$, is fundamental; it allows one, in many situations, to argue without recourse to the explicit construction of the integral expounded in n° 1 and 2. Here is an immediate consequence:

**Theorem 4.** *Let $(f_n)$ be a uniformly convergent sequence of integrable functions on a bounded interval $I$. Then the function $f(x) = \lim f_n(x)$ is integrable and*

(4.2)
$$m(f) = \int_I f(x)dx = \lim \int_I f_n(x)dx = \lim m(f_n).$$

For $r > 0$ given, and for every $n$, let us choose a step function $\varphi_n$ such that $m(|f_n - \varphi_n|) < r$, and let $N$ be an integer such that

$$n > N \Longrightarrow \|f - f_n\|_I < r$$

from the definition of uniform convergence. For $n > N$ we have

$$
\begin{aligned}
m^*(|f - \varphi_n|) &\leq m^*(|f - f_n|) + m^*(|f_n - \varphi_n|) \\
&\leq m(I)r + r,
\end{aligned}
$$

whence the integrability of $f$. Now (4.2) follows from the fact that

$$|m(f) - m(f_n)| \leq m(|f - f_n|) \leq m(I)\|f - f_n\|_I < m(I)r,$$

qed.

Proper integration theory will allow us to prove a much stronger result than the preceding: one can replace uniform convergence by simple convergence (and even much less) on condition that one assumes that there is an *integrable* function $g \geq 0$ such that

$$|f_n(x)| \leq g(x)$$

for every $n$ and every $x$ (Appendix, L19). The limit function $f$, though integrable in the modern sense of the term, need not be so in the archaic sense expounded here, even if the $f_n$ and $g$ are. Nevertheless this can happen, in which case we have a result for Riemann integrals:

**(Dominated convergence).** *Let $(f_n)$ be a sequence of functions defined and integrable on an interval $I$; assume that (i) the $f_n$ converge simply to an integrable function $f$; (ii) there exists an integrable function $g$ such that*

$$|f_n(x)| \leq |g(x)| \quad \text{for every } n \text{ and every } x \in I.$$

*Then*

$$m(f) = \lim m(f_n).$$

Since we cannot prove this very handy result yet, simple in appearance though it is – it is the analogue for "continuous sums" of the theorems on passage to the limit for sequences of normally convergent series, Chap. III, n° 13 and Chap. IV, n° 12 – , we shall not use it, except, sometimes, to show how it would greatly simplify those "elementary" proofs that require recourse to uniform convergence. The necessity of a hypothesis such as (ii) is quite clear from Figure 2: the functions $f_n$ converge simply to 0 but their integrals are all equal to 1.

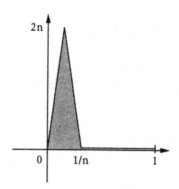

**Fig. 2.**

Theorem 4 is nevertheless prodigiously useful as we shall see immediately and in the following n°. In particular, it applies to a uniformly convergent, or *a fortiori* normally convergent, series $\sum u_n(x)$ of integrable functions: the sum of such a series is again integrable and

$$(4.3) \qquad m\left(\sum u_n\right) = \sum m(u_n),$$

i.e.

$$(4.3') \qquad \int_I \left[\sum u_n(x)\right] dx = \sum \int_I u_n(x) dx,$$

the series on the right hand side being convergent and even, in the case of normal convergence, absolutely convergent, since

$$|m(u_n)| \leq m(I)\|u_n\|_I$$

with $\sum \|u_n\|_I < +\infty$ by hypothesis.

*Example 1.* Consider a power series

$$(4.4) \qquad f(z) = \sum c_n z^n / n! = \sum c_n z^{[n]}$$

which converges on a disc $|z| < R$ of nonzero radius, and let us calculate the integral of $f(x)$ over an interval $[a, b]$ with $-R < a < b < R$. We know that the series converges normally on every disc $|z| \leq r < R$, so on the interval considered: we can therefore integrate term-by-term. But we also know (Chap. II, n° 11) that

$$(4.5) \qquad \int_a^b x^n dx = \frac{b^{n+1}}{n+1} - \frac{a^{n+1}}{n+1}$$

or, equivalently, that

$$(4.5') \qquad \int_a^b x^{[n]} dx = b^{[n+1]} - a^{[n+1]}.$$

Thus we find

$$(4.6) \qquad \int_a^b f(x) dx = F(b) - F(a)$$

where

$$F(z) = \sum c_n z^{[n+1]} = \sum c_{n-1} z^{[n]}$$

is the *primitive* power series of $f$ in the sense of Chap. II, n° 19. Since we also know that the function $F$ is differentiable (in the complex sense on $\mathbb{C}$ so *a fortiori* in the real sense on $\mathbb{R}$) and that $f$ is its derivative, (6) is just a particular case of the "fundamental theorem" which we will establish later:

$$F' = f \implies \int_a^b f(x) dx = F(b) - F(a).$$

It was this kind of calculation that led Newton and Mercator to the series

$$(4.7) \qquad \log(1+x) = x - x^2/2 + x^3/3 - \dots$$

They knew (and we know: Chap. II, n° 11) that the left hand side is the integral of the function $t \mapsto 1/(1+t)$ over the interval $[0, x]$ and they were aware of (5). The calculation is then obvious, particularly if one does not worry about Theorem 4 any more than they did. Conversely, if one first knew the formula (7), one might deduce that the integral of the function $1/(1+x)$ between $x = a$ and $x = b$ is equal to $\log(1+b) - \log(1+a)$, but this assumes that $a$ and $b$ lie in the interval of convergence of (7): a direct calculation, in Chap. II, n° 11, has already provided the result free from this restriction. The reader may amuse himself by applying (6) in the same way to the series representing $e^x$, $\sin x$, $\cos x$, etc., since their primitive series can be calculated immediately; one obtains the formulae

$$\int_a^b \cos x . dx = \sin b - \sin a, \qquad \int_a^b \sin x . dx = \cos a - \cos b,$$

$$(4.8) \qquad \int_a^b e^{tx} dx = (e^{tb} - e^{ta})/t \qquad (t \in \mathbb{C}, \ t \neq 0),$$

$$\int_a^b dx/(1+x^2) = \arctan b - \arctan a,$$

etc., which confirm the "fundamental theorem".

To conclude these trivialities on uniform limits we remark that the relation

(4.9)                $$\lim f_n = f \implies \lim m(f_n) = m(f),$$

valid in the framework of the uniform convergence, remains so under much less restricted hypotheses.

If one replaces $g$ by 1 in the Cauchy-Schwarz inequality one obtains the relation

(4.10)                $$|m(f)| \leq m(I)^{1/2}\|f\|_2.$$

From this one deduces that

(4.11)                $$\lim \|f_n - f\|_2 = 0 \implies \lim m(f_n) = m(f);$$

the same result holds for the $L^p$ norms, $p > 1$, and for the norm $\|f\|_1$ since

$$|m(f_n) - m(f)| \leq \|f_n - f\|_1$$

in this case. In other words, to obtain (9) it is enough to assume that there exists a real number $p \geq 1$ such that the integral of the function $|f_n(x) - f(x)|^p$ tends to 0.

When $\lim \|f_n - f\|_p = 0$, one says that $f_n$ converges to $f$ *in mean of order* $p$ ("in mean" for short if $p = 1$, "in quadratic mean" if $p = 2$). This is clearly the case if we have uniform convergence, but the converse is false since the value of an integral has no direct connection with that of the function at a given point or even on the neighbourhood of a point. If for example we take on $I = [0, 1]$ the functions $f_n(x) = n$ for $0 \leq x \leq 1/n^2$, $= 0$ if not, then we have $m(f_n) = 1/n$ and convergence to 0. What is essential to ensure convergence in mean is that, for $n$ large, the difference $|f_n(x) - f(x)|$ should not be $> 10^{100}$ except on intervals of total length much smaller than $10^{-100}$. All electricians know this, particularly in the case $p = 2$, since, for example, to calculate the power dissipated by an electric current of variable intensity $I(t)$ passing through a resistance during an interval of time $[a, b]$, one integrates the function $I(t)^2$ over it; "surges of current" have no influence on the result if they are concentrated on sufficiently small intervals of time.

## 5 – Application to Fourier series and to power series

For a long time it was believed, and neophytes sometimes still believe, if they trust to low level books, that integrals serve to calculate areas, volumes, centres of gravity, magnetic flux, etc. This is not false, it was even positively true in the XVII$^{th}$ century, but for ages they have served quite another purpose, namely: to do mathematics, in other words, to prove theorems. At the point where we now are in expounding the theory, we still know almost nothing. Yet nevertheless ...

Consider an *absolutely convergent Fourier series* of period 1, i.e. of the form

$$(5.1) \qquad f(x) = \sum a_n e^{2\pi i n x} \quad \text{with} \quad \sum |a_n| < +\infty,$$

where one sums over all $n \in \mathbb{Z}$ and where the factor $2\pi$ has been introduced into the exponents to simplify the formulae a little. Note in passing that the Euler relation $e^{ix} = \cos x + i.\sin x$ allows us to write (1) in the more traditional form

$$f(x) = a_0 + \sum_{n=1}^{\infty} b_n \cos 2\pi n x + c_n \sin 2\pi n x$$

which is less convenient computationally.

The first problem of the theory is to calculate the coefficients in (1) from $f$. To this purpose we remark that, for any $p, q \in \mathbb{Z}$ and $a \in \mathbb{R}$, we have[7]

$$(5.2) \qquad \int_a^{a+1} e^{2\pi i p x} \overline{e^{2\pi i q x}} dx = \int_a^{a+1} e^{2\pi i (p-q) x} dx = \begin{cases} 1 & \text{if} \quad p = q \\ 0 & \text{if} \quad p \neq q \end{cases}.$$

If $p = q$ we are integrating the constant function 1. If $p \neq q$, we can put $t = 2\pi i (p - q)$ and apply (4.8); we find the variation between $x = a$ and $x = a + 1$ of the function $e^{tx}/t$; since $t$ is a multiple of $2\pi i$ this function is of period 1, so takes the same values at $a$ and $a + 1$ – there is no point in calculating them explicitly, except to increase the chance of error – so that the integral in question is zero. If, to simplify the notation, one puts

$$(5.3) \qquad \mathbf{e}_n(x) = e^{2\pi i n x} = \exp(2\pi i n x)$$

and if one uses the notation

$$(5.4) \qquad (f \,|\, g) = \int_a^{a+1} f(x) \overline{g(x)} dx = m(f\bar{g})$$

of the end of n° 2 to denote the *scalar product* of two functions $f$ and $g$ of period 1, then the preceding formulae can be written

---

[7] The integral over an interval $[a, a+1]$ depends only on the integrand if the latter is of period 1, as we saw at the end of n° 2.

$$(5.2') \qquad\qquad (\mathbf{e}_p \,|\, \mathbf{e}_q) = \begin{cases} 1 & \text{if} \quad p = q. \\ 0 & \text{if} \quad p \neq q. \end{cases}$$

With this notation the series (1) can be written

$$f(x) = \sum a_n \mathbf{e}_n(x).$$

For $p \in \mathbb{Z}$ given, let us consider the scalar product

$$(f \,|\, \mathbf{e}_p) = m(f \overline{\mathbf{e}_p}) = m(f \mathbf{e}_{-p}).$$

Since $\sum |a_n| < +\infty$, and since the exponentials are all of modulus 1 because the exponents are purely imaginary, the series

$$f(x) \overline{\mathbf{e}_p(x)} = \sum a_n \mathbf{e}_n(x) \overline{\mathbf{e}_p(x)}$$

converges normally, so can be integrated term-by-term; in view of (2') the only term which yields a nonzero result in the integration is that for which $n = p$, so finally we find the relation

$$(5.5) \qquad\qquad a_p = (f \,|\, \mathbf{e}_p) = \int_a^{a+1} f(x) e^{-2\pi i p x} dx.$$

This formula is the basis of the theory of Fourier series: one starts from a given periodic function $f(x)$, uses (5) to *define* the coefficients $a_n$ and hopes that the function $f$ is represented by the series (1). This heavenly vision of the theory does not, alas, correspond to reality once one leaves the domain of periodic functions of class $C^1$. To begin with, the series $\sum a_n$ may well not be absolutely convergent: the case of the square waves of Chap. III, n° 2.

Let us now consider two absolutely convergent Fourier series

$$f(x) = \sum a_n \mathbf{e}_n(x), \qquad g(x) = \sum b_n \mathbf{e}_n(x)$$

and calculate their scalar product. The multiplication theorem for absolutely convergent series shows that

$$f(x) \overline{g(x)} = \sum a_p \overline{b_q} \mathbf{e}_p(x) \overline{\mathbf{e}_q(x)} = \sum a_p \overline{b_q} \mathbf{e}_{p-q}(x)$$

on using the relations

$$\mathbf{e}_p(x) \mathbf{e}_q(x) = \mathbf{e}_{p+q}(x), \qquad \overline{\mathbf{e}_n(x)} = \mathbf{e}_{-n}(x) = \mathbf{e}_n(-x).$$

The double series converges unconditionally and normally since by hypothesis the series $\sum a_n$ and $\sum b_n$ converge absolutely. We can then integrate term-by-term over $[0, 1]$, whence

$$(f \,|\, g) = \sum a_p \overline{b_q} \,(\mathbf{e}_p \,|\, \mathbf{e}_q);$$

the terms for which $p \neq q$ disappear and the *Parseval-Bessel formula*

$$(5.6) \qquad (f \mid g) = \int_a^{a+1} f(x)\overline{g(x)}dx = \sum a_n \overline{b_n}$$

remains. In particular, for any $a \in \mathbb{R}$,

$$(5.7) \qquad \|f\|_2^2 = (f \mid f) = \int_a^{a+1} |f(x)|^2 dx = \sum |a_n|^2.$$

These proofs do not apply to the square wave series of Chap. III, n° 2 and n° 11, but one can always examine what the results might mean in this case. To reduce to a Fourier series of period 1, one has to replace $x$ by $2\pi x$ in the series $\cos x - \cos 3x/3 + \cos 5x/5 - \ldots$, i.e. to consider the series

$$
(5.8) \qquad
\begin{aligned}
f(x) &= \cos 2\pi x - \cos(6\pi x)/3 + \cos(10\pi x)/5 - \ldots \\
&= [e_1(x) + e_{-1}(x)]/2 - [e_3(x)/3 + e_{-3}(x)/3]/2 + \ldots
\end{aligned}
$$

whose sum[8], if one believes Fourier, is given by

$$(5.9) \qquad f(x) = \pi/4 \text{ for } |x| < 1/4, \quad = -\pi/4 \text{ for } 1/4 < |x| < 3/4,$$

and by periodicity for the other values of $x$. If one accepts (9), the formula (5) with $a = -1/4$ here gives, up to a factor $\pi/4$ and using (4.8),

$$
\begin{aligned}
a_p &= \int_{-1/4}^{1/4} e^{-2\pi i p x} dx - \int_{1/4}^{3/4} e^{-2\pi i p x} dx = \\
&= \frac{e^{-\pi i p/2} - e^{\pi i p/2}}{-2\pi i p} - \frac{e^{-3\pi i p/2} - e^{-\pi i p/2}}{-2\pi i p} = \\
&= \left(e^{\pi i p/2} - e^{-\pi i p/2}\right)/2\pi i p - e^{-\pi i p}\left(e^{\pi i p/2} - e^{-\pi i p/2}\right)/2\pi i p = \\
&= [1 - (-1)^p] \sin(p\pi/2)/\pi p,
\end{aligned}
$$

zero if $p$ is even, and equal to $2(-1)^{(p-1)/2}/\pi p$ if $p$ is odd; since we omitted a factor $\pi/4$, we finally have

$$a_p = 0 \ (p \text{ even}) \qquad \text{or} \qquad (-1)^{(p-1)/2}/2p \ (p \text{ odd}),$$

which agrees with (8). Thus here

$$\sum |a_n|^2 = \frac{1}{2}\left(1 + 1/3^2 + 1/5^2 + \ldots\right)$$

---

[8] One might be tempted to write this series in the form $\sum(-1)^{(n-1)/2} e_n(x)/2n$ where one sums over all odd $n \in \mathbb{Z}$, but this unordered sum is no more convergent than the harmonic series; only on grouping the symmetric terms do we obtain a convergent Fourier series.

since each term is repeated twice. To apply (7), we again have to calculate the integral, which is immediate since $|f(x)|^2 = \pi^2/16$ for every $x$. Hence the formula

$$(5.10) \qquad 1 + 1/3^2 + 1/5^2 + \ldots = \pi^2/8.$$

Since one knows that

$$\pi^2/6 = \sum 1/n^2 = \sum 1/(2n)^2 + \sum 1/(2n-1)^2 = \pi^2/24 + \sum 1/(2n-1)^2,$$

it remains to observe that $1/6 - 1/24 = 1/8$ to confirm that the result is indeed correct, even if the argument is unsupported for the moment; this indicates that the hypothesis of absolute convergence in (5), (6) or (7), is probably too restricting. And this is what the theory of Fourier series will show.

Now let

$$(5.11) \qquad f(z) = \sum a_n z^n$$

be a power series which converges on a disc $|z| < R$ and therefore normally on every disc $|z| \le r < R$. Consider the function $f$ on the circle of radius $r$; again putting $e^{2\pi i t} = \mathbf{e}(t)$ with $t$ real one finds

$$(5.12) \qquad f[r\mathbf{e}(t)] = \sum a_n r^n \mathbf{e}_n(t),$$

an absolutely convergent Fourier series having exponentials only of index $n \ge 0$. Therefore, by (5),

$$(5.13) \qquad \int_0^1 f[r\mathbf{e}(t)]\overline{\mathbf{e}_n(t)}dt = \begin{cases} a_n r^n & \text{if} \quad n \ge 0, \\ 0 & \text{if} \quad n < 0. \end{cases}$$

In particular, for $n = 0$,

$$(5.14) \qquad \int_0^1 f\left(re^{2\pi i t}\right)dt = a_0 = f(0),$$

which means that the "mean value" of $f$ on the circle $|z| = r$ is equal to its value at the centre of the circle, a curious property of the analytic functions. But there is better: since (13) allows us to calculate the coefficients $a_n$ from the values of $f$ on the circle of radius $r$ it must be possible to calculate $f(z)$, and not just $f(0)$, in the same way.

Calculating formally for the moment, i.e. interchanging the $\int$ and $\sum$ signs, applying (13) for a given $r < R$ and substituting in (11) for a $z$ such that $|z| < r$, we obtain

$$f(z) \;=\; \sum_0^\infty a_n z^n = \sum (z/r)^n a_n r^n = \sum_0^\infty \int_0^1 f[re(t)][z/re(t)]^n dt =$$

$$(5.15) \quad = \int_0^1 f[re(t)] \left( \sum [z/re(t)]^n \right) dt =$$

$$= \int_0^1 \frac{re(t)}{re(t) - z} \, f[re(t)] dt \qquad \text{for } |z| < r,$$

since $z$ and $r$ do not depend on the variable of integration $t$. To justify this calculation, i.e. *Cauchy's integral formula*, which we shall write in another way below, it suffices to show that we are integrating a normally convergent series over $[0,1]$, see (4.3). The factor $f[re(t)]$, bounded because it is a continuous function of $t$, is not important. The geometric series $\sum (z/re(t))^n$ must converge, which forces $|z| < r$. If this is the case, the formula $|[z/re(t)]^n| = (|z|/r)^n = q^n$ with $q = |z|/r < 1$ implies the normal convergence of the series that we are integrating, qed.

Formula (15) shows that, on the disc $|z| < r < R$, *we can calculate $f$ from its values on the circumference* $|z| = r$ using an explicit formula of the simplest kind. One normally states it in terms of $re(t) = \zeta$, a function of $t$ whose differential is

$$d\zeta = 2\pi i re(t) dt;$$

then Cauchy's formula is written, *à la* Leibniz, in the form

$$(5.16) \qquad\qquad f(z) = \frac{1}{2\pi i} \int \frac{f(\zeta) d\zeta}{\zeta - z}$$

where one integrates along the circumference $|\zeta| = r$ and where $|z| < r$. This is, as we shall see later, a "curvilinear integral" (Chap. VIII, n°s 2 and 4).

Conversely, any function $f$ that is continuous for $|z| \leq r$ and satisfies (16) for $|z| < r$ is a power series, i.e. is analytic in $|z| < r$ : compute as in (15), but in the opposite order. This will later be used to prove that *a uniform limit of analytic functions is analytic* (Chap. VII, n° 19, where a more precise result will be found).

# § 2. Integrability Conditions

### 6 – The Borel-Lebesgue Theorem

As we have seen in Chap. II, n° 11, a very simple sufficient condition for the integrability of a real function $f$ is the existence for every $r > 0$ of a step function $\varphi$ such that

$$(6.1) \qquad |f(x) - \varphi(x)| \leq r \qquad \text{for every } x \in I;$$

for then $\varphi - r \leq f \leq \varphi + r$ and since the integrals of $\varphi - r$ and $\varphi + r$ are equal to within $2rm(I)$ the relation $m_*(f) = m^*(f)$ follows.

The preceding property means that $f$ is the uniform limit of step functions, so that the integrability of $f$ would also follow from Theorem 4. The functions possessing this property, i.e. the *regulated functions* of Chap. II, n° 11, have (Chap. III, n° 12) both left and right limits at every point of $I$. In this n° and the following, we shall show that this property characterises them, if $I$ is compact.

The idea of the proof is very simple: the whole problem is to show that, for every $r > 0$, *one can decompose $I$ into a* finite *number of subintervals $I_k$ such that the given function $f$ is constant to within $r$ on each $I_k$*. This condition is clearly necessary if the condition (1) is to be satisfied; if, conversely, it is satisfied, and if one assumes, as one may, that the $I_k$ are pairwise disjoint, one obtains (1) on taking $\varphi$ to be equal to $f(\xi_k)$ on $I_k$, where $\xi_k$ is a point chosen arbitrarily in $I_k$.

Now, given a function $f$ that has right and left limits at every $x \in I$, it is very easy to construct such intervals. Since, for every $a \in I$, the limits $f(a+)$ and $f(a-)$ exist, there is an *open* interval $]a, a + r'[$ with left end-point $a$ and an open interval $]a - r'', a[$ with right end-point $a$ on which the function is constant to within[9] $r$. And of course it is constant to within $r$ on the interval $[a, a]$. If one then considers the *open* interval $U(a) = \,]a - r'', a + r'[$ one obtains the following results: (i) each $U(a)$ is the union of at most three intervals on each of which $f$ is constant to within $r$; (ii) $U(a)$ contains $a$ for every $a \in I$. The theorem at which we aim would therefore be established if we could find a *finite* number of points $a_k$ such that

$$I \subset \bigcup U(a_k)$$

since then $I$ would be the union of its intersections with the $U(a_k)$, which are composed of at most three intervals on which $f$ is constant to within $r$.

---

[9] If $a$ is the right (resp. left) end-point of $I$, one may take any number $> 0$ for $r'$ (resp. $r''$). If the function $f$ is continuous, one can even find an open interval with centre $a$ on which $f$ is constant to within $r$, but this detail does not simplify the following proof.

This kind of question certainly set mathematicians a-thinking from about 1850 onwards, at least those who were concerned about the foundations of analysis and, in particular, about the properties of continuous functions. In their research on the "grand" integration theory, Emile Borel and Henri Lebesgue came to isolate the crucial point which their predecessors (see Theorem 8 below) had more or less used, without appreciating the generality of the statement; it was later extended, like the Bolzano-Weierstrass theorem, to much more general spaces than $\mathbb{R}$ or $\mathbb{C}$ where the notion of a compact set has meaning (see for example Dieudonné, Vol. I, Chap. III, n° 16).

**Theorem 5 (Borel-Lebesgue).** *Let $K$ be a* compact *subset of $\mathbb{R}$ (resp. $\mathbb{C}$) and $(U_i)_{i \in I}$ a family of* open *sets in $\mathbb{R}$ (resp. $\mathbb{C}$). Suppose that $K$ is contained in the union of the $U_i$. Then there is a* finite *subset $F$ of the set of indices $I$ such that $K$ is contained in the union of the $U_i$, $i \in F$. This property characterises the compact subsets of $\mathbb{R}$ (resp. $\mathbb{C}$).*

First we show that if $K$ is bounded one can, for every $r > 0$, find a *finite* number of points $x_k$ of $K$ such that $K$ is contained in the union of the open balls $B(x_k, r)$. Since $K$ is certainly contained in a compact interval or square, it is clear that one can find a finite number of open balls of radius $r/2$ which *cover* $K$, i.e. whose union contains $K$. Let us choose an $x_k \in K$ in each of those of these balls $B_k$ which actually intersect $K$. Since $B_k$ is of radius $r/2$, so of diameter $r$, we have $B_k \subset B(x_k, r)$, so that the $B(x_k, r)$ cover $K$ as desired.

To prove the existence of $F$, it thus suffices to show that there exists a number $r > 0$ possessing the following property:

(\*) for every $x \in K$ the open ball $B(x, r)$ is contained in one of the $U_i$.

If this is so, then it is enough to choose a $U_i$ containing $B(x_k, r)$ for each $k$ to obtain the first assertion of the theorem.

Suppose (\*) is false. For every $n \in \mathbb{N}$ there then must exist an $x(n) \in K$ such that the ball $B(x(n), 1/n)$ is not contained in any of the $U_i$. By BW, since $K$ is *compact*, one can extract from the sequence $x(n)$ a subsequence $x(p_n)$ which converges to an $a \in K$ (Chap. III, n° 9). Since one of the $U_i$ contains $a$, and is *open*, it contains a ball $B(a, r)$ of radius $r > 0$. For $n$ large one has both $|a - x(p_n)| < r/2$ and $1/p_n < r/2$. It follows that the ball $B(x(p_n), 1/p_n)$ is contained in $B(a, r)$ and *a fortiori* in $U_i$, a contradiction.

It remains to show that the compact sets are the only ones to have the BL property. In the first place, a set $K$ which has it is bounded; indeed, $K$ is covered by the family of the open balls $B(x, 1)$, $x \in K$, since any ball contains its centre; there is then a finite number of $x_k \in K$ such that the $B(x_k, 1)$ cover $K$, whence this property.

On the other hand, $K$ is closed, i.e. contains every adherent point $a$. Let us assume the opposite and let $a \notin K$ be an adherent point of $K$. For every

$n \in \mathbb{N}$, denote by $U_n$ the set of $x \in \mathbb{R}$ (or $\mathbb{C}$) such that $d(x,a) > 1/n$, i.e. the exterior of the ball $B(a, 1/n)$. The $U_n$ are open and cover $K$: for every $x \in K$ one has $d(x,a) > 0$ since $a \notin K$, whence, for $n$ large, $d(x,a) > 1/n$, i.e. $x \in U_n$. If then $K$ has the Borel-Lebesgue property one can cover it by a finite number of sets $U_n$; but since these form an increasing sequence this means that $K \subset U_n$ for $n$ large, in other words that the closed ball $B(a, 1/n)$ complementing $U_n$ in $\mathbb{R}$ (or $\mathbb{C}$) does not meet $K$. Contradiction, since $a$ is adherent to $K$, qed.

By a curious coincidence, the essential tool in this proof is the Bolzano-Weierstrass theorem, which, as we know (Chap. III, n° 9), characterises the compact subsets of $\mathbb{R}$ or $\mathbb{C}$. We may therefore wonder whether, conversely, it is possible to deduce BW from BL, which would allow the reader to add BL to the list of the statements equivalent to the axiom (IV) of $\mathbb{R}$ (Chap. III, end of n° 10). For a proof, see Dieudonné, *Eléments d'analyse*, Vol. I, Chap. III, n° 16.

**Corollary 1.** *Let $(K_i)_{i \in I}$ be a family of nonempty compact sets in $\mathbb{R}$ or $\mathbb{C}$. Suppose that the intersection of the $K_i$ is empty. Then there is a* finite *subset $F$ of $I$ such that the intersection of the $K_i$, $i \in F$, is empty.*

We choose any index $j$ and replace each $K_i$ by $K_i \cap K_j$. If one of these intersections is empty, the corollary is proved. So assume they are nonempty. This is equivalent to assuming that all the $K_i$ are contained in the same compact set $K$, namely $K_j$.

Let $U_i$ be the complement of $K_i$ in $\mathbb{R}$ (or $\mathbb{C}$). It is open since $K_i$ is closed. The union of the $U_i$ is the complement of the intersection of the $K_i$. If this is itself empty then the $U_i$ cover $\mathbb{R}$ (or $\mathbb{C}$) and thus $K$. By BL, there exists a finite set $F \subset I$ such that the $U_i$, $i \in F$, cover $K$. The complement of the union of these $U_i$ is the intersection of the $K_i$, $i \in F$. This cannot intersect $K$; and since it is contained in $K$ it must be empty, qed.

If we put

$$K_F = \bigcap_{i \in F} K_i$$

for every finite subset $F$ of $I$ we can reformulate the preceding corollary as follows: the $K_i$ have a point in common if and only if $K_F$ is nonempty irrespective of $F$. The case where the $K_i$ are intervals in $\mathbb{R}$ has already been treated in Chap. III, n° 9.

The reader will perhaps wonder why it is necessary to assume the $U_i$ open in the BL theorem. A trivial counterexample: cover $K$ by the closed sets $\{x\}$, $x \in K$; if $K$ is infinite it is clearly impossible to cover it by a finite number of such sets. One might prefer a less crude counterexample. Take $K = [-1, 1]$ and cover it by the intervals $]1/2, 1], ]1/3, 1/2], \ldots$ and $[-2, 0]$. Every $x > 0$ in $K$ belongs to one and only one interval $]1/n, 1/(n+1)]$, and every $x < 0$ to the interval $[-2, 0]$; the obstacle would fall if one had chosen $[-2, r]$ with an $r > 0$.

Another important consequence of BL is the *local* character of uniform convergence on a compact set:

**Corollary 2.** *Let $X$ be a subset of $\mathbb{C}$ and $(f_n)$ a sequence of scalar functions defined on $X$, converging simply to a limit $f$. Assume that, for every $a \in X$, there exists a ball $B(a)$ of centre $a$ such that the $f_n$ converge to $f$ uniformly on $B(a) \cap X$. Then the $f_n$ converge uniformly on every compact $K \subset X$ ("compact convergence" on $X$).*

We may assume the $B(a)$ open. By BL, one can cover $K$ by a finite number of balls $B(a_i)$. For $r > 0$ given, the assertion

$$(6.2) \qquad |f_n(x) - f(x)| < r \qquad \text{for every } x \in B(a_i) \cap K$$

is, for each $i$, true for $n$ large. Since, for $r$ given, these relations are finite in number, they are thus *simultaneously* true for $n$ large (Chap. II, n° 3), and since the union of the $B(a_i) \cap K$ is $K$, it follows that, for $n$ large, the inequality (2) is true for all the $x \in K$ simultaneously, qed.

Corollary 2 is particularly useful in the theory of analytic functions; $X$ is then an open subset of $\mathbb{C}$ and it is often easy to show that, for every $a \in X$, the convergence of the $f_n$ is uniform on a sufficiently small disc with centre $a$, whence compact convergence on $X$.

## 7 – Integrability of regulated or continuous functions

The arguments which led us to formulate the BL theorem at the beginning of the preceding n° lead to the following result:

**Theorem 6.** *Let $f$ be a scalar function defined on an interval $I$ of $\mathbb{R}$. The two following properties are equivalent: (i) $f$ has left and right limits at every point of $I$; (ii) there is a sequence of step functions on $I$ which converges to $f$ uniformly on every compact subset of $I$. The function $f$ is then continuous on the complement of a countable subset of $I$.*

The implication (ii) $\Longrightarrow$ (i) was established from Cauchy's criterion in Chap. III, n° 12 (Corollary of Theorem 16). The implication (i) $\Longrightarrow$ (ii) is obtained, when $I$ is compact, by observing, as at the beginning of the preceding n°, that for every $r > 0$, there exists for every $x \in I$ an open interval $U(x) = ]x - r'', x + r'[$ such that $f$ is constant to within $r$ on each of the three intervals $]x - r'', x[$, $[x, x]$ and $]x, x + r'[$; it remains only to apply BL to the $U(x)$ to obtain a finite number of intervals covering $I$ and on each of which $f$ is constant to within $r$; this argument also shows that $f$ is *bounded on every compact* $K \subset I$.

In the case of a not necessarily compact interval $I$ one clearly has to work on an arbitrary compact interval $K$ contained in $I$. One sees then that the following two properties are equivalent for a scalar function $f$ defined on $I$:

(i)  $f$ has right and left limits at every point of $I$, in other words, by definition, is *regulated on $I$*;
(ii)  for every compact interval $K \subset I$ there exists a sequence of step functions on $K$ which converges to $f$ uniformly on $K$.

One can then find a sequence $(\varphi_n)$ of step functions on $I$ (i.e. such that one can partition $I$ into a finite number of intervals on each of which the function is constant) which, for every compact $K \subset I$, converges to $f$ uniformly on $K$: choose an increasing sequence of compact intervals $K_n$ with union $I$ and, for each $n$, a step function $\varphi_n$ on $K_n$ satisfying $|f(x) - \varphi_n(x)| < 1/n$ for every $x \in K_n$, and then define $\varphi_n$ on all of $I$ by agreeing that $\varphi_n(x) = 0$ for every $x \in I - K_n$. Again $\lim \varphi_n(x) = f(x)$ for every $x \in I$ because $x \in K_p$ for $p$ large, whence $|f(x) - \varphi_n(x)| < 1/n$ for every $x \in K_p$ and every $n \geq p$ since then $K_p \subset K_n$.

It remains to prove the continuity of $f$. For every $n$, let $D_n$ be the finite set of the points of $I$ where $\varphi_n$ is discontinuous. The union $D$ of the $D_n$ is countable (Chap. I) and the $\varphi_n$, as functions defined on $I$, are all continuous at every $x \in I - D$. Similarly[10] for $f$, qed.

Note that the theorem applies to monotone functions in particular.

**Corollary.** *Every bounded and regulated function $f$ on a bounded interval $I = (a, b)$ is integrable on $I$, and*

(7.1)  $$\int_a^b f(x)dx = \lim_{u \to a+, v \to b-} \int_u^v f(x)dx.$$

Choose an $r > 0$ and a compact interval $K = [u, v]$ contained in $I$ and such that $m(I) - m(K) < r$. By Theorem 6 the function $f$ is integrable on $K$. There must therefore exist on $K$ a step function $\varphi$ such that $m_K(|f - \varphi|) < r$, where $m_K$ is the integral over $K$. We define a step function $\varphi'$ on $I$ by requiring it to be equal to $\varphi$ on $K$ and zero off $K$. Since $|f(x) - \varphi'(x)| \leq \|f\|_I$ off $K$, we see, separating the contributions over $K$ and $I - K$, that

$$\int_I |f(x) - \varphi'(x)| \, dx \leq r + [m(I) - m(K)] \|f\|_I \leq (1 + \|f\|_I) r,$$

whence $f$ is integrable on $I$. Relation (1) follows on remarking that the difference between the integrals over $I$ and $[u, v]$ is the sum of the integrals over $]a, u]$ and $[v, b[$, intervals whose lengths tend to 0, qed.

We will rediscover this in § 7 *à propos* the integration of not necessarily bounded functions over arbitrary intervals. Up to then integrals over a compact interval will almost always be sufficient for our needs, but it is good to know that in spite of its unorthodox behaviour on a neighbourhood of 0, the function $\sin(1/x)$ is integrable over $]0, 1]$ in the sense of n° 2, the most elementary that there is.

---

[10] To avoid all confusion, recall that we are dealing with the continuity of $f$ as a function on $I$ and not only on $I - D$. See n° 5 of Chap. III again.

The arguments showing that (i) $\implies$ (ii) also serve to establish the following result, already mentioned in n° 3:

**Theorem 7.** *The integral of a regulated (resp. continuous) positive function $f$ is zero if and only if the set $D = \{f(x) \neq 0\}$ is countable (resp. empty).*

The condition is sufficient. For consider a step function $\varphi \leq f$. One can have $\varphi(x) > 0$ only if $x \in D$. Since the set of points of a nonsingleton interval is uncountable (Chap. I), the function $\varphi$ is necessarily negative on all the intervals of nonzero length where it is constant. Thus $m(\varphi) \leq 0$ and, since $m(f)$ is the upper bound of these $m(\varphi)$, we also have $m(f) \leq 0$, whence $m(f) = 0$ since $f$ is positive.

To show that it is necessary, we assume first that $I$ is compact and construct a sequence $(\varphi_n)$ of step functions on $I$ such that $\|f - \varphi_n\|_I \leq 1/n$. Replacing $\varphi_n$ by $\varphi_n - 1/n$, we may assume that $\|f - \varphi_n\|_I \leq (1 + m(I))/n$ and $\varphi_n \leq f$. Since $f \geq 0$ one can even assume $\varphi_n \geq 0$ (replace them by the $\varphi_n^+$). Then $m(\varphi_n) = 0$ since $m(f) = 0$. Each $\varphi_n$ is then zero outside a finite set $D_n$. The union $D$ of the $D_n$ is countable (Chap. I) and since $f(x) = \lim \varphi_n(x)$ for every $x \in I$ it is clear that $f(x) = 0$ for every $x \notin D$.

If now $I$ is not compact it is the union of a sequence of compact $K_n$. The integral of $f$ over each $K_n$ is clearly zero; the $D \cap K_n$ are therefore countable, so $D = \bigcup D \cap K_n$ (Ch. I) is too.

If $f$ is continuous then $D$ is open and so, if not empty, must contain an interval of length $> 0$, which would have to be countable like $D$, contrary to Cantor's most famous theorem, qed.

A corollary of Theorem 7 is that if two regulated functions $f$ and $g$ are equal outside a countable set $D$ then $m(f) = m(g)$. For the function $|f - g|$ is again regulated[11] and it is positive; Theorem 7 then shows that $m(|f-g|) = 0$, whence $m(f) = m(g)$.

One might be tempted to believe that conversely, if one modifies the values of a regulated function $f$ on a countable set $D$ of points, one will again find an integrable or even regulated function. False: the constant function equal to 1 is as regulated as it is possible to be, but if you change it to have the value 0 at rational points you will obtain the Dirichlet function which is neither regulated nor Riemann integrable.

## 8 – Uniform continuity and its consequences

The principal interest of Theorem 6 is to show that every regulated function is integrable. In particular this is the case for continuous functions. The proof

---

[11] Obvious. Note, in this circle of ideas, that if $f$ is regulated and if $g$ is *continuous* then the composite function $g \circ f$ is again regulated, since if $x$ tends to $c+$ or $c-$, then $f(x)$ tends to $f(c+)$ or $f(c-)$, so that $g[f(x)]$ tends to a limit, namely $g[f(c+)]$ or $g[f(c-)]$, qed. This result may not follow if $g$ is only regulated.

of the implication (i) $\Longrightarrow$ (ii) of Theorem 6 allows us to isolate an important property that they have, namely *uniform continuity*.

Consider, generally, a scalar function $f$ defined and continuous on a subset $X$ of $\mathbb{R}$ or $\mathbb{C}$. For every $r > 0$ and every $x \in X$ there exists a number $r' > 0$ such that, for $y \in X$,

$$(8.1) \qquad d(x, y) \le r' \Longrightarrow d[f(x), f(y)] \le r.$$

The number $r'$ depends *a priori* on the choice of $r$ *and of* $x$. One says that $f$ is uniformly continuous on $X$ if, for every $r > 0$, you can choose the same $r' > 0$ for *all* $x \in X$, so that

$$(8.2) \qquad \{(x \in X) \ \& \ (y \in X) \ \& \ (d(x, y) \le r')\} \Longrightarrow d[f(x), f(y)] \le r.$$

Suppose for example that $X = \mathbb{R}$ and let us put $y = x - h$. Then (2) means that

$$(8.3) \qquad |h| \le r' \Longrightarrow |f(x - h) - f(x)| \le r \ \text{ for every } x \in \mathbb{R}.$$

Now let us introduce the *translated* functions

$$(8.4) \qquad f_h(x) = f(x - h)$$

of $f$ whose graphs are derived from the graph of $f$ by horizontal translations. This said, the fact that $d[f_h(x), f(x)] \le r$ *for every $x$* means simply, in the notation of Chap. III, n° 7, that

$$(8.5) \qquad d_{\mathbb{R}}(f, f_h) = \|f - f_h\|_{\mathbb{R}} \le r.$$

The existence, for every $r > 0$, of an $r' > 0$ satisfying (3) thus means that as $h$ tends to 0 the function $f_h(x)$ converges to $f(x)$ *uniformly on* $\mathbb{R}$. One would like to formulate uniform continuity on an arbitrary set $X$ in a similar way, but in this case the function $f_h(x)$ is defined only on the set $\neq X$ formed from $X$ by the horizontal translation of amplitude $h$, and convergence, uniform or not, no longer has a meaning.

Uniform continuity is very far from being a universal property of continuous functions. If you take the function $f(x) = e^x$ on $\mathbb{R}$ for example, when $f_h(x) = e^{-h} f(x)$, it is clear that, as $h$ tends to 0, $f_h$ converges simply to $f$ – this is continuity –, but for a given $h$ the difference $|f(x) - f_h(x)| = |e^{-h} - 1| e^x$ is not even bounded on $\mathbb{R}$, which rules out uniform convergence: in this case $\|f - f_h\|_R = +\infty$ for any $h \neq 0$.

We always have:

**Theorem 8 (Heine[12]).** *Every scalar function defined and continuous on a compact set $K \subset \mathbb{C}$ is uniformly continuous on $K$.*

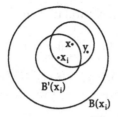

Fig. 3.

Given $r > 0$ let us choose for each $x \in K$ an open ball $B(x)$ with centre $x$ such that $f$ is constant to within $r$ in $B(x) \cap K$. Let $B'(x)$ be the open ball with centre $x$ and of radius half that of $B(x)$. Since the $B'(x)$ cover $K$, one can, by BL, find points $x_1, \ldots, x_n$ of $K$ such that the balls $B'(x_i)$ cover $K$. Let $r' > 0$ be the smallest of their radii, and let $x, y$ be two points of $K$ such that $d(x, y) < r'$. The point $x$ belongs to one of the balls $B'(x_i)$. Since the radius of $B(x_i)$ is twice that of $B'(x_i)$, itself $\geq r'$, the ball $B(x_i)$ contains $y$ too, by the triangle inequality. Since $f$ is constant to within $r$ on $B(x_i)$ we have $|f(x) - f(y)| \leq r$, qed.

**Corollary 1.** *Let $f$ be a scalar function defined and continuous on $\mathbb{R}$ (resp. $\mathbb{C}$) and zero for $|x|$ large. Then $f$ is uniformly continuous on $\mathbb{R}$ (resp. $\mathbb{C}$).*

We need only treat the case of $\mathbb{C}$. Let $K$ be a compact set outside which $f = 0$, and $H$ the set of $x \in \mathbb{C}$ such that $d(x, K) \leq 1$. Since $d(x, K)$ is a continuous function of $x$ (Chap. III, n° 10), the set $H$ is closed. It is clearly bounded like $K$, so is compact. For every $r > 0$ there is thus an $r' > 0$ such that, for $x, y \in H$, the relation $d(x, y) \leq r'$ implies $d[f(x), f(y)] \leq r$. We may assume $r' < 1$. Now let $x, y$ be two points of $\mathbb{C}$ such that $d(x, y) < r'$. If both are in $H$, the question is settled. If $x \notin H$ we have $d(x, K) > d(x, y)$, so $y \notin K$, whence $f(x) = f(y) = 0$, qed.

It is easy to understand why Theorem 8 does not apply to noncompact sets. Consider such a set $X$ and a uniformly continuous function $f$ on it; and let $a$ be an adherent point of $X$; then $f$ *tends to a limit when $x \in X$ tends to $a$*. For, take an $r > 0$; in view of Cauchy's criterion (Chap. III, n° 10, Theorem 13'), we need to prove the existence of an $r' > 0$ such that, for $x, y \in X$,

$$\{(|x - a| < r') \ \& \ (|y - a| < r')\} \Longrightarrow |f(x) - f(y)| < r.$$

But since $f$ is uniformly continuous there is an $r'' > 0$ such that the right hand inequality holds for $|x - y| < r''$; it then suffices to take $r' = r''/2$.

---

[12] Heine published in 1872, but Dugac tells us that Weierstrass had already taught the theorem in 1865, that Riemann and Dirichlet had actually used it without proof by 1854, and that it had been used implicitly by Cauchy, who had not perceived the difficulty (Chap. III, n° 6).

In these circumstances it is natural to define a function $F$ on the closure[13] $\overline{X}$ of $X$ by putting

$$F(a) = \lim_{x \to a, x \in X} f(x)$$

for every $a \in \overline{X}$; we have $F(a) = f(a)$ if $a \in X$. Let us show that *the function F is continuous on X*. For every $r > 0$ we choose an $r' > 0$ such that, for $x, y \in X$,

$$|x - y| < r' \implies |f(x) - f(y)| < r$$

and consider two points $a, b$ of $\overline{X}$ such that $|a - b| < r'$ (strict inequality). If $x, y \in X$ are sufficiently close to $a$ and $b$ respectively, we again have $|x - y| < r'$ and so $|f(x) - f(y)| < r$; since $f(x)$ and $f(y)$ tend to $F(a)$ and $F(b)$, we find in the limit that $|F(a) - F(b)| \le r$, whence the result.

This shows that the notion of uniform convergence in reality concerns only continuous functions on a *closed* set, or, equivalently, which can be extended to a closed set while remaining continuous (and even uniformly continuous). In particular:

**Corollary 2.** *Let f be a function defined and continuous on a* bounded *set* $X \subset \mathbb{C}$. *The following two properties are equivalent: (i) f is uniformly continuous on X; (ii) f is the restriction to X of a continuous function on the compact set* $\overline{X}$.

We have just seen that (i) implies (ii). The converse implication follows from Theorem 8 since $\overline{X}$ is compact.

If, for example, $X = ]0, 1]$, the function $f(x) = \sin(1/x)$ manifestly has no limit when $x$ tends to 0; this does not prevent it from being integrable since it is continuous and bounded (Corollary to Theorem 6), but does prevent it from being uniformly continuous on $X$. To verify this by a subtle use of inequalities is a gymnastic exercise in the Weierstrass tradition; Corollary 2 makes this quite unnecessary: there are enough serious occasions for dealing with inequalities that one prefers not to when one can obtain the result free. One might, otherwise, advise the amateurs to examine such functions as

$$\sin(\sin(1/x)), \quad \sin(\exp(\sin(1/x))), \quad \text{etc.}$$

"by hand".

Corollary 2 allows us to answer an approximation problem: can one approximate a given continuous function on $X$ by polynomials uniformly on $X$? We shall show in n° 28 that this is so if $X$ is a *compact* interval in $\mathbb{R}$ (or $\mathbb{C}$ so long as one uses polynomials in $x$ and $y$, and not in $z = x + iy$). But if $X$ is bounded without being compact?

Let $p$ be a polynomial satisfying $|f(x) - p(x)| \le r$ for every $x \in X$. Since the function $p$ is continuous on $\mathbb{R}$ and so on the compact closure of $X$, it

---

[13] Recall that this is the set of points that one can approximate by the $x \in X$, or, again, the smallest closed set containing $X$.

is uniformly continuous on $X$. There is therefore an $r' > 0$ such that, for $x, y \in X$, the relation $|x - y| < r'$ implies $|p(x) - p(y)| < r$ and consequently $|f(x) - f(y)| < 3r$. In other words, if $f$ is the uniform limit of polynomials (or, more generally, of uniformly continuous functions on $X$), then $f$ is uniformly continuous on $X$. Conversely, $f$ may then be extended to a continuous function on the compact set $\overline{X}$ and Weierstrass' theorem provides the desired approximation on $\overline{X}$, so *a fortiori* on $X$. The question thus lacks interest: when the answer is affirmative it results from Weierstrass' theorem for a compact set. On the other hand we have shown in Chap. III, n° 5 that if $X$ is an unbounded interval in $\mathbb{R}$ then the only uniform limits of polynomials in $X$ are the polynomials themselves. Moral: do not try to "improve" Weierstrass' theorem ...

Another consequence of Heine's theorem is the possibility of defining the integral of a continuous function $f$ over a *compact* interval $I$ by means of the standard Riemann sums.

One can, for example, like Cauchy, consider arbitrary subdivisions of $I$ and the sums $\sum f(x_k)(x_{k+1} - x_k)$ which irresistibly evoke Leibniz' notation $\int f(x)dx$ (the evocation, as concerns Cauchy, would rather go in the inverse sense ...) or even the more general sums $\sum f(\xi_k)(x_{k+1} - x_k)$ with the points $\xi_k$ chosen arbitrarily in the *closed* intervals[14] $[x_k, x_{k+1}]$. If, on each of these intervals, the function is constant to within $r$, the function $f$ is everywhere equal to within $r$ to the step function equal to $f(\xi_k)$ on $[x_k, x_{k+1}[$, so that the integral of $f$ is equal to the sum considered to within $m(I)r$. But since $f$ is uniformly continuous this condition will be satisfied so long as $|x_{k+1}-x_k| < r'$ for a suitably chosen $r' > 0$. In other words:

**Corollary 3.** *Let $f$ be a scalar function defined and continuous on a compact interval $I$. For every $r > 0$ there exists an $r' > 0$ such that*

$$(8.6) \qquad \left| \int_I f(x)dx - \sum f(\xi_k)(x_{k+1} - x_k) \right| < r$$

*for any points $\xi_k \in [x_k, x_{k+1}]$ so long as the subdivision $(x_k)$ of $I$ satisfies $|x_{k+1} - x_k| < r'$ for every $k$.*

For example one can decompose $I$ into $n$ equal intervals $I_1, \ldots, I_n$ and choose a $\xi_k \in I_k$ at random for each $k$. The corresponding Riemann sum is just

$$m(I)\frac{f(\xi_1) + \ldots + f(\xi_n)}{n} .$$

It tends to the integral of $f$ as $n$ increases indefinitely. This remark explains why the ratio $m(f)/m(I)$ between the integral of $f$ and the measure of $I$ is called the *mean value* of the function $f$ on $I$.

---

[14] For a general regulated function we have seen above that the $\xi_k$ must be *interior* to the intervals of the subdivision because the function $f$ may be discontinuous at the points $x_k$.

## 9 – Differentiation and integration under the $\int$ sign

We shall continue to explain various important consequences of uniform continuity. We have established them not only in $\mathbb{R}$, but also in $\mathbb{C}$, i.e. for functions of two real variables.

Consider a function $f(x, y)$ defined and continuous on a rectangle $K \times J$ in $\mathbb{C}$, where $K$ and $J$ are intervals in $\mathbb{R}$, and $K$ is assumed compact as its name suggests. We can integrate $f(x, y)$ with respect to $x$ for given $y$, and more generally consider the function

$$(9.1) \qquad \varphi(y) = \int_K f(x, y)\mu(x)dx$$

where $\mu$ is an arbitrary integrable function on $K$ (if not a Radon measure ...).

**Theorem 9.** *Let $K$ be a compact interval, $J$ an arbitrary interval of $\mathbb{R}$, and let $f(x, y)$ be a continuous function on $K \times J$. Then*

*(i)   the function (1) is continuous in $J$;*
*(ii)  if $f$ has a continuous partial derivative $D_2f(x, y)$ on $K \times J$ then $\varphi$ is of class $C^1$ on $J$ and*

$$(9.2) \qquad \varphi'(y) = \int_K D_2f(x, y)\mu(x)dx.$$

Continuity and differentiability at a point $y$ being local properties we can replace $J$ in what follows by a compact interval $H \subset J$ containing all points of $J$ sufficiently close to $y$.

Generally, put $\mu(f) = \int f(x)\mu(x)dx$ for every function $f$ continuous on $K$, whence, omitting the $K$ under the $\int$ sign,

$$|\mu(f)| \leq \int |f(x)|.|\mu(x)|dx \leq M(\mu)\|f\|_K$$

where $M(\mu) = \int |\mu(x)|dx$. Then $\varphi(y) = \mu(f_y)$ where $f_y(x) = f(x, y)$.

Since $f$ is continuous and so uniformly continuous on the compact set $K \times H$, one can associate with every $r > 0$ an $r' > 0$ such that, on $K \times H$,

$$(9.3)\, (|x' - x''| < r') \quad \& \quad (|y' - y''| < r') \Longrightarrow |f(x', y') - f(x'', y'')| < r.$$

For $|y'-y''| < r'$ one then has $|f(x, y')-f(x, y'')| < r$, i.e. $|f_{y'}(x) - f_{y''}(x)| < r$, for every $x \in K$; consequently,

$$(9.4) \qquad |y' - y''| < r' \Longrightarrow \|f_{y'} - f_{y''}\|_K \leq r.$$

This means that, as $y''$ tends to $y'$, the function $f_{y''}$ converges to $f_{y'}$ *uniformly on $K$.* The continuity of $\varphi$ follows from this since

$$|\varphi(y') - \varphi(y'')| = |\mu(f_{y'}) - \mu(f_{y''})| \leq M(\mu)\|f_{y'} - f_{y''}\|_K \leq M(\mu)r$$

for $|y' - y''| < r'$.

As to differentiability, we put $D_2 f = g$, $g_y(x) = g(x, y)$ and denote the right hand side of (2) by $\psi(y) = \mu(g_y)$, a continuous function of $y$ by (i) applied to $g$. Then

$$(9.5) \quad \frac{\varphi(y+h) - \varphi(y)}{h} - \psi(y) = \frac{\mu(f_{y+h}) - \mu(f_y)}{h} - \mu(g_y) =$$
$$= \mu\left[(f_{y+h} - f_y)/h - g_y\right]$$

by the linearity of $f \mapsto \mu(f)$. To show that the left hand side tends to 0 with $h$, it suffices to show that, as $h$ tends to 0, the function of $x$ to be integrated (in the third term) tends to 0 *uniformly on K* for $y$ given.

Now we proved in Chap. III, n° 16 (Corollary 4 of the Mean Value Theorem) that for every differentiable function $p$ on a compact interval $[a, b]$, we have

$$|p(b) - p(a) - p'(c)(b - a)| \leq |b - a|. \sup |p'(x) - p'(c)|$$

for every $c \in [a, b]$, the sup being taken over the $x \in [a, b]$. We apply this result to the function $y \mapsto f(x, y)$ for $x$ given; we obtain

$$|f(x, y + h) - f(x, y) - D_2 f(x, y)h| \leq |h|. \sup |D_2 f(x, y + k) - D_2 f(x, y)|,$$

the sup being taken over the $k$ lying between 0 and $h$. The function $D_2 f$ being continuous and so uniformly continuous on the compact set $K \times H$, there exists for every $r > 0$ an $r' > 0$ such that

$$|k| \leq r' \Longrightarrow |D_2 f(x, y + k) - D_2 f(x, y)| \leq r$$

for any $x \in K$ and $y \in H$. We deduce that

$$|h| \leq r' \Longrightarrow |f(x, y + h) - f(x, y) - D_2 f(x, y)h| \leq r|h|,$$

i.e. that

$$|h| \leq r' \Longrightarrow |f_{y+h}(x) - f_y(x) - hg_y(x)| \leq r|h|,$$

for any $x \in K$. On taking the sup for $x \in K$ and dividing by $|h|$, we deduce that

$$(9.6) \quad |h| \leq r' \Longrightarrow \|(f_{y+h} - f_y)/h - g_y\|_K \leq r,$$

which proves uniform convergence as announced, or, if one prefers, shows that the left hand side of (5) is $\leq M(\mu)r$, qed.

Let now $K$ and $H$ be two compact intervals, $\mu$ and $\nu$ two integrable functions on $K$ and $H$ and $f$ a continuous function in $K \times H$. We can then consider the *iterated integral* which we denote by

$$\int_H \nu(y)dy \int_K f(x,y)\mu(x)dx \text{ rather than } \int_H \left( \int_K f(x,y)\mu(x)dx \right) \nu(y)dy$$

as, in principle, one should. One can also perform these operations in the opposite order.

**Theorem 10.** *Let $K$ and $H$ be two compact intervals in $\mathbb{R}$ and $f$ a continuous function on $K \times H$. Then*

$$(9.7) \qquad \int_H \nu(y)dy \int_K f(x,y)\mu(x)dx = \int_K \mu(x)dx \int_H f(x,y)\nu(y)dy.$$

*for any integrable functions $\mu$ and $\nu$ on $K$ and $H$.*

This is the analogue of the theorem on absolutely convergent double series (Chap. II, n° 18).

To prove the equality of the two sides of (7), note that, by (3), there exist finite partitions of $K$ and $H$ into intervals $K_p$ and $H_q$ such that $f$ is constant to within $r$ on each rectangle $K_p \times H_q$. Then

$$\int_H f(x,y)\nu(y)dy = \sum \int_{H_q} f(x,y)\nu(y)dy$$

and therefore

$$(9.8) \quad \int_K \mu(x)dx \int_H f(x,y)\nu(y)dy = \sum \int_{K_p} \mu(x)dx \int_{H_q} f(x,y)\nu(y)dy.$$

Now let us choose points $\xi_p \in K_p$ and $\eta_q \in H_q$. If we replace $f(x,y)$ by $f(\xi_p, \eta_q)$ in the general term, the error is clearly bounded by

$$(9.9) \qquad r \int_{K_p} |\mu(x)|dx \int_{H_q} |\nu(y)|dy$$

The left hand side of (8) is thus equal to the "double Riemann sum"

$$(9.10) \quad \sum f(\xi_p, \eta_q) \int_{K_p} \mu(x)dx \int_{H_q} \nu(y)dy = \sum f(\xi_p, \eta_q)\mu(K_p)\nu(H_q)$$

(obvious notation!), with an error less than the sum of the expressions (9), so less than

$$r \int_K |\mu(x)|dx \int_H |\nu(y)|dy = M(\mu)M(\nu)r,$$

the product of the integrals of $|\mu|$ and $|\nu|$ over $K$ and $H$. One would find the same result on calculating in the same way from the right hand side of (7). Since $r > 0$ is arbitrary, they must be equal, qed.

The preceding theorem justifies the definition

$$(9.11) \quad \iint\limits_{K \times H} f(x,y)\mu(x)\nu(y)dxdy = \int \mu(x)dx \int f(x,y)\nu(y)dy =$$

$$= \int \nu(y)dy \int f(x,y)\mu(x)dx$$

of the *double integrals* taken over a compact rectangle $K \times H$ with respect to the "product measure" $\mu(x)\nu(y)dxdy$. On might define them in a more general framework on replacing $K \times H$ by a not too barbarous bounded subset of $\mathbb{C}$, or, equivalently, extend the theorem to discontinuous functions, but one runs quickly into great difficulties if one remains in the framework of the Riemann integral.

Consider for example the following very trivial problem: one takes a positive continuous function $\varphi(x)$ on $K$ with values in $H$ and seeks to calculate the area $A$ contained between the $x$ axis and the curve $y = \varphi(x)$ by means of a double integral rather than by the usual simple integral. Writing $E \subset \mathbb{R}^2$ for the set of $(x,y)$ such that $x \in K$ and $0 \le y \le \varphi(x)$ and $\chi_E$ for its characteristic function, equal to 1 on $E$ and to 0 elsewhere, it is "geometrically obvious" that

$$(9.12) \quad A = \iint\limits_{K \times H} \chi_E(x,y)dxdy;$$

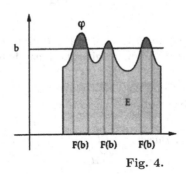

Fig. 4.

moreover, if one calculates the double integral by $\int dx \int dy$, the integral with respect to $y$, for $x$ given, involves the function equal to 1 between 0 and $\varphi(x)$ and zero elsewhere, whence $\int dy = \varphi(x)$; on integrating with respect to $x$ one thus finds the integral of the function $\varphi$, which is precisely the area in question. But let us first integrate with respect to $x$. For $y = b \in H$ given, one has $\chi_E(x,b) = 1$ if $\varphi(x) \ge b$ and $= 0$ if not; if one then considers the set $F(b) \subset K$ of $x \in K$ such that $\varphi(x) \ge b$, one has to integrate with respect to $x$ the characteristic function of $F(b)$, a compact set since $\varphi$ is continuous and $K$ compact. Now there is no reason why this function should be Riemann integrable. In fact, for *every* compact set $F \subset K$, there exists a function $\varphi$

such that $F = F(1)$; it suffices for this that $\varphi(x) = 1$ on $F$ and $\varphi(x) < 1$ for $x \notin F$. To prove the existence of $\varphi$, consider the function

$$d(x, F) = \inf_{u \in F} |x - u|;$$

it is continuous, zero on $F$ and *strictly* positive outside $F$ (Chap. III, n° 10, Example 1). Now let $M$ be the maximum of $d(x, F)$ for $x \in K$; on $K$ the function

$$\varphi(x) = 1 - d(x, F)/M$$

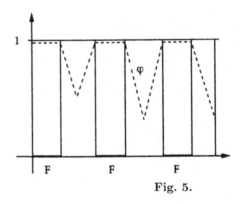

Fig. 5.

is continuous, positive, equal to 1 on $F$ and $< 1$ elsewhere. For this choice of $\varphi$ the set $\{\varphi \geq 1\}$ is just $F$. Figure 5 gives no idea of the complexity of $\varphi$ in the general case.

So we see that for it to be possible to invert the order of integration in a double integral in the Riemann theory so that

$$\iint_E f(x, y)dxdy = \iint_{K \times H} \chi_E(x, y)f(x, y)dxdy$$

for every "reasonable", for example compact or open, subset $E$, of the compact rectangle $K \times H$, as "users" unquestioningly believe, it is necessary, for a start, that the characteristic function of every compact or open subset of $\mathbb{R}$ should be integrable in the sense of this chapter. If such had been the case, no one would ever have invented the Lebesgue integral, and certainly not he himself, since this is precisely the problem which led him to his theory.

Of course, the objection does not arise for the "usual" functions: you can calculate the area of a semicircle centred on the $x$ axis by integrating first with respect to $x$ then with respect to $y$, since in this case the sets $F(b)$ are inoffensive intervals; for curves a little less convex or concave the $F(b)$ can be finite unions of closed intervals, which poses no greater problem, though again one has to justify it. But the general case lies beyond the elementary theory; we shall treat it in n° 33.

Finally we remark that all the results of this n° remain valid, with the same proofs (n° 30), when one integrates with respect to a general "Radon measure" i.e. when one replaces the integral $f \mapsto m(f)$ by a function $f \mapsto \mu(f)$ satisfying the properties of *linearity* and *continuity* of Theorem 1 which, alone, feature in all that we have just done (except for calculating the "double Riemann sums", when we have to modify a little to eliminate discontinuous functions). In other words, it is not the explicit construction of the integral which matters in these problems, but its *formal properties*. Mathematicians needed two hundred and fifty years to understand this, but we now have a century of experience.

## 10 – Semicontinuous functions[15]

To know that the regulated functions are integrable is almost always enough in elementary practice, but it is not difficult, where we now are, to anticipate the "grand" integration theory. The essential tool is a famous theorem which would have been of great use to Cauchy:

**Dini's Theorem.** [16] *Let $(f_n)$ be a* monotone *sequence of continuous real-valued functions defined on a* compact *set $K \subset \mathbb{C}$ and converging simply to a limit function $f$. Then $f$ is continuous if and* only *if the $f_n$ converge uniformly on $K$.*

We can assume that the given sequence is increasing, whence $f(x) = \sup f_n(x)$ for every $x \in K$. For every $r > 0$ and every $a \in K$, we then have

$$f(a) \geq f_n(a) > f(a) - r \text{ for } n \text{ large.}$$

If $f$ is continuous, this relation is, for $n$ given, again true on a neighbourhood of $a$. By BL, we can then find a finite number of points $a_p \in K$ and open balls $B(a_p)$ covering $K$ such that each relation

---

[15] The contents of n° 10 and 11, preparation for the Lebesgue theory, will be repeated in a more general framework in § 9, and in the Appendix to this chapter; we shall use neither the results of these two n° nor those of § 9 before the chapter devoted to them. Our aim here is to show the reader that it is not difficult to go rather further than the traditional theory, the essential being to know how far too far not to go ...

[16] Having followed the analysis courses of Joseph Bertrand and J.A. Serret at Paris in 1866, as Dugac tells us in his thesis, p. 106, and having conceived serious doubts as to the rigour of their ideas, doubts which his youth dissuaded him from making public, Ulisse Dini, professor at Pisa (where there is an "Ecole normale supérieure" which has produced a number of excellent Italian scientists), read the Germans, educated himself on Weierstrass' course, and, in 1878, published in Italian the first exposition of analysis according to Weierstrass' ideas and those of his numerous disciples, followed in 1880 by a book on Fourier series. His book was widely read because neither Weierstrass nor his disciples published anything beyond duplicated manuscript courses of very limited distribution.

(10.1)          $x \in K \cap B(a_p) \implies f(x) \geq f_n(x) > f(x) - r$

is *separately* true for $n$ large. These relations being finite in number, they are *simultaneously* true for $n$ large – unnecessary to rely on the $N_p$ and their maximum ... – and since the $B(a_p)$ cover $K$, this means that, for $n$ large, we have

$$f(x) \geq f_n(x) > f(x) - r$$

for every $x \in K$, therefore $\|f - f_n\|_K \leq r$, qed.

*Exercise.* Prove the theorem using BW.

Consider for example, for $x > 0$, the sequence $f_n(x) = n(x^{1/n} - 1)$ of Chap. II, n° 10; for $x \geq 1$, it is decreasing and tends to $\log x$; the convergence is therefore uniform on $[1, b]$ for any $b > 1$ (try to prove this "by hand" ...). The case of an interval $[a, 1]$ ($a > 0$) reduces to the preceding on putting $x = 1/y$. We therefore have uniform convergence on every compact $K \subset \mathbb{R}_+^*$. Same conclusion for the sequence $(1 + x/n)^n$ for $x \geq 0$.

Dini's Theorem holds not only for increasing *sequences* but also for what we shall call *increasing philtres* [17] of continuous functions; this terminology[18] denotes any family $(f_i)_{i \in I}$ (not necessarily countable) or set $\Phi$ of real functions defined on an arbitrary set and possessing the following property: for any functions $f$ and $g$ in the family, or the set, there exists in the family, or the set, a function $h$ that majorises $f$ and $g$ simultaneously. The most frequent case is that where

$$(f \in \Phi) \ \& \ (g \in \Phi) \implies \sup(f, g) \in \Phi.$$

The definition applies to functions defined on any set: the values of the function, not of the variable, have to be real. This is trivially the case for an increasing sequence. This is also the case, on an interval of $\mathbb{R}$ (or, more generally, in a metric space), of the set of continuous functions which are less than a given function. Similarly one defines *decreasing philtres* by reversing the sense of the inequalities.

To extend Dini's Theorem to this general case, let us consider an increasing philtre $\Phi$ of continuous real functions on the compact set $K \subset \mathbb{C}$ and assume that the function

$$\varphi(x) = \sup_{f \in \Phi} f(x),$$

the *upper envelope* of $\Phi$ (i.e., in the case of an increasing sequence, its limit), is everywhere finite and continuous. For every $r > 0$ and every $a \in K$ there

[17] Translator's note: shades of Isolde & Brangaene!

[18] A little less barbarous than N. Bourbaki's "increasing filtering sets"; I use the spelling "philtre" because the word "filter" is employed in a different sense in general topology. I have known the Bourbaki milieu well, and myself absorbed bourbachique philtres during the "grande époque" of filters enough to think that my spelling corresponds better to the psychological background of the subject ...

exists an $f \in \Phi$ such that $\varphi(a) - r < f(a)$; since $f$ and $\varphi$ are continuous this inequality is again valid on a neighbourhood of $a$ in $K$. By BL, one can then find a finite number of $a_p \in K$ and a finite number of $f_p \in \Phi$ and balls $B(a_p)$ covering $K$, such that

$$\varphi(x) - r < f_p(x) \qquad \text{on } K \cap B(a_p)$$

for every $p$. Since $\Phi$ is an increasing philtre and since the $f_p$ are finite in number, there exists a $f \in \Phi$ which majorises[19] the $f_p$. Then a fortiori $\varphi(x) - r < f(x)$ in $K \cap B(a_p)$ for any $p$, so in all $K$. Since anyhow $f(x) \leq \varphi(x)$, one finds finally that $\|\varphi - f\|_K \leq r$ and, trivially, that $\|\varphi - g\|_K \leq r$ for every $g \in \Phi$ majorising $f$. This is Dini's Theorem in this more general framework.

Since the continuous functions are integrable over a compact interval $K$ of $\mathbb{R}$ the result we have just obtained shows that, in this case,

$$m(\varphi) = \sup m(f),$$

where we revert to the notation $m(f) = \int f(x)dx$ of n° 2 for integration over $K$. The left hand side is greater than the right hand side since $\varphi$ majorises all the $f \in \Phi$; but the existence of an $f$ such that $\|\varphi - f\|_K \leq r$, hence such that the integrals of $\varphi$ and $f$ are equal to within $m(K)r$, shows that in fact the two sides are equal. This argument calls on no more than Theorem 1; the explicit construction of the integral features here no more than in the preceding n°.

Dini's Theorem serves, in the Bourbaki version which we shall follow approximately, as the point of departure on the "grand" theory of integration in view of the following result, in which connection the reader is invited to revise the generalities of Chap. II, n° 17 on infinite limits:

**Corollary 1.** *Let $K$ be a compact interval and $(f_n)$, $(g_n)$ two everywhere increasing or two everywhere decreasing sequences of real continuous functions on $K$. Assume that $\lim f_n(x) = \lim g_n(x)$ for every $x \in K$. Then*

$$\lim m(f_n) = \lim m(g_n)$$

or, in traditional notation,

$$\lim \int_K f_n(x)dx = \lim \int_K g_n(x)dx.$$

Consider for example the case of increasing sequences, put

$$\varphi(x) = \sup f_n(x) \leq +\infty$$

---

[19] If for example one has three functions $f, g, h$ in $\Phi$, then there exists a $k \in \Phi$ which majorises $f$ and $g$, then a $p \in \Phi$ which majorises $k$ and $h$, so majorising $f$, $g$ and $h$.

and consider the set $C_{\text{inf}}(\varphi)$ of all real functions $h$ defined and continuous on $K$ such that $h(x) \leq \varphi(x)$ for every $x$; we shall show that

$$(10.2) \qquad \sup m(f_n) = \sup_{h \in C_{\text{inf}}(\varphi)} m(h),$$

which will establish the corollary since the result does not involve the particular sequence $(f_n)$.

Put

$$M = \sup m(f_n) = \lim m(f_n) \leq +\infty$$

and $h_n = \inf(h, f_n)$ for every continuous function $h \leq \varphi$. The $h_n$ are $\leq h$ and form an increasing sequence of continuous functions like the $f_n$. For every $x \in K$ and every $r > 0$, we have $h(x) - r < f_n(x)$ for $n$ large since $h(x) \leq \varphi(x) = \sup f_n(x)$: this is condition (SUP 2') in the definition of an upper bound (Chap. II, n° 9). Therefore $h(x) - r \leq h_n(x)$ for $n$ large, and since $h_n(x) \leq h(x)$ we conclude that $h(x) = \sup h_n(x)$ for every $x$.

By Dini's Theorem the $h_n \leq f_n$ converge uniformly to $h$, whence $m(h) = \lim m(h_n) \leq \lim m(f_n) = M$. This inequality holding for every continuous $h \leq \varphi$ we may deduce that the right hand side of (2) is $\leq M$. But among the $h \in C_{\text{inf}}(\varphi)$ are the $f_n$ themselves, so that the right hand side of (2) majorises $m(f_n)$ for every $n$; it is therefore $\geq M$. Whence (2) and the corollary, with, moreover, the more precise result (2), qed.

It is almost obvious that the preceding corollary still holds if one substitutes increasing philtres $\Phi$ and $\Psi$ of continuous functions in place of the sequences $f_n$ and $g_n$:

$$\sup_{f \in \Phi} f(x) = \sup_{g \in \Psi} g(x) \implies \sup_{f \in \Phi} m(f) = \sup_{g \in \Psi} m(g).$$

To see this, first consider an $h \in C_{\text{inf}}(\varphi)$ and the functions $\inf(f, h)$ where $f \in \Phi$; if $f', f'' \in \Phi$ and if $f \in \Phi$ majorises $f'$ and $f''$, it is clear (sketch!) that $\inf(f, h)$ majorises $\inf(f', h)$ and $\inf(f'', h)$; the functions $\inf(f, h)$ thus form, for $h$ given, an increasing philtre of continuous functions whose upper envelope is, as above, the function $h$ itself. By Dini's Theorem for philtres, $m(h)$ is then the upper bound of the integrals of the $\inf(f, h)$, themselves majorised by the integrals of the $f \in \Phi$; we conclude that $\sup m(h) \leq \sup m(f)$; but since $\Phi \subset C_{\text{inf}}(\varphi)$, the opposite relation is obvious, whence $\sup m(f) = \sup m(h)$ and, likewise, $= \sup m(g)$.

The preceding corollary leads to a simple proof of a result which the Lebesgue theory allows one to extend to arbitrary series of integrable functions, though clearly with more work:

**Corollary 2.** *Let $\sum u_n(x)$ be a series of continuous functions on a compact interval $K$. Assume that the series converges simply to a continuous function $s(x)$ and that*

(10.3)
$$\sum \int_K |u_n(x)| dx < +\infty.$$

*Then*

(10.4)
$$\int_K s(x) dx = \sum \int_K u_n(x) dx.$$

To prove (4), one may assume $s = 0$ by replacing $u_1$ by $u_1 - s$, which is again continuous. One may also assume the $u_n$ real and then use the decomposition $u_n = u_n^+ - u_n^-$ of n° 2. These positive functions again satisfy the hypothesis (3), since $|u_n^+| \leq |u_n|$; and since $s(x) = 0$ we now have $\sum u_n^+(x) = \sum u_n^-(x) \leq +\infty$ for every $x$. Since a series with positive terms leads to an increasing sequence on considering its partial sums, Corollary 1 shows that

$$\sum m(u_n^+) = \sum m(u_n^-).$$

Since the two sides are *finite* by (3) and the inequality $m(u_n^+) \leq m(|u_n|)$, we have $\sum m(u_n) = 0$ on subtracting, qed.

Once again, only the formal properties of Theorem 1 are needed for the proof.

Condition (3) is satisfied if the given series is normally convergent on $K$, but the hypothesis (3) is weaker, even though in *elementary* practice one almost always verifies (3) by normal convergence.

Corollary 1 for increasing sequences, or its "philtrological" version, and, more precisely, the relation (2), lead us to put

(10.5)
$$m^*(\varphi) = \sup_{f \in C_{\mathrm{inf}}(\varphi)} m(f)$$

for every function $\varphi$ *which can be exhibited as* the limit of an increasing sequence of continuous functions or, more generally (?), for which

(10.6)
$$\varphi(x) = \sup_{f \in C_{\mathrm{inf}}(\varphi)} f(x)$$

for every $x \in K$; such a function takes its values in $]-\infty, +\infty]$. As we saw in Corollary 1 one might define $m^*(\varphi)$ replacing $C_{\mathrm{inf}}(\varphi)$ by any other increasing philtre $\Phi$ of continuous functions with upper envelope $\varphi$. If $m^*(\varphi) < \infty$ we shall say that $\varphi$ is *integrable* and put $m(\varphi) = m^*(\varphi)$, the *integral* of $\varphi$. As we shall see in the following n° this generalisation[20] of the Riemann integral has even simpler properties than the former, despite the fact that it applies only to very particular functions; but all these properties will be extended later to general integrable functions.

---

[20] Generalisation — since if $\varphi$ is continuous the "new" definition of $m(\varphi)$ reduces to the old, for then $\varphi \in C_{\mathrm{inf}}(\varphi)$.

First, let us elucidate a crucial point: how to characterise those functions $\varphi$ satisfying (6) by properties of an "internal" nature? These are the *lower semicontinuous* functions, or, for short, the *lsc* functions of Baire.

Let us work on a not necessarily compact interval $X$ of $\mathbb{R}$, and consider on $X$ a function $\varphi$ with values in $]-\infty, +\infty]$ satisfying (6), i.e. which is the upper envelope of a family of continuous real functions (which clearly excludes the value $-\infty$); for example the function $1/x^2(x-1)^2$ on $\mathbb{R}$, with value $+\infty$ for $x = 1$ or $0$. For every $a \in X$ and every $M < \varphi(a)$, there is, by the definition of an upper bound, a continuous function $f$ on $X$ satisfying

$$f(x) \le \varphi(x) \text{ for every } x, \qquad f(a) > M.$$

Since $f$ is continuous, we again have $f(x) > M$ on a neighbourhood of $a$, and since $\varphi$ majorises $f$, it follows that

(10.7)    $\varphi(a) > M \Longrightarrow \varphi(x) > M$  for every $x \in X$ near $a$.

This is the property which *defines* the lsc functions; equivalently, one may demand that, for every finite $M$, the set $\{\varphi > M\}$ of the $x \in X$ where $\varphi(x) > M$ must be *open in* $X$ since then, if it contains $a$, it must also contain all the points of $X$ sufficiently near $a$. Whence we deduce that the sets $\{\varphi \le M\}$ are closed[21].

If $\varphi(a)$ is finite, we may, in (7), choose $M = \varphi(a) - r$ with an arbitrary $r > 0$, whence

(10.8)    $\varphi(x) > \varphi(a) - r$     on a neighbourhood of $a$,

i.e. for every $x \in X$ such that $|x - a| < r'$, in our usual notation. Continuity would force $\varphi(x) < \varphi(a) + r$ too, but this is precisely what we do not demand of the lsc functions, whence their name. The continuous functions are characterised by the fact that both $f$ and $-f$ are lsc. For a regulated function $\varphi$ condition (8) is equivalent to saying that the right and left limit values of $\varphi$ are $\ge \varphi(a)$ at every $a \in X$.

The reader can easily check that

(i)   the sum of a finite number of lsc functions is lsc,
(ii)  if $\varphi$ and $\psi$ are lsc, then so are the functions $\sup(\varphi, \psi)$ and $\inf(\varphi, \psi)$,
(iii) the *upper* envelope $\sup \varphi_i(x)$ of a finite or infinite family $(\varphi_i)$ of lsc functions is again lsc,
(iv)  the sum, finite or not, of a series of *positive* lsc functions is again lsc.

Properties (i) and (ii) are proved by imitating what we have established for continuous functions. (iii) is a direct application of the definition of upper bounds – one can never say too often that the only useful "property" of upper

---

[21] To distinguish weak from strict inequalities is as crucial in all these questions as to distinguish open from closed sets.

bounds is their definition; (iv) follows from (i) and (iii) since the partial sums of the series form an increasing sequence. Property (ii) shows in particular that if $\varphi$ is lsc, then the functions $\inf[\varphi(x), n]$ obtained by "truncating" the graph of $\varphi$ above a horizontal $n$ are again lsc; the converse follows from (iii).

(v)  *the characteristic function $\chi_U$ of a subset $U$ of $X$, equal to 1 on $U$ and to 0 elsewhere, is lsc on $X$ if and only if $U$ is open in $X$.*

The set $\{\chi_U > M\}$ is $X$ if $M < 0$, $U$ if $0 \leq M < 1$ and empty[22] if $M \geq 1$. One must pay attention to the fact that "open in $X$" does not mean the same as "open in $\mathbb{R}$", unless $X$ itself is open.

Since the lsc functions are "half continuous", one might assume that they "half" satisfy the theorems applicable to continuous functions. This is sometimes justified:

(vi)  *let $\varphi$ be an lsc function on an interval $X$ and $K$ a compact subset of $X$; then $\varphi$ is bounded below on $K$ and there exists a point of $K$ where $\varphi$ attains its minimum.*

For every $n \in \mathbb{N}$ the set $A_n = \{\varphi \leq -n\}$ is closed in $X$, so that $A_n \cap K$ is compact; these intersections form a decreasing sequence so have a common point $a$ if they are all nonempty (Corollary 1 of BL). Absurd, since then we would have $\varphi(a) = -\infty$, an eventuality excluded by the definition of the lsc functions.

Now let $m$ be the lower bound of the $\varphi(x)$, $x \in K$. For every $n \in \mathbb{N}$ the set $K_n$ of the $x \in K$ where $\varphi(x) \leq m + 1/n$ is nonempty (definition of a lower bound) and closed (definition of the lsc functions); since the $K_n$ decrease they have a common point $c \in K$, and clearly $\varphi(c) = m$, qed.

We have seen above that every function $\varphi$ satisfying (6) on an interval $X$ is lsc; the converse holds if one assumes that there is a continuous function $f \leq \varphi$ on $X$, and thus, by (vi), if $X$ is compact. Since $\varphi - f = \varphi + (-f)$ is lsc, it suffices to treat the case of a positive function. For $a \in X$ and $M < \varphi(a)$ given, it reduces to constructing a continuous function $f \leq \varphi$ satisfying $f(a) > M$. Now there exists an $r > 0$ such that we again have $\varphi(x) > M$ for those $x \in X$ such that $|x - a| < r$. Figure 6 shows the construction of $f$, and does not require comment. We could in fact construct an increasing *sequence* of continuous functions converging to $\varphi$ [Dieudonné, Vol. 2, (12.7.8)], but this is quite unnecessary for the needs of integration theory, because of (2).

---

[22] The empty set is open because, not containing any point, it has no difficulty in satisfying the definition of an open set (all who live at least 500 years end up dying in a car accident). Moreover, since the complement of the empty set is the whole space, which is closed, it must be open. This argument also shows that the empty set is closed.

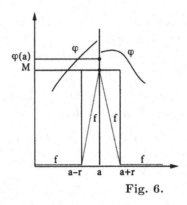

**Fig. 6.**

In all the above we have dealt with the upper envelopes of continuous functions, but of course the lower envelopes of such functions, the *upper semicontinuous* functions or *usc* functions, are no less important. These take their values in $[-\infty, +\infty[$. One passes trivially from lsc to usc by remarking that

$$\varphi \text{ is lsc } \iff -\varphi \text{ is usc.}$$

You may therefore, if it appeals to you, translate all the properties of the lsc functions into properties of the usc functions: it is enough to reverse the sense of all the inequalities and to replace the word "increasing" by the word "decreasing" everywhere. There is a theorem on the maximum, and not on the minimum, for usc functions on a compact set. Every usc function majorised by a continuous function is the lower envelope of the continuous functions which majorise it; this is always the case of a usc function on a compact interval by the maximum theorem. Likewise, the characteristic function of a set is usc if and only if the set is *closed*.

Finally, it is clear that the continuous functions are the only functions that are simultaneously lsc and usc.

For a usc function $\psi$ on a *compact* interval $K$ let $C_{\sup}(\psi)$ be the set of continuous functions $f \geq \psi$; we then put

$$(10.9) \qquad m^*(\psi) = \inf_{f \in C_{\sup}(\psi)} m(f) \geq -\infty,$$

so that $m^*(\psi) = -m_*(-\psi)$ where the right hand side is the integral of an lsc function.

## 11 – Integration of semicontinuous functions

Let us now return to the integrals of lsc functions over a compact interval $K$; these functions are bounded below but not above, so that their integrals, defined by (10.5), are $> -\infty$ but $\leq +\infty$. The essential point in the proofs is that, in (10.5), one can replace $C_{\text{inf}}(\varphi)$ by any increasing philtre of continuous functions having upper envelope $\varphi$.

(i) *Additivity of the integral:*

$$(11.1) \qquad\qquad m^*(\varphi + \psi) = m^*(\varphi) + m^*(\psi).$$

Let $\varPhi$ be the set of functions of the form $f + g$ with $f \in C_{\text{inf}}(\varphi)$ and $g \in C_{\text{inf}}(\psi)$. It is clear that $\varPhi$ is an increasing philtre of continuous functions – apply the definitions – whose upper envelope is[23] $\varphi + \psi$. Hence

$$m^*(\varphi + \psi) = \sup m(f + g) = \sup m(f) + \sup m(g) = m^*(\varphi) + m^*(\psi).$$

Similarly one can show that $m^*(\lambda\varphi) = \lambda m^*(\varphi)$ for every constant $\lambda > 0$, and even if $\lambda = 0$ so long as we define $0. + \infty = 0$. (Multiplying an lsc function by $-1$ makes it usc.)

(ii) *Passage to the limit under the $\int$ sign in an increasing sequence:*

$$(11.2) \qquad\qquad m^*(\sup \varphi_n) = \sup m^*(\varphi_n) \leq +\infty.$$

Let $\varphi(x) = \sup \varphi_n(x)$. Put $\varPhi_n = C_{\text{inf}}(\varphi_n)$ for every $n$ and let $\varPhi$ be the union of the $\varPhi_n$, i.e. the set of continuous functions $f$ satisfying $f \leq \varphi_n$ for some $n$. This is an increasing philtre: for if $f \leq \varphi_p$ and $g \leq \varphi_q$ then $\sup(f,g) = h \leq \varphi_r$ for $r \geq \max(p,q)$, and consequently $h \in \varPhi$. Finally, $\varphi$ is the upper envelope of the $f \in \varPhi$, for

$$\varphi(x) = \sup_n \varphi_n(x) = \sup_n \sup_{f \in \varPhi_n} f(x) = \sup_{f \in \bigcup \varPhi_n} f(x)$$

by the associativity of upper bounds (Chap. II, end of n° 9). We conclude that

$$m^*(\varphi) = \sup_{f \in \varPhi} m(f) = \sup_n \sup_{f \in \varPhi_n} m(f) = \sup_n m^*(\varphi_n)$$

by the definition of $m^*(\varphi_n)$, qed.

(iii) *Integration term-by-term*

$$(11.3) \qquad\qquad m^*\left(\sum \varphi_n\right) = \sum m^*(\varphi_n) \leq +\infty$$

*for every series of* positive *lsc functions.* Write $s$ and $s_n$ for the total sum and the partial sums of the series of the $\varphi_n$. Since the $\varphi_n$ are positive these partial sums form an *increasing* sequence of lsc functions of which $s$ is the limit. The integral of the left hand side is thus the limit of the integrals of

---

[23] We have already mentioned that if $A$ and $B$ are two subsets of $\mathbb{R}$ and $A + B$ is the set of $u + v$ with $u \in A$ and $v \in B$, then $\sup(A + B) = \sup A + \sup B$.

the $s_n$ by (ii), i.e., by (i), of the partial sums of the series of the integrals, qed. Note that if the $\varphi_n$ are not positive the sum of the series need not even be lsc.

If, in particular, the functions $\varphi_n$ satisfy $m^*(\varphi_n) < +\infty$, i.e. are $m$-integrable (by definition), and if $\sum m^*(\varphi_n) < +\infty$, then the sum of the series is again integrable and one may integrate it term-by-term. In particular, for *positive* lsc functions,

$$(11.4) \qquad m^*(\varphi_n) = 0 \ \text{ for every } n \Longrightarrow m^* \left( \sum \varphi_n \right) = 0.$$

Since the characteristic function of an open set $U$ in $K$ is lsc one may define the *measure of an open set* by putting

$$(11.5) \qquad m(U) = m^*(\chi_U),$$

a number clearly lying between 0 and $m(K)$; it is clear more generally that

$$(11.6) \qquad U \subset V \Longrightarrow m(U) \leq m(V)$$

since then $\chi_U \leq \chi_V$. It is easy to see that, when $U$ is an interval, $m(U)$ reduces to its usual length; for this obviously majorises $m(f)$ for every continuous function $f \leq \chi_U$ (i.e. $\leq 1$ on $U$ and $\leq 0$ elsewhere), but on replacing the discontinuities of the graph of the characteristic function at the end-points of $U$ by almost vertical line segments joining 0 to 1, one constructs functions $f$ whose integral is arbitrarily close to the length of $U$.

Properties (i), (ii) and (iii) above translate immediately:

(i') *if $U$ and $V$ are open in $K$ then $m(U \cup V) \leq m(U) + m(V)$ and*

$$(11.7) \qquad m(U \cup V) = m(U) + m(V) \ \text{ if } U \text{ and } V \text{ are disjoint.}$$

Obvious, since, in the last case, we have $\chi_{U \cup V} = \chi_U + \chi_V$.

(ii') *if $(U_n)$ is an increasing sequence of open sets then*

$$(11.8) \qquad m(\bigcup U_n) = \lim m(U_n) = \sup m(U_n).$$

Obvious since the characteristic function of the union is the limit of the sequence, increasing, of those of the $U_n$.

(iii') *if $(U_n)$ is any sequence of open sets then*

$$(11.9) \qquad m \left( \sum U_n \right) \leq \sum m(U_n)$$

*with equality if the $U_n$ are pairwise disjoint.*

Obvious, since the characteristic function of the union is less than the sum of the characteristic functions of the $U_n$ and is equal to it if the $U_n$ are pairwise disjoint.

This allows us to calculate the measure of any open $U \subset K$ explicitly. First, note that for every $a \in U$ the union of all the intervals containing $a$ and contained in $U$ is again an interval $U(a)$, clearly open in $K$ like $U$ itself, and of length $> 0$; $U$ is the union of these $U(a)$. It is immediate that, for $a \neq b$, either $U(a) = U(b)$ or $U(a) \cap U(b) = \emptyset$. The *set* (and not the family) of the $U(a)$ is *countable*, for those of these intervals which are of length $> 1/p$ are at most $m(K)p$ in number since they are pairwise disjoint. Thus we see that *every open subset $U$ of $K$* (and, in fact, of any interval, compact or not, and in particular of $\mathbb{R}$) *is the union of a finite or countable family of open pairwise disjoint intervals*. The measure of $U$ is then, by (iii'), the sum of the measures of these intervals.

We leave the reader the task of translating all these properties into terms of usc functions and of closed sets. To go further along this path would oblige us to develop all the Lebesgue theory. The reader may find these considerations insufficient: for, at this point of the exposition, (i) we are not yet able to integrate the *difference* of two lsc functions for the reason that it is neither lsc nor usc[24], (ii) we have considered integrals only over compact intervals. These limitations will be removed in the Appendix to this Chapter.

*Exercise.* We say that a set $N \subset K$ is of measure zero if, for every $r > 0$ there is an open $U \subset N$ such that $m(U) < r$. (i) Show that the union of a finite or denumerable family of sets of measure zero is of measure zero (use the fact that $r = \sum r/2^n$). Show that $\mathbb{Q} \cap K$ is of measure zero. (ii) Let $\varphi$ be a positive lsc function such that $m^*(\varphi) < +\infty$; show that the set $\{\varphi(x) = +\infty\}$ is of measure zero (use the sets $\{\varphi(x) > n\}$).

---

[24] The ingenious reader will observe that if $\varphi'$, $\varphi''$, $\psi'$ and $\psi''$ are lsc functions with finite values such that $\varphi' - \psi' = \varphi'' - \psi''$, then $\varphi' + \psi'' = \varphi'' + \psi'$, so $m(\varphi') + m(\psi'') = m(\varphi'') + m(\psi')$, so $m(\varphi') - m(\psi') = m(\varphi'') - m(\psi'')$. One may thus define $m(\theta) = m(\varphi') - m(\psi')$ without ambiguity for every function $\theta$ which can be expressed in the form of a difference of two positive lsc functions with finite values. These functions form a vector space over $\mathbb{R}$, etc. But this is not the best method for obtaining the general integrable functions: we still lack "almost everywhere zero" functions.

## § 3. The "Fundamental Theorem" (FT)

### 12 – The fundamental theorem of the differential and integral calculus

Let us return to much more elementary considerations and introduce the notion of an *oriented integral*, analogous to that of a vector on a line. To do this, first observe that we always have

$$(12.1) \qquad \int_a^b f(x)dx + \int_b^c f(x)dx = \int_a^c f(x)dx$$

if $a \leq b \leq c$. This is geometrically obvious, and has been proved in n° 2, additivity formula (2.8).

(1) shows that

$$\int_b^c = \int_a^c - \int_a^b;$$

we may then write this relation in the form

$$\int_b^c = \int_b^a + \int_a^c$$

agreeing that

$$(12.2) \qquad \int_u^v f(x)dx = - \int_v^u f(x)dx \qquad \text{if } u > v.$$

As in the case of the vectors, the relation (1) remains valid with no hypothesis on the respective positions of $a$, $b$ and $c$.

Having said this, let $f$ be a scalar function defined on an interval $I$ (of any kind), and assume that $f$ has right and left limits at each point of $I$, i.e. that $f$ is regulated. We may then, for a given $a \in I$, consider the function

$$(12.3) \qquad F(x) = \int_a^x f(t)dt$$

on $I$, with an oriented integral in the preceding sense, so the opposite of the ordinary integral for $x < a$. We have denoted the phantom variable of integration by $t$ so as not to confuse it with the variable $x$ in $F$; one can replace $t$ by $y$, $u$, $\$$ or anything one likes, except $x$, $f$ or $d$ ...

By the properties of oriented integrals

$$F(x + h) - F(x) = \int_x^{x+h} f(t)dt.$$

This relation shows first of all that $F$ is continuous: since $f$ is bounded on every compact interval $K \subset I$, the preceding integral is $O(h)$ when $h$ tends to 0.

If, on the other hand, $h$ is $> 0$ and small enough, then the function $f$ is, on $]x, x + h]$, almost equal to the limit value $f(x+)$, so that its integral is almost equal to $hf(x+)$ since the value taken by $f$ at the point $x$, or at any other individual point, has no influence on that of the integral; the quotient $[F(x + h) - F(x)]/h$ is therefore almost equal to $f(x+)$, so tends to this limit as $h > 0$ tends to 0.

Weierstrass's *Epsilontik* is missing from this argument. To introduce it, first observe that

$$\int_x^{x+h} f(x+)dt = hf(x+)$$

because $f(x+)$ is not a function of the variable of integration $t$, in other words, behaves like a constant function of $t$ for $x$ given. (Whence, once more, the necessity of not mixing the phantom or bound variables like $t$ and the free variables like $x$.) Then

(12.4)     $$\frac{F(x + h) - F(x)}{h} - f(x+) = \frac{1}{h} \int_x^{x+h} [f(t) - f(x+)]dt.$$

Now for every $r > 0$, there exists an $r' > 0$ such that

$$x < t \leq x + r' \Longrightarrow |f(t) - f(x+)| \leq r.$$

The right hand side of (4) is then, in modulus, $\leq r$. We argue similarly for $h < 0$, replacing the interval $]x, x + r'[$ by an interval $]x - r'', x[$ and $f(x+)$ by $f(x-)$. The left hand side then tends to 0, whence:

**Theorem 11.** *Let $f$ be a regulated function on an interval $I$ in $\mathbb{R}$. Then the function $F$ defined by the relation (3) is continuous and has right and left derivatives equal to $f(x+)$ and $f(x-)$ at each point $x \in I$.*

This result, for functions as then understood, is already in Newton in 1665–66 with essentially the same proof, phrased in his language of fluentes and fluxions (Chap. III, n° 14): if $y$ is the fluent which defines the curve $[y = f(x)$ in the language of today] and if $z$ is the area [i.e. the integral] between a fixed abscissa and the abscissa of the fluent $x$, then the infinitesimal increase $\dot{z}o$ of $z$ is the product of $y$ by the infinitesimal increase $\dot{x}o$ of $x$, which means that $\dot{z}/\dot{x} = y$; if one assumes that $\dot{x} = 1$, in other words that $x$ is the "time" with respect to which one derives one's fluentes to calculate their fluxions, one has $\dot{z} = y$, which, even in his conception of derivatives as "speeds of variation in time", means that the derivative of the area $z$ with respect to the variable $x$ is the ordinate $y$ of the graph at the point $x$; in Leibniz' style, $dz/dx = y$. For them, calculating the area is the same as calculating a fluent $z$ satisfying this relation. Newton justifies nothing, not even the fact that the relation $\dot{z} = y$ determines $z$ up to an additive constant, which is, however, the crucial point; a few lines suffice for him to formulate his result[25] which he illustrates by examples. *Chez* Leibniz, everything is simple too: since $F(x)$ is the "continuous sum" of the infinitely small quantities $f(t)dt$ where $t$ varies from the left hand end of the area to $x$, the infinitesimal increase in $F$ when one passes from $x$ to $x + dx$ is $f(x)dx$, whence $dF = f(x)dx$ and $f(x) = dF/dx$. Here again, the justifications awaited the XIX[th] century, but since the method worked admirably, for 150 years no one bothered to provide the rigorous proofs by "exhaustion" of Chap. II, n° 11 ...

Recall that the set $D$ of discontinuities of a regulated function $f$ is finite or countable (Theorem 6 or Chap. III, n° 12). Outside $D$ the function $F$ is therefore differentiable, with $F'(x) = f(x)$.

When the function $f$ is *continuous*, the function $F$ is even differentiable on all the interval considered, and

$$(12.5) \qquad F'(x) = f(x) \qquad \text{for every } x$$

in this case. One says then that $F$ is a *primitive* of $f$. These arguments prove the existence of a primitive for every continuous function. This result is not at all clear for a function which, though continuous, may be so savage that one cannot represent it graphically.

Now, in contrast to Newton, who did not even pose the question since he did not *see* (and one still does not see it ...) how a graph all of whose tangents are horizontal could be anything other than a line, we know (Chap. III, n° 16) that if two everywhere differentiable functions on an interval have the same derivatives everywhere, then their difference is constant. Since the addition of a constant to the function $F$ defined by (3) does not change the difference $F(x) - F(a)$, we obtain the following result:

**Theorem 12 (FT).** *Let $f$ be a scalar function defined and continuous on an interval $I$ in $\mathbb{R}$ and let $F$ be a primitive of $f$, i.e. a differentiable function such that $F'(x) = f(x)$ for every $x \in I$. Then*

$$(12.6) \qquad \int_a^b f(x)dx = F(b) - F(a)$$

*for any $a, b \in I$.*

In this way we find the results of n° 4, Example 1, again, for analytic functions, but in a much more general framework.

*Example 1.* For $x > 0$ and $s \in \mathbb{C}$, the function

$$x^s = \exp(s.\log x)$$

---

[25] *Tractatus de Methodis Serierum et Fluxionum*, pp. 195–197 and 211 of Vol. III of the *Mathematical Papers*.

has derivative $sx^{s-1}$ [Chap. IV, formula (10.10)]. The function $x^{s+1}/(s+1)$ is therefore, for $s \neq -1$, a primitive of $x^s$. Whence the formula

(12.7)    $$\int_a^b x^s dx = \frac{b^{s+1} - a^{s+1}}{s+1} \qquad (0 < a, b; \ s \in \mathbb{C}, \ s \neq -1)$$

already obtained for $s \in \mathbb{N}$ by a direct calculation of the integral (Chap. II, n° 11), but now valid for any $s \in \mathbb{C}$, $s \neq 1$.

*Example 2.* For $x > 0$ the function $\log x$ has derivative $1/x$; hence again the formula

$$\int_a^b dx/x = \log b - \log a \qquad (0 < a, b)$$

of Chap. II, n° 11.

*Example 3.* The derivative of the function $\arctan x$ is $1/(1+x^2)$; whence

$$\int_a^b \frac{dx}{1+x^2} = \arctan b - \arctan a;$$

we must pay attention to the fact that, in this calculation, we take the "principal determination" of $\arctan x$, that given by the relation

$$y = \arctan x \iff \{(x = \tan y) \ \& \ (|y| < \pi/2)\}$$

or, equivalently, the inverse function of

$$\tan : \ ]-\pi/2, \pi/2[ \ \longrightarrow \mathbb{R}.$$

*Example 4.* For $c \in \mathbb{C}$, $c \neq 0$, the derivative of $e^{cx}/c$ is $e^{cx}$; we again find the formula

$$\int_a^b e^{cx} dx = (e^{cb} - e^{ca})/c.$$

In practice, we often use the notation

(12.8)    $$F(b) - F(a) = F(x)\Big|_a^b;$$

this contradicts all the most elementary rules of mathematical logic with its $x$ which might be a $t$, a $\#$ or a $\pounds$ and which, despite its clearly phantom character, is not *linked* visibly to any of its occurrences. It would be more correct to write

$$F(x)\begin{vmatrix} x = b \\ x = a \end{vmatrix} \qquad \text{or even} \quad F(\square)\begin{vmatrix} \square = b \\ \square = a \end{vmatrix},$$

especially when $F$ depends on several variables $x$, $y$, etc. But as we have said already, one does not change society, even mathematical society, by ukase.

To denote a primitive of $f$, there is another notation, long universal, and even more catastrophic, namely

$$F(x) = \int f(x)dx,$$

omitting the limits of integration. Since the letter $x$ on the right hand side denotes a bound variable and that on the left hand side a free variable, all the tabus are violated[26]. The inventors of this system probably knew what they were doing; Leibniz printed it for the first time in 1686 in his aptly named *Geometria recondita*, and for example wrote that

$$\int xdx = x^2/2,$$

the word "integral" being introduced by Jakob Bernoulli in 1690 (Cantor, pp. 197, 218). But their principal reason is that they were much more concerned to calculate primitives rather than integrals between well defined limits, and, as we have already said, one had to wait for Fourier for the idea of displaying the integration limits in the integral notation. Imagine the confusions that this system must have provoked among less brilliant people, if not *chez* Leibniz, the Bernoullis or Euler – brains of this calibre are not born everyday. Not to ignore that relations such as

$$\int \cos x.dx = \sin x, \qquad \int dx/x = \log x,$$

etc. may induce the same confusions even nowadays...

Another way of formulating the preceding theorem starts from a function $f$ of class $C^1$, i.e. having a continuous derivative. Since $f$ is a primitive of $f'$ one finds the relation

$$(12.9) \qquad f(b) - f(a) = \int_a^b f'(x)dx,$$

as fundamental as the FT, for any $a$ and $b$ in the interval $I$ considered, or, in the language of indefinite integrals,

$$(12.10) \qquad f(x) = \int f'(x)dx,$$

---

[26] We shall use this notation frequently, and have already used it, though in quite another context, namely to denote an integral extended over an interval mentioned repeatedly in the calculations, and where no ambiguity is possible. This convention or abbreviation, which often allows us to write the integrals in the body of the text in clear language instead of using an extra line each time, economises on type and paper.

or even $f(x) = \int df(x)$. In this form, the reciprocity between derivative and integral appears clearly. In the version (9) one chooses a subdivision $a = x_1 \leq x_2 \leq \ldots \leq x_{n+1} = b$ of $[a, b]$, writes that

$$(12.11) \qquad f(b) - f(a) = \sum [f(x_{i+1}) - f(x_i)]$$

and observes that the difference $f(x_{i+1}) - f(x_i)$ is "almost" equal to $f'(x_i)(x_{i+1} - x_i)$ and, in fact, exactly equal to $f'(\xi_i)(x_{i+1} - x_i)$ for some $\xi_i \in ]x_i, x_{i+1}[$ by the mean value theorem if $f$ is real. Substituting in the above expression for $f(b) - f(a)$, one finds exactly the Riemann sums which define the integral (9). This type of argument was clearly known to Leibniz; it is linked to "the calculus of finite differences" so popular in the XVII$^{\text{th}}$ and XVIII$^{\text{th}}$ centuries. Leibniz' notation makes these results as intuitive as they must have been almost obvious in the eyes of the contemporaries who did not worry about arithmetising analysis, whence their popularity.

The formula (9) explains Theorem 19 of Chap. III, n° 17 on differentiating a uniform limit. Suppose we are given a sequence of functions $f_n$ of class $C^1$ on an interval $I$, and assume that the $f'_n$ converge to a limit $g$ uniformly on every compact $K \in I$. Then, by Theorem 4 of n° 4,

$$(12.12) \qquad \int_a^x g(t)dt = \lim \int_a^x f'_n(t)dt = \lim[f_n(x) - f_n(a)]$$

for any $a, x \in I$. If the sequence $f_n$ converges at the point $a$, it must then converge everywhere to a function $f$, which, satisfying

$$(12.13) \qquad f(x) - f(a) = \int_a^x g(t)dt$$

by (12), is a primitive of $g$; in other words, the derivative of the limit is the limit of the derivatives. An immediate estimate of the integral of $g - f'_n$ then shows that the sequence $f_n$ converges uniformly on every compact set. But Theorem 19 of Chap. III only assumes the existence of the $f'_n$, and not their continuity.

We established a theorem on "differentiation under the $\int$ sign" for integrals of the form $\int f(x, y)dx$ above. We may combine this with the FT to obtain the following occasionally useful result:

**Theorem 13.** *Let $I$ and $J$ be two intervals, $f$ a continuous function on $I \times J$ and let $\varphi, \psi : I \to J$ be two differentiable functions. Suppose that $f$ has a continuous derivative $D_1 f$ on $I \times J$. Then the function*

$$(12.14) \qquad g(x) = \int_{\varphi(x)}^{\psi(x)} f(x, y)dy$$

*is differentiable on $I$, and*

$$(12.15) \quad g'(x) = \int_{\varphi(x)}^{\psi(x)} D_1 f(x, y)dy + f[x, \psi(x)]\psi'(x) - f[x, \varphi(x)]\varphi'(x).$$

By subtracting, one need only prove this in the case where $\varphi(x) = b$ is constant. Put

$$F(x,y) = \int_b^y f(x,t)dt,$$

whence $g(x) = F[x, \psi(x)]$. Since $f$ is continuous the FT shows that

(12.16) $$D_2 F(x,y) = f(x,y).$$

Theorem 9 shows on the other hand that

(12.17) $$D_1 F(x,y) = \int_b^y D_1 f(x,t)dt.$$

By Theorem 9 (i) $D_1 F$ is continuous on $I \times J$, and since $D_2 F = f$ is also continuous we see that $F$ is a function of class $C^1$ on $I \times J$. We may thus apply the chain rule to $g(x) = F[x, \psi(x)]$ (Chap. III, n° 21), whence

$$g'(x) = D_1 F[x, \psi(x)] + D_2 F[x, \psi(x)]\psi'(x);$$

one obtains the desired formula on substituting $\psi(x)$ for $y$ in (16) and (17).

*Exercise.* By the theorem, the function

$$g(x) = \int_{x^2}^{x^3} \sin(xy)dy$$

has derivative

$$g'(x) = \int_{x^2}^{x^3} \cos(xy)ydy + 3x^2 \sin(x^4) - 2x \sin(x^3).$$

Check this result by calculating the integrals that appear in $g$ and $g'$ (for the second, integrate by parts).

*Exercise.* Prove (15) directly by writing

$$g(x+h) - g(x) = \int_b^{\psi(x+h)} f(x+h,y)dy - \int_b^{\psi(x)} f(x,y)dy.$$

Since we have just made a new incursion into the functions of two real variables, let us show how one may exploit the FT to establish one of the fundamental results of Chap. III, n° 23:

**Theorem 14.** *Let $f(x,y)$ be a function defined and continuous on $I \times J$ where $I$ and $J$ are two intervals in $\mathbb{R}$. Assume that $f$ has continuous second derivatives $D_1 D_2 f$ and $D_2 D_1 f$ on $I \times J$. Then they are equal.*

Since it is enough to verify the statement on a neighbourhood of an arbitrary point of $I \times J$ one may reduce to the case where $I = [a, b]$ and $J = [c, d]$ are compact. The FT applied to the functions $y \mapsto D_2 D_1 f(x, y)$ and $x \mapsto D_1 f(x, y)$ then shows that

$$(12.18) \quad \int_a^u dx \int_c^v D_2 D_1 f(x, y) dy =$$
$$\int_a^u [D_1 f(x, v) - D_1 f(x, c)] \, dx = f(x, v) - f(x, c) \Big|_{x=a}^{x=u}$$

for any $u \in I$ and $v \in J$. A similar calculation will show that

$$(12.19) \quad \int_c^v dy \int_a^u D_1 D_2 f(x, y) dx = f(u, y) - f(a, y) \Big|_{y=c}^{y=v},$$

a result obviously identical to that furnished by (18). But we already know that the order of integration is unimportant in these double integrals. Putting

$$g(x, y) = D_2 D_1 f(x, y) - D_1 D_2 f(x, y),$$

we thus obtain a *continuous* function on $I \times J$ such that

$$(12.20) \quad \int_a^u dx \int_c^v g(x, y) dy = 0$$

for any $u \in I$ and $v \in J$. On differentiating with respect to $u$ one finds that the integral of $g(x, y)$ between $c$ and $v$ is zero for any $v$. On differentiating with respect to $v$, one obtains $g(x, y) = 0$, qed. [Note that Chap. III, n° 23, only requires $D_1 f$ and $D_2 f$ to be differentiable at the point $(x, y)$ considered].

## 13 – Extension of the fundamental theorem to regulated functions

Let us return to the usual integral and to the "fundamental theorem". The hypothesis that $f$ is of class $C^1$ is not indispensable in justifying formula (12.9); it still holds if, for example, $f'(x)$ is a regulated function. For this, and as N. Bourbaki and also Dieudonné (*Eléments d'analyse*, Vol. 1, Chap. VIII, n° 7) do, let us say that for a regulated function $f$ defined on an arbitrary interval $I$ any *continuous* function $F$ which has a derivative equal to $f(x)$ outside a *countable* subset $D$ of $I$ is *a primitive of this regulated function*.

**Theorem 12 bis.** *Let $f$ be a regulated function on an interval $I \subset \mathbb{R}$. Then (i) $f$ has a primitive $F$ in $I$; (ii) any two primitives of $f$ are equal up to an additive constant; (iii) we have*

$$\int_a^b f(x) dx = F(b) - F(a)$$

*for any $a, b \in I$; (iv) every primitive $F$ of $f$ has right and left derivatives $F'_d(x) = f(x+)$, $F'_g(x) = f(x-)$ at every point.*

Point (i) follows from Theorem 11: the continuous function $F$ of Theorem 11 satisfies (iv) and since $f$ is continuous outside a countable set $D$, $F$ has a derivative $F'(x) = f(x)$ for $x \notin D$.

Since (iii) is valid for a particular primitive, (iii) will be established if we prove (ii), in other words that *a continuous function having a zero derivative outside a countable set $D$ is constant.* For this it is enough to show that *if the derivative is $\geq 0$ outside $D$, then the function is increasing,* for a function which is simultaneously increasing and decreasing has no choice.

We shall give two rather different and extremely ingenious proofs of this theorem[27].

It is enough to establish this result when $f'(x)$ is strictly positive everywhere on $I - D$, since if $f(x) + \varepsilon x$, which satisfies this hypothesis, is increasing for every $\varepsilon > 0$, then so clearly is $f$ in the limit. One may also restrict oneself to examining the case where $I$ is compact, since, to show that $a \leq b$ implies $f(a) \leq f(b)$, it is enough to argue on the interval $[a, b]$.

The basic idea is that the ratio $[f(x+h) - f(x)]/h$ is $> 0$ for $h$ small enough since it tends to $f'(x) > 0$; then $f(x) < f(x+h)$ for every small enough $h > 0$; it remains to pass from "locally" increasing to "globally" increasing, which the Founding Fathers considered obvious.

*First proof.* Assume that $f'(x) > 0$ for $x \in I - D$. If $f$ is not increasing, there exist points $c$, $d$ of $I$ such that $c < d$, $f(c) > f(d)$. For every number $\xi \in \,]f(d), f(c)[$, the equation $f(x) = \xi$ has at least one solution between $c$ and $d$ (intermediate value theorem). The set $E(\xi)$ of these solutions is closed since $f$ is continuous, and it is bounded since $\subset [c, d]$; it therefore contains the number $\sup E(\xi) = d_\xi$, and we have $d_\xi < d$ since $f(d_\xi) = \xi > f(d)$.

Let us show that if $h > 0$ is small enough that $d_\xi + h < d$ then $f(d_\xi + h) < f(d_\xi) = \xi$. If in fact $f(d_\xi + h) \geq \xi$, a number $> f(d)$, then the equation $f(x) = \xi$ would have a solution between $d_\xi + h$ and $d$; absurd since $\sup E(\xi) = d_\xi < d_\xi + h$.

It follows from this that, for every $\xi \in \,]f(d), f(c)[$ and every sufficiently small $h > 0$,

$$[f(d_\xi + h) - f(d_\xi)]/h < 0.$$

This is impossible if the derivative $f'(d_\xi)$ (or even only the right derivative) exists, since by hypothesis it is $> 0$; the function $f$ is thus not differentiable at any of the points $d_\xi$, which proves that $d_\xi \in D$.

Now the map $\xi \mapsto d_\xi$ of $]f(d), f(c)[$ in $D$ is injective because

---
[27] I find the first in Walter, *Analysis I*, pp. 354–359, who attributes it to L. Scheeffer, 1885, the date when arguments à *la* Cantor began to be fashionable. For the second I have simplified the method of Dieudonné, *Eléments ...*, Vol. 1, Chap. VIIII, n° 6, which treats the case of functions with values in Banach spaces, and could in fact be deduced from the result for real-valued functions with the help of the Hahn-Banach theorem. Both authors prove the mean value theorem, below, directly.

$$\xi \neq \eta \Longrightarrow f(d_\xi) = \xi \neq \eta = f(d_\eta) \Longrightarrow d_\xi \neq d_\eta.$$

If the theorem were false we would have constructed an *injective* map of an interval $]f(d), f(c)[$ of $\mathbb{R}$ into a countable set, absurd if one believes Cantor ...

As we have seen in passing, it would suffice, to obtain (2), to assume that $f$ has a positive right derivative outside $D$. The same remark applies to the proof that follows.

*Second proof.* Here again we reduce to showing that $f(a) \leq f(b)$ if $f'(x) > 0$ on $I = [a, b]$.

Let us first explain the method in the simplest case (already explained in another way in Chap. III, n° 16) where $f$ is differentiable for every $x \in I$ without exception. The set $E$ of the $x \in [a, b]$ such that $f(a) \leq f(x)$ contains $a$ and is closed since $f$ is continuous. Let $c = \sup(E) \leq b$, whence $c \in E$; assume that $c < b$. For all $x > c$ near enough to $c$, one has $f(x) > f(c) \geq f(a)$ since $f'(c) > 0$, whence $x \in E$ if $x \in I$. If $c < b$ then $E$ contains points $x > c = \sup(E)$, absurd. Whence $c = b$, qed.

Now let us pass to the general case, where the derivative exists only outside $D$. First notice that if $f(u) \geq f(a)$ at a point $u \in [a, b]$, then again $f(x) > f(u)$ [and so $> f(a)$] for every $x > u$ close enough to $u$ if $f'(u)$ *exists*; in the contrary case, one may only guarantee, for every $\varepsilon > 0$, that one has $f(x) > f(u) - \varepsilon$ [and so $> f(a) - \varepsilon$] for every $x > u$ near enough to $u$ since $f$ is continuous. This indicates that if one moves from $a$ to $b$, one is forced, in seeking to prove the inequality $f(x) \geq f(a)$, to *subtract an error term from $f(a)$ each time that one meets a point of $D$*. If one can contrive that the sum of these errors will be $\leq r$ when one arrives at the point $b$, one obtains the inequality $f(b) \geq f(a) - r$, which, valid for every $r > 0$, proves the theorem.

One must therefore, for each $\xi \in D$, allow an error $r(\xi)$ chosen so that $\sum r(\xi) \leq r$. Since $D$ is finite or countable, one need only write the points of $D$ in the form of a sequence $(\xi_n)$ and choose, for example, $r(\xi_n) = r/2^n$. We will find this technique again in the Lebesgue theory.

Let us now pass on to the formal proof. We denote by $E$ the set of $x \in [a, b]$ satisfying the relation

$$(13.1) \qquad f(x) \geq f(a) - \sum_{\xi < x} r(\xi) = g(x);$$

the series converges since it is a subseries of a convergent series of positive terms. All we have to prove is that $b \in E$.

The first claim to make is that the function $g$ defined by the right hand side of (1) is decreasing: for $y < x$, the $\xi < y$ are themselves $< x$; consequently, $g(x)$ is obtained by *subtracting* from $g(y)$ the numbers $r(\xi) > 0$ for the $\xi \in D$ such that $x \geq \xi > y$. This argument even shows that

$$(13.2) \qquad x > y \Longrightarrow g(x) \leq g(y) - r(y),$$

agreeing to put $r(y) = 0$ if $y \notin D$.

Now, as above, let $c = \sup(E)$; we show first that $c \in E$. For every $x \in E$ we have $x \leq c$ and so $g(x) \geq g(c)$, whence $f(x) \geq g(x) \geq g(c)$; since $c$ is the limit of points of $E$, we also have $f(c) \geq g(c)$ since $f$ is continuous.

It remains to prove that $c = b$. Now assume that $c < b$, and let us consider the $x \in {]c, b]}$. There are two possible cases.

(i) $f$ is differentiable at $c$. If $x$ is near enough to $c$, we have on the one hand that $f(x) > f(c)$ since $f'(c) > 0$, and on the other hand $g(x) \leq g(c)$ since $x > c$; whence, for these $x$, $f(x) > f(c) \geq g(c) \geq g(x)$ and so $x \in E$; absurd since $\sup(E) = c < x$.

(ii) $f$ is not differentiable at $c$, i.e. $c \in D$. For $c < x \leq b$, one has $f(x) < g(x)$ because $x \notin E$. Since $c < x$, we have $g(x) \leq g(c) - r(c)$ by (2). Thus $f(x) < g(c) - r(c)$. But $g(c) \leq f(c)$ since $c \in E$, see (1). Consequently

$$c < x \leq b \Longrightarrow f(x) < f(c) - r(c);$$

since $r(c) > 0$ because $c \in D$, this relation contradicts the continuity of $f$ at the point $c$ if $c < b$, qed.

As we have said, most authors deduce the preceding results from a statement which appears more general and which we have already met for everywhere differentiable functions:

**Mean value theorem.** *Let $f$ be a scalar function defined and* continuous *on an interval $I \subset \mathbb{R}$; assume that $f$ is differentiable at every point of $I - D$, where $D$ is a* countable *subset of $I$. Then, for any $a, b \in I$ with $a \leq b$,*

(13.3)                    $$|f(b) - f(a)| \leq M(b - a)$$

*where $M \leq +\infty$ is the upper bound of $|f'(x)|$ on the interval $[a, b]$.*

First assume that the function $f$ is real and that, for $x \in [a, b]$, $f'(x)$ has its values in a compact interval $[m, M]$ (there is nothing to prove if $f'$ is not bounded in $[a, b]$); we show that then

(13.4)                    $$m(y - x) \leq f(y) - f(x) \leq M(y - x)$$

for $a \leq x \leq y \leq b$. Now this means that the function $f(x) - mx$ is increasing and the function $f(x) - Mx$ is decreasing in $[a, b]$. The derivatives of these functions being respectively positive and negative outside $D$, we can write the final qed in the real case.

The case of a function $f$ with complex values reduces to the preceding case by an artifice already employed in the same context [Chap. III, n° 16, proof of (16.5)]: one applies the result already established for real functions to the function $f_z(x) = \operatorname{Re}[\bar{z}f(x)]$, where $z$ is an arbitrary complex constant; its derivative $\operatorname{Re}[\bar{z}f'(x)]$ lies, outside $D$, between $-M|z|$ and $M|z|$ where $M = \sup|f'(x)|$, so that

$$|\text{Re}[\bar{z}f(b)] - \text{Re}[\bar{z}f(a)]| \leq M|z|(b-a),$$

i.e.

$$|\text{Re}\bar{z}[f(b) - f(a)]| \leq M|z|(b-a);$$

it remains to choose $z = f(b) - f(a)$ and to divide by $|f(b) - f(a)|$.

**Corollary.** *Let $f$ be a function with complex values defined and* continuous *on an interval $I \subset \mathbb{R}$; assume $f$ differentiable at every point of $I - D$, where $D$ is a* countable *subset of $I$. If $f'(x) = 0$ for every $x \in I - D$, then $f$ is constant in $I$.*

Note that this corollary does not assume that $f'$ is a regulated function even though in practice ...

To establish *the existence* of a primitive, we have employed Theorem 11, i.e. integration theory. We are going to expound the Dieudonné method (or Bourbaki, *Functions of a real variable*) to resolve these problems without using this, a most instructive exercise[28]. The idea of the proof is to establish the result for step functions, which is easy, then to pass to uniform limits. First we assume $I = [a, b]$ compact, the general case being deduced easily as will be seen.

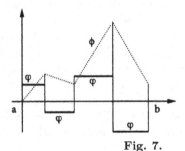

Fig. 7.

It is first of all clear that every *step* function $\varphi$ admits a primitive: using a subdivision of $I$ adapted to $\varphi$, one considers, on each interval $[x_k, x_{k+1}]$, a linear function $\Phi$ whose value at $x_k$ is equal to that of the primitive already constructed on the interval $[a, x_k]$, to ensure the continuity of the function $\Phi$ constructed piecewise in this way. $\Phi$ is a *piecewise linear* continuous function and it is not difficult to check by banal calculations that

$$(13.5) \qquad \Phi(v) - \Phi(u) = \int_u^v \varphi(x)dx$$

---

[28] The rest of this n° is more a "bonus for the reader" than an essential element of the theory.

for any $u, v \in I$. Even if one ignores, in the French or English sense, the mean value theorem, which would allow one to show that the preceding construction provides all the primitives[29] of $\varphi$, it provides, for every step function $\varphi$, at least one standard primitive defined up to an additive constant. This is the one we shall use in the course of the proof, and for good reason, since there is nothing else!

With this convention, by (5)

$$(13.6) \qquad |\Phi(u) - \Phi(v)| \leq \|\varphi\|_I . |u - v|$$

for any $u, v \in I$.

To pass from this to the case of an arbitrary regulated function $f$ one chooses a sequence of step functions $\varphi_n$ converging to $f$ uniformly on $I = [a, b]$ and, for each $n$, a primitive $\Phi_n$ of $\varphi_n$; since we obviously want the $\Phi_n$ to converge to a primitive of $f$, it is prudent to impose the condition $\Phi_n(a) = 0$. We shall show that the functions $\Phi_n$ converge uniformly on $I$ to a primitive $F$ of $f$, which will prove its existence.

Now the piecewise linear function $\Phi_{pq} = \Phi_p - \Phi_q$ is a primitive of the step function $\varphi_{pq} = \varphi_p - \varphi_q$. By (6) for $v = a$ we have

$$(13.7) \qquad |\Phi_{pq}(u)| \leq \|\varphi_{pq}\|_I . (u - a) \leq m(I)\|\varphi_{pq}\|_I$$

for every $u \in I$, whence $\|\Phi_p - \Phi_q\|_I \leq m(I)\|\varphi_p - \varphi_q\|_I$, which shows that the $\Phi_n$ satisfy Cauchy's criterion. Hence their uniform convergence to a limit function $F$, continuous like the $\Phi_n$.

We must now prove that $F'(x) = f(x)$ outside a countable set of points of $I$. Of course there is Theorem 19 of Chap. III, n° 17, but that assumes the $\varphi_n$ to be differentiable everywhere, which is not the case here except outside a finite set $D_n \subset I$ for each $n$. Outside the union $D$ of the $D_n$ one may imitate the proof of the theorem in question, Theorem 19, the essential being, in the present notation, to show that at every point $t \notin D$ one has a relation analogous to formula (17.4) of Chap. III:

$$(13.8) \qquad \lim_{x \to t} \lim_{n \to \infty} \frac{\Phi_n(x) - \Phi_n(t)}{x - t} = \lim_{n \to \infty} \lim_{x \to t} \frac{\Phi_n(x) - \Phi_n(t)}{x - t};$$

on the left hand side the limit over $n$ is $[F(x) - F(t)]/(x - t)$, so that the limit over $x$, if it exists, must be $F'(t)$; on the right hand side the limit over $x$ is the derivative in $t$ of $\Phi_n$, i.e. $\varphi_n(t)$, which exists since we are outside $D$ and *a fortiori* outside $D_n$, so that the limit over $n$ is $f(t)$; whence $F'(t) = f(t)$ modulo proving (8).

We have to argue as in Chap. III, i.e. to apply Theorem 16 of n° 12 on the "limits of limits". Again we put (for a given $t$)

---

[29] Note that while it is easy to construct a primitive of a step function "without knowing anything", to show that it is unique up to a constant reduces, even in this particularly elementary case, to proving that a function with everywhere zero derivative is constant.

$$c_n = \Phi'_n(t), \qquad u_n(x) \;=\; [\Phi_n(x) - \Phi_n(t)]/(x - t),$$
$$u(x) \;=\; [F(x) - F(t)]/(x - t)$$

and work on the set $X = I - (D \cup \{t\})$ obtained by omitting from $I$ on the one hand the points of $D$ where the $\Phi_n$ are not all differentiable, on the other hand the point $t$ where the quotients (8) are meaningless[30]. We know that $u_n(x)$ tends to $c_n$ when $x \to t$, and that $u_n(x)$ tends to $u(x)$ for every $x \in I$ when $n \to +\infty$. To deduce that the $c_n$ tend to a limit $c$ and that $u(x)$ tends to $c$ when $x \to t$, it is enough to show that the $u_n$ converge to $u$ *uniformly* on $X$ and, for this, to verify the corresponding Cauchy criterion. But again putting $\Phi_{pq} = \Phi_p - \Phi_q$, $\varphi_{pq} = \varphi_p - \varphi_q$, the general relation (6) shows that

$$|\Phi_{pq}(x) - \Phi_{pq}(t)| \leq \|\varphi_{pq}\|_I.|x - t| = \|\varphi_p - \varphi_q\|_I.|x - t|.$$

Since clearly

$$u_p(x) - u_q(x) = [\Phi_{pq}(x) - \Phi_{pq}(t)]/(x - t),$$

it follows that

$$|u_p(x) - u_q(x)| \leq \|\varphi_p - \varphi_q\|_I \qquad \text{for every } x \in X,$$

and thus $\|u_p - u_q\|_X \leq \|\varphi_p - \varphi_q\|_I$. Hence the uniform convergence of the $u_n$ on $X$. The relation (8) is thus justified at every point where the limits appearing on the *right* hand side of (8) exist, i.e. on $I - D$.

Since, what is more, the relation (5) is valid for every $\varphi_n$, it is clear that it is also valid for $f$ and $F$ by passage to the uniform limit. This finishes the proof in the case of a compact interval $I$.

The case of a noncompact interval $I$ reduces to this immediately. One chooses a point $a \in I$ and writes $I = \bigcup I_n$ where the $I_n$ are compact intervals containing $a$ and such that $I_n \subset I_{n+1}$. On each $I_n$ the function $f$ has a primitive $F_n$ such that $F_n(a) = 0$, unique since $F' = 0$ implies $F = const$ even if $F'(x)$ does not exist on the points of a countable set. Thus $F_n = F_{n+1}$ on $I_n$ for any $n$, whence the existence on $I$ of a function $F$ which, on each $I_n$, coincides with $F_n$. It is clear that $F$ satisfies Theorem 9 bis, etc.

## 14 – Convex functions; Hölder and Minkowski inequalities

Let $a$ and $b$ be two distinct points in a Cartesian space $\mathbb{R}^p$. A point $x \in \mathbb{R}^p$ lies on the line joining $a$ and $b$ if and only if the vector $x - b$ is proportional to the vector $a - b$, i.e. $x - a = t(b - a)$ or

(14.1) $$x = (1 - t)a + tb$$

---

[30] Note in passing the usefulness of defining uniform convergence for functions defined on any set, and not only on an interval of $\mathbb{R}$. Nor is it any more difficult
. . .

for some $t \in \mathbb{R}$. Since the points $a$ and $b$ correspond to the values 0 and 1 of $t$, we conclude that the points of the line segment $[a, b]$ joining $a$ to $b$ are obtained for $t \in [0, 1]$. One might even consider this statement as a definition of $[a, b]$.

We have said (Chap. III, n° 10, Example 1) that a subset $X$ of $\mathbb{R}^p$ is *convex* if

(14.2)                    $[a, b] \subset X$      for any $a, b \in X$.

In $\mathbb{R}$, the only convex sets are the intervals. In $\mathbb{C}$, the interiors (also if one includes the boundary) of a circle, of an ellipse, of a triangle, of a rectangle, etc. are convex. A circular ring is not.

One may generalise (2) and show that[31]

(14.3)                    $t_1 x_1 + \ldots + t_n x_n \in X$

for any $x_i \in X$ and $t_i$ satisfying $t_i > 0$, $\sum t_i = 1$; one shows this by induction on $n$, introducing the point

$$x = (t_1 x_1 + \ldots + t_{n-1} x_{n-1})/(t_1 + \ldots + t_{n-1}),$$

which is in $X$ by hypothesis, and observing that the point (3) is precisely $tx + (1 - t)x_n$ where $t = t_1 + \ldots + t_{n-1} = 1 - t_n$.

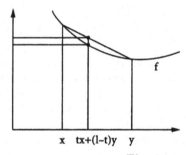

$$x \quad tx+(1-t)y \quad y$$

**Fig. 8.**

Let $f$ be a real-valued function defined on a convex set $X \subset \mathbb{R}^p$; its graph is then the set of points of $\mathbb{R}^p \times \mathbb{R} = \mathbb{R}^{p+1}$ of the form $(x, f(x))$, and the subset of $\mathbb{R}^{p+1}$ situated "above" the graph is the set of $(x, y)$ such that $x \in X$ and $y \geq f(x)$. One says that $f$ is a *convex* function if this set is convex. One sees immediately that this is so if and only if

(14.4)            $f[(1 - t)x + ty] \leq (1 - t)f(x) + tf(y)$

---

[31] The point (3) is, in mechanics, the "centre of gravity" of the "masses" $t_i \geq 0$ placed at the points $x_i$. When their sum is not equal to 1 one clearly has to divide the result by the total mass. *Exercise*: show that the medians of a triangle are concurrent.

for any $x, y \in X$ and $0 < t < 1$. Arguing as we did to establish (3), we deduce that

(14.5)        $f(t_1 x_1 + \ldots + t_n x_n) \leq t_1 f(x_1) + \ldots + t_n f(x_n)$

for any $x_i \in X$ and $t_i > 0$ of sum 1.

In the case where $X$ is an open interval of $\mathbb{R}$ one may characterise the convex functions completely by differentiability properties:

**Theorem 15.** *The real-valued function $f$ defined on an open interval $X$ of $\mathbb{R}$ is convex if and only if it is a primitive of an increasing function. The function $f$ is continuous[32], has right and left derivatives everywhere, and is differentiable outside a countable subset of $X$.*

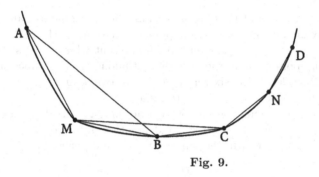

Fig. 9.

Consider Figure 9. When $B$ tends to $M$, the slope of $MB$, which decreases while remaining above that of $AM$, tends to a limit, whence the existence at every $x \in X$ of a right derivative $f'_d(x)$ and, likewise, of a left derivative $f'_s(x) \leq f'_d(x)$; whence also the continuity of $f$. The slopes of the lines $AM$, $MB$, $MC$, $BC$, $CN$ and $ND$ increase since the point $M$, for example, lies below the segment $[A, B]$. Since on the other hand the slope of $MB$ is less than that of $CN$, itself less than its limit $f'_s(y)$ when $C$ tends to $N$, so

(14.6)        $f'_s(x) \leq f'_d(x) \leq f'_s(y) \leq f'_d(y)$        for $x < y$;

the functions $f'_s$ and $f'_d$ are therefore increasing[33]. On letting $y$ tend to $x$ in (6) one finds

(14.7)        $f'_s(x) \leq f'_d(x) \leq f'_s(x+) \leq f'_d(x+)$;

on letting $x$ tend to $y$, one finds likewise

---

[32] This is not necessarily the case if $X$ is a non-open interval. On $X = [0, 1]$ the function equal to 0 for $0 < x < 1$, and to 1 at the end-points, is convex.
[33] For functions of class $C^1$ the reader may pass directly to (9).

$$f'_s(y-) \le f'_d(y-) \le f'_s(y) \le f'_d(y),$$

which, applied to $x$, allows us to complete (7) as

$$f'_s(x-) \le f'_d(x-) \le f'_s(x) \le f'_d(x) \le f'_s(x+) \le f'_d(x+).$$

Now we have $f'_d(y) \le f'_s(z)$ for $y < z$ by (6); on letting $y$ and $z$ tend to $x+$ one finds $f'_d(x+) \le f'_s(x+)$, whence

$$f'_s(x+) = f'_d(x+) \quad \text{and likewise} \quad f'_s(x-) = f'_d(x-)$$

for every $x$; finally

(14.8) $\qquad f'_s(x-) = f'_d(x-) \le f'_s(x) \le f'_d(x) \le f'_s(x+) = f'_d(x+).$

The functions $f'_s$ and $f'_d$ being increasing, and so regulated, can be discontinuous only on a countable set $D$ of points of $X$ (Chap. III, n° 12, Corollary to Theorem 16 or n° 7, Theorem 6 of the present Chap.); (8) shows moreover that the points where they are discontinuous are the same for the two functions. Outside $D$, the six terms of (8) are equal, and $f$ is differentiable since its right and left derivatives are equal.

The function $f$ being continuous, and admitting, outside $D$, a derivative equal, at one's choice, to the regulated function $f'_s$ or to $f'_d$, it must be a primitive of $f'_s$ and of $f'_d$ in the sense of n° 13; in other words,

$$f(y) - f(x) = \int_x^y f'_s(u)du = \int_x^y f'_d(u)du$$

for any $x, y \in X$.

Conversely, let us start from an increasing, so regulated, function $g$ on $X$ and let $f$ be a primitive of $g$. Consider two points $x$, $y$ of $X$ and suppose for example $x < y$. For $t \in [0,1]$ one has

(14.9)
$$\begin{aligned} f[(1-t)x + ty] - \{(1-t)f(x) + tf(y)\} = \\ = \{f[x + t(y-x)] - f(x)\} - t[f(y) - f(x)] = \\ = \int_0^{y-x} tg(x+tv)dv - t\int_0^{y-x} g(x+v)dv \end{aligned}$$

as one sees on differentiating the functions $f(x+tv)$ and $f(x+v)$ with respect to $v$ and applying the FT. But since $g$ is increasing, and since $v \ge 0$ on the interval of integration, we have

$$g(x + tv) \le g(x + v) \quad \text{for } t \in [0,1].$$

The difference between the two last integrals is thus $\le 0$ so the function $f$ is convex, qed.

**Corollary 1.** *A differentiable function f is convex on an open interval if and only if its derivative is increasing. A twice differentiable function is convex if and only if $f''(x) \geq 0$ for every x.*

For if $f'(x)$ exists everywhere, and is increasing, so regulated, then $f$ is a primitive of $f'$ and is therefore convex. If $f'$ is differentiable it is increasing if and only if $f''(x) \geq 0$ everywhere, by the mean value theorem (Chap. III, n° 16).

The case (a bonus for the reader) of a function $f$ defined on a *convex open* subset $X$ of $\mathbb{R}^p$, for example of $\mathbb{C}$, reduces to the preceding if one assumes $f$ of class $C^2$ on $X$. In the first place, the convexity of $f$ obviously means that, for any $x, y \in X$, the function $t \mapsto f[(1-t)x + ty]$, defined on a convex open subset (i.e. an open interval) of $\mathbb{R}$, is convex. Since $f$ is of class $C^1$, this function has a derivative equal to

$$\frac{d}{dt}f[(1-t)x_1+ty_1,\ldots,(1-t)x_p+ty_p] = \sum D_i f[(1-t)x+ty](y_i-x_i)$$

by the chain rule of Chap. III, n° 21. The result obtained is again differentiable if $f$ is of class $C^2$, with, for the same reason,

$$(14.10) \quad \frac{d^2}{dt^2}\ldots = \sum_{i,j} D_j D_i f[(1-t)x+ty](y_j-x_j)(y_i-x_i).$$

By Corollary 1, this result must be $\geq 0$; in particular, we must have (take $t = 0$)

$$(14.11) \qquad \sum D_j D_i f(x)u_i u_j \geq 0$$

whenever the $u_i$ can be put in the form $y_i - x_i$ for some $y \in X$. But since $X$ is open these differences can, for $x$ given, take all values sufficiently close to 0, so that, for arbitrary $u_i \in \mathbb{R}$, (11) must be satisfied by the $tu_i$ for every $t \in \mathbb{R}$ sufficiently close to 0; since this substitution multiplies (11) by $t^2 > 0$, we conclude that (11) must be satisfied for any $u_i \in \mathbb{R}$: the quadratic form (11) must be *positive semidefinite*, as one says in algebra. Conversely, it is clear that if (11) is satisfied for any $x \in X$ and $u_i \in \mathbb{R}$, then the derivative (10) is $\geq 0$ so long as $(1-t)x + ty \in X$; the function

$$t \mapsto f[(1-t)x+ty]$$

is thus convex, so $f$ is too. Consequently:

**Corollary 2.** *Let f be a real-valued function defined and of class $C^2$ in a convex open set $X \subset \mathbb{R}^p$. Then f is convex on X if and only if*

$$\sum D_i D_j f(x)u_i u_j \geq 0$$

*for any $u_i \in \mathbb{R}$ and $x \in X$.*

Theorem 15 allows us to establish the famous inequalities of Hölder and of Minkowski, which are not very useful at our level – but general culture ... The function $e^x$ has an everywhere positive second derivative and so is convex on $\mathbb{R}$, whence

$$e^{tx+(1-t)y} \leq te^x + (1-t)e^y$$

for any $x, y \in \mathbb{R}$ and $0 < t < 1$. We can also write this in the form

(14.12) $$a^t b^{1-t} \leq ta + (1-t)b$$

for $a, b > 0$ and even for $0 \leq a$, $b \leq +\infty$ on agreeing that $0^t = 0$, $(+\infty)^t = +\infty$, $t.(+\infty) = +\infty$ for $t > 0$ and that $0.(+\infty) = 0$.

Now consider an arbitrary set $X$ and suppose that to every function $f$ defined on $X$ and with values in $[0, +\infty]$ one has associated a number $\mu^*(f)$ possessing the following properties[34]:

(IS 1):  $0 \leq \mu^*(f) \leq +\infty$;
(IS 2): the relation $f \leq g$ implies $\mu^*(f) \leq \mu^*(g)$;
(IS 3):  $\mu^*(f+g) \leq \mu^*(f) + \mu^*(g)$ for any $f, g$;
(IS 4):  $\mu^*(cf) = c\mu^*(f)$ for every constant $c \geq 0$.

If $F$ and $G$ are functions on $X$ with values in $[0, +\infty]$ one has

$$F^t G^{1-t} \leq tF + (1-t)G$$

by (12); whence, using (IS 2), (IS 3) and (IS 4),

$$\begin{aligned} \mu^*(F^t G^{1-t}) &\leq \mu^*[tF + (1-t)G] \leq \mu^*(tF) + \mu^*[(1-t)G] \\ &\leq t\mu^*(F) + (1-t)\mu^*(G) \end{aligned}$$

for $0 < t < 1$. In particular we see that

$$\mu^*(F) = \mu^*(G) = 1 \Longrightarrow \mu^*(F^t G^{1-t}) \leq 1.$$

If one assumes only that $\mu^*(F)$ and $\mu^*(G)$ are finite and nonzero, and if one applies the last result to the functions $F/\mu^*(F)$ and $G/\mu^*(G)$, which is legitimate by (IS 4), the function $F^t G^{1-t}$ is divided by the constant $\mu^*(F)^t \mu^*(G)^{1-t}$, whence, using (IS 4) again,

(14.13) $$\mu^*(F^t G^{1-t}) \leq \mu^*(F)^t \mu^*(G)^{1-t}.$$

Now put

---

[34] These conditions generalise the properties of the upper integral established in n° 11 and are met in the general theory of integration. It is not really necessary to assume that $\mu^*(f)$ is defined for *every* positive function on $X$; it is enough that the formulae that we are going to write should be meaningful.

(14.14)                $$N_p(f) = \mu^* \left(|f|^p\right)^{1/p} \leq +\infty$$

for every function $f$ with complex values on $X$ and every real number $p > 0$.
If $f$ and $g$ are two such functions and if $p$ and $q$ are real numbers $> 0$, let us
choose $F = |f|^p$, $G = |g|^q$ and $t = r/p$, $1 - t = r/q$ where $r > 0$ is such that
$r/p + r/q = 1$, i.e. such that

(14.15)                $$1/p + 1/q = 1/r.$$

Then $F^t G^{1-t} = |f|^{pr/p}|g|^{qr/q} = |fg|^r$, so that (13) becomes

$$\mu^*(|fg|^r) \leq \mu^* \left(|f|^p\right)^{r/p} \mu^* \left(|g|^q\right)^{r/q},$$

whence, on raising both sides to the power $1/r$,

(14.16)        $$N_r(fg) \leq N_p(f)N_q(g) \qquad \text{for } 1/p + 1/q = 1/r.$$

Note that $N_r(f)$ *is finite if* $N_p(f)$ *and* $N_q(g)$ *are finite* because (12), applied
to $F = |f|^p$, $G = |g|^q$, $t = r/p$, $1 - t = r/q$, shows that

(14.17)                $$|fg|^r/r \leq |f|^p/p + |g|^q/q,$$

so that it remains to apply the axioms (IS).
    (16) is the inequality of (Otto) Hölder, most often used only for $r = 1$:

(14.18)        $$\mu^*(|fg|) \leq \mu^* \left(|f|^p\right)^{1/p} \mu^* \left(|g|^q\right)^{1/q} = N_p(f)N_q(g);$$

this inequality assumes that $1/p + 1/q = 1$ and thus $p, q > 1$ (*conjugate
indices*). The case where $p = q = 2$ is just the Cauchy-Schwarz inequality
extended to the functions $\mu^*$.
    From (18) one may deduce the inequality

(14.19)        $$N_p(f + g) \leq N_p(f) + N_p(g) \qquad \text{for } p > 1$$

of (Hermann) Minkowski, one of Einstein's professors at the Polytechnic of
Zürich, and, at Göttingen in 1907–1908, the inventor of the interpretation
of Relativity in $\mathbb{R}^4$ endowed with the quadratic form $x^2 + y^2 + z^2 - c^2t^2$.
He would probably have gone much further if he had not died soon after ...
There is nothing to prove if the right hand side is infinite or if the left hand
side is zero. If the right hand side is finite, so too is the left hand side, for,
the function $x^p$ being convex in $x \geq 0$ for $p > 1$, one has

$$\left(\frac{|f| + |g|}{2}\right)^p \leq \frac{1}{2}\left(|f|^p + |g|^p\right).$$

This done, we write

$$|f + g|^p \leq (|f| + |g|)^p = |f|.|f + g|^{p-1} + |g|.|f + g|^{p-1}.$$

Since $|f + g|^{(p-1)q} = |f + g|^p$, we have $N_q(|f + g|^{p-1}) < +\infty$; Hölder's inequality (18) then shows that

$$\begin{aligned}
\mu^*(|f + g|^p) &\leq [N_p(f) + N_p(g)]N_q(|f + g|^{p-1}) \\
&= [N_p(f) + N_p(g)].\mu^* \left(|f + g|^{(p-1)q}\right)^{1/q} \\
&= [N_p(f) + N_p(g)].\mu^*(|f + g|^p)^{1-1/p};
\end{aligned}$$

on multiplying both sides by $\mu^* (|f + g|^p)^{-1+1/p}$ we obtain

$$\mu^*(|f + g|^p)^{1/p} \leq N_p(f) + N_p(g),$$

i.e. (19).

It is probable that the reader has not truly understood these proofs; he may console himself in knowing that the majority of the professionals are in the same state, having only checked the argument step-by-step, registered the results, and forgotten the proofs for the rest of their lives (unless needing to expound them to students . . .).

*Example 1.* If $f$ and $g$ are regulated functions on a bounded interval $I$

$$(14.20) \qquad \left|\int_I f(x)g(x)dx\right| \leq N_p(f)N_q(f), \quad N_p(f + g) \leq N_p(f) + N_p(g)$$

where one puts

$$N_p(f) = \left(\int_I |f(x)|^p dx\right)^{1/p},$$

the $L^p$ norm of the function $f$. You might replace the traditional integral $m(f)$ by the expression

$$\mu(f) = \int |f(x)|\mu(x)dx$$

where $\mu$ is a given positive integrable function; it clearly satisfies conditions (IS 1) to (IS 4) above. One thus finds the inequalities (20) where the symbol $dx$ is replaced everywhere by $\mu(x)dx$. Without doubt we will be made to observe that this "generalisation" is illusory since it follows from the classical case on replacing $f(x)$ by $f(x)\mu(x)^{1/p}$ and $g(x)$ by $g(x)\mu(x)^{1/q}$; precisely. But the objection falls if one defines $\mu(f)$ from an arbitrary Radon measure (n° 30).

*Example 2.* Take for $X$ a finite set and $\mu^*(f) = \sum |f(x)|$. One obtains, in more traditional notation, the original versions of the inequalities:

$$(14.21) \qquad \left|\sum x_k y_k\right| \leq \left(\sum |x_k|^p\right)^{1/p} \left(\sum |y_k|^q\right)^{1/q},$$

$$(14.22) \qquad \left(\sum |x_k + y_k|^p\right)^{1/p} \le \left(\sum |x_k|^p\right)^{1/p} + \left(\sum |y_k|^p\right)^{1/p}.$$

*Example 3.* Like the preceding, but with an infinite set $X$ and, again, $\mu^*(f) = \sum |f(x)| \le +\infty$ for every function $f$ with positive values. *If the series $\sum |x_n|^p$ and $\sum |y_n|^q$ converge then the series $\sum x_n y_n$ converges absolutely and*

$$(14.23) \qquad \left|\sum x_n y_n\right| \le \left(\sum |x_n|^p\right)^{1/p} \left(\sum |y_n|^q\right)^{1/q}.$$

All this assumes $p, q > 1$ and $1/p + 1/q = 1$. The case $p = q = 2$ is the Cauchy-Schwarz inequality for series, which may be proved much more easily by passing to the limit starting from the case of a finite sum.

## § 4. Integration by parts

### 15 – Integration by parts

The chain rule shows that if $f$ and $g$ are two functions of class $C^1$ then $fg$ is a primitive of the function $f'g + fg'$; consequently, in Leibniz' style,

$$\int [f'(x)g(x) + f(x)g'(x)]dx = f(x)g(x),$$

which can be rewritten in the form

(15.1) $$\int f'(x)g(x)dx = f(x)g(x) - \int f(x)g'(x)dx;$$

this relation between *primitives* can be transformed into a relation between *definite integrals*:

(15.2) $$\int_a^b f'(x)g(x)dx = f(x)g(x)\Big|_a^b - \int_a^b f(x)g'(x)dx.$$

This is the formula for integration by parts, which allows us to calculate very many elementary integrals. It remains valid when $f$ and $g$ are the primitives of two regulated functions, since then the function $fg$ is continuous and has as derivative the regulated function $f'g+fg'$ outside a countable set of points, so is a primitive of $f'g + fg'$ (Theorem 12 bis).

*Example 1.* Let us calculate the primitive

$$\begin{aligned}
\int \log(x)dx &= \int \log(x).1.dx = \int \log(x).(x)'.dx = \\
&= \log(x)x - \int \log'(x)x dx = x\log x - \int 1dx,
\end{aligned}$$

whence

(15.3) $$\int \log(x)dx = x\log x - x.$$

It is not difficult to check that $\log x$ is indeed the derivative of $x\log x - x$.

*Example 2.* We have

$$\begin{aligned}
\int x^5 e^x dx &= \\
&= \int x^5 (e^x)'dx = x^5 e^x - \int (x^5)'e^x dx = x^5 e^x - 5\int x^4 e^x dx = \\
&= x^5 e^x - 5x^4 e^x + 5.4 \int x^3 e^x dx =
\end{aligned}$$

$$= x^5 e^x - 5x^4 e^x + 5.4x^3 e^x - 5.4.3 \int x^2 e^x dx =$$

$$= x^5 e^x - 5x^4 e^x + 5.4x^3 e^x - 5.4.3x^2 e^x + 5.4.3.2 \int x e^x dx =$$

$$= x^5 e^x - 5x^4 e^x + 5.4x^3 e^x - 5.4.3x^2 e^x + 5.4.3.2x e^x - 5.4.3.2.1 \int e^x dx =$$

$$= x^5 e^x - 5x^4 e^x + 5.4x^3 e^x - 5.4.3x^2 e^x + 5.4.3.2x e^x - 5.4.3.2.1 e^x.$$

The reader can generalise this to $x^n e^x$ for $n \in \mathbb{N}$.

And for $n < 0$? Let's try our luck.

$$\int x^{-2} e^x dx =$$

$$= \int (-x^{-1})' e^x dx = -e^x/x + \int x^{-1} e^x dx =$$

$$= -e^x/x + \int \log'(x) e^x dx = -e^x/x + e^x \log x - \int e^x \log(x) dx =$$

$$= -e^x/x + e^x \log x - \int e^x [x \log(x) - x]' dx =$$

$$= -e^x/x + e^x \log x - e^x [x \log(x) - x] + \int e^x [x \log(x) - x] dx,$$

and you may continue indefinitely without ever managing to eliminate the function $e^x \log x$. It would certainly be enough to know *one* primitive, which the theory of the power series allows us to find theoretically; but in practice one naturally tries to express the given integral in terms of a simple combination of "elementary", i.e. already known, functions. This is impossible for the functions $x^s e^x$ with $s \notin \mathbb{N}$, or $e^x \log x$, or $\exp(\exp(x))$, etc.

Moral: one does not have to go very far to find oneself face to face with elementary functions whose primitives are not elementary. To be able to calculate a primitive explicitly is almost always a miracle, and even, in teaching analysis, a contrived miracle: the calculation put to you as an exercise is feasible because the author of the exercise knows it in advance, generally from having read through the older authors to extract some from the very many exercises of the same kind.

In fact, mathematicians who have sought in vain to calculate a primitive or a definite integral of a function which is important to their applications – as happens often in mechanics or in physics – always end up changing their tactics: they give a name once and for all to the mysterious primitive or integral in question, and, instead of calculating it, try to establish its useful properties (differential equation, series expansions, asymptotic behaviour, integrals linked to these functions, etc.); these are the *special functions*. At the lowest level, this is what has always been done for the trigonometric functions: one gives them a name, and, instead of calculating them from beautiful simple algebraic formulae which do not exist, one derives their properties. This is also what the theory of the elliptic functions has allowed mathematicians to do for the primitives of the functions of the form $P(x)^{1/2}$, where $P$ is a polynomial of degree 3 or 4.

*Example 3.* Let us try to calculate the functions

$$(15.4) \qquad I_n = \int x^n \cos x.dx, \qquad J_n = \int x^n \sin x.dx.$$

We may write

$$I_n = \int x^n \sin' x.dx = x^n \sin x - n \int x^{n-1} \sin x.dx$$

and continue. It is more economical to observe that

$$I_n + iJ_n = \int x^n e^{ix} dx$$

and to calculate as in Example 2 or, for good measure, to calculate $\int x^n e^{tx} dx$ for every $t \in \mathbb{C}$. We find easily

$$(15.5) \qquad \int x^n e^{tx} dx =$$
$$= e^{tx} \left[ x^n/t - nx^{n-1}/t^2 + n(n-1)x^{n-2}/t^3 + \ldots + (-1)^n n!/t^{n+1} \right],$$

whence, for $t = i$, on separating the real and imaginary parts,

$$(15.6') \qquad I_n \quad = \quad x^n \sin x + nx^{n-1} \cos x - n(n-1)x^{n-2} \sin x - $$
$$- n(n-1)(n-2)x^{n-3} \cos x + \ldots,$$
$$(15.6'') \qquad J_n \quad = \quad -x^n \cos x + nx^{n-1} \sin x + n(n-1)x^{n-2} \cos x - $$
$$- n(n-1)(n-2)x^{n-3} \sin x - \ldots.$$

*Example 4.* Put $\log^2 x = (\log x)^2$ and calculate

$$\int x.\log^2 x.dx \quad = \quad \int \left( \frac{1}{2}x^2 \right)' \log^2 x.dx = \frac{1}{2}x^2 \log^2 x - \frac{1}{2} \int x^2 (\log^2 x)' dx =$$
$$= \quad \frac{1}{2}x^2 \log^2 x - \frac{1}{2} \int x^2 2 \log x (\log' x) dx =$$
$$= \quad \frac{1}{2}x^2 \log^2 x - \int x.\log x.dx =$$
$$= \quad \frac{1}{2}x^2 \log^2 x - \frac{1}{2}x^2 \log x + \frac{1}{2} \int x^2 \log' x.dx =$$
$$= \quad \frac{1}{2}x^2 \log^2 x - \frac{1}{2}x^2 \log x + \frac{1}{2} \int x dx =$$
$$= \quad \frac{1}{2}x^2 \log^2 x - \frac{1}{2}x^2 \log x + x^2/4.$$

One should not forget that an arbitrary constant may be added to the result in all these formulae.

## 16 – The square wave Fourier series

We are now in a position to justify the square wave series [Chap. III, eqn. (2.4) and n° 11, Example 1]

$$(16.1) \qquad s(x) = \cos x - \cos 3x/3 + \cos 5x/5 - \ldots = \pi/4 \quad \text{for} \quad |x| < \pi/2$$

which Fourier first obtained by breathtaking meaningless calculations; its value for $\pi/2 < |x| < \pi$, namely $-\pi/4$, follows from (1) on replacing $x$ by $\pi - x$. Fourier himself did not mention this explicitly in his memoir, but one has to believe that he had some doubts since he later felt the need to justify his result by patently more reasonable methods. We shall follow him.

We start quite naturally by calculating the partial sum

$$(16.2) \quad s_n(x) = \cos x - \cos 3x/3 + \ldots + (-1)^{n-1} \cos(2n-1)x/(2n-1),$$

not by an impossible direct calculation, but by differentiating it. We have

$$s_n'(x) = -\sin x + \sin 3x - \ldots + (-1)^n \sin(2n-1)x,$$

whence, putting $q = e^{ix}$ and using Euler's formulae,

$$
\begin{aligned}
-2is_n'(x) &= (q - q^{-1}) - (q^3 - q^{-3}) + \ldots + (-1)^{n-1}\left(q^{2n-1} - q^{-2n+1}\right) = \\
&= q\left[1 - q^2 + \ldots + (-1)^{n-1}q^{2n-2}\right] - q^{-1}\left[1 - q^{-2} + \ldots\right] = \\
&= q\frac{1 - (-1)^n q^{2n}}{1 + q^2} - q^{-1}\frac{1 - (-1)^n q^{-2n}}{1 + q^{-2}} = \\
&= \frac{1 - (-1)^n q^{2n}}{q^{-1} + q} - \frac{1 - (-1)^n q^{-2n}}{q + q^{-1}} = \\
&= (-1)^{n+1}\frac{q^{2n} - q^{-2n}}{q + q^{-1}} = (-1)^{n+1}2i\sin 2nx/2\cos x.
\end{aligned}
$$

Thus

$$s_n'(x) = (-1)^n \frac{\sin 2nx}{2\cos x} \; .$$

This calculation assumes $\cos x \neq 0$, which is the case in the interval of interest.

From this we deduce, by the FT, that

$$(16.3) \qquad s_n(y) - s_n(x) = (-1)^n \int_x^y \frac{\sin 2nt}{2\cos t}\, dt$$

for $x$ and $y$ in the open interval $I = ]-\pi/2, \pi/2[$. Since the sum of the series is supposed to be a constant function we ought to show that the difference (3), which tends to $s(y) - s(x)$, tends to $0$ as $n$ increases indefinitely.

This is not obvious at first sight, but let us integrate by parts; we get

$$\int_x^y \frac{\sin 2nt}{\cos t}\, dt = \left. \frac{-\cos 2nt}{2n.\cos t}\right|_x^y - \frac{1}{2n}\int_x^y \cos 2nt \frac{\sin t}{\cos^2 t}\, dt.$$

As $n$ increases indefinitely the first term of the right hand side tends to 0 because of the factor $n$ in the denominator. Likewise for the second term, for, in the interval of integration, we have

$$|\cos 2nt. \sin t| \le 1, \qquad \cos^2 t \ge m$$

where $m$ is the minimum of $\cos^2 t$ between $x$ and $y$; it is $> 0$ because a continuous function attains its minimum on a compact interval, and the interval $[x, y]$ contains no odd multiple of $\pi/2$. The factor $1/2n$ therefore provides an upper bound $O(1/n)$.

The sum $s(x)$ of the series (1) is therefore constant on the interval $|x| < \pi/2$. Its value must be

(16.4) $$s(0) = 1 - 1/3 + 1/5 - \ldots = \pi/4,$$

Leibniz' series.

We proved (4), more or less, in Chap. IV, n° 14 using the series expansion

(16.5) $$\arctan y = y - y^3/3 + y^5/5 - \ldots;$$

one obtains this by starting from the formula

$$\arctan' y = 1/(1 + y^2) = 1 - y^2 + y^4 - \ldots$$

and applying the method valid for every power series. Nevertheless one must be aware of the fact that this applies *to the interior* of the disc of convergence, i.e. for $|y| < 1$; now Leibniz' formula corresponds precisely to the value $y = 1$, where $\arctan y = \pi/4$. To justify this one has to argue as we did *à propos* the series $\log(1+x)$, which, for $x = 1$, yields the series $\log 2 = 1 - 1/2 + 1/3 - \ldots$ by passage to the limit: for $0 \le y \le 1$ the series (5) is alternating and has decreasing terms, the difference between its total sum and its $n$-th partial sum is thus majorised by its $n$-th term, so by $1/(2n + 1)$ for any $y$, whence one concludes that the partial sums of (5) converge to the total sum *uniformly* on the *closed* interval $[0, 1]$. The total sum of the series is therefore a continuous function there; likewise for the function $\arctan y$; now, if two functions are continuous for $y \le 1$, and coincide for $y < 1$, then they remain equal at $y = 1$. Whence (4).

Starting from the series (1), one may obtain others by integration. Calculating formally,

(16.6) $$\int_0^x s(t)dt = \sin x - \sin 3x/3^2 + \sin 5x/5^2 - \ldots$$

and since $s(t)$ is a step function, it will not be difficult to calculate the integral directly. But first one has to justify integrating the series (1) term-by-term.

Recall (Chap. III, n° 11) that the square wave series, though not uniformly convergent on all $\mathbb{R}$ since its sum is discontinuous, converges uniformly on every interval $[-\pi/2 + r, \pi/2 - r]$ with $r > 0$. The formula (6) is therefore legitimate for $|x| < \pi/2 - r$, so for $|x| < \pi/2$ since $r$ is arbitrarily small. Since, moreover, $s(x) = \pi/4$ on this interval, the integral is equal to $\pi x/4$. Whence

$$(16.7) \qquad \sin x - \sin 3x/3^2 + \sin 5x/5^2 - \ldots = \pi x/4 \text{ for } |x| \leq \pi/2.$$

The series is this time normally convergent on $\mathbb{R}$ because it is dominated by the series $\sum 1/(2n+1)^2$; its sum is thus continuous everywhere, and this, by passage to the limit, justifies the value $\pi^2/8$ which (7) attributes to it for $x = \pi/2$.

For $\pi/2 \leq |x| \leq 3\pi/2$, one puts $x = y + \pi$, which brings one back to the preceding case:

$$(16.8) \qquad \sin x - \sin 3x/3^2 + \sin 5x/5^2 - \ldots \; = \; \pi(\pi - x)/4$$
$$\text{for } \pi/2 \leq |x| \leq 3\pi/2.$$

The left hand side being periodic, it is unnecessary to continue the calculation; it would be better to sketch the "curve", displayed by Fourier himself:

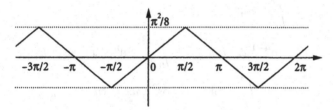

Fig. 10.

Note that for $x = \pi/2$ we again find the relation

$$1 + 1/3^2 + 1/5^2 + \ldots = \pi^2/8$$

obtained in n° 5, eqn. (5.10) by applying the Parseval-Bessel formula to the square wave series even though we had not yet justified formula (1).

Let us continue on the same track. Being normally convergent the series (7) can be integrated term-by-term, for example on $[-\pi/2, x]$, which produces the series $-\cos x + \cos 3x/3^3 + \ldots$. For $-\pi/2 \leq x \leq \pi/2$ the formula (7) yields an integral equal to $\pi x^2/8 - \pi^3/32$, whence

$$(16.9) \qquad \cos x - \cos 3x/3^3 + \ldots = \frac{\pi}{8}(\pi/2 + x)(\pi/2 - x) \text{ for } |x| \leq \pi/2.$$

For $x = 0$ one finds

$$(16.10) \qquad\qquad 1 - 1/3^3 + 1/5^3 - \ldots = \pi^3/32.$$

On continuing to integrate the reader would find infinitely many other formulae of the same kind, the first already in Euler, by other methods, of course. One may also examine what the Parseval-Bessel formula of (5.6) or (5.7) of this Chap. V may provide; there is no problem since the successive primitive series of the square wave series are (more and more) absolutely convergent. This kind of exercise is appreciably more instructive than calculating a primitive of a rational fraction chosen only to make the candidate stumble over calculations of no other interest.

### 17 – Wallis' formula

Let us put

$$(17.1) \qquad\qquad I_n = \int_0^{\pi/2} \sin^n x.dx$$

for $n \in \mathbb{N}$. First,

$$(17.2) \qquad\qquad I_0 = \pi/2, \qquad I_1 = 1.$$

Next, integrate by parts, whence

$$
\begin{aligned}
I_n &= -\int_0^{\pi/2} \sin^{n-1} x. \cos' x.dx = \\
&= -\sin^{n-1} x. \cos x \Big|_0^{\pi/2} + \int_0^{\pi/2} (\sin^{n-1} x)' \cos x.dx = \\
&= (n-1)\int_0^{\pi/2} \sin^{n-2} x. \cos^2 x.dx = (n-1)(I_{n-2} - I_n).
\end{aligned}
$$

Finally we obtain $I_n = I_{n-2}(n-1)/n$, whence

$$I_{2n} = \frac{(2n-1)(2n-3)\ldots 1}{2n(2n-2)\ldots 2} I_0 \qquad \text{with } I_0 = \pi/2$$

and

$$I_{2n+1} = \frac{2n(2n-2)\ldots 2}{(2n+1)(2n-1)\ldots 3} I_1 \qquad \text{with } I_1 = 1.$$

Since $0 \le \sin x \le 1$ on the interval of integration it is clear that $I_{n+1} \le I_n$, thus that $I_{2n+1} \le I_{2n} \le I_{2n-1}$, whence

$$1 \le I_{2n}/I_{2n+1} \le I_{2n-1}/I_{2n+1} = 1 + 1/2n.$$

The ratio

$$I_{2n}/I_{2n+1} = \frac{(2n+1)(2n-1)^2(2n-3)^2\ldots 3^2}{(2n)^2(2n-2)^2\ldots 2^2} \frac{\pi}{2}$$

thus tends to 1, whence *Wallis'* famous *formula*

$$(17.3) \qquad \frac{2}{\pi} = \lim \frac{1.3.3.5.5.7\ldots(2n-1)(2n+1)}{2.2.4.4.6.6\ldots 2n.2n}.$$

Fourier, who used it, wrote it in the form

$$\frac{2}{\pi} = 3.3.5.5.7.7.\ldots/2.2.4.4.6.6.\ldots;$$

though neither the numerator nor the denominator makes any sense, one may try to interpret it as an infinite product (Chap. IV, n° 17). In the form $3/2.3/2.5/4.5/4\ldots$ it is clearly divergent since the general term is of the form $(p+1)/p = 1 + 1/p$. One may then try the product of all the expressions $3.3/2.2$, etc., i.e. the infinite product with general term $(2n+1)^2/4n^2 = (1+1/2n)^2$; but

$$(1+1/2n)^2 = 1 + 1/n + 1/4n^2 = 1 + u_n$$

where the series $\sum u_n$ is divergent since $u_n \sim 1/n$; new catastrophe. One might prefer the groupings $3.3/4.4$, etc., which leads to the product of the

$$(2n+1)^2/(2n+2)^2 = [1 - 1/(2n+2)]^2 = 1 - 1/(n+1) + 1/4(n+1)^2 = 1 + u_n,$$

which diverges as much as the preceding ones, and for the same reason, namely that $u_n \sim -1/n$. The solution is to write (3) in the form

$$2/\pi = \lim \frac{1.3}{2.2} \frac{3.5}{4.4} \frac{5.7}{6.6} \cdots \frac{(2n-1)(2n+1)}{2n.2n},$$

or

$$(17.4) \qquad 2/\pi = \prod_{n=1}^{\infty}(1 - 1/4n^2) = (1 - 1/4)(1 - 1/16)(1 - 1/36)\ldots$$

The infinite product is this time absolutely convergent like the series $\sum 1/n^2$ and is greatly preferable to the formula (3), which would lead you to perdition as we have just seen.

A very fast proof Wallis' formula comes from starting from the infinite product

$$\sin \pi x = \pi x \prod(1 - x^2/n^2);$$

one recovers Wallis when $x = 1/2$.

# § 5. Taylor's Formula

## 18 – Taylor's Formula

Chap. II, n° 19, has shown us that if a function $f(z)$ is analytic on an open set in $\mathbb{C}$ containing a point $a$, then the power series which represents it on a neighbourhood of $a$ has the form

$$(18.1) \qquad f(z) = f(a) + f'(a)(z-a)/1! + f''(a)(z-a)^2/2! + \ldots.$$

This result applies in particular to functions which are defined and analytic on an interval in $\mathbb{R}$.

One obviously cannot hope for so much for every function defined on such an interval, even if it is indefinitely differentiable, since all the derivatives of such a function may well be zero at $a$ without the function being zero on a neighbourhood of $a$, as Cauchy showed (Chap. IV, end of n° 5). The situation is still more desperate when we deal with functions which are not indefinitely differentiable.

But let us start from the formula

$$f(t)\Big|_a^x = \int_a^x f'(t)dt = \int_a^x f'(t)P_0(t)dt$$

where $P_0(t) = 1$, and let $P_1(t)$ be a primitive of $P_0$. An integration by parts shows that

$$f(t)\Big|_a^x = f'(t)P_1(t)\Big|_a^x - \int_a^x f''(t)P_1(t)dt.$$

If $P_2(t)$ is a primitive of $P_1$ and if one integrates again by parts, one finds

$$f(t)\Big|_a^x = f'(t)P_1(t) - f''(t)P_2(t)\Big|_a^x + \int_a^x f'''(t)P_2(t)dt.$$

If $f$ is of class $C^{n+1}$, i.e. has continuous derivatives of order $\leq n+1$ on the interval $I$ of $\mathbb{R}$ where it is defined, one may continue up to the integral involving $f^{(n+1)}$. The result is clearly that, if one chooses polynomials $P_k(t)$ satisfying

$$(18.2) \qquad\qquad P'_k = P_{k-1}, \qquad P_0 = 1,$$

then

$$f(t)\Big|_a^x = f'(t)P_1(t) - f''(t)P_2(t) + \ldots + (-1)^{n-1}f^{(n)}(t)P_n(t)\Big|_a^x +$$

$$(18.3) \qquad\qquad + (-1)^n \int_a^x f^{(n+1)}(t)P_n(t)dt.$$

Since we would like to express $f(x)$ in terms of its derivatives at $a$ and of the factors $(x-a)^k$, we ought to choose the $P_k$ so as to make the terms

$f'(x), \ldots, f^{(n)}(x)$ that appear in the integrated parts disappear from the right hand side of (3); to do this we have to choose the $P_k$ to vanish for $t = x$, in other words

(18.4) $$P_k(t) = (t - x)^k/k! = (t - x)^{[k]}.$$

After this calculation we obtain the following result:

**Theorem 16.** *Let $f$ be a function defined and of class $C^{n+1}$ on an interval $I$ of $\mathbb{R}$. Then, for any $a, x \in I$,*

(18.5) $$\begin{aligned} f(x) &= f(a) + f'(a)(x - a) + f''(a)(x - a)^{[2]} + \ldots \\ &\quad \ldots + f^{(n)}(a)(x - a)^{[n]} + r_n(x) \end{aligned}$$

*where*

(18.6) $$r_n(x) = \int_a^x f^{(n+1)}(t)(x - t)^{[n]}dt.$$

This is *Taylor's formula with integral remainder* for functions of class $C^{n+1}$, and is due, with this very proof, to Cauchy.

An expression for the remainder that is sometimes more useful comes from putting $x = a + h$ and replacing the function $f(x)$ by the function $g(u) = f(a + uh)$, where $u$ now varies in $[0, 1]$. The limits $a$ and $x$ become 0 and 1 and we have $g^{(p)}(u) = h^p f^{(p)}(a + uh)$ by the most trivial form of the chain rule. On applying the formula to $g$ between $u = 0$ and $u = 1$ one finds the following version of Taylor's formula for $f$:

(18.5') $$f(a + h) = f(a) + f'(a)h + \ldots + f^{(n)}(a)h^n/n! + r_n(a + h),$$

where

(18.6') $$r_n(a + h) = \frac{h^{n+1}}{n!}\int_0^1 f^{(n+1)}(a + uh)(1 - u)^n du.$$

The simplest and most useful particular case is

(18.6") $$f(x + h) - f(x) - f'(x)h = \frac{1}{2}h^2 \int_0^1 f''(x + uh)(1 - u)\,du.$$

For $a = 0$, (5) becomes a formula of Maclaurin type:

(18.7) $$f(x) = f(0) + f'(0)x + f''(0)x^2/2! + \ldots + f^{(n)}(0)x^n/n! + r_n(x)$$

with

(18.8) $$\begin{aligned} r_n(x) &= \int_0^x f^{(n+1)}(t)(x - t)^{[n]}dt = \\ &= \frac{x^{n+1}}{n!}\int_0^1 f^{(n+1)}(ux)(1 - u)^n dt. \end{aligned}$$

This expresses $f$ as the sum of a polynomial

$$p(x) = f(0) + f'(0)x + \ldots + f^{(n)}(0)x^{[n]}$$

of degree $n$ having at $x = 0$ (or, in the general case, at $x = a$) the same derivatives of order $\leq n$ as $f$, and a "remainder" given by (8). It is called the "remainder" because, if $f$ is analytic, $r_n(x)$ is actually the $n$-th remainder of the Taylor series of $f$; but here there is no series. Moreover, when one speaks of the partial sums and of the remainders of a series, one very much hopes that these will be increasingly negligible as $n$ increases.

The true situation is less impressive. If you apply (7) to an indefinitely differentiable function all of whose derivatives are zero at the origin, you will find $f(x) = r_n(x)$. Far from being negligible, the "remainder" is then the predominant term.

All the same, one can estimate it. Since $f^{(n+1)}$ is continuous it is bounded on every compact set contained in the interval $I$ where $f$ is defined. If then $x$ remains in a compact interval $K$ contained in $I$ and containing $a$ (for example, an interval of centre $a$ if $a$ is interior to $I$), and if, as always, one puts

(18.9)    $$\left\| f^{(p)} \right\|_K = \sup_{x \in K} \left| f^{(p)}(x) \right|,$$

then $\left| f^{(n+1)}(t)(x-t)^n \right| \leq \left\| f^{(n+1)} \right\|_K \cdot \left| (x-t)^n \right|$ on the interval of integration; but on this interval we have $\left| (x-t)^n \right| = \pm(x-t)^n$ with the $+$ sign if $x > a$ and the $-$ sign if $x < a$, in which case the $+$ sign is reestablished from the fact that the *oriented* integral from $a$ to $x$ is accounted negatively. The function $t \mapsto (x-t)^{[n]}$ having as a primitive the function

$$t \mapsto -(x-t)^{[n+1]}$$

which vanishes for $t = x$, one finds finally that

(18.10)    $$|r_n(x)| \leq \left\| f^{(n+1)} \right\|_K \cdot |x-a|^{n+1}/(n+1)! \quad \text{for } x \in K.$$

(6) makes the result even more obvious.

Modifying the notation, in particular we have

(18.11)    $$f(a+h) = f(a) + f'(a)h + f''(a)h^{[2]} + \ldots$$
$$\ldots + f^{(n)}(a)h^{[n]} + O(h^{n+1})$$

as $h$ tends to 0. If, for example, a function $f$ of class $C^\infty$ vanishes together with all its successive derivatives at $x = a$, then

$$f(x) = O\big((x-a)^n\big) \quad \text{for every } n$$

as $x$ tends to $a$, which shows that, even in this case, Taylor's formula does provide information: $f(a+h)$ tends to 0 more rapidly than every power of $h$.

We say then that the function $f$ is *flat* at the point $a$. The aeroplane which will never be built follows a trajectory of this kind on landing.

*Exercise.* Prove that

$$p(h) = f(a) + f'(a)h + \ldots + f^{(n)}(a)h^{[n]}$$

is the only polynomial satisfying

$$f(a + h) = p(a) + o(h^n), \qquad d^\circ(p) \leq n.$$

These results assume that $f^{(n+1)}$ exists and is continuous. One can in fact prove the formula otherwise, without this last hypothesis, in the case of a real-valued function; the practical usefulness of the result in relation to the preceding one is weak, to speak moderately, but the proof is ingenious. To simplify we restrict to the Maclaurin case, and, for $x = b$ given, let us consider the functions

$$
\begin{aligned}
g(t) &= f(b) - f(t) - f'(t)(b - t) - \ldots - f^{(n)}(t)(b - t)^{[n]}, \\
h(t) &= g(t) - g(0)(b - t)^{[n+1]}/b^{[n+1]}
\end{aligned}
$$

like Cauchy and his uniformed students, most of whom were probably dozing. By hypothesis, $g$, and so $h$ also, have first derivatives, not necessarily continuous, between $0$ and $b$. We have $h(0) = 0$ by a direct calculation, and $h(b) = g(b) = f(b) - f(b) = 0$. Therefore there is a number $\xi$ between $0$ and $b$ where $h'(\xi) = 0$ (mean value theorem, or even Rolle's theorem, Chap. III, n° 16). Now a calculation similar to the one that led us to (3) shows that

$$g'(t) = -f^{(n+1)}(t)(b - t)^{[n]},$$

whence

$$
\begin{aligned}
h'(t) &= -f^{(n+1)}(t)(b - t)^{[n]} + g(0)(b - t)^{[n]}/b^{[n+1]} \\
&= \left[ g(0) - f^{(n+1)}(t)b^{[n+1]} \right] (b - t)^{[n]}/b^{[n+1]}.
\end{aligned}
$$

Since $h'(\xi) = 0$ we therefore have $g(0) = f^{(n+1)}(\xi)b^{[n+1]}$ for a $\xi$ between $0$ and $b = x$, and since

$$g(0) = f(x) - \left\{ f(0) + f'(0)x + \ldots + f^{(n)}(0)x^{[n]} \right\},$$

finally we have

(18.12)   $f(x) = f(0) + f'(0)x + \ldots + f^{(n)}(0)x^{[n]} + f^{(n+1)}(\xi)x^{[n+1]};$

this is Maclaurin's formula with *Lagrange remainder*. In the general case one obtains

(18.13)   $f(a+h)$   $=$   $f(a) + f'(a)h + f''(a)h^{[2]} + \dots$
$$\dots + f^{(n)}(a)h^{[n]} + f^{(n+1)}(a+\theta h)h^{[n+1]}$$

where, yet again, $f$ has real values, is $(n+1)$ times differentiable, and where the number traditionally denoted by $\theta$ is between 0 and 1 so that the point $\xi = a + \theta h$ lies between $a$ and $x = a + h$.

Applied to a function of class $C^\infty$ these results sometimes allow us to pass to a Taylor *series*.

*Example 1.* Take $a = 0$ and $f(x) = \sin x$. For $n = 2p$ one finds $|r_n(x)| \leq |x|^{2p}/(2p)!$ since the successive derivatives are everywhere less than 1 in modulus. On passing to the limit one thus recovers the formula

$$\sin x = \lim \left[ x - x^3/3! + \dots + (-1)^{p-1}x^{2p-1}/(2p-1)! \right].$$

This argument can be generalised. For a $C^\infty$ function the Taylor formula with remainder applies for any $n$. To pass from this to a power series expansion, it suffices – but this is the crucial point – to know that the remainder tends to 0 as $n$ increases indefinitely, in other words to have a suitable estimate of the successive derivatives. If for example there exist constants $M$ and $q$ such that $|f^{(n)}(x)| \leq Mq^n$ for any $n$, then the formula (10), for $a = 0$, shows that

$$|r_n(x)| \leq Mq^n|x|^n/n! = M(q|x|)^n/n!$$

and the passage to the limit is justified. In other cases one has to argue in some other way.

*Example 2.* Take $f(x) = e^x$. All the derivatives are equal to $f$, so that, for $K$ given, the factor $\left\| f^{(n+1)} \right\|_K$ in (10) does not depend on $n$. Thus $\lim r_n(x) = 0$ for every $x$ and one recovers the power series for the exponential function.

*Example 3.* Take $f(x) = (1+x)^s$ with $s \in \mathbb{C}$ and $-1 < x$, whence

$$f^{(n+1)}(x) = s(s-1)\dots(s-n)(1+x)^{s-n-1}.$$

Here, by (8),

$$r_n(x) = s(s-1)\dots(s-n)\frac{x^{n+1}}{n!} \int_0^1 (1+ux)^{s-n-1}(1-u)^n dt,$$

or

$$r_n(x) = \frac{s(s-1)\dots(s-n)}{n!} x^{n+1} \int_0^1 (1+xu)^{s-1} \left( \frac{1-u}{1+xu} \right)^n du.$$

For $x > -1$ we have $1 + xu > 1 - u$ on the interval of integration, so that $0 < (1-u)^n/(1+xu)^n < 1$ for every $n$. The modulus of the integral is therefore less than a number independent of $n$. Since the series with general term $s(s-1)\ldots(s-n)x^{n+1}/n!$ converges for $|x| < 1$ (use the $u_{n+1}/u_n$ criterion, this is not exactly the binomial series), its general term tends to 0. Likewise therefore for $r_n(x)$. On passing to the limit in Taylor's formula of order $n$ one thus recovers Newton's *series*

$$(1+x)^s = 1 + sx + s(s-1)x^2/2! + \ldots$$

for $|x| < 1$, $x$ real, $s$ complex.

We said above that if $f$ is of class $C^{n+1}$ then the remainder in the Taylor formula is $O\left((x-a)^{n+1})\right)$ as $x$ tends to $a$. One may improve this result for $f$ real, assuming $f^{(n+1)}(a) \neq 0$. The ratio $r_n(x)/(x-a)^{n+1} = f^{(n+1)}(\xi)$ then tends to $f^{(n+1)}(a)$ since the derivative of order $n+1$ is continuous and $\xi$ lies between $a$ and $x$, so tends to $a$. If one assumes $f^{(n+1)}(a) \neq 0$, then, in the Maclaurin case, for simplicity,

$$(18.14) \quad f(x) - f(0) - f'(0)x - \ldots - f^{(n)}(0)x^{[n]} \sim f^{(n+1)}(0)x^{[n+1]}$$

as $x$ tends to 0, the $\sim$ sign signifying, we recall, that the *ratio* between the two sides of (14) tends to 1.

This formula sometimes allows one to calculate the limits of quotients $f(x)/g(x)$ as $x$ tends to a point $a$ where $f(a) = g(a) = 0$. For $f$ and $g$ of class $C^n$, suppose that we have

$$f(a) = f'(a) = \ldots = f^{(n-1)}(a) = 0,$$
$$g(a) = g'(a) = \ldots = g^{(n-1)}(a) = 0, \qquad g^{(n)}(a) \neq 0.$$

On a neighbourhood of $a$ one has $g(x) \sim g^{(n)}(a)(x-a)^{[n]}$. If $f^{(n)}(a) \neq 0$, then likewise $f(x) \sim f^{(n)}(a)(x-a)^{[n]}$. Consequently

$$(18.15) \qquad \lim f(x)/g(x) = f^{(n)}(a)/g^{(n)}(a).$$

If $f^{(n)}(a) = 0$ then $f(x) = o((x-a)^n)$ and the result remains valid. For example, to find the limit of the ratio

$$\frac{\tan x - x}{x - \sin x}$$

as $x$ tends to 0. Here $f(0) = f'(0) = f''(0) = 0$, $f'''(0) = 2$ and $g(0) = g'(0) = g''(0) = 0$, $g'''(0) = 1$. The limit is therefore $1/2$.

This is the famous *l'Hôpital's rule* (he proved it for $n = 1$, the general case being apparently due to Maclaurin), named for a Parisian marquis who, in 1696, earned himself an enviable mathematical reputation by publishing a book entitled *Analyse des infiniment petits, pour l'intelligence des lignes courbes*, the first exposition to the public of what one then knew as the differential calculus *chez* Leibniz and the Bernoullis. The author made no mystery of the fact that he had learned everything from his correspondence and conversations with them: "I have simply appropriated their discoveries and those of Mr. Leibnis [*sic*]", he wrote, skimping the name of his hero, not even giving him a francophone *t*. This did not prevent Johann Bernoulli from accusing him of having plagiarised one of his manuscripts and from attacking him in 1742, which was 38 years after the death of the presumed culprit. Since Johann and Jakob Bernoulli did not hesitate to accuse each other of plagiarism in the years from 1695, it is probably better to reserve one's judgement, as does Moritz Cantor, pp. 222–228 of his Vol. III.

"Science is a cruel game. One shoots one's adversaries down in flames, one grills one's competitors, one demolishes one's rivals", Loup Verlet tells us, p. 97 of *La malle de Newton*. One should not disdain disputes over priority; they reveal one of the most permanent and pervasive traits of the psychology of scientists: the defence of their intellectual property. They are an obligatory corollary of the following principle: the second person to prove a theorem or to discover an AIDS vaccine will not enhance his reputation; he has lost his time, except insofar as the necessary work may have taught him techniques of future application. One has to be the first[35]. Moreover, since scientists, and particularly "pure" mathematicians, derive no capital from their work other than *the recognition of their prowess by their peers*[36], those who value this have no choice. Those who patent their discoveries – again not the case of the pure mathematicians, except recently in cryptology[37] – or depend on

---

[35] The subject is expounded with raw clarity in an interview with a Franco-American biologist, *Portrait of a biologist as capitalist savage*, to be found in Bruno Latour, *Petites leçons de sociologie des sciences* (Paris, La Découverte, 1993). The subject had inspired Robert K. Merton much earlier. In the real world it happens that a new result is obtained almost simultaneously by researchers working independently of one another; questions of priority are of more interest to them than to spectators. A particularly extreme case is described by Nicholas Wade, *The Nobel Duel* (Doubleday, 1981). In an other domain, let us quote the immortal declaration of the heroic Frenchman, who, on account of a young German, missed first place in the cyclists' Tour de France of 1997: "What is missing, I think, is the opportunity to chance on a year without someone better than oneself."

[36] This is the function of prizes, medals of gold or of chocolate, colloquia in honour of, seats in academies, etc. which scientists dish out for mutual reward and admiration. In mathematics, it has even happened, over the last fifteen years, that the complete works of great men have been published before their deaths, despite having to add supplementary volumes later. This facilitates colleagues' work, but the psychological effect on the person so honoured is probably not negligible. The *Science Citation Index*, moreover, allows every scientist to know how many times his papers have been cited each year. When the total exceeds a hundred (for Einstein, the record holder, the total even exceeded nine hundred some years ago), one might think that the "reward" is sufficient.

[37] Neal Koblitz, *A Course in Number Theory and Cryptography* (Springer, 2d. ed., 1994), propounds the hypothesis that publications in certain parts of number theory may some day have to be submitted to preliminary censoring by the National Security Agency. This idea is not so wild, since, when the public key systems invented by mathematicians – their own enterprise – appeared some

subventions from organisations advised by scientists have no more: if the hormone you are trying to synthesise may bring you fifty thousand dollars a year for seventeen years it is urgent not to be second.

With the passion for their profession, which, throughout history, has led scientists – not all, of course – to accept unworthy or mediocre conditions of life or work, this is the reason why you will see first class people spend sixty hours per week in the laboratory, for forty years, for a CNRS "director of research" salary, which any Polytechnic student or former pupil of an *Haute Ecole des Etudes Commerciales Libérales et Avancées* can obtain in industry or in a bank two years after leaving the institution. And so one lauds the disinterestedness of these "heroes of science" who have no choice.

No choice? After all, as a garage owner who presented an unconscionably overinflated account explained to the present author (who, scion of the *lower middle class*, was not a customer of great standing and did not care to accept it), "with your brain, you should have chosen another profession". Garage owner, for example. The garage owners could, in exchange, write the mathematics which sells itself cheap, teach the chemistry of polymers or the history of the Middle Ages to two hundred students every year, or try to understand Alzheimer's disease. Then we might praise the disinterestedness of the garage owners.

Others say that the ill-paid scientists are recompensed in that they enjoy themselves; this may often be true while they are working, but their families do not always appreciate this, and the argument does not have a universal validity. Marcel Dassault, the famous French aeroplane manufacturer, said, and repeated during his life, that what "amused him", was to "make aeroplanes" and one does not doubt this. This did not stop him from going to his office in a Rolls Royce, and this did not stop his son, who also amuses himself as he can and continues to make man-hunting fighter planes, from

---

dozen years ago, the reactions of the NSA have been (i) to try to forbid them (in France, their use is subject to prior authorisation), (ii) to impose limits on the degree of security, (iii) to take charge of research contracts in this field. Remember that the NSA, created in 1952 and with an enormous budget, has the mission of ensuring the security of the telecommunications of the American government and the insecurity of those of others; see James Bamford, *The Puzzle Palace* (Houghton Mifflin, 1982) and *Body of Secrets* (Arrow Books, 2003). In "militarily sensitive" areas, access to an American thesis, or to certain courses or seminars, may be limited to people who have submitted to a "security clearance" guaranteeing their "loyalty". In the USSR, *all* scientific or technical publications were, in principle, subject to prior censorship. Koblitz notes that, up to very recently, number theory had never lent itself to any application outside pure mathematics. The interest of some mathematicians in cryptology is however longstanding – Viète and Wallis for example – and one knows the part played during the War by the Turing team; as for those of our contemporaries who are involved, they do not have to proclaim it *urbi et orbi*. The novelty is the recourse to very advanced mathematics, with, necessarily, the cooperation of professionals in the theory, simply to understand it; the bridge players of the Enigma project are probably not enough.

setting himself as the aim of his animal hunting for 1998 to kill, from a tur-
reted (armoured?) $4 \times 4$, one hundred and eighty-five head of large game in
his modest property of eight hundred hectares near Paris, which landed him
in trouble[38]. Mr. Gates, despite his $N(t)10^{10}$ dollars, also seems to amuse
himself.

Another long trivial theme – though not in the XVII[th] century for Fermat –, the question of the *maxima and minima of functions*. If a differentiable
function $f$ has a *local* maximum or minimum at a point $a$ i.e. if, on a neighbourhood of $a$, one has $f(x) \leq f(a)$ in the first case and $f(x) \geq f(a)$ in
the second, it is clear that $f'(a) = 0$. But this condition does not suffice,
as shown by the function $x^3$ at $x = 0$. To elucidate the question one has to
examine the higher derivatives. If

$$f'(a) = f''(a) = \ldots = f^{(n)}(a) = 0, f^{(n+1)}(a) \neq 0,$$

then $f(x) - f(a) \sim f^{(n+1)}(a)(x-a)^{[n+1]}$ as we have seen; consequently, the
difference $f(x) - f(a)$ has the sign of the right hand side for $x$ sufficiently
near $a$, whence the conclusion: maximum if $n$ is even and $f^{(n+1)}(a) < 0$,
minimum if $n$ is even and $f^{(n+1)}(a) > 0$, while the graph of $f$ crosses a
horizontal tangent (point of inflexion) if $n$ is odd.

---

[38] *Le Monde* of 23 April 1997; the officials of the National Water and Forests Office
apparently do not appreciate that the methods honoured in military aviation
since the Great War are being applied to deer and boar.

# § 6. The change of variable formula

## 19 – Change of variable in an integral

As we said above, the rules for calculating derivatives can be interpreted in the language of integrals. We have seen the interpretation of the relation $(fg)' = f'g + fg'$ in terms of primitives. The other fundamental rule of calculus, namely that the derivative of a composite function $f[u(x)]$ is $f'[u(x)].u'(x)$, likewise leads to an almost trivial integral formula, but it facilitates many explicit evaluations. It also has very important applications, mainly in the theory of line integrals (Vol. III).

**Theorem 17.** *Let $u$ be a real function defined and of class $C^1$ on a compact interval $I = [a, b]$, and $f$ a function defined and continuous on the interval $J = u(I)$. Then*

$$(19.1) \qquad \int_{u(a)}^{u(b)} f(y)dy = \int_a^b f[u(x)]u'(x)dx.$$

For, let $F$ be a primitive of $f$ on $J$, so that the left hand side is equal to $F[u(b)] - F[u(a)]$. The function $G(x) = F[u(x)]$ is differentiable on $I$ and $G'(x) = F'[u(x)]u'(x) = f[u(x)]u'(x)$. The right hand side of (1) is thus equal to $G(b) - G(a) = F[u(b)] - F[u(a)]$, i.e. to the left hand side, qed.

Note that we are dealing with *oriented* integrals in this formula, since, even if $a < b$, one may well have $u(a) > u(b)$.

In the Leibniz notation for primitives one would write

$$(19.2) \qquad \int f(y)dy = \int f'[u(x)]u'(x)dx.$$

The formula is self-explanatory: one replaces $y$ by its expression as a function of $x$, simultaneously in $f(y)$ and in $dy$ (Chap. III, n° 14 and 15). This is one of the great advantages of Leibniz' system, with the analogous formula $dy/dx = dy/du.du/dx$.

One may widen the hypotheses of Theorem 17 a little when applying it to regulated functions, but the reader new to the subject would be better to keep to the very simple Theorem 17 for the moment.

The whole question is to reassure oneself that, $F$ being a primitive of the regulated function $f$ and $u$ a primitive of the regulated function $u'$ in the sense of n° 13, then $G(x) = F[u(x)]$ is again a primitive of $f[u(x)]u'(x)$. Since $G$ is continuous, because $u$ and $F$ are, it is enough to convince oneself that (i) the function $f[u(x)]u'(x)$ is regulated, (ii) $G'(x)$ exists and is equal to $f[u(x)]u'(x)$ outside a countable set of values of $x$.

Point (i): assume that $u'(x)$ and $f[u(x)]$ are regulated. The first condition is satisfied if $u$ is a primitive of a regulated function. So is the second if $f$ is continuous since then $f \circ u$ is continuous; if $f$ is only regulated one has to check that $f \circ u$ has right and left limit values; now, as $h$ tends to 0 through, let us say, positive values, $u(x+h)$ tends to $u(x)$, though not necessarily in a monotone way; to be certain that $f[u(x)]$ is regulated it is therefore prudent to assume $u$ monotone, which is the most general case in practice.

Point (ii): the derivative $G'(x)$ exists provided that $u'(x)$ and $F'[u(x)]$ exist. The derivative $u'(x)$ exists outside a countable set $D$ and the derivative $F'(y)$ exists either everywhere if $f$ is continuous, or outside a countable set $D'$ if $f$ is only regulated. If $f$ is continuous then $G'(x) = f[u(x)]u'(x)$ outside $D$, and since, in this case, the function $f[u(x)]u'(x)$ is regulated as we have seen à propos the point (i), everything works. If $f$ is only regulated, in which case we had better assume $u$ monotone (as we have seen in point (i) for $f[u(x)]u'(x)$ to be regulated), the existence of $G'(x)$ assumes $x \notin D$ and $u(x) \notin D'$, i.e. $x \notin D \cup u^{-1}(D')$. If $u$ is strictly monotone, so injective, the inverse image $u^{-1}(D')$ is countable like $D'$, so also is its union with $D$; then $G'(x) = f[u(x)]u'(x)$ outside a countable set and everything works again. If $u$ is not strictly monotone there are intervals on which $u$ is constant, so also $f[u(x)]$ and the relation $G'(x) = f[u(x)]u'(x)$ can be written on these intervals as $0 = 0$, which shows that it is not false ...

In conclusion, we see that the change of variable formula is valid in the two following cases: (a) *f is continuous and u is a primitive of a regulated function*; (b) *f is regulated and u is a monotone primitive of a regulated function*.

In practice, one most often assumes the function $u$ to be strictly monotone, or, almost equivalently, that its derivative is always $> 0$ or always $< 0$. On writing $a$ and $b$ for what we wrote as $u(a)$ and $u(b)$ in (1), we then find the relation

$$(19.3) \qquad \int_a^b f(y)dy = \int_{u^{-1}(a)}^{u^{-1}(b)} f[u(x)]u'(x)dx$$

where $u^{-1} : J \longrightarrow I$ denotes the inverse map of $u$.

*Example 1.* Calculate the indefinite integral $\int (x^2 + 1)^3 x dx$. Putting $u(x) = x^2 + 1$ we have $u'(x)dx = 2xdx$, so we need to calculate $\frac{1}{2} \int u(x)^3 u'(x)dx$; this is situation (2) with $f(y) = y^3$. Thus

$$\int (x^2 + 1)^3 x dx = \frac{1}{2} \int y^3 dy = y^4/8 \qquad \text{with } y = x^2 + 1.$$

To calculate the given integral between the limits $x = 2$ and $x = 3$, for example, one notes that, in this case, $u(a) = 5$ and $u(b) = 10$, whence

$$\int_2^3 (x^2 + 1)^3 x dx = y^4/8 \Big|_5^{10} = (10^4 - 5^4)/8.$$

In practice, one puts it as follows: make the change of variable $y = x^2 + 1$; then $dy = 2xdx$ and $(x^2 + 1)^3 = y^3$ and consequently

$$\int (x^2 + 1)^3 x dx = \frac{1}{2} \int y^3 dy = y^4/8 = (x^2 + 1)^4/8.$$

*Example 2.* Let $f$ be a real function of class $C^1$ on an interval $I$, *not vanishing* on $I$. To calculate $\int f'(x)/f(x).dx$ one performs the change of variable $y = f(x)$, whence $dy = f'(x)dx$ and

$$\int \frac{f'(x)}{f(x)} dx = \int dy/y.$$

It remains to find a primitive of the function $1/y$ on the interval $J = f(I)$. Since $f$ does not vanish it has constant sign on $I$. If it is positive, the function $\log y$ will serve. If it is negative, it is the function $\log |y|$ which suits, since for $y < 0$ the derivative of the latter, i.e. of $\log(-y)$, is $-\log'(-y) = 1/y$. In conclusion, one finds

(19.4) $$\int \frac{f'(x)}{f(x)} dx = \log |f(x)|.$$

For example,

$$\int_{1/2}^{4} \frac{dx}{x.\log x} = \int_{1/2}^{4} \frac{\log' x}{\log x} dx = \log(|\log x|)\Big|_{1/2}^{4} = \log \frac{\log 4}{\log 2} = \log 2$$

since $4 = 2^2$.

Likewise

$$\int \tan x.dx = -\int \cos' x/\cos x.dx = -\log |\cos x|,$$

so long as one works on an interval where the function $\cos x$ does not vanish, say $]-\pi/2, \pi/2[$. Whence for example

$$\int_{0}^{\pi/4} \tan x.dx = -\log(\cos x)\Big|_{0}^{\pi/4} = -\log\left(1/\sqrt{2}\right) = \frac{1}{2}\log 2.$$

*Example 3.* To calculate $\int dx/\sin x$ on an interval where the sine function does not vanish, for example on $]0, \pi[$. If one is inspired, or if one has read all the books, one observes that

$$1/\sin x = 1/2\sin(x/2)\cos(x/2) = 1/2\tan(x/2)\cos^2(x/2) = f'(x)/f(x)$$

where $f(x) = \tan x/2$ and $f'(x) = 1/2\cos^2(x/2)$, whence

$$\int dx/\sin x = \log|\tan x/2|.$$

This kind of recourse to Providence will not take one very far if one has no general procedure for calculating integrals of the form

(19.5) $$\int \frac{\sum a_{pq} \cos^p x.\sin^q x}{\sum b_{pq} \cos^p x.\sin^q x} dx$$

with a finite number of nonzero coefficients $a_{pq}$ and $b_{pq}$, in other words a rational function of $\cos x$ and of $\sin x$. The method is to perform the change of variable

$$(19.6) \qquad\qquad x = 2\arctan y, \qquad y = \tan(x/2).$$

Trigonometry then shows that

$$(19.7) \qquad \sin x = 2y/(y^2 + 1), \qquad \cos x = (1 - y^2)/(y^2 + 1)$$

using the relations

$$\sin 2t = 2\sin t.\cos t = 2\tan t.\cos^2 t = 2\tan t/(\tan^2 t + 1),$$
$$\cos 2t = 2\cos^2 t - 1 = 2/(\tan^2 t + 1) - 1 = (1 - \tan^2 t)/(\tan^2 t + 1).$$

Further,

$$dx = 2dy/(y^2 + 1).$$

On substituting in (5) one reduces to calculating an integral of a rational function of $y$, which will be the aim of the following n°.

*Example 4.* To calculate

$$\int \frac{x^4 + 1}{\sqrt{(x+1)(x-5)}}\, dx;$$

we have to work in the interval $x < -1$, or in the interval $x > 5$ to obtain a real result. We have $(x+1)(x-5) = (x-3)^2 - 4$, which suggests the change of variable $x = 3 + 2y$, whence $dx = 2dy$ and reduction to

$$\int \frac{(2y+3)^4 + 1}{\sqrt{y^2 - 1}}\, dy.$$

A second change of variable $y = \cosh z$ reduces us to

$$\int \frac{(2\cosh z + 3)^4 + 1}{\sinh z}\, \sinh z.dz = z + \int (2\cosh z + 3)^4 dz$$

and to calculating the primitives of the functions $\cosh^n x$, which can be done in several ways, the banal method – expanding as exponentials – most often being the best.

If, in this example, the denominator had been the square root of a trinomial without real roots we would have put it into the standard form $(x - a)^2 + b^2$ and the change of variable $x - a = by$ would have led us to $(y^2 + 1)^{1/2}$, in which case it is the change of variable $y = \sinh z$ which leads us to the result.

There are also cases where, in the given trinomial, the coefficient of $x^2$ is $< 0$. The same changes of variable lead this time to integrals in $(1 - y^2)^{1/2}$ or in $(-1 - y^2)^{1/2} = i(1 + y^2)^{1/2}$. The second case is treated by putting $y = \sinh z$ as above. In the first the change of variable $y = \sin z$ leads us to the result.

*Example* 5 (Darboux, 1875). Consider the integral

$$\int_0^1 2n^2x.\exp(-n^2x^2)dx;$$

the change of variable $y = n^2x^2$, for which $2n^2xdx = dy$, transforms it into the integral of $e^{-y}$ taken from 0 to $n^2$. The result, $1 - \exp(-n^2)$, tends to 1 as $n$ increases indefinitely, although the function being integrated tends to 0. How do you explain this strange phenomenon which Gaston Darboux was, apparently, the first to discover?

## 20 – Integration of rational fractions

One does need these from time to time in real mathematics; but very rarely. In teaching they are useful only to (i) accustom students to algebraic calculation, which will always be useful elsewhere, (ii) provide examiners with an inexhaustible reservoir of exercises built up over the generations, so enabling them, according to point (i), to test the candidate's virtuosity. They may also be needed in certain electrotechnical calculations, for example, but this is surely not the principal motivation of the subject, inaugurated by Leibniz who was not thinking of the XIX$^{\text{th}}$ and XX$^{\text{th}}$ century students who were obliged to suffer the fallout ...

Let $f(x) = P(x)/Q(x)$ be a rational function of $x$, where $P$ and $Q$ are polynomials. By using the d'Alembert-Gauss theorem which we shall prove later, and a few ideas from algebra, we may write $Q$ in the form $Q(x) = Q_1(x)\ldots Q_r(x)$ where each of the $Q_i$ is, up to a constant factor, of the form

$$Q_k(x) = (x - a_k)^{n_k};$$

the $a_k$ are the various distinct roots of $Q$, perhaps complex, and the integers $n_k$ are their orders of multiplicity, by definition. It is shown in all the algebra textbooks that one may write $f$ in the form

(20.1) $$f(x) = p(x) + \sum \frac{p_k(x)}{(x - a_k)^{n_k}}$$

with a polynomial $p$, the quotient of the Euclidean division of $P$ by $Q$, and polynomials $p_k$ of degrees $< n_k$. On writing $p_k$ as a polynomial in $x - a_k$ one finally finds a *decomposition into simple elements* of the form

(20.2) $$f(x) = p(x) + \sum_{k,n} A_{kn}/(x - a_k)^n$$

with a finite number of constants $A_{kn}$ [39]. The search for a primitive of $f$ thus reduces to that of a primitive of the polynomial $p$ – immediate calculation – and of functions of the form $(x - a)^{-n}$ where $n$ is an integer $\geq 1$. The result

---

[39] Let $p$ and $q$ be two polynomials in one variable with coefficients in $K = \mathbb{Q}, \mathbb{R}, \mathbb{C}$ or any other field.

$$(20.3) \qquad \int \frac{dx}{(x-a)^n} = -\frac{1}{(n-1)(x-a)^{n-1}}$$

is obvious if $n \neq 1$, but less so for $n = 1$.

It is first of all prudent – even if $n > 1$ – to work in an interval $I$ of $\mathbb{R}$ not containing $a$. If $a$ is real, then

$$(20.4) \qquad \int \frac{dx}{x-a} = \log|x-a| \qquad \text{if } a \in \mathbb{R}$$

since, for $y \neq 0$, the function $\log|y|$ has derivative $1/y$. In (4) we therefore have to take $\log(a-x)$ for the primitive if $a$ is to the right of the interval $I$ and $\log(x-a)$ if it is to the left.

If $a$ is complex the business is more complicated.

Recall first (Chap. IV, n° 14, section (x) and § 4) that, for nonzero $z \in \mathbb{C}$, one defines the expression $\mathcal{L}og\ z$ by

$$\mathcal{L}og\ z = w \iff z = e^w.$$

There are infinitely many possible values, differing by a multiple of $2i\pi$. Putting $w = u + iv$, we have $z = e^u e^{iv}$, whence $u = \log|z|$ and $v = \arg z$, i.e.

---

(i) Consider the set of polynomials of the form $up + vq$, where $u$ and $v$ are arbitrary polynomials with coefficients in $K$. Among those which are not identically zero let $d = u_0 p + v_0 q$ be a polynomial of minimal degree, not greater than the degrees of $p$ and $q$ since $p = 1p + 0q$, $q = 0p + 1q$. Every polynomial which divides $p$ and $q$ divides all the $up + vq$, so divides $d$. On the other hand, $d$ itself divides $p$ and $q$, because the Euclidean division algorithm yields a relation of the form $p = du + d'$, with $d'$ of degree strictly less than the degree of $d$, and the relation $d' = (1 - uu_0)p - uv_0 q$ shows then that $d' = 0$ since $d$ is of minimum degree among the *nonzero* polynomials which can be written in the form $up+vq$. In brief, $d$ is the gcd of $p$ and $q$.

(ii) If $d$ is constant, i.e. if $p$ and $q$ have no common nonconstant divisor, i.e. if $p$ and $q$ are *mutually prime*, one may assume $d = 1$ whence

$$r/pq = r(u_0 p + v_0 q)/pq = rv_0/p + ru_0/q;$$

every rational fraction with denominator $pq$ is thus the sum of two rational fractions whose denominators are respectively $p$ and $q$. More generally, every rational fraction whose denominator is a product $p_1 \ldots p_k$ of pairwise mutually prime polynomials is the sum of rational fractions having only one of the $p_i$ in its denominator: $p_1$, for example, is prime to $p_2 \ldots p_k$, which allows us to simplify the denominators step-by-step.

(iii) Suppose $q(X) = (X - a_1)^? \ldots (X - a_k)^?$ with pairwise distinct roots $a_i$ and integer exponents. The polynomials $(X - a_i)^?$ are pairwise mutually prime since their divisors are obvious. Every rational fraction of the form $p/q$ thus decomposes as a sum of fractions of the form $p_i(X)/(X - a_i)^?$. On writing $p_i(X)$ as a polynomial in $X - a_i$ one obtains the desired decomposition into "simple elements", qed.

(20.5) $$\mathcal{L}og\ z = \log|z| + i\arg z.$$

If $z = x + iy$ we conclude that

(20.6) $$\mathcal{L}og\ z = \frac{1}{2}\log(x^2 + y^2) + i\arg z,$$

all this up to $2ki\pi$. For example, for $x$ real,

(20.7) $$\mathcal{L}og(x - i) = \frac{1}{2}\log(x^2 + 1) + i\arg(x - i).$$

Now we saw in Chap. IV, § 4 (v) that in the open set $G = \mathbb{C} - \mathbb{R}_-$ obtained by removing the real half axis $x \leq 0$ from $\mathbb{C}$ there are *uniform branches* of the pseudofunction $\mathcal{L}og$; such a branch is a (true) *continuous* function (and in fact analytic) $L(z)$ which, on $G$, satisfies the relation $z = e^{L(z)}$ for every $z$; every other solution is obtained by adding a constant multiple of $2\pi i$ to $L(z)$, and the simplest solution, which one generally calls the *principal determination* of the log in $G$, is that for which

(20.8) $$L(z) = \log|z| + i\arg z \qquad \text{with } |\arg z| < \pi;$$

this function is even analytic and satisfies

(20.9) $$L'(z) = 1/z,$$

the derivative being taken, of course, in the complex sense (Chap. II, n° 19).

To extend the formula (4) to the case where $a$ is complex it is enough to consider the function $x \mapsto L(x - a)$ on $\mathbb{R}$. Since $a$ is not real, the points $x - a$, situated on the horizontal through $a$, all lie in the open set $G = \mathbb{C} - \mathbb{R}_-$; the function

(20.10) $$L(x - a) = \log|x - a| + i\arg(x - a) \quad \text{with } |\arg(x - a)| < \pi,$$

obtained by composing $x \mapsto x - a$ and the analytic function $L$, is *ipso facto* of class $C^\infty$ in $\mathbb{R}$ and its derivative is the function $L'(x - a) = 1/(x - a)$ by Chap. III, n° 21, Example 1, where we showed generally that if $g(z)$ is holomorphic and $f(t)$ is differentiable then

$$\frac{d}{dt}g[f(t)] = g'[f(t)]f'(t).$$

From this we deduce that up to an additive constant

(20.11) $$\int \frac{dx}{x - a} = L(x - a) = \log|x - a| + i\arg(x - a), \qquad a \notin \mathbb{R},$$

where the argument must be chosen between $-\pi$ and $+\pi$ so that the right hand side will, at least, be a continuous function of $x$, and so in fact $C^\infty$.

Suppose for example that we are to integrate $1/(x-i)$ from $x = 1$ to $x = 2$. We have to calculate the variation of the function $\log|x - i| + i\arg(x - i)$ between these values. That of the log is

$$\log|(2 - i)/(1 - i)| = \frac{1}{2}\log(5/2).$$

The points $x - i$ lie below the real axis, so that one *has to* choose their arguments between $-\pi$ and $0$; then $\arg(2-i) = -\pi/6$ and $\arg(1-i) = -\pi/4$. Finally the desired integral equals

$$\frac{1}{2}(\log 5 - \log 2) + i\pi/12.$$

Let us now give two examples of application to primitives of rational functions.

*Example 1.* Calculate

$$\int \frac{dx}{(x^2 + 1)^2}.$$

We write

(20.12) $\qquad 1/(x^2 + 1)^2 \;=\; A/(x - i)^2 + B/(x - i) +$
$$+ B'/(x + i) + A'/(x + i)^2$$

with coefficients to be determined. Multiplying through by $(x - i)^2$ one finds $1/(x + i)^2$ on left hand side and, on the right hand side, $A$ plus terms containing factors $x - i$, so zero for $x = i$. Consequently $A = 1/(2i)^2 = -1/4$. Similarly, $A' = -1/4$. The terms in $A$ and $A'$ have sum

$$-i\left[(x - i)^2 + (x + i)^2\right]/4(x^2 + 1)^2 = -\frac{1}{2}(x^2 - 1)/(x^2 + 1)^2;$$

on substituting this in the left hand side of (12) one obtains the relation

$$\frac{1}{2}(x^2 + 1) = B/(x - i) + B'/(x + i);$$

here again, one multiplies through by $x - i$ and puts $x = i$ in the result; this gives $B = 1/4i = -i/4$ and likewise $B' = i/4$. Whence finally

$$1/(x^2 + 1)^2 = -1/4(x - i)^2 - i/4(x - i) + i/4(x + i) - 1/4(x + i)^2.$$

Consequently,

$$I = \int \frac{dx}{(x^2 + 1)^2} = \frac{1}{4}\left(\frac{1}{x - i} + \frac{1}{x + i}\right) + \frac{1}{4}\left[L(x + i) - L(x - i)\right].$$

The rest of the problem is now to express this result in real terms. Now $L(x + i) = \frac{1}{2}\log(x^2 + 1) + i\arg(x + i)$ with a similar formula for $L(x - i)$;

it is clear on the other hand (sketch!) that $\arg(x - i) = -\arg(x + i)$ if one chooses the arguments between $-\pi$ and $+\pi$. The expression between [ ] is therefore equal to $2i\arg(x + i)$, the argument being chosen between $0$ and $\pi$ since $x + i$ is above the real axis. Thus, up to an additive constant (we are calculating a primitive),

$$I = x/2(x^2 + 1) - \frac{1}{2}\arg(x + i) \qquad \text{with } 0 < \arg(x + i) < \pi.$$

To reduce to a more familiar expression we note that the argument $t$ of $x + i$ satisfies $\tan t = 1/x$, whence

$$\arg(x + i) = \arctan(1/x) = \pi/2 - \arctan x + 2k\pi.$$

The left hand side having to be a continuous function of $x \in \mathbb{R}$ and the function $\arctan x$ also being so, if one insists on values between $-\pi/2$ and $+\pi/2$, the integer $k$ must be independent of $x$; for $k = 0$, one actually finds for the right hand side a value between $0$ and $\pi$, as it must be. The constant $\pi/2$ being unimportant in calculating the desired primitive, we conclude that

$$I = x/2(x^2 + 1) + \frac{1}{2}\arctan x + const.$$

The reader may, as an elementary prudence, check the result by calculating its derivative; the fact that it is real is already a good sign ...

The reader will find very many examples of this technique in all the textbooks, although the great majority of authors recoil from the complex log. Moreover one is not always forced to use these when the denominator $Q$ of the given real rational function has complex roots. For then, taking together the conjugate imaginary terms of the decomposition (2) in the case of a real function, one is led to sums of expressions of the form $(Ax+B)/(x^2+px+q)^n$ where the trinomial $x^2 + px + q$, with $p$ and $q$ real, has no real roots, i.e. can be written $(x - a)^2 + b^2$ with $a$, $b$ real and $b \neq 0$. The change of variable $x = ay + b$ then reduces it to calculating integrals of the form

$$I_n = \int dx/(x^2 + 1)^n, \qquad J_n = \int x dx/(x^2 + 1)^n.$$

Since 1 is the derivative of the function $x$, an integration by parts gives

$$\begin{aligned} I_n &= x/(x^2 + 1)^n - 2n\int x^2 dx/(x^2 + 1)^{n+1} = \\ &= x/(x^2 + 1)^n - 2nI_n + 2nI_{n+1} \end{aligned}$$

since $x^2 = (x^2 + 1) - 1$; on replacing $n$ by $n - 1$, this relation can again be written as

$$(2n - 2)I_n = (2n - 1)I_{n-1} - x/(x^2 + 1)^{n-1},$$

which allows us to calculate step-by-step, starting from $I_1 = \arctan x$; one can even calculate the general formula for $I_n$ directly, but it is clearly not worth doing. A similar method applies to the $J_n$.

*Example 2.* Leibniz having believed he would flabbergast Newton with his series for $\pi/4$, the latter retorted that he knew others, and better ones, especially the formula

(20.13)    $\pi/2\sqrt{2} = 1 + 2(1/3.5 - 1/7.9 + 1/11.13 - \ldots)$,

but of course without presenting the proof.

But we know that he derived it from integrating the function

$$\frac{1+x^2}{1+x^4} = \frac{1}{2}\frac{1}{x^2 - x\sqrt{2}+1} + \frac{1}{2}\frac{1}{x^2 + x\sqrt{2}+1}.$$

Putting $\varepsilon = \pm 1$, one has $x^2 - \varepsilon x\sqrt{2}+1 = \left(x - \varepsilon/\sqrt{2}\right)^2 + 1/2$, which suggests the change of variable $x - \varepsilon/\sqrt{2} = y/\sqrt{2}$; then $dx = dy/\sqrt{2}$ and $x^2 - \varepsilon x\sqrt{2} + 1 = \frac{1}{2}(y^2 + 1)$; consequently,

$$\sqrt{2}\int dx/\left(x^2 - \varepsilon x\sqrt{2}+1\right) = \int dy/(y^2+1) = \arctan y = \arctan\left(x\sqrt{2} - \varepsilon\right).$$

One deduces

$$\sqrt{2}\int \frac{1+x^2}{1+x^4}\,dx = \arctan\left(x\sqrt{2}+1\right) + \arctan\left(x\sqrt{2}-1\right).$$

But the addition formula

$$\tan(u+v) = \frac{\tan u + \tan v}{1 - \tan u.\tan v}$$

shows that

$$\arctan x + \arctan y = \arctan\frac{x+y}{1-xy} + k\pi.$$

An easy calculation then shows that

(20.14)    $\sqrt{2}\int \dfrac{1+x^2}{1+x^4}\,dx = \arctan\left[x\sqrt{2}/(1-x^2)\right],$

whence

(20.15)    $\sqrt{2}\int_0^t \dfrac{1+x^2}{1+x^4}\,dx = \arctan\left[t\sqrt{2}/(1-t^2)\right]$

for $0 \le t < 1$. As $t$ tends to 1, $t\sqrt{2}/(1-t^2)$ tends to $+\infty$, its arctan tends to $\pi/2$ and finally one finds

(20.16)    $\displaystyle\int_0^1 \frac{1+x^2}{1+x^4}\,dx = \pi/2\sqrt{2}.$

On the other hand, the integrand is represented by the power series

$$(1+x^2)\left(1 - x^4 + x^8 - x^{12} + \ldots\right) = (1-x^4+x^8-\ldots)+\left(x^2 - x^6 + x^{10} - \ldots\right);$$

Can one integrate this term-by-term between 0 and 1?

There is no problem in integrating over $[0, t]$ with $t < 1$. One finds

$$(t - t^5/5 + t^9/9 - \ldots) + (t^3/3 - t^7/7 + t^{11}/11 - \ldots).$$

It remains to pass to the limit in each series as $t$ tends to 1. Once again we have alternating series with decreasing terms: by Leibniz' estimate of the remainder, namely $t^n/n < 1/n$, the partial sums $s_n(t)$, clearly continuous for $|t| \leq 1$, converge to the total sum $s(t)$ uniformly on the *closed* interval $|t| \leq 1$, so that it is a continuous function of $t$ for $t \leq 1$. One may then write that

$$
\begin{aligned}
\lim_{t \to 1-0} s(t) &= \lim_{t \to 1-0} \lim_{n} s_n(t) \quad [\text{by definition of } s(t)] = \\
&= \lim_{n} \lim_{t} s_n(t) \quad [\text{Chap. III, n° 12, Theorem 16}] = \\
&= \lim_{n} s_n(1) \quad [\text{since } s_n \text{ is continuous}] = s(1).
\end{aligned}
$$

By (16), one finally finds

$$\pi/2\sqrt{2} = (1 - 1/5 + 1/9 - \ldots) + (1/3 - 1/7 + 1/11 - \ldots)$$

and has only to rearrange the terms to obtain Newton's series.

# § 7. Generalised Riemann integrals

Up to now we have attempted to integrate only *bounded* functions on a bounded and, most often, *compact* interval. To get further we have to free ourselves from these restrictions. The method is quite similar to that for passing from the partial sums to the total sum of a series. To avoid complications which would not be helpful at this level, we restrict ourselves to *regulated* functions, i.e. those having right and left limits at each point, and so are integrable on every compact interval (or even bounded interval, if they are themselves bounded: n° 7, corollary to Theorem 6); if we really wanted to generalise, we would have to go to the grand integration theory (Appendix). This would moreover allow us to simplify many proofs very appreciably, also the somewhat artificial statements needed in order to remain at the "elementary" level.

## 21 – Convergent integrals: examples and definitions

Suppose for example that we wish to assign a meaning to the integral

$$\int_0^b dx/x.$$

It is natural, if only for a geometric reason, to consider it as the limit of the integral over $(u, b)$ as $u > 0$ tends to 0. This is equal to $\log b - \log u$ and, hard luck, therefore tends to $+\infty$, which is not the result hoped for, even if, after all, we have attributed the sum $+\infty$ to divergent series with positive terms. One could also attribute to $1/x$ the value $+\infty$ at $x = 0$, so obtaining a lower semicontinuous function on $[0, b]$ to which one applies the definition of the integral given in n° 11; the result is the same, as one sees on calculating the integrals over $[0, b]$ of the functions $\inf(n, 1/x)$ and then passing to the limit as $n \to +\infty$. Replacing $1/x$ by $x^s$ with $s$ real, a function of which a primitive is $x^{s+1}/(s+1)$, one obtains the same result if $s + 1 < 0$. For $s + 1 > 0$ the integral over the interval $[u, b]$, equal to $b^{s+1}/(s+1) - u^{s+1}/(s+1)$, tends to $b^{s+1}/(s+1)$; whence "clearly", i.e. by definition,

(21.1)     $$\int_0^b x^s dx = b^{s+1}/(s+1) \qquad \text{if } s > -1 \text{ and } b > 0.$$

Consider next the integral

$$\int_a^{+\infty} x^s dx$$

with $a > 0$ to eliminate a possible difficulty at $x = 0$, and where $s$ is real. It is natural to consider this as the limit of the integral over $(a, v)$ as $v$ increases indefinitely. If $s = -1$, one finds $\log v - \log a$, which tends to $+\infty$.

If $s \neq -1$, one finds $v^{s+1}/(s+1) - a^{s+1}/(s+1)$. If $s > -1$, the result increases indefinitely. If $s < -1$, it tends to $-a^{s+1}/(s+1)$, whence the formula

$$(21.2) \qquad \int_a^{+\infty} x^s dx = -a^{s+1}/(s+1) \quad \text{for } s < -1, \ a > 0.$$

We remark in passing that the hypotheses on $s$ that give a meaning to the integrals (1) and (2) are mutually exclusive; in other words, the integral

$$\int_0^{+\infty} x^s dx \qquad \textbf{IS NEVER FINITE.}$$

Let us generalise. Let $f$ be a regulated function on a noncompact interval $X = (a, b)$, either not bounded, or bounded but not closed. If one argues as we did in defining convergence of a series, one is led to associate a "partial integral"

$$(21.3) \quad s(K) = \int_K f(x)dx = \int_u^v f(x)dx = s(u,v) = F(v) - F(u),$$

to every *compact* interval $K = [u, v]$ contained in $X$, where $F$ is a primitive of $f$ in $X$. We then say that the integral

$$(21.4) \qquad\qquad s(X) = \int_X f(x)dx = \int_a^b f(x)dx$$

*converges* if $s(K)$ tends to a limit – which, by definition, will be the integral (4) – as $K$ "tends to" $X$, i.e. when $u$ and $v$ tend respectively to[40] $a$ and $b$. As in the case of a series [Chap. II, eqn. (15.4)], this means[41] that for every $r > 0$ there exists a compact interval $K \subset X$ such that, for every compact interval $K' \subset X$,

$$(21.5) \qquad\qquad K \subset K' \Longrightarrow |s(K') - s(X)| < r.$$

One could also then say that $f$ is "integrable" on $X$, but it is better to abstain carefully from this when the integral of $|f|$ does not converge, this term having been reserved, in the only theory which has counted for a long time, that of Lebesgue, for *absolutely* integrable functions. It would be better to speak of *semiconvergent* integrals when $\int |f(x)| \, dx$ does not converge.

(5), again, means that, for every $r > 0$, one has $|s(X) - s(u,v)| < r$ once $u$ is close enough to $a$ and $v$ close enough to $b$. If, for example, $a$ is finite

---

[40] If $X$ and $f$ are bounded, in which case $f$ is integrable on $X$ in the sense of n° 2 (n° 7, corollary to Theorem 6), this definition is compatible with that of n° 2.

[41] One can define this type of limit precisely. Let $\varphi(K)$ be a function of a variable compact set $K \subset X$. One says that it tends to a limit $u$ as $K$ tends to $X$ if for every $r > 0$ there exists a compact $K \subset X$ such that $K \subset K' \subset X \Longrightarrow |u - \varphi(K')| < r$. This is definition (21.5).

and $b = +\infty$, this means that, for every $r > 0$, there exists an $r' > 0$ and an $N > 0$ such that

$$\{(u - a < r') \ \& \ (v > N)\} \Longrightarrow |s(X) - s(u, v)| < r.$$

In terms of primitives,

(21.6)    $$\int_a^b f(x)dx = \lim_{v \to b, v < b} F(v) - \lim_{u \to a, u > a} F(u),$$

so that *the integral converges if and only if $F$ has finite limit values at $a$ and $b$.* These limits actually exist if the integral converges, for, in this case, one has $|s(u', v') - s(u'', v'')| < r$ if $u'$ and $u''$ are close enough to $a$ and $v'$ and $v''$ close enough to $b$; taking $v' = v''$, one sees then that $|F(u') - F(u'')| < r$, so that Cauchy's criterion is satisfied by $F(u)$ as $u$ tends to $a$. It is clear conversely that the integral converges if $F$ has limits at the end-points of $X$.

This is exactly what we have verified in the preceding examples. The method also works for integrating an exponential function $e^{cx}$ with $c$ real and nonzero, since the behaviour of the primitive $e^{cx}/c$ as $x$ tends to $+\infty$ or $-\infty$ has been elucidated in Chap. IV. In particular, one may integrate $e^x$ from $-\infty$ to any finite limit, but one cannot integrate from $-\infty$, nor from a finite limit, to $+\infty$.

But in general one does not know $F$, so the usefulness of (6) is heavily constrained; it is better, in most cases, to examine the order of magnitude of $f(x)$ as $x$ tends to $a$ or to $b$, just as one does for series.

## 22 – Absolutely convergent integrals

The theory of series is particularly simple when they are absolutely convergent. Likewise here. We shall say that the integral (21.4) is *absolutely convergent* if the integral of $|f(x)|$ is convergent, in which case one may say that $f$ is *integrable* over $X$ (or, to reassure the reader who is starting these topics, *absolutely integrable*) without risking a collision with the Lebesgue theory.

**Theorem 18.** *(i) Let $f$ be a positive regulated function defined on an interval $X = (a, b)$. Then $f$ is integrable on $X$ if and only if the integrals over the compact subsets $K \subset X$ are bounded above; and then*

(22.1)    $$\int_X f(x)dx = \sup_{K \subset X} \int_K f(x)dx;$$

*(ii) let $f$ be a regulated function defined on an interval $X$; if the integral $\int f(x)dx$ extended over $X$ is absolutely convergent (i.e. if $f$ is absolutely integrable on $X$) then it is convergent and*

(22.2)    $$\left| \int_X f(x)dx \right| \leq \int_X |f(x)|dx.$$

To prove point (i) we observe that, $f$ being positive, $s(K)$ is an increasing function of $K$:

$$K \subset K' \implies s(K) \leq s(K').$$

The arguments of Chap. II, n° 9 on increasing sequences transpose immediately to here without our needing to expound them all again. One might also observe that, if $f$ is positive, its primitives are increasing functions; these tend to finite limits at the end-points of $X$ if and only if they are bounded on $X$.

To establish the assertion (ii) one can reduce to the case of a real function, then to that of a positive function by writing $f = f^+ - f^-$. Since $f^+$ and $f^-$ are majorised by $|f|$, point (i) shows that the integrals of these functions converge, so that of $f$ too. One could also use directly one of the numerous variants of Cauchy's criterion adapted to the situation, namely that the partial integrals $s(K)$ tend to a limit if and only if for every $r > 0$ there exists a compact subset $K \subset X$ such that

$$|s(K') - s(K'')| < r \quad \text{for any } K' \supset K \text{ and } K'' \supset K;$$

but, since $K'$ and $K''$ contain $K$, we have[42]

$$
\begin{aligned}
|s(K') - s(K'')| &= \left| \int_{K'-K} f(x)dx - \int_{K''-K} f(x)dx \right| \\
&\leq \int_{(K'-K)\cup(K''-K)} |f(x)|dx,
\end{aligned}
$$

the integral of $|f|$ being extended over the set $(K' \cup K'') - K$; if we write $S(K)$ for partial integrals relative to $|f|$ we have

$$|s(K') - s(K'')| \leq S(K' \cup K'') - S(K),$$

an arbitrarily small quantity for any $K', K'' \supset K$ for $K$ "large enough" if the $S(K)$ are bounded above.

The inequality (2) is obvious when integrating over a compact interval $K \subset X$, hence propagates to the limit, qed.

Theorem 18 has some trivial consequences, which we use constantly.

**Corollary 1.** *Let $f$ be a bounded regulated function on an interval $X$, and $\mu$ an absolutely integrable regulated function on $X$. Then the function $f(x)\mu(x)$ is absolutely integrable on $X$ and*

$$\int |f(x)\mu(x)dx| \leq \|f\|_X \int |\mu(x)|dx.$$

---

[42] In what follows, we will have occasion to integrate over a set which is a finite union of intervals having, pairwise, at most one point in common; the integral will clearly, by definition, be the sum of the integrals extended over these intervals. This, furthermore, is equivalent to multiplying the integrand by the characteristic function, a step function, of this union.

Obvious. As we shall do on various occasions in the rest of this §, we have used the $\int$ sign to denote integrals extended over $X$.

*Example*: the *Fourier transform*

$$(22.3) \qquad \hat{\mu}(y) = \int_{\mathbb{R}} e^{-2\pi i x y} \mu(x) dx$$

of an absolutely integrable function on $X = \mathbb{R}$ is defined for every $y \in \mathbb{R}$.

In the following statement one says that a function $f$ is *square* (understood: absolutely) *integrable* on an interval $X$ if the function $|f|^2$ is integrable on $X$. One may then generalise Cauchy-Schwarz:

**Corollary 2.** *Let $f$ and $g$ be two regulated square integrable functions on an interval $X$; then the function $f(x)\overline{g(x)}$ is absolutely integrable on $X$ and*

$$\left| \int f(x)\overline{g(x)}dx \right|^2 \leq \int |f(x)|^2 dx . \int |g(x)|^2 dx.$$

One replaces $f$ and $g$ by $|f|$ and $|g|$, writes the Cauchy-Schwarz inequality for every compact interval $K \subset X$ and notes that the left hand side is, for any $K$, majorised by the right hand side of the inequality to be established, whence the result on passage to the limit. Or else, see the end of n° 14, which will prove more generally that if the functions $|f|^p$ and $|g|^q$ are integrable for $1/p + 1/q = 1$, then $fg$ is also integrable, and a Hölder inequality is valid.

The convergence conditions for the integrals involving $x^s$ established above for $s$ real can be transformed immediately into conditions for absolute convergence in the case of a complex exponent, since

$$|x^s| = x^{\text{Re}(s)} \qquad \text{for } x > 0.$$

On the other hand, point (ii) of Theorem 18 shows that if, on a neighbourhood of one of the limits of integration, one has a relation of the form

$$f(x) = O(g(x)),$$

then *absolute* convergence (on a neighbourhood of this end-point) of the integral of $g(x)$ implies that of the integral of $f(x)$; if one has the more precise relation $f(x) \asymp g(x)$ then the integrals are of the same nature as concerns absolute convergence. (It might, on the other hand, happen that one of them converges non-absolutely and that the other diverges, as can occur with series.)

In elementary practice the *absolute* convergence of an integral is almost always shown by comparing the behaviour of the integrand with that of a combination of classical functions: exponentials, powers, logarithms, etc. It is worth having these results permanently available, and to have understood their reasons.

First of all, putting $X = (a, b)$, one may always choose a $c$ such that $a < c < b$ and decompose the integral into integrals extended over $(a, c)$ and over $(c, b)$. If the integrand is regulated there will be no problem of convergence on a neighbourhood of $c$, and this allows us to isolate the difficulties. One may always, in the case where $c$ is the right endpoint, reduce to the case where $c = +\infty$ by the change of variable $c - x = 1/y$. If $c$ is the left endpoint, one may reduce to the case where $c = 0$ by $x - c = y$, or to $c = -\infty$ by $x - c = -1/y$.

Now consider the prototype integral

$$(22.4) \qquad \int_a^{+\infty} \log^m x . x^n e^{-sx} dx \qquad (a > 0)$$

where $m$, $n$, $s$ are *a priori* complex but can in fact be assumed real, since the modulus of the integrand is obtained by replacing the exponents by their real parts. In view of the orders of increase of the three functions involved, it is pretty clear that the convergence of the integral is, for $s \neq 0$, governed by the exponential factor. It follows immediately from Chap. IV, n° 5, that

$$\log^m x . x^n = o(e^{rx}) \quad \text{as } x \to +\infty \text{ for every } r > 0;$$

the integral will thus converge if there exists an $r > 0$ making the integral of $e^{(r-s)x}$ converge, i.e. if $r - s < 0$, whence convergence for $s > 0$, strict inequality. For $s < 0$, the integrand grows indefinitely, whence divergence.

In the case where $s = 0$, the change of variable $x = e^y$ leads us to integrate the function $y^m e^{(n+1)y}$ on a neighbourhood of infinity; the integral is thus convergent for $n < -1$ and divergent for $n > -1$.

If, finally, we have $n = -1$, so that we are dealing with the integral of $x^{-1} \log^m x$, the same change of variable reduces to the function $y^m$, whence convergence for $m < -1$ and divergence for $m \geq -1$.

In conclusion, convergence is governed by the exponential function if this is actually present or, if it is absent ($s = 0$), by the power function if that is actually present; if these two functions are absent, the integral converges if and only if $m < -1$.

One might study integrals similar to (4), but containing more simple factors, by the same method; you could, for example, introduce the factors $\log \log x$, or $\log \log \log x$, etc. ..., which grow more and more slowly and do not affect the result so long as there are factors present which decrease much more rapidly than them. You might also insert a factor $x^{-x}$, which tends to 0 fast enough to annihilate even the most vertical exponential functions ... Etc.

The study of an integral such as

$$(22.5) \qquad \int_0^b |\log x|^m x^n dx \qquad (0 < b < +\infty)$$

reduces to the preceding case; the change of variable $x = 1/y$ transforms it into the integral of the function $\log^m y.y^{-n-2}$ on a neighbourhood of $+\infty$. The integral (5) is thus convergent if $-n - 2 < -1$, i.e. if $n > -1$; it diverges if $n < -1$. If $n = -1$, the integral converges if $m < -1$ and diverges in the contrary case. Note in passing that (5) *always converges* for $n = 0$, i.e. *when the term $x^n$ is absent*, a result due to the fact that on a neighbourhood of 0 a power of the log grows less quickly than any negative power of $x$, for example than the function $x^{-1/2}$ whose integral converges (primitive: $2x^{1/2}$). For $n = 0$, $m = 1$, note that $\log x$ has primitive $x \log x - x$, a function which tends to a limit, namely 0, when $x \to 0$.

*Exercise*: extend this calculation to the case of an arbitrary integer $m > 0$ by integrating by parts.

The case of rational functions is particularly simple: if $p$ and $q$ are polynomials and $q$ has no real roots then the *absolute convergence* of the integral

$$\int_{-\infty}^{+\infty} \frac{p(x)}{q(x)} \, dx$$

depends only on the integer $n = d^\circ(q) - d^\circ(p)$ since the function is equivalent to $1/x^n$ to within a constant factor; so absolute convergence is equivalent to the condition $n \geq 2$.

*Example 1.* Consider Euler's ubiquitous Gamma function

(22.6) $$\Gamma(s) = \int_0^{+\infty} e^{-x} x^{s-1} dx.$$

Absolute convergence at infinity is automatic, but, at 0, requires $\mathrm{Re}(s) > 0$. An integration by parts[43] then shows that

$$\Gamma(s+1) = \int_0^{+\infty} e^{-x} x^s \, dx = -e^{-x} x^s \Big|_0^{+\infty} + s \int_0^{+\infty} e^{-x} x^{s-1} dx$$

and since the integrated-out part is clearly zero, we have

(22.7) $$\Gamma(s+1) = s\Gamma(s).$$

Since it is clear that $\Gamma(1) = 1$ we deduce that

(22.8) $$\Gamma(n) = (n-1)!$$

for every integer $n > 1$. This is Euler's method for defining the "factorial" of an arbitrary complex number. We shall see in n° 25 that the $\Gamma$ function is holomorphic in the half plane $\mathrm{Re}(s) > 0$ and that it is even the restriction of a function holomorphic on $\mathbb{C} - \{0, -1, -2, \ldots\}$ .

---

[43] The formula $\int f'g = fg - \int fg'$ for integration by parts applies to the integrals considered here, on condition that we check that the function $fg$ has limit values at the end-points of the interval of integration $X$: integrate over a compact $K \subset X$ and pass to the limit.

*Example 2.* Euler also studied the integral

$$(22.9) \qquad B(x,y) = \int_0^1 t^{x-1}(1-t)^{y-1}dt$$

where $x$ and $y$ are *a priori* complex (and rational for him). Absolute convergence on a neighbourhood of 0 requires $\mathrm{Re}(x) > 0$ and, on a neighbourhood of 1, $\mathrm{Re}(y) > 0$. Clearly

$$(22.10) \qquad B(x,y) = B(y,x)$$

(change of variable $t \mapsto 1 - t$). The change of variable $t \mapsto \sin^2 t$ shows that

$$(22.11) \qquad B(x,y) = 2 \int_0^{\pi/2} \sin^{2x-1} t . \cos^{2y-1} t.dt.$$

We shall see later (n° 26, Example 1) that

$$(22.12) \qquad B(x,y) = \Gamma(x)\Gamma(y)/\Gamma(x+y),$$

a famous formula due to Euler with, as always, a proof which posterity, principally Jacobi, has rectified. It immediately provides the explicit value of (11) for $x,y \in \mathbb{N}$.

## 23 – Passage to the limit under the $\int$ sign

For generalised or "improper" Riemann integrals there are theorems on passage to the limit which the Lebesgue theory has rendered obsolete, but remain usable at a more elementary level. For example:

**Theorem 19 (Poor man's dominated convergence).** *Let $(f_n)$ be a sequence of regulated functions, absolutely integrable on an interval $X \subset R$. Assume that*
   *(i) the $f_n$ converge to a limit $f$ uniformly on every compact $K \subset X$,*
   *(ii) there exists a positive function $p$, integrable on $X$, and such that $|f_n(x)| \le p(x)$ for any $x$ and $n$.*
   *Then the function $f$ is absolutely integrable on $X$ and*

$$(23.1) \qquad \int f(x)dx = \lim \int f_n(x)dx.$$

First of all, it is clear that $f$, being regulated like the $f_n$, is absolutely integrable on $X$ since $|f(x)| \le p(x)$ for every $x$. Since $p$ is positive and integrable, for every $r > 0$ there exists a compact interval $K \subset X$ such that

$$(23.2) \qquad \int_{X-K} p(x)dx < r$$

(and even for every $K' \supset K$), whence the same relation for each $|f_n(x)|$ and for $|f(x)|$. On the other hand

(23.3) $$\int_K |f(x) - f_n(x)| dx < r \qquad \text{for } n \text{ large}$$

since $f_n$ converges uniformly to $f$ on $K$. Now

$$\left| \int_X f(x) dx - \int_X f_n(x) dx \right| \le \int_X |f(x) - f_n(x)| dx$$

is the sum of the analogous integrals extended over $K$ and $X - K$; by (2), the second is $< 2r$ since $|f(x) - f_n(x)| \le 2p(x)$ for any $x$ and $n$; the first is $< r$ for $n$ large by (3). The left hand side of the preceding relation (and even the right hand side) thus tends to 0, qed.

*Example 1.* Consider the function

$$\Gamma(s) = \int_0^{+\infty} e^{-x} x^{s-1} dx, \qquad \text{Re}(s) > 0,$$

again, and observe that

$$e^{-x} x^{s-1} = \lim(1 - x/n)^n x^{s-1}.$$

We cannot just bluntly apply Theorem 19 since the functions on the right hand side are not integrable between 0 and $+\infty$: convergence at 0 presupposes $\text{Re}(s) > 0$ and convergence at infinity $\text{Re}(s) < -n$. For $x < n$ we always have [Chap. III, n° 16]

$$\log[(1 - x/n)^n] = n.\log(1 - x/n) = -x - x^2/2n - \ldots < -x$$

and so $(1 - x/n)^n < e^{-x}$. Now consider the functions

(23.4) $$f_n(x) = \begin{cases} (1 - x/n)^n x^{s-1} & \text{for} \quad 0 < x \le n, \\ 0 & \text{for} \quad x > n; \end{cases}$$

they converge to the absolutely integrable function $e^{-x} x^{s-1}$ and satisfy $|f_n(x)| \le |e^{-x} x^{s-1}|$. To be able to apply Theorem 19 it therefore suffices to show that convergence is uniform on every compact subset of $]0, +\infty[$. If we accept this point provisionally we then find

$$\Gamma(s) = \lim \int_0^n (1 - x/n)^n x^{s-1} dx = \lim n^s \int_0^1 (1 - u)^n u^{s-1} du;$$

on integrating by parts à la Leibniz, i.e. without limits of integration, we find that

$$\int (1 - u)^n u^{s-1} du = (1 - u)^n u^s/s + \frac{n}{s} \int (1 - u)^{n-1} u^s du$$

and since the integrated-out part is zero for $u = 0$ [because $\mathrm{Re}(s) > 0$] and $u = 1$, we find

$$\int_0^1 (1-u)^n u^{s-1} du = \frac{n}{s} \int_0^1 (1-u)^{n-1} x^s dx;$$

whence, iterating,

(23.5) $$\int_0^1 (1-u)^n u^{s-1} du = \frac{n!}{s(s+1)\ldots(s+n)}.$$

A little less than two centuries after 1812 and Gauss, who did not know that Euler, ever present, had preceded him along this path about 1776, as Remmert, *Funktionentheorie 2*, pp. 34–36, tells us, we find that

(23.6) $$\Gamma(s) = \lim n! n^s / s(s+1) \ldots (s+n)$$

for $\mathrm{Re}(s) > 0$. We can derive an expansion of $\Gamma$ as an infinite product, but we still lack the necessary "Euler's constant" which will appear in Chap. VI, n° 18.

We still have to show that the functions (4) converge uniformly on every compact interval $K = [a, b]$ with $0 < a < b < +\infty$. The factor $x^{s-1}$ being bounded on $K$ since $a > 0$, it is enough to examine the factor $(1 - x/n)^n$. For $n > b$ we have $|x/n| < 1$ in $K$ and thus

$$\log\left[(1-x/n)^n\right] = n.\log(1-x/n) = -\left(x + x^2/2n + x^3/3n^2 + \ldots\right);$$

the sequence of functions $\log\left[(1-x/n)^n\right]$, thus also that of the functions $(1-x/n)^n$, is therefore increasing on $K$, and even on $[0, b]$, for $n > b$. Since it converges to the continuous function $e^{-x}$ uniform convergence on $K$ follows from Dini's Theorem of n° 10.

More elementarily, so more complicated: first remark that $\log\left[(1-x/n)^n\right] = -x - x^2/2n - \ldots$ converges uniformly to $-x$ on $[0, b]$, since, for $n > b$,

$$\left|x^2/2n + x^3/3n^2 + \ldots\right| \le \frac{b^2}{n}\left(1 + b/n + b^2/n^2 + \ldots\right) = \frac{b}{n-b}$$

for every $x \in [0, b]$. Since $(1-x/n)^n = \exp[n.\log(1-x/n)]$, it remains either to "dirty one's hands" by calculating (exercise!), or to establish a general lemma to bypass the calculations:

**Lemma.** *Let $K$ be a compact subset of $\mathbb{C}$, let $(f_n)$ be a sequence of functions which converges uniformly on $K$ to a bounded limit function $f$ on $K$, and let $g$ be a function defined and continuous on an open set $U$ containing the closure of $f(K)$. Then the composite function $g_n = g \circ f_n$ is defined on $K$ for $n$ large and converges uniformly to $g \circ f$.*

The limit $f$ being bounded, the closure $H$ of $f(K)$ is compact so that, for every integer $p$, the set $H_p$ of the $z \in \mathbb{C}$ such that $d(z, H) \leq 1/p$ is also compact. Since $U$ is open and contains $H$, it contains[44] an $H_p$. Since $\|f - f_n\|_K \leq 1/p$ for $n$ large, we therefore have $f_n(K) \subset H_p \subset U$, which allows us to define $g_n(x) = g[f_n(x)]$. Now the function $g$ is *uniformly* continuous on the compact $H_p$; for every $r > 0$ there is therefore an $r' > 0$ such that $|g(z') - g(z'')| \leq r$ if $z', z'' \in H_p$ satisfy $|z' - z''| \leq r'$; now this, for $n$ large, is the case for any $x \in K$ if one takes $z' = f(x)$ and $z'' = f_n(x)$. Hence $\|g \circ f - g \circ f_n\|_K \leq r$ for $n$ large, qed.

Note in passing that the lemma makes no hypothesis as to the nature of the $f_n$; in particular, they are not assumed continuous; it is $g$ which must be. But if the $f_n$ are continuous, the function $f$ is so too, and the closure $H$ of $f(K)$ is in fact the compact set $f(K)$ itself.

If, instead of integrating a sequence of functions one integrates a series, one has to consider the partial sums $s_n(x)$ of the series and apply the preceding theorem to them. The simplest result is the following:

**Theorem 20.** *Let $\sum u_n(x)$ be a series of absolutely integrable regulated functions on an interval $X$. Assume that (i) the series converges uniformly on every compact $K \subset X$; (ii) there exists a positive function $p(x)$, integrable on $X$, such that $\sum |u_n(x)| \leq p(x)$ for every $x \in X$. Then the function $s(x) = \sum u_n(x)$ is absolutely integrable on $X$ and*

$$(23.7) \qquad \int s(x)dx = \sum \int u_n(x)dx.$$

The hypothesis (i) shows that the partial sums $s_n(x)$ converge uniformly on every compact $K \subset X$; since (ii) shows that $|s_n(x)| \leq p(x)$, one need only apply the preceding theorem to the $s_n$.

Condition (ii) is analogous to normal convergence on all of $X$, but more restrictive. In fact, and in contrast to the case of integrals extended over a compact interval, *normal convergence in $X$ is not enough to assure (7) if $X$ is not bounded.* We do know then, of course, that for $n$ large the difference between the total sum $s(x)$ and the partial sum $s_n(x)$ is $< r$ for any $x \in X$ since this is majorised by the $n$-th remainder of the series $\sum v_n$ which dominates the series $\sum u_n(x)$. But we cannot extract any estimate for the difference between their integrals over $X$ if $X$ is not bounded.

One may however establish a useful result whose formulation is very close to that of one of the fundamental results of the Lebesgue theory:

---

[44] The $H_p \cap (\mathbb{C} - U)$ are closed and bounded and form a decreasing sequence of compacta; their intersection, contained simultaneously in $H$ (because $H = \cap H_p$ for every compact $H$) and in $\mathbb{C} - U$, is empty; thus $H_p \cap (\mathbb{C} - U) = \emptyset$, i.e. $H_p \subset U$, for $p$ large: Chap. III, n° 9 or corollary 1 of BL.

**Theorem 21.** *Let $X$ be an interval and $u_n(x)$ a series of regulated functions which converges normally on every compact $K \subset X$. Assume that*

$$(23.8) \qquad \sum \int |u_n(x)| dx < +\infty.$$

*Then the function $s(x) = \sum u_n(x)$ is absolutely integrable on $X$ and*

$$\int s(x)dx = \sum \int u_n(x)dx.$$

Let us put, generally,

$$m_I(f) = \int_I f(x)dx$$

and consider a compact interval $K \subset X$. Since the given series converges normally on $K$ ("uniformly" would suffice) and one may integrate term-by-term on a compact set (n° 4), the relation $|s(x)| \leq \sum |u_n(x)|$ shows that

$$m_K(|s|) \leq m_K\left(\sum |u_n|\right) = \sum m_K(|u_n|) \leq \sum m_X(|u_n|) = M < +\infty$$

by (8). The regulated function $s$ is therefore absolutely integrable on $X$ [Theorem 18, (i)], with $m_X(|s|) \leq \sum m_X(|u_n|)$. Omitting the first $N$ terms from the series, one finds in the same way that

$$m_X\left(\left|s - \sum_{p=1}^{N} u_p\right|\right) \leq \sum_{p=N+1}^{\infty} m_X(|u_n|).$$

The result is $\leq r$ for $N$ large since $\sum m_X(|u_n|) < +\infty$. It follows that

$$\left|m_X(s) - \sum_{p=1}^{N} m_X(u_p)\right| \leq m_X\left(\left|s - \sum_{p=1}^{N} u_p\right|\right) \leq r \quad \text{for } N \text{ large,}$$

whence the theorem.

In the Lebesgue theory, the two preceding theorems are valid without the hypothesis of uniform or normal convergence, which would considerably simplify the arguments of Example 1; simple (or even only "almost everywhere") convergence is enough to assure the result; in fact, the hypothesis (8) even *implies* "almost everywhere" absolute convergence of the series $\sum u_n(x)$, as we shall see. On the other hand, hypothesis (ii), "dominated convergence", is essential even in the "grand" integration theory, where one ignores (in the English sense) the "semiconvergent" integrals, so specific to $\mathbb{R}$.

Instead of integrating a function of $x$ depending on an integer $n$ one may consider more generally an integral of the form

$$\int_X f(x,y)dx$$

where $y$ varies in an arbitrary subset $Y$ of $\mathbb{R}$ or even of $\mathbb{C}$, and examine what happens when $y$ tends to a closure point $b$ of $Y$. The hypotheses to make are obvious:

(i)    the function $x \mapsto f(x,y)$ is regulated for every $y$;
(ii)   $\lim_{y \to b} f(x,y) = g(x)$ exists for every $x \in X$ and the limit is uniform on every compact $K$ of $X$, i.e. for every $r > 0$ there is an $r' > 0$ such that

$$|f(x,y) - g(x)| \le r \ \text{ for every } x \in K$$

for every $y$ such that $|y - b| \le r'$;
(iii)  there exists a positive integrable function $p$ on $X$ such that $|f(x,y)| \le p(x)$ on $X$ for every $y \in Y$ close enough to $b$.

Then we may write

$$\lim_{y \to b} \int f(x,y)dx = \int dx \lim_{y \to b} f(x,y).$$

The hypotheses (i), (ii) and (iii) show that $g$ is regulated and absolutely integrable, after which it is enough to copy the proof of Theorem 21, replacing $f_n(x)$ everywhere by $f(x,y)$ and the expression "for $n$ large" by "for $y$ close enough to $b$" (or "for $y$ large" if $y$ tends to infinity). We could have established this general result directly; the theorem for sequences can be deduced from it on taking $Y = \mathbb{N}$ and $b = +\infty$.

In the most frequent applications of this result one seeks to show that the integral is a continuous function of $y$:

**Theorem 22.** *Let $X$ be an interval, $H$ a compact subset of $\mathbb{C}$, $f$ a function defined and continuous on $X \times H$ and $\mu$ a function defined and regulated in $X$. Assume that there is a positive function $p$ on $X$ such that $|f(x,y)| \le p(x)$ on $X \times H$ and $\int p(x)|\mu(x)|dx < +\infty$. Then the function $y \mapsto \int f(x,y)\mu(x)dx$ is continuous on $H$.*

Hypotheses (i) and (iii) above are clearly satisfied by $f(x,y)\mu(x)$ when $y$ tends to a $b \in H$. If $K$ is a compact subset of $X$, then the function $f$ is uniformly continuous on the compact $K \times H$; consequently, the hypothesis (ii) is satisfied also[45]. Thus $\lim \int f(x,y)\mu(x)dx = \int f(x,b)\mu(x)dx$, qed.

---

[45] Recall why. For every $r > 0$ there exists an $r' > 0$ such that the values of $f$ at two points of $K \times H$ distant at most $r'$ from each other are equal to within $r$. It follows that $|y - b| < r' \implies |f(x,y) - f(x,b)| < r$ for every $x \in K$, which means that, as $y$ tends to $b$, $f(x,y)$ tends to $f(x,b)$ uniformly on $K$. The factor $\mu(x)$, which is bounded on every compact set, like every regulated function, does not change the conclusion. Note in passing that if we have introduced a *function* $\mu(x)$, it is because we do not yet know how to treat an integral $\int f(x,y)d\mu(x)$ with respect to an arbitrary *measure* on a noncompact interval.

In practice, the continuous function $f$ is defined on $X \times Y$ where $Y \subset \mathbb{C}$ is not necessarily compact: the case of an arbitrary interval in $\mathbb{R}$ or of an open subset of $\mathbb{C}$ for example. To apply the theorem, it is enough to work on an arbitrarily small neighbourhood of a point $b \in Y$ since continuity is a property of local nature. So everything works *if every* $b \in Y$ *has a compact neighbourhood in* $Y$. Now a neighbourhood of $b$ in $Y$ contains, by definition, all the points of $Y$ whose distance to $b$ is sufficiently small. The hypothesis in question thus means that *there exists an* $r > 0$ *such that the set of* $y \in Y$ *such that* $d(b, y) \leq r$ (weak inequality) *is compact.* A subset of $\mathbb{C}$ having this property at each of its points is said to be *locally compact*. This is the case if $Y = F \cap U$ with $F$ closed and $U$ open[46]: choose $r$ so that the closed disc $d(b, y) \leq r$ is in $U$, then take for a neighbourhood of $b$ in $Y$ the intersection of this disc with $F$: it is closed in $\mathbb{C}$, so compact. In $\mathbb{R}$, every interval is locally compact; $\mathbb{Q}$ is not (exercise!). In $\mathbb{C}$, the union $Y$ of the open disc $D : |z| < 1$ and of the compact interval $[1, 3]$ is not, even though both these two sets are: the intersection of $Y$ with a closed disc of centre 1 is never closed. We might have said all this in Chap. III, but the reader is perhaps grateful to have been spared this at the beginning of the theory ...

In conclusion, Theorem 22 remains valid if one assumes $H$ only *locally* compact. By a happy coincidence, the locally compact sets are, among the subsets of $\mathbb{C}$, those on which one may construct an integration theory *à la* Lebesgue and, for a start, give a reasonable definition of Radon measures, as we shall see in n° 31.

### 24 – Series and integrals

One may sometimes compare an integral to a series, and vice-versa, to decide on its convergence or divergence. If, for example, $f$ is a regulated function on an interval $X = [a, +\infty[$ with $a$ finite, then $f$ has a limit value at $a$ and the convergence of the integral on a neighbourhood of $a$ poses no problem; it is then clear that

$$\int_X |f(x)| dx < +\infty \iff \sum_{n \geq a} \int_n^{n+1} |f(x)| dx < +\infty$$

because the partial sums of the series are, more or less, the partial integrals over the intervals $[a, n]$.

Consider now a function $f$ defined for $x \geq a$ finite, *positive, decreasing, and tending to* 0 *at infinity* (without this, for a decreasing function, the integral has no chance of converging); being monotone, $f$ is regulated (and, in applications, is always continuous). For every $n \geq a$ the integral of $f$ over the interval $[n, n+1]$ lies between $f(n)$ and $f(n+1)$ since $f$ is positive and decreasing. The series (1) is therefore of a similar nature to the series $\sum f(n)$

---

[46] One can prove the converse, but it is hardly worthwhile.

and consequently, *the integral of $f$ on the interval $[a, +\infty]$ converges if and only if the series $\sum f(n)$ converges*, with

$$(24.1) \qquad \sum_{n \geq a} f(n) \leq \int_a^{+\infty} f(x)dx \leq f(a) + \sum_{n > a} f(n);$$

the term $f(a)$ comes from the interval $[a, p]$ where $p$ is the smallest integer $\geq a$. A sketch will make the result obvious.

If for example $f(x) \asymp c/x^s$ with $s$ real, the integral converges like the Riemann series $\sum 1/n^s$, i.e. for $s > 1$.

There are also the integrals of "oscillating" functions. Consider for example the integral

$$(24.2) \qquad I = \int_a^{+\infty} f(x)\sin(\pi x)dx$$

where $f$ is again positive, decreasing, and tends to 0 at infinity. The integral between $n$ and $n+1$ this time lies *up to sign* between $f(n)$ and $f(n + 1)$ since, on the interval considered, $\sin(\pi x)$ is either everywhere between 0 and 1, or everywhere between $-1$ and 0. This suggests comparison with the alternating series $\sum (-1)^n f(n)$, which converges since $f$ decreases and tends to 0. But it is better to compare with the series with general term

$$u_n = \int_n^{n+1} f(x)\sin(\pi x)dx.$$

It is clear that the $u_n$ are alternately positive and negative. On the other hand

$$u_{n+1} = -\int_n^{n+1} f(x + 1)\sin(\pi x)dx$$

thanks to the change of variable $x \mapsto x+1$. Since $f(x+1) \leq f(x)$, we conclude that $|u_{n+1}| \leq |u_n|$. Finally, and as we have seen, $|u_n|$ always lies between $f(n)$ and $f(n+1)$, so tends to 0. The alternating series $u_n$ is therefore convergent. Now let $p$ be the smallest integer $\geq a$. For $p \leq n \leq v < n + 1$ we have

$$\int_a^v f(x)\sin(\pi x)dx = \int_a^p \ldots + (u_p + \ldots + u_{n-1}) + \int_n^v f(x)\sin(\pi x)dx.$$

Since the last integral is, in modulus, $\leq f(n)$ and so tends to 0, and since the series $u_n$ converges, it is clear that the left hand side tends to a limit as $v \to +\infty$, namely

$$I = \int_a^p f(x)\sin(\pi x)dx + \sum_{n \geq p} u_n.$$

The "remainder" of an alternating series being, in absolute value, less than the first term neglected, one thus obtains, for every $n > a$, the inequality

(24.3)
$$\left| I - \int_a^n f(x)\sin(\pi x)dx \right| \le f(n).$$

From this one can deduce an important result on the Fourier transform:

**Theorem 23.** *Let $f$ be a positive regulated function, defined for $x \ge a > -\infty$, decreasing, and tending to 0 at infinity. Then the integral*

$$\varphi(y) = \int_a^{+\infty} f(x)\sin(2\pi xy)dx$$

*converges for any $y \ne 0$, and is a continuous function of $y$.*

To see this, assume $y > 0$ and perform the change of variable $2xy = u$, whence

$$2y\varphi(y) = \int_{2ay}^{+\infty} f(u/2y)\sin(\pi u)du.$$

Convergence is clear, and (3) can now be written

(24.4)
$$\left| 2y\varphi(y) - \int_{2ay}^n f(u/2y)\sin(\pi u)du \right| \le f(n/2y).$$

Let us work on an interval $y \ge b > 0$. We have $f(n/2y) \le f(n/2b)$, so that $f(n/2y)$ converges uniformly to 0 on this interval. Thus it remains to show that the integral in (4) is a continuous function of $y$ for any $n$; for then $2y\varphi(y)$ will be the uniform limit of continuous functions on $y > b$. Now, returning to the initial variable of integration $x = u/2y$, the integral in question can be written

$$\int_a^{n/2y} \sin(2\pi xy)f(x)dx,$$

and its continuity as a function of $y$ is clear, even though the upper limit of integration depends on $y$. The reader may provide the $\varepsilon$, inspired by Theorem 13 of n° 12.

Dirichlet's integral

$$\int_0^{+\infty} \sin(2\pi xy)dx/x$$

fits into this framework, for the function $\sin(2\pi xy)/x$ tends to $2\pi y$ at the origin, so that it is enough to examine its behaviour at infinity, given by the preceding theorem. (Note that in fact the integral does not depend on $y$). Same remark for the Fresnel integrals of the kind

$$\int_0^{+\infty} \cos(2\pi x^2 y)dx, \qquad y \ne 0;$$

the change of variable $x^2 = t$ leads to

$$\int_0^{+\infty} \cos(2\pi yt)t^{-1/2}dt;$$

there is no problem at $t = 0$ since $-\frac{1}{2} > -1$. The problem is to calculate the integral explicitly.

The preceding theorem applies also to the Fourier integrals

$$\hat{f}(y) = \int_{-\infty}^{+\infty} f(x)e^{-2\pi ixy}dx,$$

which, by Euler's formulae, reduce to four integrals of the preceding type. Theorem 25 thus applies here when $f(x)$ tends to 0 at infinity in a monotone fashion for $|x|$ large: the integral converges for $y \neq 0$ and $\hat{f}(y)$ is continuous on $\mathbb{R}^* = \mathbb{R} - \{0\}$. This is the case, for example, if $f(x) = p(x)/q(x)$ is a rational function for which $d^\circ(q) = d^\circ(p) + 1$; the function $f$ tends to 0 monotonely at infinity because its derivative has only a finite number of roots, so has constant sign on a neighbourhood of $+\infty$ or $-\infty$. We should not forget that the integral does not converge absolutely.

## 25 – Differentiation under the $\int$ sign

To extend the theorem on differentiation under the $\int$ sign to "improper" integrals one considers as in n° 9 a continuous function $f$ on a rectangle $X \times J$ where, this time, $X$ is no longer compact, and one assumes that $D_2f$ exists and is continuous on $X \times J$. In ignorance of what a measure is one may always examine an integral of the form

$$(25.1) \qquad g(y) = \int f(x,y)\mu(x)dx,$$

where $\mu$ is a regulated function on $X$, and seek hypotheses to assure that

$$(25.2) \qquad g'(y) = \int D_2f(x,y)\mu(x)dx.$$

We assume of course that (1) and (2) are convergent integrals for any $y \in J$. In problems of this kind the principle is the same as for the analogous problems for series: one replaces $X = (a,b)$ by a compact interval $K = [u,v]$ contained in $X$, applies Theorem 9 of n° 9 to the function[47]

$$(25.3) \qquad g_K(y) = \int_K f(x,y)\mu(x)dx,$$

---

[47] We remarked at the end of n° 9 that Theorem 9 does not rest on the explicit construction of the usual integral, but only on its properties of linearity and continuity. These would be equally valid if one defined the integral by the formula $\mu(f) = \int f(x)\mu(x)dx$. Theorem 9 does not apply directly to the function $f(x,y)\mu(x)$, since it is no longer necessarily continuous, but the result still holds.

then one passes to the limit as $u$ and $v$ tend respectively to $a$ and $b$, i.e. as $K$ tends to $X$. Since

$$(25.4) \qquad g'_K(y) = \int_K D_2 f(x,y)\mu(x)dx$$

by Theorem 9 of n° 9, we have to show that the derivative of a limit is the limit of the derivatives, a problem which Theorem 19 of Chap. III, n° 17 is there to resolve: since $g_K(y)$ tends to $g(y)$ for every $y$, it is enough to show that $g'_K(y)$ converges to the integral (2) *uniformly on every compact subset $H$ of $J$* as $K$ tends to $X$. Equivalently, that for every $r > 0$ there exists a compact interval $K \subset X$ such that

$$(25.5) \qquad \left| \int_{X-K'} D_2 f(x,y)\mu(x)dx \right| \le r \text{ for every } y \in H \text{ and every } K' \supset K;$$

the integral (5) is actually the difference between $g'_K(y)$ and the integral (2) to which it must converge [and which we will have the right to denote $g'(y)$ *after* having justified the passage to the limit]. A brutal way of guaranteeing (5) is to assume the existence of a positive function $p_H(x)$ such that

$$(25.6) \qquad |D_2 f(x,y)| \le p_H(x) \text{ with } \int_X p_H(x)|\mu(x)|dx < +\infty$$

for any $x \in X$ and $y \in H$; the left hand side of (5) is then majorised by the integral of $p_H|\mu|$ over $X - K'$, so is $< r$ for any $y \in H$ if $K'$ contains a large enough compact interval $K \subset X$.

This is the argument used to show that if, for a series of differentiable functions the series of its derivatives converges *normally*, then one may differentiate it term-by-term. The integrals of $D_2 f$ on the compact sets play the rôle of the partial sums of the derived series; the existence of a function $p_H$ satisfying (3) plays the rôle of normal convergence and guarantees that for $K \subset X$ large enough the "remainder" of the "sum" of the $D_2 f(x,y)$, i.e. the integral on $X - K$, is in modulus $\le r$ for any $y$. One cannot recommend the reader too strongly to let himself be guided by these analogies between "continuous sums", i.e. integrals, and "discrete sums", i.e. series.

We thus obtain a simple but useful result:

**Theorem 24.** *Let $X$ and $J$ be two intervals in $\mathbb{R}$, let $\mu$ be a regulated function on $X$ and $f$ a function defined and continuous on $X \times J$. Assume that*

*(i)    the integral*

$$g(y) = \int_X f(x,y)\mu(x)dx$$

*converges for every $y \in J$;*

*(ii)   the function $f$ has a continuous partial derivative $D_2 f(x,y)$ on $X \times J$;*

*(iii) for every compact $H \subset J$ there exists a positive function $p_H$ on $X$ such that $|D_2 f(x,y)| \le p_H(x)$ for every $x \in X$ and every $y \in H$, and $\int p_H(x)|\mu(x)|dx < +\infty$.*

*Then the function $g$ is differentiable and*

$$(25.7) \qquad g'(y) = \int_X D_2 f(x,y)\mu(x)dx.$$

*Example 1.* If $X = Y = \mathbb{R}$, if $\mu$ is an absolutely integrable regulated function on $\mathbb{R}$ and if $f(x,y) = e^{-2\pi ixy}$, then the function $g(y)$ is just the Fourier transform $\hat{\mu}$ of $\mu$. Here

$$D_2 f(x,y) = -2\pi ix e^{-2\pi ixy}$$

and so $|D_2 f(x,y)| = 2\pi|x| = p(x)$, and this is clearly the smallest positive function which dominates $x \mapsto D_2 f(x,y)$ for a (or for all) $y \in Y$. Conclusion: if

$$\int_\mathbb{R} |x\mu(x)|dx < +\infty,$$

then $\hat{\mu}$ is differentiable and

$$\hat{\mu}'(y) = -2\pi i \int_\mathbb{R} x\mu(x)e^{-2\pi ixy}dx$$

is the Fourier transform of $-2\pi ix\mu(x)$.

*Example 2.* In particular choose $\mu(x) = \exp(-\pi x^2)$, an integrable function on $\mathbb{R}$ since it decreases at infinity more rapidly than $|x|^{-n}$ for any $n > 0$. We have $-2\pi ix\mu(x) = i\mu'(x)$, whence, integrating by parts,

$$\hat{\mu}'(y) = i \int_{-\infty}^{+\infty} \mu'(x)\exp(-2\pi ixy)dx = -2\pi y \int_{-\infty}^{+\infty} \mu(x)\exp(-2\pi ixy)dx$$

since the integrated-out part is zero because of the decrease of $\mu$ at infinity. One obtains the relation

$$\hat{\mu}'(y) = -2\pi y\hat{\mu}(y),$$

a relation satisfied equally by $\mu$. The function $\hat{\mu}/\mu$ therefore has derivative zero, whence

$$(25.8) \qquad \hat{\mu}(y) = c\mu(y) = c\exp(-\pi y^2)$$

with a constant

$$(25.9) \qquad c = \hat{\mu}(0) = \int_\mathbb{R} \exp(-\pi x^2)dx.$$

It will emerge later that $c = 1$, and this without the least calculation, thanks to the general Poisson summation formula

$$\sum \mu(n) = \sum \hat{\mu}(n),$$

where the sums are over $\mathbb{Z}$. In part this explains the rôle of the function $\exp(-\pi x^2)$ in the calculus of probabilities (Gauss' normal law).

The preceding theorem can be used to show that a function is holomorphic:

**Theorem 24 bis.** *Let $X$ be an interval in $\mathbb{R}$, $U$ an open subset of $\mathbb{C}$, $\mu$ a regulated function in $X$, and $f$ a function defined and continuous on $X \times U$ satisfying the following conditions:*

(i)   *the integral*

$$g(z) = \int_X f(t, z)\mu(t)dt$$

   *converges absolutely for every $z \in U$;*
(ii)  *the function $z \mapsto f(t, z)$ is holomorphic on $U$ for every $t \in X$ and its derivative $f'(t, z)$ with respect to $z$ is continuous on $X \times U$;*
(iii) *for every compact $H \subset U$, there exists on $X$ a positive function $p_H(t)$ such that $|f'(t, z)| \leq p_H(t)$ for every $t \in X$ and every $z \in H$ and $\int p_H(t)|\mu(t)|dt \leq +\infty$.*

   *Then $g$ is holomorphic on $U$ and*

(25.10)   $$g'(z) = \int_X f'(t, z)\mu(t)dt.$$

Putting $z = x + iy$, Theorem 24 shows that one may differentiate under the $\int$ sign, either with respect to $x$ for $y$ given, or with respect to $y$ for $x$ given. Since $z \mapsto f(t, z)$ satisfies the Cauchy's holomorphy condition [Chap. II, eqn. (19.10)], so clearly $g$ does too, qed.

*Example 3.* If $\mu$ is a regulated function on the closed interval $[0, +\infty[$ and is $O(t^N)$ at infinity for some $N$, its *Laplace transform* or *complex Fourier transform*

$$L_\mu(z) = \int_0^{+\infty} e^{2\pi itz}\mu(t)dt$$

is defined on $U : \mathrm{Im}(z) > 0$ since then $|e^{2\pi itz}\mu(t)| = O(e^{-2\pi ty}t^N)$ at infinity. Here $f(t, z) = e^{2\pi itz}$, whence $|f'(t, z)| = 2\pi te^{-2\pi ty}$. Since every compact subset $H$ of $U$ is contained in a half plane $\mathrm{Im}(z) \geq \sigma > 0$, we have, in $H$, that $|f'(t, z)| \leq 2\pi te^{-2\pi \sigma t} = p_H(t)$ with $\int p_H(t)|\mu(t)|dt < +\infty$ since the function $p_H(t)\mu(t) = O(e^{-2\pi \sigma t}t^{N+1})$ is absolutely integrable on $\mathbb{R}_+$. The function $L_\mu$ is therefore holomorphic on $U$.

This calculation, iterated, shows further that the (complex) derivatives of $L_\mu$ are given by

$$(25.11) \qquad L_\mu^{(n)}(z) = (2\pi i)^n \int_0^{+\infty} e^{2\pi itz} t^n \mu(t)dt.$$

*Example 4.* The function $\Gamma(s) = \int e^{-x} x^{s-1} dx$ is holomorphic in the half plane $\mathrm{Re}(s) > 0$ where it is defined. It is clear that

(i)  the function $s \mapsto e^{-x} x^{s-1} = e^{-x} \exp[(s-1)\log x]$ is holomorphic for every $x > 0$ since it is the composite of two holomorphic functions;

(ii)  its complex derivative[48] $e^{-x} x^{s-1} \log x$ is continuous;

(iii)  if $s$ remains in a compact subset $H$ of the half plane $\mathrm{Re}(s) > 0$, strict inequality, then $s$ is subject to conditions $a \le \mathrm{Re}(s) \le b$ with $0 < a < b < +\infty$, so that

$$\left| e^{-x} x^{s-1} \log x \right| \le p_H(x) = \begin{cases} e^{-x} x^{a-1} |\log x| & \text{if } 0 < x \le 1, \\ e^{-x} x^{b-1} \log x & \text{if } 1 < x < +\infty \end{cases}.$$

Now the integral of $x^{a-1} \log x$ converges at 0 for $a > 0$ and that of $e^{-x} x^{b-1} \log x$ converges at infinity for any $b$. Whence dominated convergence and the result.

One sees at the same time that

$$(25.12) \qquad \Gamma'(s) = \int_0^{+\infty} e^{-x} x^{s-1} \log x.dx.$$

Example 5. Let us write

$$(25.13) \qquad \Gamma(s) = \int_0^1 e^{-x} x^{s-1} dx + \int_1^{+\infty} e^{-x} x^{s-1} dx.$$

The second integral converges for any $s \in \mathbb{C}$. So, as in the preceding example, is a holomorphic function of $s$ in all of $\mathbb{C}$. In the first integral, term-by-term integration of the exponential series gives, for $\mathrm{Re}(s) > 0$,

$$(25.14) \quad \int_0^1 e^{-x} x^{s-1} dx = \sum \frac{(-1)^n}{n!} \int_0^1 x^{n+s-1} dx = \sum_{\mathbb{N}} \frac{(-1)^n}{n!(s+n)};$$

the operation is justified because (i) the series $\sum (-1)^n x^{n+s-1}/n!$ to be integrated over $X = ]0,1]$ converges normally on every compact $K \subset X$, (ii) the series

$$\sum |(-1)^n x^{n+s-1}/n!| = e^x x^{\mathrm{Re}(s)-1} = p(x)$$

---

[48]  Recall (Chap. III, n° 21, Theorem 22) that the chain rule valid for functions of a real variable is also valid for holomorphic functions of a complex variable.

is integrable on $X$ since $\mathrm{Re}(s) > 0$: so Theorem 20 applies.

The result (14) is a series of holomorphic functions on the open set

$$(25.15) \qquad U = \mathbb{C} - \{0, -1, -2, \dots, \},$$

a series which converges normally on every compact[49] $H \subset U$. If we knew generally that the sum of such a series is again holomorphic, we could deduce from this and from (13) that $\Gamma(s)$ is the restriction to the half plane $\mathrm{Re}(s) > 0$ of a holomorphic function on $U$. We do not know this yet, even though we know (Chap. III, n° 22) that if a sequence or series of holomorphic functions on an open subset $U$ of $\mathbb{C}$ converges uniformly on every compact set *and so does its derived series*, then the sum of the given series is holomorphic, and its derivative is obtained by differentiating term-by-term. We stated then that this result, a trivial consequence of the Cauchy equation and of the much more general theorem on sequences or series of $C^1$ functions in the plane, is much too weak to be of interest, but here it will suffice for our needs. It all reduces to showing that the derived series

$$\sum (-1)^{n+1}/n!(s+n)^2$$

of (14) converges normally on every compact $H \subset U$, which is clear.

Another procedure. An integration by parts shows immediately that

$$\int_0^1 e^{-x} x^{s-1} dx = \frac{1}{s} + \frac{1}{s} \int_0^1 e^{-x} x^s dx$$

for $\mathrm{Re}(s) > 0$; but the integral obtained converges for $\mathrm{Re}(s) > -1$ and depends holomorphically on $s$ in this half plane as in Example 4; this allows us to *extend* the function[50] $\Gamma$ *analytically* (it would be better to say *holomorphically* at this stage of the exposition) to the half plane $\mathrm{Re}(s) > -1$ minus the point 0. This done, a new integration by parts yields the relation

$$(25.16) \qquad \int_0^1 e^{-x} x^{s-1} dx = \frac{1}{s} + \frac{1}{s(s+1)} + \frac{1}{s(s+1)} \int_0^1 e^{-x} x^{s+1} dx$$

with an integral converging now for $\mathrm{Re}(s) > -2$ and so holomorphic in this half plane. Pursuing the calculations, one defines $\Gamma(s)$ in all the half planes

---

[49] It suffices to prove the existence of an $r > 0$ such that $|s + n| \geq r$ for any $s \in H$ and $n \in \mathbb{N}$. This is equivalent to saying that the distance $d(H, -\mathbb{N})$ between the closed set $H$ and $-\mathbb{N}$ is $> 0$, which follows from the fact that they are disjoint, with $H$ compact. One may also argue directly.

[50] Given two open sets $U \subset V$ in $\mathbb{C}$ and an analytic (resp. holomorphic) function on $U$, to extend $f$ to $V$ analytically (resp. holomorphically) consists of constructing an analytic (resp. holomorphic) function on $V$ coinciding with $f$ on $U$. If $V$ is connected the analytic extension, if one exists, is unique (Chap. II, n° 20). Recall also (Chap. VII, n° 14) that the terms "analytic" and "holomorphic" are synonymous.

$\text{Re}(s) > -n$, apart from the points $0, -1$, etc. where the rational fractions $1/s$, $1/s(s+1)$, $1/s(s+1)(s+2)$ etc. appear.

The final result is that one may extend the function $\Gamma$ holomorphically to the open set (15) and that it is given there by the formula

$$(25.17) \qquad \Gamma(s) = \int_1^{+\infty} e^{-x}x^{s-1}dx + \sum_{n=0}^{\infty} \frac{(-1)^n}{n!(s+n)}$$

in which everything converges for any $s \neq 0, -1, \ldots$. As we shall see in Chap. VII, n° 20, Example 4, these various methods of defining $\Gamma(s)$ beyond the half plane $\text{Re}(s) > 0$ all lead to the same function.

## 26 – Integration under the $\int$ sign

We saw in n° 9, Theorem 10, that if $f$ is a continuous function on $K \times H$, where $K$ and $H$ are compact intervals, then

$$\int dx \int f(x,y)dy = \int dy \int f(x,y)dx.$$

Does this result extend to arbitrary intervals? Yes, on condition that, as always, one imposes hypotheses of domination by fixed integrable functions.

**Theorem 25 (Poor man's Lebesgue-Fubini).** *Let $X$ and $Y$ be two intervals and $f$ a continuous function on $X \times Y$. Suppose that the following conditions are satisfied:*

*(i)  for every compact $K \subset X$ there exists a positive integrable function $q_K(y)$ on $Y$ such that $|f(x,y)| \leq q_K(y)$ in $K \times Y$;*

*(ii)  for every compact $H \subset Y$ there exists a positive and integrable function $p_H(x)$ on $X$ such that $|f(x,y)| \leq p_H(x)$ on $X \times H$;*

*(iii)  one of the two relations*

$$(26.1) \qquad \int_X dx \int_Y |f(x,y)|dy < +\infty, \qquad \int_Y dy \int_Y |f(x,y)|dx < +\infty,$$

*is satisfied.*

*Then the two relations (1) are also satisfied, and*

$$(26.2) \qquad \int_X dx \int_Y f(x,y)dy = \int_Y dy \int_X f(x,y)dx.$$

In what follows we shall put

$$(26.3) \qquad g_J(x) = \int_J f(x,y)dy, \qquad h_I(y) = \int_I f(x,y)dx$$

for any intervals $I \subset X$ and $J \subset Y$. We shall also employ the notation $m_I$ to denote an integral over $I$.

First we note that, by Theorem 22 of n° 23 for $\mu = 1$, and by the hypotheses (i) and (ii), the functions (3) are continuous for any $I$ and $J$. Let us start from the relation

$$(26.4) \quad m_K(g_H) = \int_K dx \int_H f(x,y)dy = \int_H dy \int_K f(x,y)dx = m_H(h_K)$$

valid for any compact[51] $K \subset X$ and $H \subset Y$ (n° 9, Theorem 9). The whole problem is to pass to the limit under the $\int$ signs as $K$ and $H$ tend to $X$ and $Y$.

(a) First we show that we may pass to the limit with respect to $H$ for $K$ fixed. Hypothesis (i) shows that $f(x,y)$ is absolutely integrable on $Y$ for every $x \in X$ and that, further,

$$(26.5) \quad \left| \int_Y f(x,y)dy - \int_H f(x,y)dy \right| \leq \int_{Y-H} q_K(y)dy \ \text{ for every } x \in K,$$

a result $\leq r$ for $H$ large enough since $q_K$ is integrable on $Y$. Since the left hand side can also be written as $|g_Y(x) - g_H(x)|$, (5) shows that

$$\|g_Y - g_H\|_K \leq r$$

for $H$ large enough. Since we may pass to the limit under the $\int$ sign when we integrate uniformly convergent continuous functions on a compact set, we obtain

$$(26.6) \quad \lim_{H \to Y} m_K(g_H) = m_K(g_Y).$$

(b) Next we have to pass to the limit along $K$. By (4) and the definition of an integral extended to $X$ or $Y$, we have

$$(26.7) \quad m_Y(h_K) = \lim_{H \to Y} m_H(h_K) = \lim_{H \to Y} m_K(g_H);$$

there is no problem of convergence since, on the left hand side, the integral on $K$, i.e. the function $h_K(y)$, is majorised in modulus by $m(K)q_K(y)$ by hypothesis (i), whence the absolute convergence of the integral over $Y$.

Comparing (6) and (7), we find

$$(26.8) \quad m_Y(h_K) = m_K(g_Y)$$

for every compact interval $K \subset X$, which would be (2) if $X$ were compact. Likewise we find

---

[51] The reader will already no doubt have observed that we often omit the word "interval".

(26.9)                              $$m_X(g_H) = m_H(h_X)$$

for every compact $H \subset Y$. It remains to pass from here to (2), a relation which can again be written

(26.9')                             $$m_X(g_Y) = m_Y(h_X).$$

(c) These results do not rely on hypothesis (iii) of the statement and remain valid if one replaces $f$ by the function $|f|$, which satisfies (i) and (ii) like $f$. Suppose now that $\int dx \int |f(x,y)|dy < +\infty$. Applying (9) to $|f|$, we obtain

(26.10)
$$\int_H dy \int_X |f(x,y)|dx = \int_X dx \int_H |f(x,y)|dy \le$$
$$\le \int_X dx \int_Y |f(x,y)|dy$$

for every compact $H \subset Y$. Taking the upper bound of the left hand side as $H$ varies in $Y$, we obtain (Theorem 18, (i))

(26.11)
$$\int_Y dy \int_X |f(x,y)|dx \le \int_X dx \int_Y |f(x,y)|dy < +\infty.$$

So we see that if the first integral (1) is finite, the second is too. But now one may argue starting from second as we have just done, starting from the first. Obviously one obtains the reverse inequality to (11), which must then in fact be an *equality*. Whence (2) for the function $|f|$.

(d) It remains to obtain (2) for the function $f$ itself. An easy method is to reduce to a real function by considering $\mathrm{Re}(f)$ and $\mathrm{Im}(f)$, functions which, bounded in modulus by $|f|$, again satisfy the hypotheses of the theorem. The function $f$ being now assumed real, we write $f = f^+ - f^-$ as always; these two positive functions, majorised by $|f|$, also satisfy the hypotheses of the theorem, and since they are identical to their absolute values (11) reduces to (2) for these two functions; whence (2) for $f$.

Another method, which, unlike the preceding, could be applied to functions with values in Banach spaces, even of infinite dimension, consists of starting from (9) and showing that, as $H \to Y$, the two sides of (9) converge to the two sides of (9'). This is the case of the right hand side by definition of $m_Y(h_X)$. To examine the left hand side, first note that

(26.12)
$$|m_X(g_Y) - m_X(g_H)| \le m_X(|g_Y - g_H|) \le$$
$$\le \int_X dx \int_{Y-H} |f(x,y)|dy$$

for every compact $H \subset Y$. Since we are already able to invert the integrations for the function $|f|$, we can apply it to the "intervals" $X$ and $Y - H$ (the fact that $Y - H$ is the union of two disjoint intervals does not change anything).

One may thus interchange the order of the integrations in the third term of (12); and thus obtain the integral on $Y - H$ of a positive integrable function on $Y$, a result $< r$ for $H$ large enough. The first term of (12) therefore tends to 0 as $H \to Y$, so that the left hand side of (9) tends to that of (9'), qed.

Given the conditions of the preceding theorem, one often says that $f(x, y)$ is *absolutely integrable on $X \times Y$* and puts

$$(26.13) \quad \iint_{X \times Y} f(x, y)dxdy = \int_Y dx \int_Y f(x, y)dy = \int_Y dy \int_X f(x, y)dx.$$

In practice, one may often substitute the following condition for the hypotheses (i), (ii) and (iii): *there exist positive, regulated and integrable functions $p$ and $q$ on $X$ and $Y$ respectively such that $|f(x, y)| \leq p(x)q(y)$ on $X \times Y$.* The hypotheses (i) and (ii) are satisfied on choosing

$$p_H(x) = \|q\|_H p(x) \qquad \text{and} \qquad q_K(y) = \|p\|_K q(y)$$

for any $K$ and $H$ (the uniform norms are finite since $p$ and $q$ are regulated). The hypothesis (iii) is also satisfied, since, for example, $\int |f(x, y)|dy \leq Mp(x)$ where $M = \int q(y)dy$, whence the absolute convergence of the repeated integrals.

The hypotheses (i) and (ii) are unnecessary in the complete Lebesgue-Fubini theorem and one contents oneself with hypothesis (iii), but one cannot again obtain a Rolls for the price of a VW. N° 33 of § 9 will provide, an inevitable intermediate stage, the LF theorem for semicontinuous functions, thanks to which one can justify what all the users do instinctively when they integrate a continuous function on a simple compact set in $\mathbb{C}$.

*Exercise.* Extend Theorem 10 of n° 9 to noncompact intervals $X$ and $Y$.

*Example 1.* First note that if the continuous functions $f(x)$ and $g(y)$ are defined and absolutely integrable on the intervals $X$ and $Y$ then the function $f(x)g(y)$ is absolutely integrable on $X \times Y$, and clearly

$$\iint_{X \times Y} f(x)g(y)dxdy = \int_X f(x)dx. \int_Y g(y)dy,$$

the product of the integrals of $f$ and $g$. Now let us choose $X = Y = ]0, +\infty[$, $f(x) = e^{-x}x^{a-1}$ and $g(y) = e^{-y}y^{b-1}$, with $\text{Re}(a) > 0$ and $\text{Re}(b) > 0$. We obtain the relation

$$(26.14) \quad \Gamma(a)\Gamma(b) = \iint_{X \times Y} e^{-x-y}x^{a-1}y^{b-1}dxdy = \int dx \int \dots dy.$$

If, for each $x$, one effects the change of variable $y = (u^{-1} - 1)x$ in the $y$-integration, one finds

$$(26.15) \quad \Gamma(a)\Gamma(b) \;=\; \int_0^{+\infty} dx \int_0^1 e^{-u/x} x^{a+b-1} (1-u)^{b-1} u^{-b-1} du \;=\;$$

$$= \int_0^1 (1-u)^{b-1} u^{-b-1} du \int_0^{+\infty} e^{-u/x} x^{a+b-1} dx;$$

the change of variable $x = tu$ in the $x$ integral for $u$ given then yields

$$\Gamma(a)\Gamma(b) \;=\; \int_0^1 (1-u)^{b-1} u^{-b-1} du \int_0^{+\infty} e^{-t} t^{a+b-1} u^{a+b} dt \;=\;$$

$$= \int_0^1 (1-u)^{b-1} u^{a-1} du \int_0^{+\infty} e^{-t} t^{a+b-1} dt,$$

whence the famous formula

$$(26.16) \qquad\qquad \Gamma(a)\Gamma(b) = \Gamma(a+b)B(a,b)$$

announced above.

# § 8. Approximation Theorems

### 27 – How to make $C^\infty$ a function which is not

In about 1926–1927 the physicist Paul-Adrien-Maurice Dirac, in Dublin, had
the idea of introducing a function $\delta(x)$ on $\mathbb{R}$ (and even on $\mathbb{R}^3$ or $\mathbb{R}^4$) possessing
two supernatural properties: on the one hand

$$\delta(x) = 0 \quad \text{for} \quad x \neq 0, \qquad \delta(0) = +\infty,$$

while on the other hand[52]

$$\int f(x)\delta(x)dx = f(0)$$

for every just-a-little-reasonable function $f$; a little later, Dirac and the
theoretical physicists juggled, in Relativity space, with "functions" such as
$\delta(c^2t^2 - x^2 - y^2 - z^2)$ and, as the inventor of the theory of distributions[53]
has written, "lived in a fantastic universe which they knew how to manage admirably, practically faultlessly, though never able to justify it at all".
Dirac had, to be sure, explained how one could "approximate" his function
by considering, for $\varepsilon > 0$, the function equal to $1/2\varepsilon$ for $|x| < \varepsilon$ and zero else-
where, or the "bell curve" functions $\exp(-\pi x^2/\varepsilon)/\sqrt{\varepsilon}$, whose integral over $\mathbb{R}$
is equal to 1 and whose graph, as $\varepsilon \to 0$, more and more closely resembles
a sky-scraper of infinite height and of zero base representing the function $\delta$,
but since furthermore he allowed himself to differentiate his "function" and
to write formulae such as

$$\int f(x)\delta'(x)dx = -f'(0),$$

those mathematicians who tried to understand him understood nothing. It
was twenty years later that the distributions of which we speak below gave a
meaning to these calculations, and it was 1954 when the formulae in several
variables of theoretical physics were at last justified – but not for nothing ...
– by the Swiss mathematician Paul Méthée.

   Leaving differentiation aside for the moment, Dirac's idea raises the question of whether one may approximate the value of a function $f$ at a point, say
$x = 0$, with the help of integrals involving $f$, for example by using functions
$u_n(x)$ such that

$$(27.1) \qquad \lim \int f(x)u_n(x)dx = f(0)$$

---

[52] In this n° and in the following, we write a simple $\int$ for an integral extended over
$\mathbb{R}$.

[53] Laurent Schwartz, *Un mathématicien aux prises avec le siècle* (Paris, Odile Jacob, 1997), pp. 230–231 (trans. *A Mathematician Grappling with his Century*,
Birkhäuser, 2001).

for every "reasonable" function $f$.

Since as yet we know how to integrate only regulated functions, we shall suppose in what follows that $f$ and the $u_n$ are such. To give a meaning to (1) for every "reasonable" function $f$ – let us say bounded on $\mathbb{R}$ –, we impose on them the condition

(D1) the $u_n(x)$ are absolutely integrable on $\mathbb{R}$,

a superfluous condition if, as is often the case, the $u_n$ are zero outside a compact interval.

The most "reasonable" functions being the constants, we must consequently impose on the $u_n$ the condition

$$(D\,2) \qquad\qquad \lim \int u_n(x)dx = 1$$

if we want (1) to hold. Then trivially

$$(27.2) \qquad\qquad f(0) = \lim \int f(0)u_n(x)dx,$$

so, to obtain (1), we are led to examine the difference

$$(27.3) \qquad \int [f(x) - f(0)]u_n(x)dx = \int_{|x|\leq r} + \int_{|x|\geq r} \qquad (r > 0).$$

Dirac's idea, that "almost all" the mass of the measure $u_n(x)dx$ is concentrated on a neighbourhood of the origin for $n$ large, leads us to introduce the condition

(D 3) for any $r > 0$,

$$(27.4) \qquad\qquad \lim_{n\to\infty} \int_{|x|\geq r} |u_n(x)|dx = 0.$$

If this is the case, and if, as always, one writes $\|f\|$ for the uniform norm of $f$ on $\mathbb{R}$, the second integral on the right hand side of (3) is majorised by $2\|f\|\varepsilon$ for $n$ large.

Assume now that $f$ is continuous at the origin, and consider, in (3), the integral over the interval $|x| \leq r$. For $\varepsilon > 0$ given one may choose $r$ so that

$$(27.5) \qquad\qquad |f(x) - f(0)| \leq \varepsilon \qquad \text{for } |x| \leq r.$$

The integral is then, in absolute value, majorised by $\varepsilon \int |u_n(x)|dx$ where one integrates over $|x| \leq r$ and *a fortiori* if one integrates over all $\mathbb{R}$. To be sure that the result is arbitrarily small, it is thus enough to assume that

$$(D\,4) \qquad\qquad \sup \int |u_n(x)|dx = M < +\infty,$$

a superfluous condition, by (D 2), if the $u_n$ are positive.

To recapitulate: $\varepsilon > 0$ being given, one chooses $r > 0$ to ensure (5); on the right hand side of (3) the first integral is then $\leq M\varepsilon$ for any $n$, and since, for $r$ given, the second tends to 0 as $n$ increases, by (4), we see finally that the integral (3) tends to 0, whence (1), in view of (2).

A sequence of regulated functions satisfying the conditions (D 1) to (D 4) is called a *Dirac sequence* on $\mathbb{R}$. We have established the following result:

**Dirac's lemma.** *Let $(u_n)$ be a Dirac sequence. For every regulated function $f$ that is defined and bounded on $\mathbb{R}$, and is continuous at the origin, we have*

$$(27.6) \qquad f(0) = \lim \int f(x)u_n(x)dx.$$

In practice, the Dirac sequences that one uses often satisfy more restrictive conditions, namely

(i)    $u_n$ is positive;
(ii)   for every $r > 0$, $u_n$ is zero outside $[-r, r]$ for $n$ large;
(iii)  the integral of $u_n$ is equal to 1 for every $n$.

Conditions (i) and (iii) imply properties (D 1), (D 2) and (D 4) imposed in the general case, and (ii) implies (D 3). Condition (ii) is not satisfied in some important cases, as is shown by the exponential functions that we shall use in the following n°.

If the $u_n$ satisfy (i), (ii) and (iii), if, more generally, they are all zero for $|x| \geq A$, it is unnecessary to assume $f$ bounded in the above statement since nothing changes if one replaces $f(x)$ by 0 for $|x| \geq A$.

*Example 1.* Consider on $\mathbb{R}$ a function $u(x)$ which is *regulated, positive, with total integral 1*, and put $u_n(x) = nu(nx)$. The condition (D 1) is satisfied, also (D2) (change of variable $nx = y$ in the integral) and condition (D 3) is satisfied because

$$\int_{|x| \geq r} u_n(x)dx = \int_{|x| \geq nr} u(x)dx,$$

a result which tends to 0 for every $r > 0$ since $u$ is integrable on $\mathbb{R}$. Consequently,

$$(27.7) \qquad f(0) = \lim n \int f(x)u(nx)dx$$

for every regulated function $f$ which is continuous at the origin.

If $u$ is of compact support the function $u_n(x) = nu(nx)$ is zero for $|x| \geq A/n$, i.e. outside an ever-shrinking interval with centre 0; the factor $n$ in its definition shows that, on the other hand, it takes very large values on a neighbourhood of 0, an indispensable condition if its integral is to remain equal to 1. The lemma which we have generously attributed to Dirac shows

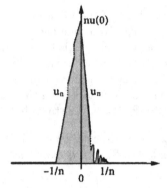

Fig. 11.

that in a certain sense Dirac's $\delta$ "function" is the "limit" of the functions $u_n(x)$; indeed, if $u$ is zero for $|x| > A$, then

$$\lim u_n(x) = 0 \qquad \text{for } x \neq 0$$

since $u_n$ is zero outside $[-A/n, A/n]$. If one has chosen $u$ so that $u(0) > 0$ it is also clear that $u_n(0) = nu(0)$ increases indefinitely.

Most authors choose ultraregular positive functions for the $u_n$, with pretty bell-shaped graphs symmetric with respect to the origin, growing higher and higher, and whose base shrinks more and more, so that the area contained between the graph and the $x$ axis remains equal to 1. One may, for example, choose $u_n(x) = c_n(1 - x^2)^n$ for $|x| < 1$, $= 0$ for $|x| > 1$, the constant $c_n$ being chosen so that $\int u_n = 1$; the method of Example 1 would lead to the functions $u_n(x) = cn(1 - x^2/n^2)$ for $|x| < 1/n$, $= 0$ elsewhere, with $c = 3/4$. One often also chooses $u_n(x) = c_n \exp(-nx^2)$, with the appropriate choice of $c_n$; it would be better to take

$$u_n(x) = n \exp(-\pi n^2 x^2)$$

since the integral over $\mathbb{R}$ of the function $\exp(-\pi x^2)$ is equal to 1, as we shall see later; these functions do not have compact support but nevertheless form Dirac sequences, and, up to notation, had already appeared, not only in Dirac but also, a half-century earlier, in Weierstrass, in the proof of his theorem on approximation by polynomials (following n°). In fact, none of this is necessary because Dirac's lemma, which assumes nothing as to the "elementary", "classical", or other nature of the $u_n$, generalises to all measures defined over all locally compact spaces, i.e. to situations where polynomials, exponentials and other curiosities of the real line are unknown. Let us add that if the reader were to restrict himself to proving Dirac's lemma for the functions $n \exp(-\pi n^2 x^2)$ for example, he might possibly be tempted to perform explicit calculations offering no benefit other than the risk of error.

The most important consequence of Dirac's lemma is provided by the following statement:

**Theorem 26.** *Let $(u_n)$ be a Dirac sequence. For every function $f$ defined and continuous on $\mathbb{R}$ one has*

$$(27.8) \qquad f(x) = \lim \int f(x-y)u_n(y)dy$$

*uniformly on every compact subset of $\mathbb{R}$ if $f$ is bounded or just if the $u_n$ vanish outside the same compact set.*

To establish (8) for $f$ bounded it is enough to apply the lemma to the function $y \mapsto f(x-y)$. The little calculation in Dirac's lemma shows moreover that

$$(27.9) \qquad \left| f(x)\int u_n(y)dy - \int f(x-y)u_n(y)dy \right| \le$$
$$\le \int |f(x) - f(x-y)|\, |u_n(y)|\, dy$$

and since the integral of $u_n$ tends to 1, we reduce to proving that the right hand side converges uniformly to 0 when $x$ remains in a compact subset $K$ of $\mathbb{R}$.

Take an $\varepsilon > 0$. Since $f$ is *uniformly* continuous on every compact $K$ in $\mathbb{R}$ there is an $r > 0$ such that $|f(x-y) - f(x)| \le \varepsilon$ for $x \in K$ and $|y| \le r$. By virtue of property (D 4) of Dirac sequences, the contribution of the interval $|y| \le r$ is thus $\le M\varepsilon$ for any $n$ and $x \in K$.

For such a choice of $r$ the contribution of the set $|y| \ge r$ to the right hand side of (9) is less than the product of $2\|f\|$ by the integral of $|u_n|$ extended over $|y| \ge r$; for $n$ large it is therefore $< \varepsilon$ *for any $x \in \mathbb{R}$*, by (D 3).

Finally one obtains an estimate for (9) valid for all the $x \in K$ simultaneously, whence the theorem in this case.

If the $u_n$ are all zero for $|y| \ge A$, the first part of the argument survives unchanged. In the second, one remarks that the contribution of the set $|y| \ge r$ is in fact an integral over $r \le |y| \le A$; if $x$ remains in the compact $|x| \le B$, the integral (9) involves only the pairs $(x, y)$ satisfying $|x| \le B$, $|y| \le A+B$, a set on which the difference $|f(x) - f(x-y)|$ is bounded, which allows one to conclude as in the preceding case.

The interest of Theorem 26 is that it allows one to approximate the function $f$ by $C^\infty$ functions, or even polynomials, much more "regular" than itself. The idea had already been met in 1926 *chez* the American Norbert Wiener, the future inventor of "cybernetics", whom Dirac no doubt had not read.

In the first case, one remarks for a start, that thanks to the function

$$u(x) = \exp(-1/x) \text{ for } x > 0, \quad = 0 \text{ for } x \le 0,$$

which is $C^\infty$ on $\mathbb{R}$ as we saw a long time ago, there are "many" $C^\infty$ functions *of compact support*[54] on $\mathbb{R}$: to obtain one it suffices to multiply a $C^\infty$

---

[54] Recall that the "support" of a function $f$ is the smallest closed set outside which it is zero, i.e. the closure of the set $\{f \ne 0\}$.

function vanishing for $x < a$ by a $C^\infty$ function zero vanishing for $x > b$. The set of these functions is a vector space which, since Schwartz, has been denoted $\mathcal{D}(\mathbb{R})$ or simply $\mathcal{D}$. There are clearly even positive $C^\infty$ functions, zero outside arbitrarily given intervals; on dividing by their integral over $\mathbb{R}$ one may assume that the integral is equal to 1. Thus there exist Dirac sequences formed of functions $\varphi_n \in \mathcal{D}$. Using the function $u(x)$ above, one may for example take $\varphi_n(x) = c_n u(x + 1/n)u(1/n - x)$ with a constant $c_n$ such that $\int \varphi_n(x)dx = 1$.

The functions (8) which, for every continuous function $f$ on $\mathbb{R}$, converge to $f$, are then $C^\infty$, as we shall see. It all amounts to showing that for every $\varphi \in \mathcal{D}$ zero for $|x| \geq A$ and every continuous function $f$, the *convolution product*

$$(27.10) \qquad f \star \varphi(x) = \int f(x - y)\varphi(y)dy = \int_{-A}^{A} f(x - y)\varphi(y)dy$$

is $C^\infty$; this is the method of *regularising* an "arbitrary" function; it even provides a $C^\infty$ result for every *regulated* function $f$ on $\mathbb{R}$.

We remark that the change of variable $x - y = t$ transforms (10) into

$$(27.10') \; f \star \varphi(x) = \int f(t)\varphi(x - t)dt = \int \varphi(x - t)f(t)dt = \varphi \star f(x)$$

(one has $dy = -dt$, but the integral changes orientation and the factor $-1$ is eliminated on reestablishing the natural orientation). To check that the integral is a $C^\infty$ function of $x$ one may restrict to a compact interval $J = [-A, A]$. Since $\varphi(x-t)$ is zero for $|x-t| \geq B$, because $\varphi$ is of compact support, the integral (10') is, for every $x \in J$, extended over the interval $K = [-A-B, A+B]$. We are then in the situation of Theorem 9 of n° 9: the function $\varphi(x-t)$ plays the rôle of the function $f(x, y)$ of the theorem and $f(t)$ that of $\mu$.

The convolution product is therefore differentiable, so, returning to (10'),

$$(27.11) \qquad (f \star \varphi)' = f \star \varphi' \;\; \text{or} \;\; D(f \star \varphi) = f \star D\varphi.$$

Since the derivative $D\varphi$ of a function in $\mathcal{D}$ is again in $\mathcal{D}$, one may iterate (11), which leads to the general relation

$$(27.12) \qquad (f \star \varphi)^{(n)} = f \star \varphi^{(n)} \;\; \text{or} \;\; D^n(f \star \varphi) = f \star D^n\varphi$$

hypothesising only that $f$ is regulated on $\mathbb{R}$. It is not the differentiability or the continuity of $f$ which matters, it is that of $\varphi$.

**Theorem 27.** *For every function $f$ defined and continuous on $\mathbb{R}$ there exists a sequence $f_n$ of $C^\infty$ functions which converges to $f$ uniformly on every compact subset of $\mathbb{R}$. If $f$ is of compact support one may assume that the $f_n$ are zero outside a fixed compact set.*

Obvious: one applies Theorem 26 to a Dirac sequence formed by functions of $\mathcal{D}$, bearing in mind what we want to establish. If $f$ is zero for $|x| > A$ and if one assumes, for example, that the $\varphi_n$ vanish for $|x| \geq 1/n$, it is clear that the $f_n$ are all zero for $|x| \geq A + 1$, qed.

If the function $f$ is $C^1$, one can, in formula (10), differentiate directly with respect to $x$, again thanks to Theorem 9 of n° 9, which shows that

$$(27.13) \qquad D(f \star \varphi) = Df \star \varphi$$

in this case. Replacing $\varphi$ by the $\varphi_n \in \mathcal{D}$ of a Dirac sequence one sees that the $Df_n$ converge to $Df$ uniformly on every compact set. Therefore:

**Corollary.** *Let $f$ be a function of class $C^p$ ($p \leq +\infty$) on $\mathbb{R}$. There exists a sequence $f_n$ of $C^\infty$ functions such that*

$$\lim f_n^{(r)}(x) = f^{(r)}(x)$$

*uniformly on every compact for every finite $r \leq p$.*

All these results extend, with the same proofs, to functions defined on $\mathbb{R}^p$.

## 28 – Approximation by polynomials

We shall now prove Weierstrass' theorem on the uniform approximation of continuous functions by polynomials on a compact set. The proof we shall give of this – Weierstrass', up to a few details – also uses approximation by convolution products.

(i) We start from a function $u$ which is positive, integrable, has integral 1 over $\mathbb{R}$, but is not zero for $|x|$ large because we shall choose for $u$ an everywhere convergent power series. For every function $f(x)$ which is continuous on $\mathbb{R}$ and zero for $|x| \geq A$ given, we put

$$(28.1) \qquad f_n(x) = n \int f(y) u(nx - ny) dy.$$

These functions being the convolution products of $f$ by the functions $u_n(x) = nu(nx)$, which form a Dirac sequence, the $f_n$ converge to $f$ uniformly on every compact set.

(ii) We assume that $u(x) = \sum a_p x^p$ is the sum of a power series that converges for every $x \in \mathbb{C}$; then

$$f_n(x) = n \int f(y) dy \sum_p a_p n^p (x - y)^p.$$

For $x$ and $n$ given, the power series in $x - y$ converges normally on every compact set, and in particular on the interval $|y| \leq A$ outside which $f(y) = 0$. Since $f$ is bounded we may integrate term-by-term, whence

$$(28.2) \qquad f_n(x) = \sum_p na_p n^p \int (x-y)^p f(y) dy = \sum f_{n,p}(x),$$

where we integrate over $|y| \leq A$. For $|x| \leq A$ we have $|x - y| \leq 2A$, whence

$$(28.3) \qquad |f_{n,p}(x)| \leq 2An\|f\|.|a_p|(2nA)^p,$$

the factor 2A coming from the integration over $|y| < A$. Since the power series $u(x)$ converges for any $x$, so for example at the point $2nA$, the right hand side of (3) is the general term of a convergent series. The series (2) therefore converges normally on $|x| \leq A$.

(iii) The general term of the series (2) is a *polynomial* in $x$, as we see on expanding $(x - y)^p$. Since it converges normally on $|x| \leq A$, its partial sums converge to $f_n$ uniformly on this interval. Since the $f_n$ converge to $f$ uniformly in $|x| \leq A$, by (i), we can, replacing them by partial sums of sufficiently high order, obtain a sequence of polynomials converging to $f$ uniformly on $|x| \leq A$.

(iv) We still have to show the existence of $u$. The function $u(x) = c.\exp(-\pi x^2)$ meets the requirements. It is clearly positive, integrable, has total integral 1 for $c$ suitably chosen ($c = 1$, in fact) and is expandable as an everywhere convergent power series.

**Theorem 28 (Weierstrass, 1885).** *Let $f$ be a real function defined and continuous on a* compact *interval $K \subset \mathbb{R}$. Then there exists a sequence of polynomials which converges to $f$ uniformly on $K$.*

It is enough to observe that $f$ can be extended to a continuous function on all $\mathbb{R}$, zero for $|x|$ large: complete the graph of $f$ by linear functions.

For a function $f$ defined and continuous on all $\mathbb{R}$, or, more generally, on an unbounded interval $I$, it is impossible to find a sequence of polynomials $p_n$ which converges uniformly to $f$ on $I$ except in the trivial case where $f$ is itself a polynomial (Chap. III, n° 5, end). As we then noted, it is always possible to demand that the $p_n$ should converge to $f$ uniformly on every compact $K \subset I$.

If the interval $I$ is bounded but not compact, we have seen in n° 8, as a consequence of Corollary 2 of the uniform continuity theorem, that approximation by polynomials is possible only if $f$ is uniformly continuous on $I$; $f$ then extends to be a continuous function on the compact interval obtained by adjoining the endpoints to $I$, and Weierstrass' theorem applied to this compact interval yields the result. In short, Weierstrass' theorem is best possible. Having said this, there are more difficult approximation theorems, where, instead of considering polynomials, one considers for example, linear combinations of exponential functions $\exp(a_n x)$ with given $a_n$ (suitably chosen ...).

Up to details, the proof of Weierstrass' theorem which we have presented is that of Weierstrass himself; his aim was to show that, though it is certainly impossible to represent every continuous function by simple analytic formulae (algebraic, power series, etc.), it is however possible to find series of simple functions – as it happens, polynomials and not only monomials – which represent them.

There are many other proofs of the theorem; Hairer and Wanner, *Analysis by Its History*, p. 264, cited a dozen (from G. Meinardus, 1964), the latest dating from 1934. This leads me to suspect, without having checked, that perhaps one should add to their list the one and only model proof, applicable in the much more general framework of compact topological spaces, namely the Stone-Weierstrass theorem, a generalisation obtained in the 1930s by a Chicago mathematician who notably wrote in this period the first systematic exposition of the theory of "abstract" Hilbert spaces; he did much after 1945 to invite or recruit foreign mathematicians – I benefitted in 1950 – and moreover had a pronounced taste for gastronomy in general and French in particular; this was a good sign in an American, but one has to say that he was the son of a Chief Justice of the Supreme Court and not of a corn farmer from the Bible Belt. The theorem is as follows. Assume given on a *compact* space $X$, (Appendix to Chap. III, n° 7) a set $A$ – the initial letter of the word "algebra" – of continuous functions with complex values satisfying the following conditions:

(a) the complex constants are in $A$;
(b) the sum and the product of two functions of $A$ are again in $A$;
(c) for every $f \in A$, the complex conjugate function $\bar{f}$ is in $A$;
(d) for any distinct points $x, y \in X$ there exists an $f \in A$ such that $f(x) \neq f(y)$.

Then *every continuous function on $X$ is the uniform limit on $X$ of a sequence of functions $f_n \in A$*. For a proof with hardly any calculations, see Dieudonné, Vol. 1, Chap. VII, n° 3, or Serge Lang, *Analysis I* (Addison-Wesley, 1968), Chap. VIII, n° 5. If $X$ is a compact subset of $\mathbb{C}$, one may take for $A$ the set of polynomials in $x$ and $y$ [but not just the polynomials in $z$: they do not satisfy condition (c), without mentioning the fact that, in an open subset of $\mathbb{C}$, a uniform limit of polynomials in $z$ is holomorphic, as we shall see], whence Weierstrass' theorem for two variables, or for $p$ variables on taking $X$ in $\mathbb{R}^p$.

In particular let us take for $X$ the circle $|z| = 1$, the set of complex numbers of the form $z = e^{2\pi i t}$ with $t \in \mathbb{R}$ defined modulo $\mathbb{Z}$. A function $f$ defined and continuous on $X$ is transformed into a function $g(t) = f(e^{2\pi i t})$ defined, continuous and of period 1 on $\mathbb{R}$, and vice-versa as is easy to see – one has only to check continuity. On $X$, a polynomial in $x = (z + \bar{z})/2$ and $y = (z - \bar{z})/2i$, i.e. in $z$ and $\bar{z}$, clearly reduces to a finite sum of the form $\sum a_n e^{2\pi i n t}$ with the $n \in \mathbb{Z}$, in other words to a trigonometric polynomial (Chap. III, n° 5). Corollary: every function defined, continuous and periodic on $\mathbb{R}$ is the uniform limit of trigonometric polynomials, as announced in Chap. III, n° 5. This is the result on which one may base the whole theory of Fourier series. One may also, we shall see this in the chapter dedicated to the subject, prove it by explicit elementary calculations, as did Weierstrass himself. But Stone's method applies to generalisations of Fourier series (harmonic analysis on compact groups for example) where direct and explicit calculations would be impossible. Moreover, in this way one has no need of providential functions like $\exp(-\pi x^2)$.

This kind of "nonconstructive" proof naturally does not satisfy the calculators. There is a proof with calculations in Hairer and Wanner, pp. 264–268, and, moreover, the graphs showing the approximations they provide.

## 29 – Functions having given derivatives at a point

To end this section with a nonobvious theoretical exercise we shall prove the following result[55]:

**Theorem 29 (Emile Borel, 1895).** *For every sequence $(a_n)$ of complex numbers there exists an indefinitely differentiable function $f$ of compact support on $\mathbb{R}$ such that $f^{(n)}(0) = a_n$ for every $n \in \mathbb{N}$.*

Our first move, faced by this theorem, is to put

$$(29.1) \qquad f(x) = \sum a_n x^n / n!$$

in accordance with Maclaurin's formula. Bad idea: the series has every chance of diverging for $x \neq 0$.

All the same, (1) contains the germ of an idea for a proof. The function $a_n x^n / n! = a_n x^{[n]}$ has the virtue that, at the origin, all its derivatives are zero except for the $n$-th, which is equal to $a_n$. We shall replace it by a function $f_n \in \mathcal{D}$, the space of $C^\infty$ functions of compact support on $\mathbb{R}$, possessing the same properties but making the series converge. And since we will have to calculate the successive derivatives of the series it will be necessary to differentiate it term-by-term, i.e. to apply Theorem 20 of Chap. III, n° 17. In other words, the function $f$ will be given by the formula

$$(29.2) \qquad f(x) = \sum f_n(x)$$

where the $f_n \in \mathcal{D}$ satisfy, for example,

$$(29.3) \qquad f_n(x) = 0 \qquad \text{for} \qquad |x| \geq 1$$

to yield a result of compact support, satisfy also

---

[55] The proof which follows (H. Mirkil, 1956) develops the one that one finds for example in Lars Hörmander, *The Analysis of Linear Partial Differential Operators* (Vol. 1, Springer-Verlag, 1983, p. 16), where it takes sixteen lines. Borel's complete result is much stronger but requires difficult results on analytic functions: one may assume $f$ to be of class $C^\infty$ on a neighbourhood of 0 and *real-analytic* apart from at the origin. See Remmert, *Funktionentheorie II*, p. 237.

$$(29.4) \qquad f_n^{(r)}(0) = \begin{cases} a_n & \text{if} \quad r = n \\ 0 & \text{if} \quad r \neq n \end{cases}$$

and finally, and this is the crucial point, are chosen so that, for every $r \geq 0$, the series of the derivatives $\sum f_n^{(r)}(x)$ converges uniformly on $|x| \leq 1$ [or on $\mathbb{R}$, which comes to the same by (3)]. The proof divides into several parts.

(i) *Construction of* $f_0$ or, equivalently, of a function $h$ which is zero for $|x| \geq 1$, equal to 1 at $x = 0$ and all of whose derivatives are zero at the origin. We start with a function of the type

$$(29.5) \qquad g(x) = \begin{cases} 1 & \text{for} \quad |x| \leq A \\ 0 & \text{for} \quad |x| > A \end{cases}$$

and "regularise" it with the help of a convolution product

$$(29.6) \qquad h(x) = \int \varphi(x - y)g(y)dy$$

with a $\varphi \in \mathcal{D}$, positive, of total integral equal to 1, and zero for $|x| > C$, where $C$ will be chosen later. The integral is taken over the interval $|y| \leq A$ and cannot be $\neq 0$ for a given $x$ unless there exists a $y$ satisfying $|y| \leq A$ and $|x - y| \leq C$ simultaneously, which requires $|x| \leq A + C$. Then

$$h(x) = \int_{-A}^{+A} \varphi(x - y)dy = \int_{x-A}^{x+A} \varphi(z)dz.$$

The support $[-C, C]$ of $\varphi$ is contained in $[x - A, x + A]$ so long as $x - A \leq -C < C \leq x + A$, i.e. $C - A \leq x \leq C + A$; choosing $A = 3/4$ and $C = 1/4$, one sees then that the function $h$ is zero for $|x| > 1$ and equal to $\int \varphi(z)dz = 1$ for $|x| < 1/2$. Its graph is of the type below (fig. 12).

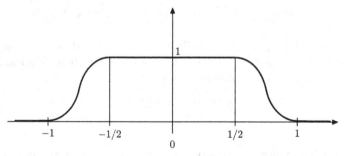

**Fig. 12.**

(ii) *Choice of the* $f_n$. We put

(29.7) $$f_n(x) = h(b_n x)a_n x^{[n]}.$$

with the $b_n > 0$ to be chosen later. Since we want the $f_n$ to be zero for $|x| \geq 1$, it is prudent to impose the condition

(29.8) $$b_n > 1$$

on them. Let us calculate the derivatives of the $f_n$ at the origin. They are obtained from Leibniz' formula: $(fg)^{[r]} = \sum f^{[r-p]}g^{[p]}$. At $x = 0$ all the derivatives of $x^{[n]}$ are zero except the $n$-th, equal to 1. Those of $h(b_n x)$ are all zero at the origin starting from the first. The derivative of order $r$ of $f_n$ cannot be $\neq 0$ unless there exists a $p$ such that $p = n$ and $p = r$ simultaneously, in other words if $r = n$. In this case, there remains $f_n^{(n)}(0) = a_n$, so that the $f_n$ do satisfy condition (4), and this whatever the $b_n$.

(iii) *Convergence of the series* $\sum f_n^{(r)}(x)$. We shall show that one may choose the $b_n$ so that

(29.9)    $|f_n^{(r)}(x)| \leq 1/2^n$  for every $x$, every $r$, and every $n > r$,

the sole significance of the numbers $1/2^n$ being that they form a convergent series. As this is the case, it is clear that the series $\sum f_n(x)$ will be normally convergent as will be all the derived series, the sum of the series will therefore be in $\mathcal{D}$ and its derivatives, calculated by differentiating the series term-by-term, will be the $a_n$ at the origin, by the relations (4).

It remains to choose the $b_n$. By (7) and Leibniz, we have

(29.10) $$f_n^{(r)}(x) = \sum ?h^{(r-p)}(b_n x)b_n^{r-p}a_n x^{[n-p]}$$

with numerical coefficients denoted ? and whose exact values are of little importance. Since $h(b_n x) = 0$ for $|x| > 1/b_n$, it suffices, to evaluate the result, to work on the interval $|x| \leq 1/b_n$, which allows us to estimate the monomials appearing in (10). In this interval we have $|b_n^{r-p}x^{n-p}| \leq b_n^{r-n}$, an expression independent of $p$, whence, passing to the uniform norms,

(29.11)    $\left\|f_n^{(r)}\right\| \leq b_n^{r-n}|a_n| \sum ? \left\|h^{(r-p)}\right\| = M_{r,n}b_n^{r-n} \leq M_{r,n}/b_n$

for $r < n$. Since, for $n$ given, the conditions (9) to be satisfied involve only the $r < n$, so are finite in number, it then suffices, to satisfy them simultaneously, to choose $M_{r,n}/b_n \leq 1/2^n$ for every $r < n$, i.e.

$$b_n \geq \max_{0 \leq r < n} 2^n M_{r,n},$$

qed.

This proof is typical of the current techniques in analysis. All the work consists of rigorously controlling the orders of magnitude of the numbers or functions that one is manipulating. Nothing is calculated explicitly. We are at the antipodes of the analysis of the Founders.

# § 9. Radon measures in $\mathbb{R}$ or $\mathbb{C}$

## 30 – Radon measures on a compact set

As we have already observed on several occasions, many of the results of integration theory use only a few quasi-algebraic properties of the integral $m(f)$ of a function: linearity, positivity and continuity with respect to uniform convergence, in other words, Theorem 1 of n° 2.

We have also observed on occasion that there are curious analogies between integrals and series. Let us work on a compact interval of $\mathbb{R}$ and consider, for example, the two following situations:

(i) Theorem 9 of n° 9 and Theorem 24 of n° 25 which allow one to calculate the derivative of a "continuous" sum

$$(30.1) \qquad \varphi(y) = \int f(x, y)\mu(x)dx$$

of functions by the formula

$$(30.2) \qquad \varphi'(y) = \int D_2 f(x, y)\mu(x)dx,$$

(ii) Theorem 19 of Chap. III, n° 17 which, translated into the language of series, allows one to differentiate a "discrete" sum

$$(30.3) \qquad \varphi(y) = \sum f_n(y)$$

of functions by the formula

$$(30.4) \qquad \varphi'(y) = \sum f_n'(y).$$

The analogy would be even clearer if, starting from a finite or denumerable set $D$ of points of $\mathbb{R}$, a scalar function $\mu(\xi)$ on $D$ satisfying $\sum |\mu(\xi)| < +\infty$, and a function $f(\xi, y)$ defined on $D \times Y$, one put

$$(30.3') \qquad \varphi(y) = \sum f(\xi, y)\mu(\xi)$$

when one would have

$$(30.4') \qquad \varphi'(y) = \sum D_2 f(\xi, y)\mu(\xi)$$

of course under suitable hypotheses as in the case (i).

One may unify the two cases formally by writing, for every reasonable function $f$ of a "continuous" or "discrete" variable, $\mu(f) = \int f(x)\mu(x)dx$ in the first case and $\mu(f) = \sum f(\xi)\mu(\xi)$ in the second; using the notation $f_y(x) = f(x, y)$ one then has

$$\varphi(y) = \mu(f_y) \qquad \text{and} \qquad \varphi'(y) = \mu[(D_2 f)_y]$$

in both cases.

We are led to introduce, in a general way, functions $f \mapsto \mu(f)$ in which the variable $f$ is a more or less arbitrary *function* on a given interval $X$ (or a metric space, or even, in the "abstract" theory of integration, an arbitrary set) possessing properties formally analogous to those of integrals and series. Clearly one has to impose some restrictions on the category of functions $f$ considered: it is not possible to define the expression $\int f(x)dx$ in a natural way for *every* function $f$ on $\mathbb{R}$. On the other hand, the problem, whether dealing with series or integrals, has always been the following: one is given $\mu(f)$ for particularly simple functions $f$ (finite sums in the discrete case, step functions in the continuous case) and one hopes to extend the construction in a natural way to more complicated functions (series in the discrete case, integrals of regulated or semicontinuous functions, or even more general in the Lebesgue theory, in the case of continuous sums).

In the simplest case, of a *compact* interval $K \subset \mathbb{R}$, the constructions in n° 1 and 2 of this Chapter led us to associate to each *step* function $\varphi$ a number $\mu(\varphi)$ possessing the following properties:

(i) *linearity*: $\mu(\alpha\varphi + \beta\psi) = \alpha\mu(\varphi) + \beta\mu(\psi)$ for any constants $\alpha$ and $\beta$ and step functions $\varphi$ and $\psi$;

(ii) *continuity* with respect to uniform convergence: there exists a constant $M(\mu) \geq 0$ such that

(30.5) $$|\mu(\varphi)| \leq M(\mu)\|\varphi\|_K$$

for any $\varphi$. If, for every interval $I \subset K$, we write $\chi_I$ for the function equal to 1 on $I$ and to 0 in $K - I$, we may then associate a "measure"

$$\mu(I) = \mu(\chi_I)$$

to $I$, which manifestly has the additivity property (M 2) of n° 1:

$$I = I_1 \cup \ldots \cup I_n \implies \mu(I) = \mu(I_1) + \ldots + \mu(I_n)$$

if the $I_k$ are pairwise disjoint, because $\chi_I$ is then the sum of the characteristic functions of the $I_k$. From this one can calculate the integral $\mu(\varphi)$ of every step function using a finite *partition* of $K$ into intervals $I_k$ on which $\varphi$ is constant; on choosing points $\xi_k \in I_k$, one has

$$\mu(\varphi) = \sum \varphi(\xi_k)\mu(I_k)$$

since

$$\varphi(x) = \sum \varphi(\xi_k)\chi_{I_k}(x)$$

for every $x \in K$, whence the formula by the linearity of $\varphi \mapsto \mu(\varphi)$.

Starting from this data one may then define $\mu(f)$ for every regulated function $f$ on $K$ by passing to uniform limits: if the step functions $\varphi_n$ converge uniformly to $f$ then the relation (5), which implies

$$|\mu(\varphi_p) - \mu(\varphi_q)| \leq M(\mu)\|\varphi_p - \varphi_q\|_K,$$

shows that the integrals $\mu(\varphi_n)$ form a Cauchy sequence, so converge; the limit depends only on $f$ since, if $(\psi_n)$ is another uniform approximation to $f$, the relation (5) shows that $\mu(\varphi_n) - \mu(\psi_n)$ tends to 0. (Compare the construction of the real numbers starting from Cauchy sequences of rational numbers.) Whence $\mu(f)$, with, quite clearly, the two usual properties of linearity and continuity:

$$|\mu(f)| \leq M(\mu)\|f\|_K.$$

Note in passing that this construction, which does not involve the "lower" and "upper" integrals of n° 1, but applies – no big deal! – only to regulated functions, does not use the hypothesis of *positivity* of $\mu$, namely that

(30.6) $$\varphi \geq 0 \Longrightarrow \mu(\varphi) \geq 0.$$

If this is satisfied then the relation $\varphi \leq \psi$ implies $\mu(\varphi) \leq \mu(\psi)$ and all the arguments of n° 1 concerning the lower and upper integrals of a bounded function apply unchanged. As we have observed since n° 1, the three conditions imposed on our functions $\mu(I)$ would be satisfied if one put $\mu(I) = \mu(v) - \mu(u)$ for $I = (u, v)$, where $\mu(x)$ is an increasing function on $K$. Note also that this construction mingles the discrete and the continuous sums: for example, put

$$\mu(\varphi) = \int_K \varphi(x)\mu(x)dx + \sum \varphi(\xi)c(\xi)$$

where one integrates over $K$ and where $\sum |c(\xi)| < +\infty$; if $\varphi$ is the characteristic function of an interval $I \subset K$ of any type, one finds clearly

$$\mu(I) = \int_I \mu(x)dx + \sum_{\xi \in I} c(\xi).$$

Physically, this comes down to considering that one has, on the one hand, a distribution of masses on $K$ whose density in the usual sense (the ratio of the mass of an "infinitely small" segment of $K$ to its length) is given by the regulated function $\mu(x)$ and, on the other hand, a countable set of "point" masses $c(\xi)$. One also meets this kind of situation in the most modern physics: in the spectrum of the radiation emitted by the Sun, there are "bands", whose intensity is a continuous function of the frequency, and "lines" which concentrate a nonzero intensity on an interval consisting of a single frequency. Nothing very artificial here; Newton would have said that one meets this in Nature . . .

In elementary practice one is above all interested in integrating *continuous* functions; when, in the theorems on differentiation under the $\int$ sign for example, we have introduced an arbitrary regulated function $\mu(x)$ in front of the symbol $dx$ of *Lebesgue measure*

$$f \longmapsto m(f) = \int_X f(x)dx$$

on an interval $X$, this was not for the pleasure of integrating discontinuous functions; it was in order to obtain a theorem applicable to the measure

$$f \longmapsto \mu(f) = \int_X f(x)\mu(x)dx.$$

There are other more technical reasons to think that in a "good" integration theory the starting point is not the measure $\mu(I)$ of an arbitrary interval $I \subset K$, but rather the integral $\mu(f)$ of an arbitrary continuous function $f$ on $K$; anyway, and as we have just seen, the passage from measures of intervals to the integration of continuous (or even regulated) functions is quasi-instantaneous, once one has understood the construction of the classical integral.

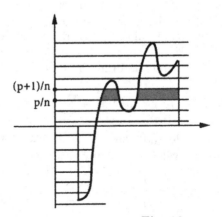

$(p+1)/n$
$p/n$

**Fig. 13.**

The real problem, solved by Lebesgue, is to integrate functions much more general than the continuous or regulated functions (start with the semicontinuous functions). *Chez* Lebesgue, a century ago, one started by extending the concept of the measure of an interval to that of the measure of a much more complicated set $E \subset K$, for example to the sets which are countable unions of countable intersections of countable unions of countable intersections of open intervals (it can get even worse, but this is not important). Starting from this one integrates a function $f$ – assumed bounded for simplicity – as follows: for every integer $n \geq 1$, consider for any $p \in \mathbb{Z}$ the set

$E_{n,p} = \{p/n \leq f(x) < p/n + 1/n\}$ on which $f$ is equal to $p/n$ to within $1/n$; these sets form a finite partition of $K$; if they belong to the category of those for which one can define $m(E)$ [or $\mu(E)$ in the case of an arbitrary measure], one may consider the *Lebesgue sum*, $\sum_p m(E_{p,n})p/n$ (as against the Riemann sum); geometrically, one considers the set of the points $(x, y)$ in the plane lying between $K$ and the graph of $f$, cuts it by the lines $y = p/n$ into *horizontal* slices having the $E_{p,n}$ as bases, and approximates the required area $\int f(x)dx$ by the sum of the areas of these horizontal slices (fig. 13). Lebesgue's genius was not just to have replaced the decomposition into vertical slices by decomposition into horizontal slices; it was to have understood that this innocent modification of the traditional procedure provided a method formidably more powerful than that of Riemann. It has been generalised *ad libitum*, but no one has ever progressed in a way that would be useful beyond certain *ad hoc* problems. The mode of exposition has only been modified from that chosen by Lebesgue in an age when the concepts of vector space, of linear form and of norm had not yet been isolated: analysis had been *arithmetised* and was now, in Germany rather more than in France, as rigorous as number theory, but it had not yet been *algebraised*; in fact, integration theory was probably the impetus that forced the analysts to learn, even to invent, what a vector space of infinite dimension ought to be, starting with Hilbert spaces.

Since, for every compact set $K \subset \mathbb{C}$ (or in $\mathbb{R}^p$, or for every metric compact space), one has available the space[56] $L(K) = C^0(K)$ of the scalar continuous functions on $K$ and the norm

$$\|f\|_K = \sup |f(x)|$$

of uniform convergence on $K$, one is led to define a *Radon measure* on $K$ in the following way: it is a map

$$\mu : L(K) \longmapsto \mathbb{C}$$

which is *linear* in the general sense of algebra and *continuous* in the general sense of the theory of normed vector spaces: there exists a constant $M(\mu)$ such that

$$|\mu(f)| \leq M(\mu)\|f\|_K$$

for every $f \in L(K)$. Such a measure is said to be *positive* if

$$f \geq 0 \Longrightarrow \mu(f) \geq 0.$$

One shows without much difficulty (Chap. XI, n° 17, Theorem 29) that every measure $\mu$ can be put in the form $\mu(f) = \mu_1(f) - \mu_2(f) + i\mu_3(f) - i\mu_4(f)$ with

---

[56] The notation $L(X)$ was introduced in André Weil, *L'integration dans les groupes topologiques et ses applications* (Paris, Hermann, 1940), a book from which many of my generation learned integration and generalised Fourier analysis. I assume that Weil chose it not only in homage to Lebesgue, but also because he composed directly onto the typewriter ...

positive measures $\mu_k$. It is generally worth restricting oneself – or reducing – to the case of positive measures, by far the most important, and the only ones for which a grand theory has been constructed.

Leibniz' notation having amply proved its usefulness, one imitates it by writing

$$\mu(f) = \int_K f(x)d\mu(x),$$

a notation which will justify itself better below, but which one may use without understanding its origin. So, by definition, we have the relation

$$(30.7) \quad \int_K (3\sin x - 5\log x)\, d\mu(x) = 3\int_K \sin x\, d\mu(x) - 5\int_K \log x\, d\mu(x)$$

for any constants 3 and $-5$ and continuous functions sin and log on $K$, also that

$$(30.8) \quad \left| \int_K f(x)d\mu(x) \right| \le M(\mu)\|f\|_K$$

for every continuous function $f$ on $K$, with a constant $M(\mu)$ to be chosen as small as possible; this is the *norm of the measure* $\mu$, notation $\|\mu\|$; for the usual measure $m$ on an interval of $\mathbb{R}$, we have $\|m\| = m(K)$. The object of this condition is to guarantee that Theorem 4 concerning uniform limits applies again here: if a sequence of continuous functions $f_n$ converges uniformly on $K$ to a necessarily continuous limit $f$ one has

$$(30.9) \quad |\mu(f) - \mu(f_n)| = |\mu(f - f_n)| \le M(\mu)\|f - f_n\|_K,$$

whence $\mu(f) = \lim \mu(f_n)$. If, likewise, a series $s(x) = \sum u_n(x)$ of continuous functions on $K$ converges normally, one may integrate term-by-term:

$$(30.10) \quad \mu\left(\sum u_n\right) = \sum \mu(u_n).$$

In short, *we have transformed Theorems 1 and 4 of § 1 into definitions.*

*Example 1.* Choose a function $\mu(x)$, integrable (in the usual sense) on an interval $K \subset \mathbb{R}$, and put

$$(30.11) \quad \mu(f) = \int f(x)\mu(x)dx$$

for every $f \in L(K)$. Linearity is obvious and continuity follows from the inequalities

$$|\mu(f)| \le \int |f(x)|.|\mu(x)|dx \le \|f\|_K . \int |\mu(x)|dx.$$

Here $\|\mu\| \le \int |\mu(x)|dx$ (and, in fact, we have equality).

The simplest case is obtained by choosing $\mu(x) = 1$; this is the measure $m$ which appears in all of classical analysis and has been the subject of this chapter. It is called the *Lebesgue measure* on $K$, not, of course, because poor Henri Lebesgue had discovered that the length of an interval $(a, b)$ is equal to $b - a$, but because it is for this particularly important measure that he invented the grand integration theory that was later extended to all measures defined on all reasonable topological spaces. In the general case (11) one speaks of the *measure of density* $\mu(x)$ with respect to Lebesgue measure: obvious physical interpretation.

*Example 2.* Choose a countable set $D$ of points of $K$ and, for every $\xi \in D$, a number $c(\xi) \in \mathbb{C}$; assuming $\sum |c(\xi)| < +\infty$ one may define

$$(30.12) \qquad \mu(f) = \sum c(\xi) f(\xi)$$

for every continuous function $f$ on $K$, the series being taken over $D$. No hypothesis on the compact set $K$ is necessary here.

*Example 3.* Take $K = A \times B$ where $A$ and $B$ are compact intervals in $\mathbb{R}$ and put

$$m(f) = \iint\limits_{A \times B} f(x, y) dx dy$$

for every continuous function $f$ on $K$ (n° 9, Theorem 10).

*Example 4.* Choose for $K$ the set $\mathbb{T} : |z| = 1$ of complex numbers of modulus 1 (unit circle); the "parametric representation" $z = \exp(2\pi i t) = \mathbf{e}(t)$ of the points of $\mathbb{T}$ transforms any function $f \in L(\mathbb{T})$ into a continuous function $f[\mathbf{e}(t)]$ of period 1 on $\mathbb{R}$. One thus obtains a very privileged measure on $\mathbb{T}$ by putting

$$(30.13) \qquad m(f) = \int_0^1 f[\mathbf{e}(t)] dt = \frac{1}{2\pi} \int_0^{2\pi} f(e^{it}) dt$$

for every $f \in L(\mathbb{T})$; this is what dominates the theory of Fourier series. One could clearly replace $\mathbb{T}$ by any other parametrised curve, but it is better to defer this type of example to when we shall need it (mainly line integrals of holomorphic functions).

We shall see later how, in the case of an *interval*, one may construct all the measures on $K$ by a procedure analogous to that of n° 1 and 2 concerning Lebesgue measure.

We shall now show quickly how some of the theorems on Lebesgue measure extend to Radon measures.

*Differentiation under the $\int$ sign.* Theorem 9 of n° 9 extends trivially (i.e. with the same proof) to the functions

$$(30.14) \qquad g(y) = \int f(x, y) d\mu(x),$$

where we have not indicated that the integral is extended over $K$; $f$ is a continuous function on $K \times J$ and $J$ an interval of $\mathbb{R}$. The result is a continuous function and it is of class $C^1$ if $D_2 f$ exists and is continuous, with, in this case,

$$(30.15) \qquad g'(y) = \int D_2 f(x, y) d\mu(x).$$

To reproduce the proofs *verbatim* here would be to waste our time and that of the reader. The simplest case is that of an interval $K \subset \mathbb{R}$, but in fact the argument and the result apply to every compact set $K$ if one knows that, on every compact set (for example, here, $K \times H$ where $H \subset J$ is a compact interval), a continuous function is uniformly continuous.

    *Exercise.* The function $f \star \mu(x) = \int f(x - y) d\mu(y)$ is $C^\infty$ if $f$ is $C^\infty$ and of compact support on $\mathbb{R}$.

*Double integrals.* A slightly less easy exercise consists of generalising Theorem 10 to measures:

**Theorem 30.** *Let $K$ and $H \subset \mathbb{R}$ be two compact sets, $\mu$ and $\nu$ measures on $K$ and $H$, and $f$ a continuous function on $K \times H$. Then*

$$(30.16) \qquad \int_K d\mu(x) \int_H f(x, y) d\nu(y) = \int_H d\nu(y) \int_K f(x, y) d\mu(x).$$

Inspired by the proof of Theorem 10 one is led, for a given $r > 0$, to take partitions $(K_p)$ of $K$ and $(H_q)$ of $H$, and to compare each member of (16) to the sum of the analogous expressions obtained on replacing $K$ and $H$ by $K_p$ and $H_q$, i.e. by multiplying $f(x, y)$ by $\chi_p(x)\theta_q(y)$ where $\chi_p$ and $\theta_q$ are the characteristic functions of $K_p$ and $H_q$. Since

$$(30.17) \qquad \sum \chi_p(x) = 1 \ \text{ on } K, \qquad \sum \theta_q(y) = 1 \ \text{ on } H,$$

we have $f(x, y) = \sum f(x, y)\chi_p(x)\theta_q(y)$ on $K \times H$, which explains why the sum of the integrals over the products $K_p \times H_q$ is the integral of $f$ over $K \times H$.

    This method is not directly applicable here: for an arbitrary measure we don't yet know how to integrate discontinuous functions. The solution is to replace the discontinuous functions which frustrate us by *continuous* positive functions $k_p$ and $h_q$ on $K$ and $H$ still satisfying (17) and zero outside sets $A_p$ or $B_q$ which are small enough that the function $f$ is constant to within $r$ on each product $A_p \times B_q$. By (17) and the linearity of $\mu$ and $\nu$, we will then have

$$(30.18) \qquad \int d\mu(x) \int f(x, y) d\nu(y) =$$
$$= \sum \int d\mu(x) \int f(x, y) k_p(x) h_q(y) d\nu(y),$$

this is all meaningful, and we can imitate the proof of Theorem 10.

Let us be more precise. Choose an $r > 0$ and an $r' > 0$ such that (uniform continuity: n° 8)

(30.19) $(|x' - x''| \leq r')$ & $(|y' - y''| \leq r') \Longrightarrow |f(x', y') - f(x'', y'')| \leq r.$

Cover $K$ by a finite number of *closed* sets $F_p$ of diameters[57] $\leq r'/2$ (we are not dealing with a partition) and, for each $p$, let us choose an *open* set $U_p \supset F_p$ of diameter $\leq r'$; we may, but there is no point, assume that these sets are intervals. For every $p$ there exists a continuous positive function $\varphi_p$ on $K$ which is strictly positive on $F_p$ and zero outside $U_p$, for example $\varphi_p(x) = d(x, K - U_p \cap K)$, the distance from the point $x$ to the complement of $U_p \cap K$ in $K$, closed and disjoint from $F_p$. The continuous function $k(x) = \sum \varphi_p(x)$ being $> 0$ at all $x \in K$, since the $F_p$ cover $K$, we need only put $k_p(x) = \varphi_p(x)/k(x)$ to obtain the functions we seek.

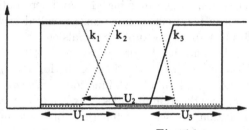

Fig. 14.

Similarly we construct continuous positive functions $h_q$ on $H$, with sum 1 and zero outside open sets $V_q$ of diameters $< r'$. Finally, we choose points $\xi_p \in A_p = U_p \cap K$ and $\eta_q \in B_q = V_q \cap H$.

In the general term of the sum (18) the integrand cannot be $\neq 0$ at a point $(x, y)$ unless $g_p(x)$ and $h_q(y)$ are so, i.e. if $(x, y) \in A_p \times B_q$. Then $|x - \xi_p| \leq r'$ and $|y - \eta_q| \leq r'$ and thus, by (19),

(30.20)    $|f(x, y)k_p(x)h_q(y) - f(\xi_p, \eta_q)k_p(x)h_q(y)| \leq rk_p(x)h_q(y).$

In fact, this inequality is valid for any $(x, y) \in K \times H$ since outside $A_p \times B_q$ either $k_p(x) = 0$ or $h_q(y) = 0$. On adding the inequalities (20) and remembering that $\sum k_p(x)h_q(y) = 1$ one finds

$$\left| f(x, y) - \sum f(\xi_p, \eta_q)k_p(x)h_q(y) \right| \leq r$$

for any $x$ and $y$. Put $g(x, y) = \sum f(\xi_p, \eta_q)k_p(x)h_q(y)$; then

(30.21)                     $\|f - g\|_{K \times H} \leq r.$

---

[57] The diameter of a set $X \subset \mathbb{C}$ is the number $\sup d(x, y)$ where $x$, $y$ vary in $X$.

Denote the two sides of (16) by $\lambda'(f)$ and $\lambda''(f)$. These are measures on $K \times H$: linearity is obvious, and continuity follows from $|\lambda'(f)| \leq M(\mu)M(\nu)\|f\|_{K \times H}$, with the same inequality for $\lambda''$. By (21), we then have $\lambda'(f) = \lambda'(g)$ and $\lambda''(f) = \lambda''(g)$ to within $M(\mu)M(\nu)r$. Since $r > 0$ is arbitrary it is then enough to show that $\lambda'(g) = \lambda''(g)$ to establish (16).

But this is obvious: for any $k \in L(K)$ and $h \in L(H)$ we have

$$\int d\mu(x) \int k(x)h(y)d\nu(y) = \int d\mu(x)k(x) \int h(y)d\nu(y) =$$
$$= \int d\mu(x)k(x)\nu(h) = \nu(h) \int k(x)d\mu(x) = \mu(k)\nu(h)$$

and the same calculation, with the same result, on interchanging the order of the integrations; since $g$ is a linear combination of functions of the form $k(x)h(y)$ we have $\lambda'(g) = \lambda''(g)$, qed.

This proof generalises fully: in every metric space one may find systems of continuous positive functions satisfying (17) and zero outside arbitrarily small given open sets; such systems of functions are called *partitions of unity*. The method applies to triple, quadruple, integrals etc.

The two essential points in the proof are that (i) the identity (16) is obvious if $f(x, y)$ is a finite sum of functions of the form $g(x)h(y)$, (ii) every $f \in L(K \times H)$ is, by (21), the limit of such functions, *uniformly on $K \times H$*. The general Stone-Weierstrass theorem provides (ii) without the least calculation: it is enough to check that the set $A$ of functions of the form $\sum g_p(x)h_q(y)$, with $g_p \in L(K)$ and $h_q \in L(H)$ complex-valued, satisfies the conditions (a), (b), (c), (d) of n° 28; a very easy exercise.

Finally one has to observe that the map

$$f \longmapsto \lambda(f) = \int\int f(x, y)d\mu(x)d\nu(y)$$

– it is now unnecessary to specify the order of the integrations – is a *measure* on the compact set $K \times H$, the *product measure* or *Cartesian product* of $\mu$ and $\nu$. On choosing $d\mu(x) = dx$ and $d\nu(y) = dy$ one recovers Lebesgue measure in the plane.

All the theory expounded in n°s 10 and 11 extends to general positive measures, with the very same proofs: we have established what we need and only have to replace $m(f)$ everywhere by $\mu(f)$ and $m(K)$ by $M(\mu)$ or $\|\mu\|$.

## 31 – Measures on a locally compact set

Since Lebesgue measure allows one to integrate over intervals which are neither compact nor even bounded we should be able to extend the definition of

Radon measures to this case. First we examine the classical situation more closely, it being a little less simple than the compact case.

If $X = (a, b) \subset \mathbb{R}$ is an arbitrary interval one cannot define the integral $\int f(x)dx$ over $X$ for just any continuous function $f$ on $X$: there must be convergence conditions at the endpoints of $X$ if these do not belong to $X$ or are infinite. A radical method to eliminate them is to assume the function $f(x)$ to be zero when $x$ is close enough to $a$ or $b$, i.e. that there exists a *compact*[58] interval $K = [u, v] \subset X$ such that $f(x) = 0$ for $x \notin K$. Then

$$(31.1) \qquad \int_{K'} f(x)dx = \int_K f(x)dx \qquad \text{for } K \subset K' \subset X,$$

so that the integral taken over $X$ converges absolutely for a trivial reason and reduces to the integral over any compact set outside which $f$ is zero. The functions of this kind are called *of compact support on $X$* (n° 27); the set of these is a vector space which we again denote by $L(X)$ and which many other authors denote $C_c^0(X)$, with a suffix $c$ whose significance is obvious. For every compact $K \subset X$ the set of $f \in L(X)$ which vanish outside $K$ is a vector subspace $L(X, K)$ of $L(X)$.

In this way Lebesgue measure on $X$ gives us a (positive) linear form $f \mapsto m(f)$ on $L(X)$. We have a norm on $L(X)$

$$(31.2) \qquad \|f\|_X = \sup_{x \in X} |f(x)|,$$

but if $X$ is not bounded the linear form $m$ is not continuous relative to this norm, in other words, there is no finite constant $M$ such that

$$(31.3) \qquad |m(f)| = \left| \int_u^v f(x)dx \right| \leq M . \sup_{x \in X} |f(x)| = M\|f\|_X$$

for every continuous function $f$ on $X$ which is zero outside some compact interval $K = [u, v] \subset X$ and otherwise arbitrary. For example take $X = \, ]0, +\infty[$, with $0 < u < v < +\infty$; you can clearly find a function $f$ with values everywhere between 0 and 1, equal to 1 on $K$ and zero outside a compact $K' \subset X$ a little larger than $K$ (make the characteristic function of $K$ continuous by replacing its discontinuities at $u$ and $v$ by line segments); for such a function the left hand side of the preceding inequality is $> m(K) = v - u$ and the right hand side is equal to $M$; impossible if $u$ and $v$ can take arbitrary values between 0 and $+\infty$. For $X = [0, +\infty[$, one may take $u = 0$, but the difficulty remains.

Failing (3), one may all the same observe that

---

[58] If $a$ is finite one must have $a \leq u$ if $a \in X$, $a < u$ if not; if $b$ is finite similarly one must have $b \geq v$ if $b \in X$, $b > v$ if not. In the case where $a$ or $b$ is infinite both $u$ and $v$ must be finite. If for example $X = [0, +\infty[$, a continuous function on $X$ is so in particular at 0, so that the only difficulty in integrating it over $X$ comes from the other endpoint of $X$.

(31.4)         $|m(f)| \leq m(K)\|f\|_X$      for every $f \in L(X, K)$

since in fact the integral over $X$ involves only the compact $K \subset X$ off which $f$ is zero. This is the result we can generalise.

We shall call a *Radon measure on X* any linear form $\mu$ on the vector space $L(X)$ having the following property: *for every compact $K \subset X$ there exists a constant $M_K(\mu)$ such that*

(31.5)                    $|\mu(f)| \leq M_K(\mu)\|f\|_X$

*for every $f \in L(X, K)$.* In other words, we assume that the linear form $\mu$ is continuous on each subspace $L(X, K)$, though not necessarily on all of $L(X) = \bigcup L(X, K)$.

*Example 1.* Take for $X$ the *open* interval $]0, +\infty[$ and

(31.6)                    $\mu(f) = \int_0^{+\infty} f(x)dx/x.$

There is no problem with convergence for $f \in L(X)$ since $f(x)$ is zero on a neighbourhood of 0 and for $x$ large. If $f$ is zero outside $K = [u, v]$ then

$$|\mu(f)| \leq (\log v - \log u)\|f\|_X,$$

whence continuity.

One could clearly replace the function $1/x$ by any regulated function $p$ on $X$; if the function $p$ is absolutely integrable in $X$ then

$$|\mu(f)| \leq M(\mu)\|f\|_X$$

where $M(\mu) = \int |p(x)|dx$ no longer depends on the support $K$ of $f$. In this particular case the linear form $\mu$ is continuous on $L(X)$ and not just on the subspaces $L(X, K)$. A measure possessing this property is said to be *bounded* or *of finite total mass* on $X$.

*Example 2.* If one replaces the open interval $]0, +\infty[$ by the closed interval $[0, +\infty[$ the formula (6) is no longer meaningful, since, in this case, a function $f \in L(X)$ is required to be zero for $x$ large but not on a neighbourhood of 0, so allowing every chance of making the integral (6) divergent.

But one can replace $1/x$ by a function that poses no problem at the origin, and, for example, put

(31.7)        $\mu(f) = \int_0^{+\infty} f(x)x^s dx$      with $\mathrm{Re}(s) > -1$.

If $f$ is zero outside $K = [0, v]$ then

$$|\mu(f)| \leq M_K(\mu)\|f\|_X$$

with

$$M_K(\mu) = \int_0^v |x^s| dx = \frac{v^{\mathrm{Re}(s)+1}}{\mathrm{Re}(s) + 1}.$$

One obtains a bounded measure on replacing $x^s$ by an absolutely integrable function on $X$, for example $x^s e^{-x}$ with $\mathrm{Re}(s) > -1$.

*Example 3.* Choose a compact interval $K \subset X$, a measure $\mu$ on $K$ and consider the linear form $f \mapsto \int f(x) d\mu(x)$, where one integrates over $K$, so involving only the values of $f$ on this fixed compact set. This example shows that *a measure on $K$ may also be considered as a measure on $X$*: all the mass is supported by $K$.

*Example 4.* For $X = \mathbb{R}$ put

(31.8) $$\mu(f) = \sum f(n),$$

summing over $\mathbb{Z}$. If $f$ is zero outside a compact $K$, only the $n \in K$ count, whence $|\mu(f)| \leq M_K(\mu)\|f\|_X$ where $M_K(\mu)$ is the number of integers in $K$.
    More generally one may put

(31.9) $$\mu(f) = \sum c(\xi) f(\xi)$$

assuming only that for every compact $K \subset X$ the series taken over the $\xi \in K$ converges absolutely. Then, for $f \in L(X, K)$,

$$|\mu(f)| \leq M_K(\mu)\|f\|_X \qquad \text{where} \qquad M_K(\mu) = \sum_{\xi \in K} |c(\xi)|$$

since the $\xi \notin K$ do not appear in (9). If the total series $\sum |c(\xi)|$ converges one obtains a bounded measure as in Example 2, for example

$\mu(f) = \sum n^{-2} \sin(1/n) f(1/n)$ for $X = [0, +\infty[$, where one sums over the integers $> 0$.

In all this, the fact that $X$ is an interval of $\mathbb{R}$ hardly plays a rôle. One could assume that $X$ is a *locally compact* subset of $\mathbb{C}$ (end of n° 23) – an open set, a closed set, or, the general case, the intersection of an open and of a closed[59] set – and consider the vector space $L(X)$ of continuous functions on $X$ which are zero outside some compact subset of $X$, with its obvious vector subspaces $L(X, K)$. A Radon measure on $X$ is then again a linear form on $L(X)$ whose restriction to each $L(X, K)$ satisfies a relation (5).
    The assumption that $X$ is locally compact is needed to assure the existence of "many" functions $f \in L(X)$. The proof rests on the following lemma:

---

[59] *Exercise.* Show from the definition that the intersection of two locally compact subsets of $\mathbb{C}$ is locally compact.

**Lemma.** *Let $X$ be a subset of $\mathbb{C}$, $K$ a compact subset of $X$ and $F$ a subset of $X$ disjoint from $K$ and closed in[60] $X$. There exists a continuous function $f$ on $X$ such that*

(31.10)    $f = 1$ *on* $K$,    $0 \leq f < 1$ *on* $X - K$,    $f = 0$ *on* $F$.

*If $X$ is locally compact one may assume $f \in L(X)$.*

Consider the function $d(x, K)$; it is continuous, zero on $K$ and $> 0$ outside $K$. Let us show that there exists an $r > 0$ such that $d(x, K) \geq r$ for every $x \in F$. If not, there exist points $x_n \in F$ and $y_n \in K$ such that $d(x_n, y_n)$ tends to 0. Since $K$ is compact, one can (Bolzano-Weierstrass) assume that $y_n$ tends to a limit $b \in K \subset X$; it is clear that then the $x_n \in F$ tend to $b$. But since $F$ is closed in $X$ this implies $b \in F \cap K = \emptyset$, absurd.

This done, we put $f(x) = \varphi[d(x, K)]$ with a function $\varphi(t)$ defined and continuous for $t \geq 0$. For $f$ to satisfy the conditions (10) it is enough that $\varphi$ should satisfy the following conditions: (i) $\varphi(0) = 1$; (ii) $0 \leq \varphi(t) < 1$ for any $t > 0$; (iii) $\varphi(t) = 0$ for $t \geq r$. The existence of such functions is clear.

Now assume $X$ locally compact. Since $F$ is closed in $X$ for every $a \in X - F$ there exists an open ball $B_X(a)$ of $X$ [the set of $x \in X$ such that $d(a, x) < r$] such that the corresponding closed ball $\overline{B_X}(a)$ [the set of $x \in X$ such that $d(a, x) \leq r$] does not meet $F$. Since $X$ is locally compact one may assume $\overline{B_X}(a)$ compact by choosing $r$ small enough. Since $K$ is also compact one may (Borel-Lebesgue) cover it with a finite number of balls $B_X(a_p)$ [these are the intersections of $X$ with open balls of $\mathbb{C}$]. The union $U$ of these balls is open in $X$ and contained in the compact union $H$ of the $\overline{B_X}(a_p)$, which does not meet $F$. If one applies the lemma to $K$ and $X - U \supset X - H$ one obtains a function $f$ which satisfies (10) and is zero outside the compact subset $H$ of $X$, qed.

*Exercise* – Let $X$ be the (not locally compact) union of the open disc $|z| < 1$ and of the interval $[1, 2]$ of $\mathbb{R}$. Show that $f(1) = 0$ for every function $f$ continuous in $X$ and zero outside a compact $K \subset X$.

Examples of measures on a locally compact subset of $\mathbb{C}$ are not always as obvious as in $\mathbb{R}$. Certainly there are discrete measures like those of Example 4 of n° 31. It is easy to obtain measures analogous to those of Examples 1 and 2 if $X$ is *open*: choose an arbitrary continuous function $\rho$ on $X$ (for we do not yet know how to integrate anything else) and put

$$\mu(f) = \int\!\!\int f(x, y)\rho(x, y)dxdy$$

for every $f \in L(X)$; for every compact $K \subset X$ the set $X - K$ is open in $\mathbb{C}$ so that on agreeing to attribute to $f(x, y)\rho(x, y)$ the value 0 outside $X$ one

---

[60] This means that every limit *in* $X$ of points of $F$ is in $F$, or again, that $F$ is the intersection of $X$ and of a closed set in $\mathbb{C}$. This is the general concept of closed set in the metric space obtained by endowing the set $X$ with its usual distance.

defines a continuous function of compact support on $\mathbb{C}$; the integral $\mu(f)$ is then obtained by integrating over any compact rectangle $A \times B$ containing the support of $f$.

If $X = I \times J$, where $I$ and $J$ are arbitrary intervals of $\mathbb{R}$, and if one has measures $\mu$ and $\nu$ on $I$ and $J$, one can define a product measure

$$f \longmapsto \int d\mu(x) \int f(x,y)d\nu(y) = \int d\nu(y) \int f(x,y)d\mu(x)$$

as in the case where $I$ and $J$ are compact; to legitimise this construction one has to prove an analogue of Theorem 30, which is hardly necessary since $f$, being zero outside a compact subset of $I \times J$, is zero outside a rectangle $K \times H$ where $K \subset I$ and $H \subset J$ are compact[61]; the arguments of Theorem 30 are thus directly applicable here. Starting from these products of measures, you can choose a continuous function $\rho$ on $X = I \times J$ and consider the linear form

$$f \longmapsto \int\!\!\int f(x,y)\rho(x,y)d\mu(x)d\nu(y)$$

as we did for Lebesgue measure. For example, take $I = ]0, +\infty[$ open, $J = [0, +\infty[$ closed and

$$\mu(f) = \int f(x)dx/x, \qquad \nu(f) = \int f(x)x^{-1/2}dx;$$

on the square $I \times J \subset \mathbb{C}$, neither open nor closed in $\mathbb{C}$ but nevertheless locally compact, one obtains the measure

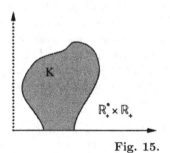

**Fig. 15.**

$$f \longmapsto \int\!\!\int f(x,y)dxdy/xy^{1/2}.$$

The integral is defined since the points of a compact subset of $I \times J$ cannot approach indefinitely close to the subset $x = 0$ of the frontier of $I \times J$ where

---

[61] Proof: the projections $(x,y) \mapsto x$ and $(x,y) \mapsto y$ are continuous maps of $\mathbb{C}$ into $\mathbb{R}$ and, in particular, of $I \times J$ into $I$ and $J$ respectively; they therefore transform every compact subset of $I \times J$ into compacta $K \subset I$ and $H \subset J$ (Chap. III, n° 9, Theorem 11), so that the given compact set is contained in $K \times H$.

the integration in $x$ would diverge. In particular, a compact subset remains separated from the point $(0,0)$, which does not belong to $I \times J$.

*Exercise* (less easy). Choose $X = \mathbb{C}$ and try to define a measure on $X$ by the formula

$$\mu(f) = \int\!\!\int f(x,y)\,(x^2 + y^2)^{-s}\,dxdy;$$

find the real values of $s$ for which this is meaningful.

*The integration of semicontinuous functions* in the case of a noncompact interval $X$, more generally of a locally compact subset of $\mathbb{C}$, proceeds more or less as in the compact case. Dini's Theorem shows that if an increasing sequence of continuous functions $f_n$ has a continuous function $f$ for its upper envelope, then convergence is uniform on every compact $K \subset X$. As in the case where $X$ is compact, one deduces that if $f$ and the $f_n$ are in $L(X)$ then $\mu(f) = \lim \mu(f_n) = \sup \mu(f_n)$ for every positive measure $\mu$ on $X$.

On replacing the $f_n$ by $f_n - f_1$, so replacing $f$ by $f - f_1$ and $\mu(f_n)$ by $\mu(f_n) - \mu(f_1)$, we may assume the $f_n$ positive. Since they are majorised by $f$ they are all zero outside a compact $K$ of $X$ independent of $n$. Thus

$$|\mu(f) - \mu(f_n)| \le M_K(\mu)\|f - f_n\|_X,$$

which yields the result, clearly applicable to increasing philtres as in n° 10.

To pass from this to the lsc functions we shall restrict ourselves to functions $\varphi$ which are *positive outside a compact* $K \subset X$ for otherwise one cannot hope that $L(X)$ contains an $f \le \varphi$. Being lsc, such a function is bounded below on the compact set $K = \{\varphi \le 0\}$ [n° 10, (vi)], whence the existence of a $f_0 \in L(X)$ majorised by[62] $\varphi$. If one writes $L_{\inf}(\varphi)$ for the set of $f \in L(X)$ satisfying $f \le \varphi$ one sees, as in n° 10, on considering $\varphi - f_0$ which is lsc and positive, that $\varphi$ is the upper envelope of the $f \in L_{\inf}(\varphi)$: in the case of an arbitrary locally compact $X \subset \mathbb{C}$, the lemma above, applied taking for $K$ the set $\{a\}$ and for $F$ the set of $x \in X$ such that $d(a,x) \ge r$, replaces the figure of n° 10. Then one puts

(31.11) $$\mu^*(\varphi) = \sup_{f \in L_{\inf}(\varphi)} \mu(f).$$

The crucial point, as in n° 10, is that *one may calculate $\mu^*(\varphi)$ using any increasing philtre* $\Phi \subset L_{\inf}(\varphi)$ *having $\varphi$ for upper envelope.* First, the sup of the $\mu(f)$ for $f \in \Phi$ is clearly $\le \mu^*(\varphi)$. Oppositely, for every $h \in L_{\inf}(\varphi)$ the set of functions $\inf(f,h)$, where $f \in \Phi$, is an increasing philtre (obvious) whose upper envelope is $h$ (obvious); the version of Dini's Theorem just obtained then shows that $\mu(h)$ is the upper bound of the integrals of the $\inf(f,h)$;

---

[62] Let $-m, m \ge 0$, be the minimum of $\varphi$ on $K$, i.e. on $X$. By the lemma above there exists an $f \ge 0$ in $L(X)$ which is equal to $m$ on $K$. The function $-f$ is the one we need.

since $\inf(f, h) \leq f$ we conclude that the upper bound of the $\mu(f)$, $f \in \Phi$, majorises $\mu(h)$ for every $h \in L_{\text{inf}}(\varphi)$, so majorises $\mu^*(\varphi)$, whence the result.

From here, the machine turns on its own and yields the three properties (i), (ii) and (iii) of n° 11. No need to reproduce the proofs: you replace $m$ by $\mu$ and $C_{\text{inf}}$ by $L_{\text{inf}}$.

One may also consider on $X$ the usc functions $\psi$ which are lower envelopes of continuous functions of compact support; this assumes these $\psi$ *negative outside a compact* (so zero outside a compact if they are positive everywhere), and since they are bounded above on it, the existence in $L(X)$ of functions which majorise it is obvious as in the other case. Writing $L_{\text{sup}}(\psi)$ for the set of $f \in L(X)$ which majorise $\psi$, one denotes by $\mu^*(\psi)$ the lower bound of the $\mu(f)$ when $f$ runs through $L_{\text{sup}}(\psi)$. Replacing $\psi$ by $-\psi$ would bring us back to the previous case.

These constructions apply in particular to the everywhere continuous functions on $X$. Such a function $\varphi$ is, in many ways, the difference of two positive continuous functions $\varphi'$ and $\varphi''$, for example $\varphi^+$ and $\varphi^-$, to which the definition of the integral of an lsc function applies. It is therefore natural to define $\mu(\varphi) = \mu(\varphi') - \mu(\varphi'')$, but the definition has no meaning if $\mu(\varphi') = \mu(\varphi'') = +\infty$, and depends *a priori* on the choice of $\varphi'$ and $\varphi''$. To eliminate the first objection one limits oneself to *integrable* continuous functions for $\mu$ (understood: absolutely) for which one may choose $\varphi'$ and $\varphi''$ to have finite integrals; since $0 \leq \varphi^+ \leq \varphi'$ we have the same for $\varphi^+$ and $\varphi^-$, so that $|\varphi| = \varphi^+ + \varphi^-$ is also integrable. The second objection is not one: for

$$\varphi'' - \varphi' = \psi'' - \psi' \Longrightarrow \varphi'' + \psi' = \psi'' + \varphi',$$

whence the result thanks to the additivity of the integral.

In the case of Lebesgue measure on an interval $X$ we recover the definitions of n° 22. We may restrict ourselves to the case of a positive continuous function $\varphi$. N° 22 defines convergence of the integral $\int \varphi(x)dx$ by insisting that the integrals on the compact sets $K \subset X$ be bounded above; here we assume that the integrals of the functions $f \in L(X)$ majorised by $\varphi$ are bounded above. Since each of these $f$ is zero outside a compact of $X$ it is clear that convergence in the sense of n° 22 implies convergence in the present sense. Moreover, it is clear, since $\varphi$ is continuous, that, for every compact interval $K \subset X$, there exists an $f \in L(X)$ equal to $\varphi$ on $K$ and $\leq \varphi$ everywhere elsewhere: multiply $\varphi$ by a continuous function of compact support with values in $[0, 1]$ and equal to 1 on $K$. Whence the implication in the reverse sense, and the equality of the integrals of $\varphi$ on $X$ defined by the two methods.

## 32 – The Stieltjes construction

Returning to the case of ℝ, let us take an *increasing* function $\mu(x)$ (in the wide sense) on $X = (a, b)$ and show how, in generalising the definition of the usual integral, one may associate with it a measure on $X$ which we shall again denote by $\mu$. One can show that in this way one obtains all the positive measures on $X$. The results of this n° are rarely useful and will not be used in this volume, so the reader may go directly on to the following n°; but the arguments brought into play are excellent exercises in analysis.

(i) *Definition of the measure.* Instead of defining $\mu(f)$ directly for every $f \in L(X)$ we shall first do so for the step functions which vanish outside a compact subset[63] of $X$. As in n° 1, this is equivalent to attributing a "measure" to each bounded interval $I = (u, v)$ such that $[u, v] \subset X$; this condition is superfluous if $X$ is compact, but if $X$ is open at one of its endpoints where the function $\mu(x)$ may tend to $+\infty$ or $-\infty$, it avoids infinite measures. This said, one puts

$$
\begin{aligned}
(32.1) \qquad \mu(]u, v[) &= \mu(v-) - \mu(u+), \\
\mu(]u, v]) &= \mu(v+) - \mu(u+), \\
\mu([u, v[) &= \mu(v-) - \mu(u-), \\
\mu([u, v]) &= \mu(v+) - \mu(u-),
\end{aligned}
$$

agreeing that $\mu(a-) = \mu(a)$ at the left endpoint of $X$ and $\mu(b+) = \mu(b)$ at the right endpoint when $X$ contains $a$ or $b$. Then $\mu[(u, v)] = \mu(v) - \mu(u)$ if the function $\mu$ is *continuous*, but, as one sees immediately, the definition chosen in the general case permits point masses at the points where $\mu$ is discontinuous: the measure of a singleton interval $[u, u]$ is equal to the jump

$$
(32.2) \qquad \mu(u+) - \mu(u-) = \Delta\mu(u)
$$

of the function $\mu$ at this point. One may note in passing that these formulae involve only the right and left limit values of $\mu$; one could for example assume $\mu$ continuous on the right, replacing $\mu(x)$ by $\mu(x+)$ for every $x$.

The main merit of these definitions is that, if one has a *partition* of $I = (u, v)$ into pairwise disjoint intervals $I_1, \ldots, I_n$, then

$$
(32.3) \qquad \mu(I) = \sum \mu(I_p).
$$

One may actually assume that $I_p = (x_p, x_{p+1})$ with weak inequalities

$$
u = x_1 \leq x_2 \leq \ldots \leq x_{n+1} = v
$$

to allow for the possibly singleton intervals. In calculating $\mu(I_1) + \mu(I_2)$, for example, two cases are possible; if $x_2$ belongs to $I_2$, then it does not belong to $I_1$, and the sum is equal to

$$
[\mu(x_2-) - \mu(u?)] + [\mu(x_3??) - \mu(x_2-)] = \mu(x_3??) - \mu(u?),
$$

---

[63] The reader may assume $X$ compact to start with, which simplifies the arguments a little.

where ? is the sign $-$ or $+$ as $I$ contains $x_1 = u$ or not, and where ?? is likewise $+$ or $-$ as $I_2$ contains $x_3$ or not. If, on the contrary, $x_2$ does not belong to $I_2$, and so belongs to $I_1$, one finds

$$[\mu(x_2+) - \mu(u??)] + [\mu(x_3??) - \mu(x_2+)] = \mu(x_3??) - \mu(u?).$$

On pursuing these small calculations step-by-step one finds finally that the right hand side of (3) is equal to $\mu(v??) - \mu(u?)$, i.e. to $\mu(I)$.

This being so, one defines the integral of a step function $\varphi$ by the obvious formula: one chooses intervals $I_p \subset X$, pairwise disjoint, finite in number, and with compact union $K \subset X$, such that $\varphi$ is constant on the $I_p$ and zero outside $K$, and puts

$$(32.4) \qquad \mu(\varphi) = \sum \varphi(\xi_p)\mu(I_p)$$

where $\xi_p \in I_p$ for every $p$. The additivity formula (3) guarantees as in n° 1 that the integral depends only on $\varphi$ and not on the partition chosen; it is then obvious that the map $\varphi \mapsto \mu(\varphi)$ is linear, that $\mu(\varphi) \geq 0$ for every $\varphi \geq 0$, and that it is continuous in the sense that if $\varphi$ is zero outside a compact interval $K \subset X$ then

$$(32.5) \qquad |\mu(\varphi)| \leq \mu(K)\|\varphi\|_X$$

since the $I_p$ are, or may be assumed to be all contained in $K$, so that $\sum \mu(I_p) \leq \mu(K)$.

With these conventions, one may construct a Riemann integration theory as in n° 2 of this chapter. To define $\mu(f)$ for an $f \in L(X, K)$ for instance, choose a sequence of step functions $\varphi_n$ which vanish outside $K$ and converge uniformly to $f$; then $\mu(f) = \lim \mu(\varphi_n)$ exists (Cauchy's criterion) and depends only on $f$.

To proceed further, let us consider a finite *partition* of $K$ into intervals $I_p$ sufficiently small for $f$ to be constant to within $r$ on each $I_p$ and let us choose an $x_p$ in each $I_p$. Then $\|f - \sum f(x_p)\chi_{I_p}\|_K \leq r$, whence

$$(32.6) \qquad |\mu(f) - \sum \mu(I_p)f(x_p)| \leq \mu(K)r.$$

From here, the fact that the map $f \mapsto \mu(f)$ of $L(X, K)$ into $\mathbb{C}$ satisfies Theorem 1 is too obvious to deserve yet another model proof.

(ii) *History.* The measures that we have just described were published in 1895 by the Dutchman Thomas Stieltjes (1856–1894), then professor at Toulouse, in a long memoir on certain analytic functions which he represents by a formula

$$f(z) = \int \frac{d\mu(t)}{t - z}$$

and also studies series and true integrals. In the Louis XIV typography[64] of the *Annales de la Faculté des Sciences de Toulouse* of the period, he devoted no more than two pages to explaining the construction of his integral with respect to a monotone function; although remarking on the analogy with a distribution of masses and the fact that the discontinuities of the function $\mu$ correspond to discrete masses, Stieltjes restricts himself to saying that, to integrate a function $f(x)$ over a bounded interval $(a, b)$, one considers a subdivision $a = x_1 < x_2 < \ldots < x_n = b$ (strict inequalities) of it, chooses points $\xi_p$ such that $x_p \leq \xi_p \leq x_{p+1}$ (weak inequalities) and one calculates the sum

$$\sum f(\xi_p) \left[\mu(x_{p+1}) - \mu(x_p)\right]$$

which, according to him, converges to the integral $\int f(x)d\mu(x)$ as the subdivision becomes ever finer; he proves nothing, refrains from detailing the rôle of the discontinuities of $\mu$, and says only that one argues as in the usual case, which is a little optimistic if one starts from his formula; after this he returns to his analytic functions.

This first generalisation of the Riemann integral provoked no interest, notably not even on the part of Lebesgue who passed over it in silence in his 1903 book, where he expounds his own work. But in 1909, a Hungarian mathematician, Frigyes Riesz (1880–1956), one of the creators of functional analysis, showed in a note in the *Comptes rendus de l'Académie des sciences de Paris* that, if $K$ is a compact interval, then every continuous linear form on $L(K)$ is a difference of Stieltjes integrals $f \mapsto \int f(x)d\mu(x)$ (i.e. is defined by a not necessarily positive measure $\mu$ [65]); at the same period, Hilbert, Hellinger and Toeplitz, who began to generalise the classical "diagonalisation" in finite dimensions to the linear operators on a Hilbert ... space, proclaimed the usefulness of Stieltjes integrals in their theory; we shall explain why in volume IV [Chap. XI, n° 22, (iv)]. As a result, Lebesgue remarked in the second edition of his book, rather roundaboutly, that his theory extends to the Stieltjes integrals, which was incontestable but a little delayed. In 1913 Radon generalised the Stieltjes integral to the case of several variables starting in $\mathbb{R}^n$ from a function $\mu(E)$ defined on reasonable subsets $E$ of the space and satisfying additivity conditions analogous to those of Lebesgue measure[66]; he

---

[64] Many French administrations still have a tendency to use it; it minimises the import of the text, i.e. the information made public. The comparison with official American documents, parliamentary reports for example, is edifying: several thousand pages of dense text every year for the discussion in committee of the defence budget, several dozen with wide margins in France.

[65] For a proof, see Walter Rudin, *Real and Complex Analysis* (McGraw Hill, 1966), Chap. 2.

[66] This is the method that one finds in Hans Grauert and Ingo Lieb, *Differential-und Integralrechnung III* (Springer, 1968), Chap. I. Playing with half open and semi closed parallelipipeds is not exactly relaxing. It is easier in Rudin, but since he first devotes a whole chapter to "abstract" measures which he never uses in his book, the energy expended in pure loss is about the same.

shows how to integrate with respect to such a measure. One could then easily liberate oneself from the hypothesis that $X \subset \mathbb{R}^n$ (Maurice Fréchet, 1915), after which the general theory would occupy a generation of mathematicians, not to say two[67], which vied to generalise them, even though the results are not always of great use, or put to work principles very different from those of Lebesgue or Radon.

(iii) *The increasing function associated with a discrete measure.* Consider the discrete measure (31.9) of the preceding n°, assuming that the $c(\xi)$, $\xi \in D$, are positive so as to obtain a positive measure. To obtain the increasing function $\mu(x)$ which, according to F. Riesz (there was also a Marcel Riesz, his brother, a first class analyst and great lover of spirits, at Lund, while Frigyes served his whole career in Hungary) defines it, let us choose once and for all a $c \in X$ (it is simplest to choose $c = a$ if $a \in X$) and put[68]

$$(32.7) \qquad \mu(x) = \begin{cases} \sum_{\xi \in [c,x]} c(\xi) & \text{if } c \leq x, \\ -\sum_{\xi \in ]x,c[} c(\xi) & \text{if } x < c. \end{cases}$$

The series is the partial sum of an unconditionally convergent series, so converges. It is an increasing function since, as $x$ increases, the sum (7) contains more and more positive terms if $x \geq c$ and fewer and fewer negative terms if $x < c$. It is even right continuous. For consider two points $x$ and $x + h > x$ and assume first that $c \leq x$; now $\mu(x + h) - \mu(x)$ is the sum of the masses $c(\xi)$ contained in the interval $]x, x + h]$; since the series $\sum c(\xi)$ converges unconditionally there is, for every $r > 0$, a finite subset $F$ of $D \cap ]x, x + h]$ such that the sum of the $c(\xi)$ for $\xi \notin F$ is $< r$; the elements of $F$ being finite in number, $]x, x + h]$ contains no element of $F$ if $h$ is small enough. The difference $\mu(x + h) - \mu(x)$ is thus $< r$ for $h > 0$ small enough, whence the right continuity of $\mu$ in this case. If $x < x + h < c$, the difference $\mu(x + h) - \mu(x)$ is again the sum of the masses contained in $]x, x + h]$ and the argument is the same. Note in passing the importance of not confusing the signs [ and ], the classical pitfall in this subject …

To calculate $\mu(x-)$, note that for $h > 0$, $\mu(x) - \mu(x - h)$ is the sum of the masses contained in $]x - h, x]$, a sum which always contains the mass $c(x)$, possibly zero, placed at the point $x$ and which, for $h$ small enough, is arbitrarily close as above. In other words, $\mu(x) - \mu(x-) = c(x)$, whence

$$(32.8) \qquad \mu(x+) = \mu(x) = \mu(x-) + c(x),$$

---

[67] The book of Jean-Paul Pier, *Histoire de l'intégration* (Masson, 1996) contains a very rich and useful bibliography, but cites numerous not always illuminating commentaries, especially when written in the "academic eulogy" style; no one ever needed Darboux to know that Riemann was a great mathematician (and, moreover, not on account of his integral …).

[68] Compare with the function (13.1) used in extending the FT to the primitives of regulated functions, in which the weak inequality is replaced by a strict inequality.

so that $\mu(x)$ has a discontinuity of amplitude $\Delta\mu(\xi) = c(\xi)$ at each point $\xi \in D$ and is continuous at the other points of $I$.

(Since the $\xi$ might well be all the rational points of $X$, this shows in passing that the increasing functions are not just those suggested by the usual naive sketches.)

This done, it remains to apply the definitions (1) and to check that one recovers the formula

$$(32.9) \qquad \mu(I) = \sum_{\xi \in I} c(\xi)$$

for every bounded interval $I = (u, v)$ such that $[u, v] \subset X$.

Next we have to check that the integral of a function $f \in L(X, K)$ is well defined by formula (31.9) of the preceding n°. For this, let us consider a partition of $K$ into intervals $I_p$ on which $f$ is constant to within $r$ and apply the approximation (6); the only intervals which count are those which contain points of $D$ and (9) shows that

$$\mu(f) = \sum_p f(x_p) \sum_{\xi \in I_p} c(\xi)$$

to within $\mu(K)r$. But for $\xi \in I_p$ one has $f(x_p) = f(\xi)$ to within $r$; on replacing $f(x_p)$ by $f(\xi)$ for all the $\xi \in I_p$ in the preceding formula, one commits, for each $p$, an error less than $r$ times the sum of the $c(\xi)$ for $\xi \in I_p$, so a total error less than $\mu(K)r$. Since $f$ is zero outside $K$ we have

$$(32.10) \qquad \mu(f) = \sum_{\xi \in D} c(\xi)f(\xi)$$

to within $2\mu(K)r$, whence the equality since $r > 0$ is arbitrary.

(iv) *The discrete and continuous components of a measure.* The sums (9) appear again in the general case of an arbitrary increasing function $\mu$ when one wants to make the rôle of the discontinuities of $\mu$ in calculating $\mu(f)$ more precise. Furthermore we shall see that as well as a "continuous sum", as in the classical case, $\mu(f)$ includes a "discrete sum", namely the sum

$$(32.11) \qquad \mu_d(f) = \sum \Delta\mu(\xi)f(\xi)$$

extended over all the discontinuities of the increasing function $\mu$; recall that $\Delta\mu(\xi) = \mu(\xi+) - \mu(\xi-)$.

First note that the set $D$ of these points of discontinuity is *countable*[69] and that

$$\sum_{\xi \in K} \Delta\mu(\xi) \leq \mu(K)$$

---

[69] This also results from the fact that a monotone function is regulated.

for every compact $K = [u, v] \subset X$. To see this, one first remarks that the partial sum extended over a finite subset $F$ of $D \cap K$ is $\leq \mu(K)$; for if one orders the points $\xi_1 < \ldots < \xi_n$ of $F$ one has

$$\mu(\xi_1 -) < \mu(\xi_1 +) \leq \ldots \leq \mu(\xi_n -) < \mu(\xi_n +);$$

the sum in question is thus majorised by $\mu(\xi_n +) - \mu(\xi_1 -)$ and so by $\mu(v+) - \mu(u-) = \mu(K)$ since $\mu(x)$ is increasing. For every integer $n \geq 1$, the $\xi \in K$ where $\mu(\xi) > 1/n$ are therefore finite in number, whence simultaneously the denumerability of $D \cap K$, so of $D$ since $X$ is the union of a sequence of compact sets; hence the desired inequality, which makes the series (11) absolutely convergent for every bounded function $f$ vanishing outside $K$, so for $f \in L(X, K)$.

Now the expression (11) reenters the preceding framework (iii). One is thus led to associate to the expression (11), a discrete measure on $X$, the corresponding increasing function (7), here with $c(\xi) = \Delta\mu(\xi)$. Now assume $\mu(x)$ *right continuous*, which we may, as we have seen above, and put

$$(32.12) \qquad \mu(x) = \mu_d(x) + \mu_c(x);$$

we thus define a *continuous* function $\mu_c$ since $\mu$ and $\mu_d$ have the same points of discontinuity $\xi \in D$, are both right continuous, and have the same "jumps" at these points. The function $\mu_c$ is moreover *increasing*. To see this, we have to check that, for $x \leq y$, we have $\mu(x) - \mu_d(x) \leq \mu(y) - \mu_d(y)$, i.e.

$$\mu_d(y) - \mu_d(x) \leq \mu(y) - \mu(x) = \mu(y+) - \mu(x+),$$

which, by (7), means that the sum of the jumps of the function $\mu$ at the points of the interval $]x, y]$ where it is discontinuous is less than its variation between $x$ and $y$, which is clear as we saw in proving (8).

This done, the functions $\mu(x)$, $\mu_d(x)$ and $\mu_c(x)$ define Radon measures on $X$ and it is obvious – use Riemann sums – that, for every continuous function $f \in L(X)$, one has

$$(32.13) \qquad \mu(f) = \mu_d(f) + \mu_c(f) = \mu_c(f) + \sum \Delta\mu(\xi)f(\xi).$$

Since the function $\mu_c(x)$ is continuous, the formulae (1) simplify:

$$(32.14) \qquad \mu_c(I) = \mu_c(v) - \mu_c(u) \qquad \text{if } I = (u, v),$$

whether $I$ is open, or closed, or ..., and it is vain, in the "Riemann sums" for $\mu_c$, to allow for the nonexisting discontinuities of $\mu_c(x)$. The measure $\mu_d$ provides the "discrete sum" and the measure $\mu_c$ the "continuous sum" to which we alluded above.

The ideal case is where the function $\mu_c(x)$ is of class $C^1$. The mean value theorem then shows that if $I = (u, v)$

$$\mu_c(I) = \mu'_c(w)(v - u) = \mu'_c(w)m(I)$$

for a $w \in I$, where $m(I)$ is Euclidean length. The Riemann sum $\sum \mu_c(I_p)f(x_p)$ relative to a sufficiently fine partition of $K$, and in which $x_p$ can be assumed equal to $w_p$, becomes $\sum f(w_p)\mu'_c(w_p)m(I_p)$; on decorating this calculation with the inevitable $\varepsilon$ and $\delta$ and calling $\mu$ what we would call $\mu_c$, one then finds that *for every increasing function $\mu$ of class $C^1$ (so continuous), one has*

(32.15)
$$\int f(x)d\mu(x) = \int f(x)\mu'(x)dx;$$

in 1697, but not in 1997, Leibniz would have said to you: obvious since $d\mu(x) = \mu'(x)dx$ ... But here like elsewhere, it is not the notation which makes the formula obvious; it the formula that explains the notation.

Consider for example on $X = ]0, +\infty[$ the measure

$$f \longmapsto \int f(x)dx/x$$

where one integrates over X. It corresponds to the monotone function $\mu(x) = \log x$: this is of class $C^1$ and has derivative $1/x$, whence the result by (15). For this measure, the measure of an interval $I = (u, v)$ with $0 < u < v < +\infty$ is $\mu(v) - \mu(u) = \log(v/u)$. One could extend this argument to any positive continuous function $p$ other than $1/x$: the increasing function defining the Radon measure

$$f \longmapsto \int f(x)p(x)dx$$

is any primitive $P(x)$ of $p$; the measure $P(v) - P(u)$ of an interval $I$ is the integral of $p$ extended over $I$.

*Exercise.* Prove (15) assuming that $\mu(x)$ is a primitive of a regulated function $\geq 0$.

Finally we remark that there are much more complicated monotone functions than the preceding, and for which the nondiscrete component of the corresponding integral $\int f(x)d\mu(x)$ is *not* of the form $\int f(x)p(x)dx$, even if you permit "densities" $p(x)$ that are Lebesgue integrable on every compact set ("singular" measures concentrated on sets of measure zero). The general theory of integration treats them exactly like the others.

## 33 – Application to double integrals

As we saw at the end of n° 9 and in n° 30, one may integrate any continuous function $f(x, y)$ on a rectangle $X \times Y$ with respect to Lebesgue measure $dxdy$ or, more generally, with respect to a product measure $d\mu(x)d\nu(y)$, where $X$ and $Y$ are compact intervals, and even if $X$ and $Y$ are noncompact, provided that $f \in L(X \times Y)$. But to integrate over an arbitrary compact set $K \subset X \times Y$ – in other words, and by definition, to integrate the function equal to $f$ on $K$

and to 0 elsewhere over $X \times Y$ – is not as simple, unless one imposes *ad hoc* hypotheses on $K$ as the physicists and engineers do, or, like a mathematician of the last century, declares that "the possibility of inverting the integrations rests on the obvious principle that a sum remains the same for any order in which one adds the parts[70]". The end of n° 9 has demonstrated the difficulty of the general problem.

The first problem is to *define* an integral over a compact set $K \subset X \times Y$, in other words to integrate the function $\varphi$ equal to $f$ on $K$ and to 0 at the other points of $X \times Y$. Most luckily, it is *upper semicontinuous* if $f$ is *positive*. For, consider a point $a \in X \times Y$. If $a \notin K$, then $\varphi(x) = 0$ on a neighbourhood of $a$ in $X \times Y$ since $K$ is closed, whence the continuity of $\varphi$ at $a$. At a point $a \in K$, since $f$ is continuous on $K$, for every $r > 0$ there exists an open disc $B(a)$ such that $\varphi(x) = f(x) < f(a) + r$ on $K \cap B(a)$; since $\varphi(x) = 0$ for $x \notin K$ and $f(a) \geq 0$, we have $\varphi(x) < f(a) + r = \varphi(a) + r$ for every $x \in B(a)$, whence the result. If $f$ is negative, the function $\varphi$ is on the contrary lsc since it is the negative of the usc function constructed starting from $-f$. If $f$ changes sign, catastrophe: we do not know how to integrate a function which is neither lsc nor usc. But one may always write $f = f^+ - f^-$ and deal with $f^+$ and $f^-$, for lack of a less crude method.

In these circumstances, as much to generalise and establish the standard result:

**Theorem 31.** *Let $X$ and $Y$ be two intervals, $\mu$ and $\nu$ positive measures on $X$ and $Y$, and $\lambda$ the product measure on $X \times Y$. Let $\varphi$ be an lsc (resp. usc) function on $X \times Y$ which is[71] the upper (resp. lower) envelope of the $f \in L(X \times Y)$ which minorise (resp. majorise) it. Then we have the following properties:*

*(i)   For every $x \in X$, the function $y \mapsto \varphi(x, y)$ is lsc (resp. usc);*
*(ii)  the function*

$$(33.1) \qquad x \longmapsto \int_Y \varphi(x, y) d\nu(y),$$

*with values $> -\infty$ (resp. $< +\infty$), is lsc (resp. usc);*

*(iii) we have*

$$(33.2) \qquad \lambda(\varphi) = \int_X d\mu(x) \int_Y \varphi(x, y) d\nu(y),$$

*the two members being simultaneously finite or infinite;*

---

[70] Joseph Bertrand in his *Traité de calcul différentiel et intégral* (1870), cited by Jean-Paul Pier, *Histoire de l'intégration* (Masson, 1997), p. 104. Bertrand taught analysis at the Ecole polytechnique from 1856 to 1894, which indicates the capacity for renewal of the institution at this period. There also, in parallel, were Charles Hermite (1869–1876) and Camille Jordan (1876–1911), who introduced some notions of set theory before 1900.

[71] a condition always satisfied if $X$ and $Y$ are compact or if $\varphi$ is positive (resp. negative), etc.

*(iv) we may invert the order of the integrations in (2).*

We shall examine the case of an usc function, the other being trivial to deduce from it: multiply the function by $-1$. As always, the two crucial points will be that (a) a lower envelope of continuous functions is usc; (b) one may calculate the integral of an usc function from any decreasing philtre of continuous functions having $\varphi$ as lower envelope. It remains to combine these tools with the *definitions*; there is nothing more to the very short proof.

That the function $y \mapsto \varphi(x, y) = \varphi_x(y)$ is usc on $Y$ for every $x \in X$ may be seen either by using the definition in terms of inequalities, or from the fact that $\varphi$ is the lower envelope of the set $\Phi = L_{\text{sup}}(\varphi)$ of the functions $f \in L(X \times Y)$ which majorise it.

One may therefore integrate $\varphi_x$, the result being $< +\infty$, though it can be $-\infty$ for an usc function. For $x$ given, it is clear that the set of functions $f_x(y) = f(x, y)$, where $f \in \Phi$, is a decreasing philtre of continuous functions on $Y$ whose lower envelope is $\varphi_x$. Thus

$$(33.3) \qquad\qquad \nu(\varphi_x) = \inf \nu(f_x).$$

Let us now put $F_f(x) = \nu(f_x) = \int f(x, y) d\nu(y)$ and $F_\varphi(x) = \nu(\varphi_x)$. Since the $f$ vary in a decreasing philtre, it is the same for the $F_f$, because

$$f \leq g \Longrightarrow f_x \leq g_x \Longrightarrow F_f(x) \leq F_g(x)$$

since $\nu$ is positive. Now the functions $F_f(x) = \nu(f_x)$ are continuous on $X$ because, $f$ being in $L(X \times Y)$, the $f_x$ are zero outside the same compact subset of $Y$, which allows us to argue as in n° 9, Theorem 9. Since

$$F_\varphi(x) = \nu(\varphi_x) = \inf \nu(f_x) = \inf F_f(x),$$

one concludes both that $F_\varphi$ is usc – point (ii) of the statement – and that

$$(33.4) \qquad\qquad \mu(F_\varphi) = \inf \mu(F_f).$$

But (4) can again be written

$$(33.5) \int d\mu(x) \int \varphi(x, y) d\nu(y) = \inf \int d\mu(x) \int f(x, y) d\nu(y) = \inf \lambda(f)$$

where the inf is relative to the $f \in \Phi$. Since, in (5), $f$ runs through the set of $f \in L(X \times Y)$ which majorise $\varphi$, the right hand side of (5) is, by definition of the integral of an usc function, equal to the integral $\lambda(\varphi)$ of $\varphi$ with respect to $\lambda$. The relation (5) therefore establishes (2). Point (iv) of the statement is then obvious, qed.

To complete the enunciation of Theorem 31, assume that the function $\varphi$ is lsc, positive and integrable on $X \times Y$, i.e., that $\lambda(\varphi) < +\infty$. By (2) we then have

$$\int_X d\mu(x) \int_Y \varphi(x,y) d\nu(y) < +\infty;$$

adopting the notation of the proof, this means that the function $F_f(x) = \int \varphi(x,y) d\nu(y)$, which is lsc and positive, has finite integral and so is integrable with respect to $\mu$. As we have seen (n° 11, *Exercise*) for the case of Lebesgue measure on a compact interval – but it generalises immediately – we can deduce that $F_\varphi$ is finite outside a set $N$ of $\mu$-measure zero, and therefore the function $y \longmapsto \varphi(x,y)$, which is lsc, is integrable with respect to $\nu$ for every $x \in X - N$.

In the case we started from at the beginning of this n°, where $\varphi$ is equal to a continuous positive function $f$ on a compact set $K \subset X \times Y$ and to 0 outside $K$, which ought, by definition, lead to the integral of $f$ on $K$, it is helpful to make (2) a little more explicit, where everything is now finite since $\varphi$ is bounded above *and* below. Now, to integrate $\varphi(x,y)$ with respect to $y$ for $x$ given is equivalent to integrating over $Y$ the function equal to $f(x,y)$ if $(x,y) \in K$ and to 0 elsewhere. If one denotes by $K(x)$ the compact set of the $y \in Y$ such that $(x,y) \in K$ – the "section" of $K$ by the vertical through the point $x$ – the number $\int \varphi(x,y) dy$ is then obtained by integrating over $Y$ the function equal to $f(x,y)$ for $y \in K(x)$ and to 0 elsewhere, and this is usc (same argument as in $X \times Y$). It is natural to denote this integral by

$$(33.6) \qquad \int_{K(x)} f(x,y) dy;$$

the Lebesgue-Fubini formula is then written

$$(33.7) \qquad \iint_K f(x,y) dx dy = \int_X dx \int_{K(x)} f(x,y) dy$$

according to tradition. But as we have already observed at the end of n° 9 the sets $K(x)$ can be arbitrary compact sets in $Y$, so that, as intuitive as it may seem, formula (7) masks an integration theory already much more advanced than that of Riemann. And we have had to assume $f$ positive to establish it! Of course, this restriction, which may seem ridiculous, will be eliminated by the complete Lebesgue theory (Chap. XI, § 4, n° 10).

# § 10. Schwartz distributions

### 34 – Definition and examples

We showed in the preceding § how one may introduce the concept of the integral, classical or not, into a particularly simple general framework: that of linear continuous forms on a vector space endowed with a "topology".

In analysis there is another operation bearing on functions and possessing analogous linearity and continuity properties: derivation. This is not defined for every continuous function but it is easy to introduce it into the present framework. Consider for simplicity a compact interval $K \subset \mathbb{R}$ and the vector space $C^1(K)$ of the functions of class $C^1$ in $K$, endpoints included. Choose measures $\mu$ and $\nu$ in $K$ and, for every $f \in C^1(K)$, put

$$(34.1) \qquad T(f) = \int_K f(x)d\mu(x) + \int_K f'(x)d\nu(x).$$

It is clear that $f \mapsto T(f)$ is linear and that

$$(34.2) \quad |T(f)| \leq M(\mu)\|f\|_K + M(\nu)\|f'\|_K \leq M(\|f\|_K + \|f'\|_K)$$

where $M$ is a constant independent of $f$, whence, for any $f, g \in C^1(K)$,

$$|T(f) - T(g)| \leq M(\|f - g\|_K + \|f' - g'\|_K).$$

If a sequence of functions $f_n \in C^1(K)$ converges uniformly to a limit $f$ *and if* the $f'_n$ converge uniformly to a limit $g$, in which case $f \in C^1(K)$ and $g = f'$ (Theorem 19 of Chap. III, n° 17), then $T(f) = \lim T(f_n)$.

These results can be interpreted on defining a norm on $C^1(K)$ by the formula

$$\|\|f\|\| = \|f\| + \|f'\|$$

where the norms on the right hand side are the uniform norms on $K$. The relation (2) shows that $T$ is a continuous linear form on $C^1$ for this norm. Theorem 19 of Chap. III, on the other hand, shows that Cauchy's criterion holds in $C^1(K)$: if indeed one has $\|\|f_p - f_q\|\| < r$ for $p$ and $q$ large, then also $\|f_p - f_q\| < r$ and $\|f'_p - f'_q\| < r$; consequently, the $f_n$ and their derivatives converge uniformly to functions $f$ and $g$ and it is clear, as above, that $f$ is the limit of the $f_n$ in the sense that $\lim \|\|f - f_n\|\| = 0$. In other words, $C^1(K)$ is a *complete* normed vector space – in short, a *Banach space*. It can be shown that there are no continuous linear forms on $C^1(K)$ other than the expressions of the form (1).

This kind of analogy between measures and derivations directly inspired Laurent Schwartz, the inventor of the theory of distributions. In the period when he created his theory one already knew of similar, but limited, attempts, due, for example, to the German Salomon Bochner who emigrated to the USA before 1940, and much more to the Soviet Sergei Sobolev[72] in his work on partial differential equations. But it was Schwartz, during and after W.W. II, who understood the enormous generality of the concept of distribution and formulated it in a perfectly clear way, placing it in the framework of the theory of topological vector spaces[73].

To obtain a satisfactory theory it is clearly necessary to allow for derivations of any order. One therefore has to replace the $C^1$ functions envisaged above by $C^\infty$ functions; when one works on $\mathbb{R}$, to which we confine ourselves in this §, we have, as in integration theory, to restrict ourselves to $C^\infty$ functions of compact support, i.e. zero for $|x|$ large: this is the vector space $\mathcal{D}$ or $\mathcal{D}(\mathbb{R})$ we have already used in n° 27 to "regularise" i.e. make $C^\infty$, with the help of convolution products, functions which are not. For Schwartz, a *distribution* on $\mathbb{R}$ is a linear form on $\mathcal{D}$, i.e. a map $T$ of $\mathcal{D}$ into $\mathbb{C}$ such that[74]

$$T(\alpha\varphi + \beta\psi) = \alpha T(\varphi) + \beta T(\psi)$$

and satisfying a *continuity* condition analogous to that imposed on measures in n° 31, but distinctly less obvious.

In the first place, one wants the measures to be particular distributions. This indicates that a distribution cannot have a continuity property unless one restricts to the vector subspace $\mathcal{D}(\mathbb{R}, K) = \mathcal{D}(K)$ of the functions which vanish outside a given compact subset $K$ of $\mathbb{R}$.

Since every $\varphi \in \mathcal{D}$, and all its successive derivatives, are continuous and so bounded (compact support), for every integer $r \geq 0$ one may define a norm

(34.3)     $$\|\varphi\|^{(r)} = \|\varphi\| + \|\varphi'\| + \ldots + \|\varphi^{(r)}\|$$

on $\mathcal{D}$, where, as in all the rest of this §, the notation

$$\|\varphi\| = \sup|\varphi(x)| = \|\varphi\|_{\mathbb{R}}$$

denotes the norm of uniform convergence on $\mathbb{R}$; definition (3) directly generalises the norm $\||\varphi\||$, which we introduced temporarily above, on $C^1(K)$. Clearly

$$\|\varphi + \psi\|^{(r)} \leq \|\varphi\|^{(r)} + \|\psi\|^{(r)},$$

so that the expression $d_r(\varphi, \psi) = \|\psi - \varphi\|^{(r)}$ satisfies the triangle inequality; to say that $d_r(\varphi, \psi) \leq \varepsilon$ implies that, for every $h \leq r$, one has $\left|\varphi^{(h)}(x) - \psi^{(h)}(x)\right| \leq \varepsilon$ for every $x \in \mathbb{R}$. Also

---

[72] S. Bochner *Vorlesungen über Fouriersche Integrale* (Leipzig, Akademie Verlagsgesellschaft, 1932), S. Sobolev *Méthode nouvelle à résoudre le probléme de Cauchy* ... (Mat. Sbornik, 1936, 1(43), pp. 39-71). According to a recent Russian book on the history of Soviet nuclear weapons, Sobolev was in charge of the mathematical and computational part of the project in 1943-1953. (Courtesy of Jean-Marie Kantor.)

[73] Schwartz describes his discovery in detail in Chapter 6 of his memoirs, *Un mathématicien aux prises avec le siècle* (Paris, Odile Jacob, 1997) (trans. *A Mathematician Grappling with his Century*, Birkhäuser, 2001).

[74] It has been standard usage since Schwartz to denote the elements of $\mathcal{D}$ by Greek letters and to denote "arbitrary" functions by Roman letters.

(34.4) $$\|\varphi\|^{(r)} \leq \|\varphi\|^{(r+1)}$$

for any $r$ and $\varphi$, and $\|c\varphi\|^{(r)} = |c|.\|\varphi\|^{(r)}$ for every constant $c \in \mathbb{C}$.

The introduction of these norms $\|\varphi\|^{(r)}$ on $\mathcal{D}$ allows one to verify the continuity of linear functions $T(\varphi)$ much more general than (1): choose $p+1$ measures $\mu_i$ $(0 \leq i \leq p)$ on $\mathbb{R}$ and put

(34.5) $$T(\varphi) = \sum \int \varphi^{(i)}(x)d\mu_i(x) = \sum \mu_i(\varphi^{(i)}).$$

Then
$$\|\varphi^{(i)}\| \leq \|\varphi\|^{(p)}$$

for every $i \leq p$ and consequently

(34.6)     $|T(\varphi)| \leq M_K(T)\|\varphi\|^{(p)}$     where $M_K(T) = \sum M_K(\mu_i)$

by definition (31.5) of a measure.

One might wonder whether it is possible to construct more sophisticated linear forms on $\mathcal{D}$, involving, maybe, infinitely many derivatives of $\varphi$. This is sometimes possible:
$$T(\varphi) = \sum \varphi^{(n)}(n),$$

the series taken over $\mathbb{N}$. There is no problem of convergence since, for a function zero outside a compact set, all the terms of large enough rank are zero; and in a given subspace $\mathcal{D}(K)$ the preceding formula is just a particular case of (5), with Dirac measures at the points $n$ situated in $K$. Another attempt: put
$$T(\varphi) = \sum c_n\varphi^{(n)}(0)$$

with the coefficients $c_n$ chosen to make the series converge for any $\varphi$. Now the derivatives at a point of a $\varphi \in \mathcal{D}$ can be chosen arbitrarily (n° 29); the series $\sum c_n a_n$ would therefore have to converge for any $a_n \in \mathbb{C}$, for example if $a_n = 1/c_n$; absurd.

These remarks and, of course, a now more extensive experience, show that, in a given subspace $\mathcal{D}(K)$, one should not hope to go further than expressions of the form (5); in fact, one of the first theorems proved by Schwartz – and very easy thanks to the theory of Banach spaces[75] – was that, in his theory, *every* distribution reduces, on a compact $K$, to the form (5), with measures $\mu_i$ depending on $K$.

This indicates that the continuity condition to impose on a distribution $T$ is the following: for every compact $K \subset \mathbb{R}$ there exist a $p \in \mathbb{N}$ and a constant $M_K(T)$ such that

---

[75] The corresponding theorem for distributions on $\mathbb{T}$ ($C^\infty$ functions of period 1) can be proved elementarily with the help of Fourier series, as we shall see in Chap. VII, end of n° 10.

(34.7)        $|T(\varphi)| \le M_K(T).\|\varphi\|^{(p)}$    for every $\varphi \in \mathcal{D}(K)$.

Why is this a "continuity" property ? Because it is natural to define the notion of convergence of a sequence on each $\mathcal{D}(K)$ in the following way:

a sequence of functions $\varphi_n \in \mathcal{D}(K)$ converges to a $\varphi \in \mathcal{D}(K)$ if for every $r \in \mathbb{N}$ the sequence of derivatives $\varphi_n^{(r)}$ converges uniformly to $\varphi^{(r)}$.

This definition is inspired by Theorem 19 of Chap. III, n° 17 and presents the advantage that, as in $C^1(K)$, there is an analogue of Cauchy's criterion for this concept of convergence: to verify that a sequence of functions $\varphi_n \in \mathcal{D}$ converges in the preceding sense it suffices (and is necessary) to check that

$$\left\|\varphi_p^{(r)} - \varphi_q^{(r)}\right\| \le \varepsilon \qquad \text{for } p, q \text{ large,}$$

for every $r \in \mathbb{N}$ and every $\varepsilon > 0$; the $\varphi_n$ and all their successive derivatives then converging uniformly to functions vanishing outside $K$, the theorem in question assures that the limit $\varphi$ of the $\varphi_n$ is $C^\infty$ and that $\varphi_n^{(r)}$ converges uniformly to $\varphi^{(r)}$ for any $r$; in other words, the sequence $\varphi_n$ converges to $\varphi$ in $\mathcal{D}(K)$ in the preceding sense.

This said, it is clear that this concept of convergence means that the uniform distance $\|\varphi^{(r)} - \varphi_n^{(r)}\|$ tends to 0 for any $r$; it is clearly equivalent to requiring

(34.8)        $$\lim \|\varphi - \varphi_n\|^{(r)} = 0 \qquad \text{for every } r.$$

The condition (4) imposed on the distributions then implies that, if a sequence $\varphi_n \in \mathcal{D}(K)$ converges to a limit $\varphi \in \mathcal{D}(K)$ in the preceding sense, one has

$$T(\varphi) = \lim T(\varphi_n).$$

Conversely, one may prove – it is not totally obvious – that this property forces the existence of a majorisation (7) for every compact $K \subset \mathbb{R}$.

One should pay attention to the fact that to formulate this continuity condition one must work in the vector subspaces $\mathcal{D}(K)$, failing which one would restrict considerably the definition of distributions as in integration theory. The distribution – in fact, a measure – $T(\varphi) = \sum \varphi(p)$, where one sums over $\mathbb{Z}$, provides a counterexample: take for $\varphi_n$ a $C^\infty$ function with values in $[0, 1/n]$, vanishing outside $[n-1, 2n]$, and equal to $1/n$ on $[n, 2n-1]$; the $\varphi_n$ converge to 0 uniformly on $\mathbb{R}$, but $T(\varphi_n) = 1$ for every $n$.

*Example 1.* Every absolutely integrable function $f$ on every compact interval of $\mathbb{R}$ (for example $\log|x|$ despite its singularity at the origin) defines a distribution[76] which is in fact a measure

---

[76] In all the rest of this chapter, the $\int$ sign will denote an integral extended over all $\mathbb{R}$.

$$T_f(\varphi) = \int \varphi(x)f(x)dx.$$

More generally, it is clear that every measure, restricted to $\mathcal{D}(\mathbb{R}) \subset L(\mathbb{R})$, is a distribution.

*Example 2.* Choose an $a \in \mathbb{R}$, an integer $k \in \mathbb{N}$ and put

$$T(\varphi) = \varphi^{(k)}(a).$$

For $k = 0$ one obtains the Dirac measure at the point $a$, denoted by $\delta_a$ or $\varepsilon_a$:

$$\delta_a(\varphi) = \varphi(a).$$

One could consider more generally distributions of the form

$$T(\varphi) = \sum c_n \varphi^{(k_n)}(a_n)$$

with a finite number of points $a_n$, given constants $c_n$, and arbitrary orders of differentiation $k_n$, for example

$$T(\varphi) = \varphi(0) + 3\varphi''(4) - \varphi'''(\pi).$$

One may even allow a countable infinity of points $a_k$ subject to a few precautions. A formula such as

$$T(\varphi) = \sum c_n \varphi^{(k)}(n),$$

where one sums over the $n \in \mathbb{Z}$, with a $k$ independent of $n$, defines a distribution because, for every compact $K$, the "series" in reality reduces to a finite sum for the $\varphi$ vanishing outside $K$. More subtly, let us choose an arbitrary sequence of points $a_n \in \mathbb{R}$, also $c_n$ such that $\sum |c_n| < +\infty$, and orders of differentiation $k_n$ *all less than the same integer $k$* and let us put

$$T(\varphi) = \sum c_n \varphi^{(k_n)}(a_n).$$

Since the derivatives involved are all of order $\leq k$, the general term of the series is, in modulus, less than $|c_n|.\|\varphi\|^{(k)}$, whence $|T(\varphi)| \leq M\|\varphi\|^{(k)}$ where $M = \sum |c_n|$, which proves continuity.

These elementary examples and the formula (5) are enough to show how *the theory of distributions allows one to unify the differential and integral calculus.*

*Exercise.* Given an interval $X \subset \mathbb{R}$ write $\mathcal{D}(X)$ for the set of functions defined on $X$, indefinitely differentiable (including at the endpoints of $X$ if they belong to $X$), and zero outside some compact subset of $X$. Find a reasonable definition for distributions on $X$. Find a distribution on $X =\,]0,1]$ which is not the restriction to the $\varphi \in \mathcal{D}(X)$ of a distribution on $\mathbb{R}$.

## 35 – Derivatives of a distribution

One of the conjuring tricks (it is nothing else, despite its usefulness, and it is already in Sobolev) which the concept of distribution allows is to *attribute derivatives to functions which do not have them*. To understand this, let us start from a function $f$ of class $C^1$ on $\mathbb{R}$ and consider the distribution

$$T_{f'}(\varphi) = \int \varphi(x) f'(x) dx$$

associated to its derivative. In view of the fact that $\varphi(x) = 0$ for $|x|$ large, an integration by parts shows that

$$T_{f'}(\varphi) = - \int \varphi'(x) f(x) dx.$$

If, as above, one associates the distribution

(35.1) $$T_f : \varphi \longmapsto \int \varphi(x) f(x) dx$$

to any $f \in C^1(\mathbb{R})$, then

(35.2) $$T_{f'}(\varphi) = -T_f(\varphi').$$

Starting from this, one *defines* the derivative $T'$ of a distribution $T$ by putting

(35.3) $$T'(\varphi) = -T(\varphi') \qquad \text{for every } \varphi \in \mathcal{D}.$$

Since obviously

$$\|\varphi'\|^{(r)} \leq \|\varphi\|^{(r+1)}$$

for every $r$, the continuity of $T$ propagates immediately to $T'$. One may then repeat this operation and define the successive derivatives of $T$, clearly given by

(35.4) $$T^{(p)}(\varphi) = (-1)^p T(\varphi^{(p)}).$$

If then you associate to each regulated function $f$ on $\mathbb{R}$ the distribution $T_f$ given by (1), you can define the "derivative" of $f$ ... But this is no longer a *function* in the usual sense – miracles exist even less in Nature than infinitely small numbers ... –, it is a *distribution* which can be fearfully complicated and which, for this reason, one refrains in general from calculating explicitly. The difficulty does not appear if $f$ is $C^\infty$; in this case, one may apply the traditional formula for integration by parts *ad libitum* and obtain the formula

(35.5) $$(T_f)^{(k)} = T_{f^{(k)}},$$

which shows that the definition of the successive derivatives of a distribution is compatible with that of the successive derivatives of a $C^\infty$ function.

*Example 1.* Take for $T$ the Dirac measure

$$T(\varphi) = \varphi(0)$$

at the origin. We find $T'(\varphi) = -\varphi'(0)$ in accordance with Dirac's baroque formula to which we alluded at the beginning of n° 27. One might, like Dirac himself, continue:

$$T''(\varphi) = +\varphi''(0), \qquad T'''(\varphi) = -\varphi'''(0), \quad \text{etc.}$$

One may ask why it was twenty years before a mathematician justified this "obvious" calculation. In Dirac's time, some theoretical physicists had begun to understand what was already called a *functional space*, i.e. a vector space of infinite dimension whose elements are, not the usual vectors in Euclidean or relativistic space or in the configuration space of a system of particles, but functions of one or several real variables; and in which one has a concept of "convergence" coming from a "norm". From 1930 on, it was clear that the probabilistic interpretation of the Schrödinger equation of quantum mechanics associates with every system of physical particles a square integrable function (in the sense of the Lebesgue theory) on a Cartesian space $E$ of finite dimension whose points correspond to all the possible configurations of the system; when one integrates the square of this function over a subset $M$ of the space of possible configurations one obtains the probability that the configuration of the system at the instant considered is one of those in $M$. This is what the mathematicians already called a Hilbert space, an infinite dimensional generalisation of Euclidean spaces endowed with a scalar product[77] (see the Appendix to Chap. III, n° 5, and the chapter on Fourier series). But the right functional space, which would have allowed one to understand Dirac's acrobatics, namely Schwartz' space $\mathcal{D}$, had not yet been invented, either because no one had thought of it, or, more probably, because no one

---

[77] A very important part of quantum mechanics was invented by physicists working either permanently or temporarily at Göttingen or nearby, or regularly appearing as participants at the international meetings that took place in Copenhagen, Cambridge, Münich, Hamburg, Zürich, but not in France – the "travelling seminar" of the physicists of the period, where many Americans of the Second World War learned their trade; see Donald Fleming and Bernard Bailyn, *The Intellectual Migration* (Harvard UP, 1969), Daniel J. Kevles, *The Physicists* (MIT Press, 1971) or Richard Rhodes, *The Making of the Atomic Bomb* (Simon & Schuster, 1986, 886 pp.), who explains the system and where you will find much other information. It happens that Hilbert and other well-informed mathematicians in the latest progress in "modern" mathematics were also professors at Göttingen or nearby; the famous *Methoden der Mathematischen Physik* of Courant and Hilbert appeared at this time, the "abstract" theory of Hilbert spaces was constructed in 1927–1930 by von Neumann at Göttingen and Hamburg, and was integrated into quantum mechanics in his *Grundlagen der Quantenmechanik* (Springer, 1929). Von Neumann was in the USA from 1933, like Richard Courant who founded an institute of applied mathematics at the University of New York, which, after 1945, prospered in the regular American way of the time: military contracts.

had had the idea of considering vector spaces in which convergence is defined
not by a single distance, as in the Hilbert or Banach spaces already known
(the second ones only by few mathematicians), but by infinitely many functions $d_r(f, g)$ as in Schwartz' space $\mathcal{D}$. To understand this kind of situation
it would have been necessary simultaneously to master integration theory,
partial differential equations in several variables to have really interesting
examples, the "abstract" algebra linear then in development, and the general theory of topological vector spaces where one is given in advance a family
of "distances" that determine convergence.

This was much too much for the physicists and even for the immense majority of the mathematicians of the time, the main exception being possibly
Sobolev in the 1930s. The theory of partial differential equations was in an almost total state of chaos. That of "locally convex topological vector spaces",
a natural extension of the work of Stephan Banach and of the Polish school
between the two Wars, was not invented until during the War by George
W. Mackey in the USA and, independently, by Dieudonné and Schwartz a
little later[78], then powerfully developed by Alexandre Grothendieck; so it
was constructed at the same time as the distributions were, and under their
influence. Afterwards Schwartz' theory spread everywhere, including in the
USSR where I. M. Gelfand and his school published several volumes on the
subject filled with examples, until it became a fundamental tool in the theory
of partial differential equations; see for example the formidable volumes of
Lars Hörmander, one of the principal proponents of the theory. Even more
extraordinary, the general theory of distributions itself, which brought the
first French Fields Medal to Schwartz in 1950, contained no truly "profound"
theorem – not, though, its applications – and required "only" the ability to
detect analogies between a dozen disparate domains and to isolate the general
principle which unified all. The philosophers of science call this a paradigm,
a new vision which not only puts order and clarity into chaos, but also and
above all allows one to pose new problems. Universal gravity, the analysis of
Newton and Leibniz, the atomic theory in chemistry, the theories of evolution of Darwin, of heredity of Mendel, the bacteria of Pasteur, relativity and
quantum mechanics, etc.

*Example 2.* Take for $f$ the function equal to 1 for $x > 0$ and to 0 for $x < 0$.
Then
$$T_f(\varphi) = \int_0^{+\infty} \varphi(x)dx,$$

whence

---

[78] In 1943–1945 one did not yet have the Internet and, in 1946, while my thesis was
almost complete, I discovered it, and in Russian, in a Soviet article of 1943 which
had just arrived in paris after an inexplicable delay. Being naturally curious, and
its authors being not entirely unknown to me, thanks to some work from before
1940, I had the good idea of reading it (i.e., then, of having it translated).

$$T_f'(\varphi) = -T_f(\varphi') = -\int_0^{+\infty} \varphi'(x)dx = \varphi(0)$$

since the primitive $\varphi$ of $\varphi'$ is zero for $x$ large. In other words, the derivative of the distribution associated to $f$ is the Dirac measure at the origin, and not a function in the usual sense. Obvious extension to the distributions defined by a step function; its derivatives are linear combinations of Dirac measures at the points where the function is discontinuous. For an arbitrary regulated function $f$, the uniform limit on every compact of a sequence of step functions $f_n$, one has

$$T_{f'}(\varphi) = -T_f(\varphi') = -\int f(x)\varphi'(x)dx = -\lim \int f_n(x)\varphi'(x)dx$$

since in fact one integrates over a compact interval, whence

$$T_{f'}(\varphi) = \lim T_{f_n'}(\varphi),$$

a result practically impossible to explain in the general case.

*Example 3.* Consider the distribution

$$T(\varphi) = \int_0^{+\infty} \varphi(x)\log x.dx;$$

the integral converges absolutely on a neighbourhood of 0 (no problem at infinity) since $|\varphi(x)\log x| \leq \|\varphi\|.|\log x|$. To calculate $T'$ one integrates naively by parts (if necessary passing to the limit on the interval $[u, +\infty[$ as $u \to 0+)$:

$$-T'(\varphi) = \int_0^{+\infty} \varphi'(x)\log x.dx = \varphi(x)\log x\Big|_0^{+\infty} - \int_0^{+\infty} \varphi(x)x^{-1}dx$$

and one obtains infinite expressions. This shows that it would be better to use a primitive of $\varphi$ which vanishes at the origin in order to neutralise the logarithm; simplest would be to choose $\varphi(x) - \varphi(0)$, which is $O(x)$, but then the difficulty reemerges at infinity because of the term $\varphi(0)\log x$ in the integrated-out part and of the term $\varphi(0)/x$ in the last integral[79]. To cut the Gordian knot one divides the integral in two; first

$$\int_0^1 \varphi'(x)\log x.dx = [\varphi(x) - \varphi(0)]\log x\Big|_0^1 - \int_0^1 [\varphi(x) - \varphi(0)]dx/x =$$

$$= -\int_0^1 [\varphi(x) - \varphi(0)]dx/x$$

since the integrated-out part is zero by $\varphi(x) - \varphi(0) = O(x)$ and $\log 1 = 0$. The integral obtained converges since $[\varphi(x) - \varphi(0)]/x$ is bounded on a neighbourhood of 0 (and even tends to a limit as $x \to 0$). On the other hand, without any problem of convergence,

---

[79] It is very rare to see Dieudonné deceive himself, but he does so in his *Eléments d'analyse*, Vol. 3, p. 247 à *propos* the same example.

$$\int_1^{+\infty} \varphi'(x)\log x.dx = \varphi(x)\log x\Big|_1^{+\infty} - \int_1^{+\infty}\varphi(x)dx/x = -\int_1^{+\infty}\varphi(x)dx/x$$

since $\varphi(x)$ is zero for $x$ large. Finally,

$$T'(\varphi) = \int_0^1 [\varphi(x) - \varphi(0)]dx/x + \int_1^{+\infty}\varphi(x)dx/x,$$

which shows that the derivative *in the sense of distributions* of the function $\log x$ is not what one might have believed. We recommend the reader to redo the calculation using an arbitrary point $a > 0$ as intermediate, and to verify that one finds the same result.

*Exercise.* For every $\varphi \in \mathcal{D}$ put

$$T(\varphi) = \lim_{\varepsilon \to 0} \int_{|x|>\varepsilon}\varphi(x)dx/x.$$

Show that the limit exists and that $T$ is a distribution. Calculate its derivative, and a "primitive".

# Appendix to Chapter V

# Introduction to the Lebesgue Theory

In Volume IV, Chapter XI we shall expound in detail the theory invented in about 1900 by Henri Lebesgue, which all mathematicians have used for a long time. One can, nevertheless, give a good idea of it in fifteen or so pages since, apart from Dini's Theorem, the arguments used are technically very simple: the usual operations of set theory (including the generalities of Chap. I on countable sets), very simple inequalities (the triangle inequality, the generalities of Chap. II, n° 17 on infinite limits), the definition and properties of upper bounds and of absolutely convergent series. The difficulty is not in proving the theorems; it is to present them in a logically coherent order, as in any theory when one wants to reach difficult theorems starting from almost nothing. We shall adopt the method perfected about fifty years ago by N. Bourbaki.

In what follows we shall write $X$ for the set on which we intend to develop the integration theory; $X$ is then a *locally compact* subset of $\mathbb{C}$, for example an interval of any kind in $\mathbb{R}$ or, in the general case, the intersection of an open and a closed set in $\mathbb{C}$. We shall write $L(X)$ for the set of complex continuous functions defined on X and zero outside a compact subset of X; a positive measure on X is thus, by definition, a linear map

$$\mu : L(X) \longrightarrow \mathbb{C}$$

such that $\mu(f) \geq 0$ for every $f \geq 0$. The most important case at our level is naturally that of the usual Lebesgue measure on an interval of $\mathbb{R}$, but to restrict oneself to this does not simplify anything in the proofs or their statements. Recall that for every function $f$ with complex values the symbol $|f|$ denotes the function $x \mapsto |f(x)|$.

(i) *Integration of lsc functions* . As we saw in n° 31, we can immediately define the upper integral $\mu^*(\varphi)$ of a real lsc function on $X$ so long as we assume $\varphi$ positive outside a compact $K \subset X$; we shall write $\mathscr{I}(X) = \mathscr{I}$ for the set of these functions. The upper integral

$$\mu^*(\varphi) = \sup_{f \leq \varphi, f \in L(X)} \mu(f)$$

is always defined for $\varphi \in \mathscr{I}$, and we have $-\infty < \mu^*(\varphi) \leq +\infty$. The whole Lebesgue theory can be constructed with the help of these functions and of their upper integrals. Their properties are strictly the same as in n° 11:

(L 1) *We have* $\mu^*(\varphi + \psi) = \mu^*(\varphi) + \mu^*(\psi)$ *for any* $\varphi, \psi \in \mathscr{I}$.

(L 2) *If* $\Phi \subset \mathscr{I}$ *is an increasing philtre*

$$(1) \qquad\qquad \mu^*(\sup \varphi) = \sup \mu^*(\varphi)$$

and in particular $\mu^*(\sup \varphi_n) = \sup \mu^*(\varphi_n)$ for every *increasing* sequence.

(L 3) *We have* $\mu^*(\sum \varphi_n) = \sum \mu^*(\varphi_n)$ *for any* positive $\varphi_n \in \mathscr{I}$.

(L 1) and (L 2) are proved as in n° 11, using Dini's theorem (n° 30).

(ii) *Measure of an open set.* For every open set $U$ in $X$ one puts

$$\mu^*(U) = \mu^*(\chi_U)$$

where the function $\chi_U$, equal to 1 on $U$ and to 0 on $X - U$, is lsc since $U$ is open in $X$. The statements (i'), (ii') and (iii') of n° 11 are still valid here because they are mere translations of (L 1), (L 2) and (L 3). Note that if $X$ is not compact then $\mu^*(U)$ can take the value $+\infty$.

(iii) *Upper integral of a positive function* Now consider a function $f$ on $X$, with values in $[0, +\infty]$. There exist functions $\varphi \in \mathscr{I}$ such that $\varphi \geq f$, for example the function everywhere equal to $+\infty$. So we define the *upper integral* of $f$ by putting

$$(2) \qquad\qquad \mu^*(f) = \inf_{\varphi \geq f} \mu^*(\varphi) \leq +\infty.$$

Despite the notation, this definition *is not* identical to that of n° 1: for the Dirichlet function one has $\mu^*(f) = 1$ in the Riemann theory, but $\mu^*(f) = 0$ in the Lebesgue theory as we shall see below. If $f$ is lsc and *a fortiori* continuous, the definition (2) provides the same value as (1) since $f$ appears among those $\varphi \in \mathscr{I}$ which majorise $f$. Another trivial point, always useful, is that

$$f \leq g \Longrightarrow \mu^*(f) \leq \mu^*(g).$$

For every set $A \subset X$ one similarly puts

$$(3) \qquad\qquad \mu^*(A) = \mu^*(\chi_A),$$

the upper integral of the characteristic function of A; as in the case of open sets, the properties of the measures of sets will be obtained immediately, at the end of this Appendix, by applying to their characteristic functions the statements valid for arbitrary functions. For the moment let us just note that

$$A \subset B \Longrightarrow \mu^*(A) \leq \mu^*(B).$$

For a function $f$ with complex (or even vector) values one puts[80]

$$(4) \qquad\qquad N_1(f) = \mu^*(|f|) \leq +\infty.$$

We always have

$$(5) \qquad\qquad N_1(f + g) \leq N_1(f) + N_1(g);$$

this is clear if one of the terms on the right hand side is infinite; in the opposite case, and by definition, for any $r > 0$ there exist lsc functions $\varphi$ and $\psi$ such that

$$|f| \leq \varphi, \quad \mu^*(\varphi) \leq \mu^*(|f|) + r, \quad |g| \leq \psi, \quad \mu^*(\psi) \leq \mu^*(|g|) + r;$$

then $|f + g| \leq \varphi + \psi$, whence, from (L 1),

$$\mu^*(|f + g|) \leq \mu^*(\varphi + \psi) = \mu^*(\varphi) + \mu^*(\psi) \leq \mu^*(|f|) + \mu^*(|g|) + 2r,$$

qed.

We also have

$$(6) \qquad\qquad N_1(\lambda f) = |\lambda| N_1(f)$$

for every scalar $\lambda$ so long as we agree, as everywhere in this context, that

$$0. + \infty = 0$$

since $N_1(0) = 0$. This convention must particularly be respected when calculating a product $fg$ of two functions: if for example we have $f(x) = 0$ and $g(x) = +\infty$ for an $x \in X$ we must agree that $fg$ is zero at the point $x$.

If one denotes by $\mathscr{F}^1(X; \mu) = \mathscr{F}^1$ the set of complex functions such that $N_1(f) < +\infty$ one obtains a vector space on which the function $N_1$ is a norm – up to one "detail": the relation $N_1(f) = 0$ does not imply $f = 0$.

The first important statement is the following:

(L 4) *If $f(x) = \sum f_n(x) \leq +\infty$ is the sum of a series of positive functions, then*

$$(7) \qquad\qquad \mu^*(f) \leq \sum \mu^*(f_n).$$

There is nothing to prove if the right hand side is infinite. If it is finite there exists, for $r > 0$ given, and for every $n$, a function $\varphi_n \in \mathscr{I}$ satisfying $f_n \leq \varphi_n$, $\mu^*(\varphi_n) \leq \mu^*(f_n) + r/2^n$: this is definition (2); the function $\varphi = \sum \varphi_n$ is lsc, it majorises $f$, and, from (L 3),

---

[80] $N_1(f)$ is often written $\|f\|_1$ in spite of the fact that $N_1(f) = 0$ does not imply $f = 0$. We shall also use the notation $N_1(f) = \mu^*(f)$ for functions with values in $[0, +\infty]$.

$$\mu^*(f) \le \mu^*(\varphi) = \sum \mu^*(\varphi_n) \le \sum [\mu^*(f_n) + r/2^n] = r + \sum \mu^*(f_n),$$

qed. Here we note the appearance of a "trick" apparently unknown before Borel and Lebesgue, despite its simplicity: to use the formula $r = \sum r/2^n$ to estimate the sum of a *series* in a controlled way. The only importance of the numbers $1/2^n$ is that their sum is 1. (7) implies

(7')    $$\mu^*(\bigcup A_n) \le \sum \mu^*(A_n)$$

for any sets $A_n \subset X$, because the characteristic function $\chi$ of the union of the $A_n$ is majorised by the sum of the characteristic functions $\chi_n$ of the $A_n$. Do not believe that you will obtain an equality if you assume that the $A_n$ are pairwise disjoint: for this you have also to assume that the $A_n$ are "measurable", as we shall see.

(iv) *Sets of measure zero.* The relation (15') suggests the notion of a set of measure zero or *negligible*, i.e., such that

(8)    $$\mu^*(N) = \mu^*(\chi_N) = 0.$$

This is equivalent to requiring that, for every $r > 0$, there exists an *open* set $U$ in $X$ such that

(8')    $$N \subset U \quad \& \quad \mu^*(U) \le r.$$

First of all, (8') implies (8) since $\mu^*(N) \le \mu^*(U)$. If, conversely, $N$ satisfies (8), there exists for every $r > 0$ a $\varphi \in \mathscr{I}$ such that $\varphi \ge \chi_N$ and $\mu^*(\varphi) \le r$; the relation $\varphi(x) > 1/2$ defines an open $U \supset N$ for which $\chi_U \le 2\varphi$, whence

$$\mu^*(U) = \mu^*(\chi_U) \le 2\mu^*(\varphi) \le 2r$$

and (8').

(L 5) *Every subset of a negligible set is negligible; the union of a finite or countable family of negligible sets is negligible.*

Use (3) and (7').

For the usual Lebesgue measure (8') shows that a singleton set is negligible. Then so is every *countable* set, for example the set of $\{x \in X\}$ with rational coordinates, even though this set is everywhere dense in X. But the converse is false. The most famous counterexample (see the end of this (viii)) is *Cantor's triadic set* in $X = [0,1]$ consisting of the $x \in X$ which can be written in base 3 enumeration without using the digit 1. For all that, one cannot deduce from this that every union of negligible sets is negligible: if such were the case, every set would be negligible, being the union of singleton sets! Countability is essential in (L 5).

When one has a relation $P\{x\}$ depending on a variable $x \in X$, for example $f(x) \ge g(x)$, one says that $P\{x\}$ is *true almost everywhere* (a.e.) if the set of $x$ such that $P\{x\}$ is not true is of measure zero. If one has a *finite or countable* family of statements $P_n\{x\}$ and if each of them, taken separately, is true almost everywhere, i.e., outside a negligible set $N_n$, then they are simultaneously true outside $N = \bigcup N_n$, so almost everywhere, from (L 5). For example, the sum of a series of functions that are zero almost everywhere is zero almost everywhere, similarly the product of two almost everywhere zero functions, their upper envelope, etc.

The set of complex-valued functions that are almost everywhere zero, or, as one says, *negligible,* is thus a vector subspace $\mathcal{N}$ of $\mathscr{F}^1$. One might prefer never to meet these functions, known to *exist* only in the sense of mathematical logic, but it is not generally the theory of integration which allows one to eliminate, even less to exhibit them. They serve essentially to camouflage the horrors "of no importance" since they do not count in calculating integrals: even in the simple Riemann theory one knows that if two regulated functions are equal outside a countable set their integrals are equal (n° 7, Theorem 7).

(L 6) *Let $f$ be a function with complex values; then*

$$(9) \qquad\qquad N_1(f) = 0 \Longleftrightarrow f(x) = 0 \text{ a.e.}$$

*If $f$ is a function with values in $[0, +\infty]$ then*

$$(10) \qquad\qquad N_1(f) < +\infty \Longrightarrow f(x) < +\infty \text{ a.e.}$$

Assuming $\mu^*(|f|) = 0$, for every integer $p \geq 1$ consider the set $N_p = \{|f(x)| > 1/p\}$ and write $\chi_p$ for its characteristic function. We have $|f| > \chi_p/p$, whence $\mu^*(N_p) = \mu^*(\chi_p) \leq p\mu^*(|f|) = 0$. Being the union of the $N_p$ the set $N = \{f(x) \neq 0\}$ is of measure zero, by (L 5). Assume conversely that $N$ is of measure zero, and now put

$$N_p = \{p < |f(x)| \leq p+1\} \subset N$$

for every $p \geq 0$, and let $\chi_p$ be the characteristic function of $N_p$. It is clear that $|f|\chi_p \leq (p+1)\chi_p$ whence $\mu^*(|f|\chi_p) \leq (p+1)\mu^*(\chi_p) = 0$ since $N_p$, contained in $N$, is of measure zero, by (L 5). Since $|f| = \sum |f|\chi_p$ we have $\mu^*(|f|) = 0$ by (L 4), whence (9).

To prove (10), put $A_p = \{f(x) > p\}$ for every $p \geq 1$ and again let $\chi_p$ be the characteristic function of $A_p$. We have $f > p\chi_p$, whence $\mu^*(\chi_p) \leq \mu^*(f)/p$. Since the set $N = \{f(x) = +\infty\}$ is the intersection of the $A_p$, we obtain $\mu^*(N) \leq \mu^*(f)/p$ for every $p$, whence $\mu^*(N) = 0$ if $\mu^*(f) < +\infty$, qed.

(L 6) shows that if two functions $f$ and $g$ are equal almost everywhere then $N_1(f) \leq N_1(g) + N_1(f-g) = N_1(g)$, by (9); by symmetry, one sees that

$$(11) \qquad\qquad f = g \text{ a.e.} \Longrightarrow N_1(f) = N_1(g).$$

The number $N_1(f)$ thus depends only on the *equivalence class* of $f$ modulo the vector subspace $\mathcal{N}$ of negligible functions; on writing $\tilde{f}$ for this class, i.e.

the *set* of all the functions almost everywhere equal to $f$ (Chap. I, end of n° 4), we put (but see footnote[80])

$$(12) \qquad ||\tilde{f}||_1 = N_1(f).$$

The relations (5) and (6) are valid for these classes, and as we have done what was strictly necessary for $||\tilde{f}||_1 = 0$ to imply $f(x) = 0$ a.e., i.e. $\tilde{f} = 0$, we see that the quotient space

$$F^1 = \mathscr{F}^1/\mathcal{N}$$

is a true *normed* vector space (Appendix to Chap. III, n° 5).

(v) *Integrable functions* . This done, a function $f$ with complex values will be said to be *integrable* on $X$ for the measure considered if, for every $r > 0$, there exists a *continuous* function $g \in L(X)$ such that

$$(13) \qquad N_1(f - g) = \mu^*(|f - g|) < r$$

or, equivalently, if there exists a sequence of continuous functions $f_n \in L(X)$ such that

$$(13') \qquad \lim N_1(f - f_n) = 0;$$

if so one defines the integral of $f$ by

$$(14) \qquad \mu(f) = \int f(x)d\mu(x) = \lim \mu(f_n).$$

As in n° 2 of § 1, this limit exists and does not depend on the sequence $(f_n)$ chosen. For

$$\begin{aligned} |\mu(f_p) - \mu(f_q)| &= |\mu(f_p - f_q)| \le \mu(|f_p - f_q|) = N_1(f_p - f_q) \\ &\le N_1(f_p - f) + N_1(f - f_q); \end{aligned}$$

the sequence $\mu(f_n)$ thus satisfies Cauchy's criterion. If another sequence $(g_n)$ of continuous functions satisfies (13'), then similarly

$$\begin{aligned} |\mu(f_n) - \mu(g_n)| &\le \mu(|f_n - g_n|) = N_1(f_n - g_n) \\ &\le N_1(f_n - f) + N_1(f - g_n), \end{aligned}$$

whence $\lim \mu(f_n) - \mu(g_n) = 0$, qed.

Integrable functions have properties which are almost trivial, and others which are less so. Let us start with the first.

It is immediately obvious that every function $f \in L(X)$ is integrable [take $g = f$ in (13)] and that its Lebesgue integral is equal to the number $\mu(f)$, the value at $f$ of the linear form $\mu : L(X) \longrightarrow \mathbb{C}$.

If $f$ and $g$ are equal almost everywhere and if $f$ is integrable, then $g$ is integrable and $\mu(f) = \mu(g)$ because, in (13'), $N_1(f - f_n)$ is unchanged if we replace $f$ by $g$. This allows us to say that a function $f$ defined outside a negligible set $N \subset X$ is *integrable* if any function $g$ everywhere defined and such that $f = g$ outside $N$ is so; one then defines $\mu(f) = \mu(g)$.

(13') shows further that $N_1(f) = \lim N_1(f_n) = \lim \mu(|f_n|)$. Now we know that $|\mu(f_n)| \leq \mu(|f_n|)$ since we are dealing with continuous functions (same proof as in n° 2, Theorem 1). Thus

$$|\mu(f)| = \lim |\mu(f_n)| \leq \lim \mu(|f_n|) = \lim N_1(f_n),$$

whence

(15) $$|\mu(f)| \leq N_1(f)$$

for every integrable function $f$.

(L 7) *If $f$ is integrable then so is $|f|$ and*

(16) $$N_1(f) = \mu(|f|) = \int |f(x)| d\mu(x).$$

To prove this we go back again to (13') and (14). Using the fact that $\left||u| - |v|\right| \leq |u - v|$ for any $u, v \in \mathbb{C}$ we obtain the inequality $\left||f| - |f_n|\right| \leq |f - f_n|$; this shows that $\lim N_1(\left||f| - |f_n|\right|) = 0$; since the functions $|f_n|$ are in $L(X)$, $|f|$ is integrable and we have $\mu(|f|) = \lim \mu(|f_n|)$ by definition of the integral. Now $N_1(f_n)$ converges to $N_1(f)$. Like (16), the definition of $N_1(f)$ for every *continuous* function $f$ applies to the $f_n$ so

$$N_1(f) = \lim N_1(f_n) = \lim \mu(|f_n|) = \mu(|f|),$$

qed.

(L 8) *If $f$ and $g$ are integrable then so are $\alpha f + \beta g$ for any $\alpha, \beta \in \mathbb{C}$ and*

(17) $$\mu(\alpha f + \beta g) = \alpha\mu(f) + \beta\mu(g).$$

Let $(f_n)$ and $(g_n)$ be sequences in $L(X)$ such that $\lim N_1(f - f_n) = \lim N_1(g - g_n) = 0$. The triangle inequality

$$N_1[(\alpha f + \beta g) - (\alpha f_n + \beta g_n)] \leq |\alpha| N_1(f - f_n) + |\beta| N_1(g - g_n)$$

proves the integrability of $\alpha f + \beta g$. Moreover, by the definition of the integral,

$$\mu(\alpha f + \beta g) = \lim \mu(\alpha f_n + \beta g_n) = \lim \alpha\mu(f_n) + \beta\mu(g_n),$$

whence (17).

(L 9) *Let $(f_n)$ be a sequence of integrable functions and $f$ a function such that $\lim N_1(f - f_n) = 0$. Then $f$ is integrable and*

(18)
$$\int f(x)d\mu(x) = \lim \int f_n(x)d\mu(x).$$

For every $r > 0$ we have $N_1(f - f_n) < r$ for all sufficiently large $n$; on the other hand, for every $n$ there exists a function $g_n \in L(X)$ such that $N_1(f_n - g_n) < 1/n$; for $n$ large we thus have $N_1(f - g_n) < r + 1/n < 2r$, whence the integrability of $f$. Moreover, from (L 8) and (15),

$$|\mu(f) - \mu(f_n)| = |\mu(f - f_n)| \le N_1(f - f_n),$$

whence (18).

(L 10) *If $f$ and $g$ are integrable and real-valued then* $\inf(f, g)$ *and* $\sup(f, g)$ *are integrable.*

Since we know that $f - g$ and $|f - g|$ are integrable, it is enough, as in n° 2, Theorem 2, to show that $f^+$ is integrable. To do this one uses (13') and the inequality $|f^+ - g^+| \le |f - g|$, valid for any real $f$ and $g$, qed.

(L 11) *A function $\varphi \in \mathscr{I}$ is integrable if and only if $\mu^*(\varphi) < +\infty$. If so, $\mu(\varphi) = \mu^*(\varphi)$.*

Whether $\varphi$ is or is not lsc, the condition $\mu^*(\varphi) < +\infty$ is necessary. If it is satisfied there exists for every $r > 0$ a continuous function $f \le \varphi$ such that

$$\mu(f) \le \mu^*(\varphi) \le \mu(f) + r;$$

since $\varphi = f + (\varphi - f) = f + |\varphi - f|$ and since $f$ and $\varphi - f$ are lsc since $f$ is continuous, we have $\mu^*(\varphi) = \mu(f) + \mu^*(|\varphi - f|)$ and so

$$N_1(\varphi - f) = \mu^*(|\varphi - f|) = \mu^*(\varphi) - \mu(f) \le r$$

from (L 1); whence the integrability of $\varphi$. The preceding relation also shows that

$$|\mu(\varphi) - \mu(f)| \le N_1(\varphi - f) = |\mu^*(\varphi) - \mu(f)| \le r,$$

whence $\mu(\varphi) = \mu^*(\varphi)$, qed.

(vi) *Convergence in mean: Cauchy's criterion.* We say that the functions $f_n$ *converge in mean* to a function $f$ when $\lim N_1(f - f_n) = 0$, i.e.

$$\lim \int |f(x) - f_n(x)|d\mu(x) = 0$$

from (L 7); sometimes one writes the preceding relation in the simpler form

$$f(x) = \text{l.i.m.} f_n(x),$$

the *limit in mean*.

We shall show that *Cauchy's criterion is valid for convergence in mean;* this is not so in the Riemann theory and this is one of the fundamental breakthroughs accomplished by Lebesgue (or rather by his immediate successors).

First we prove the following result:

(L 12) *Let $(f_n)$ be a series of integrable functions such that*

$$\sum \int |f_n(x)|d\mu(x) = \sum N_1(f_n) < +\infty.$$

*Then $\sum |f_n(x)| < +\infty$ a.e., every function $f$ such that $f(x) = \sum f_n(x)$ a.e. is integrable, and*

$$\lim N_1(f - f_1 - \ldots - f_n) = 0,$$
$$\int f(x)d\mu(x) = \sum \int f_n(x)d\mu(x),$$
$$\sum \left| \int f_n(x)d\mu(x) \right| < +\infty.$$

We put $F(x) = \sum |f_n(x)| \leq +\infty$; from (L 4) we have $N_1(F) \leq \sum N_1(f_n) < +\infty$, so $F(x) < +\infty$ a.e. from (L 6), so that the series $\sum f_n(x)$ converges absolutely almost everywhere. Let $f$ be a function almost everywhere equal to the sum of this series. Since $|f(x)| \leq F(x)$ a.e. we have

$$N_1(f) = \mu^*(|f|) \leq \mu^*(|F|) = N_1(F) \leq \sum N_1(f_n).$$

If we suppress the first $p$ terms of the given series we replace $f$ by a function equal almost everywhere to $f - (f_1 + \ldots + f_p)$, whence, by the same argument,

$$(19) \qquad N_1(f - f_1 - \ldots - f_p) \leq \sum_{q>p} N_1(f_q),$$

which is arbitrarily small for $p$ large. Since $f_1 + \ldots + f_p$ is integrable, so is $f$, by (L 9), and

$$(20) \qquad \mu(f) = \lim \mu(f_1 + \ldots + f_p) = \sum \mu(f_p).$$

The series is absolutely convergent since $|\mu(f_p)| \leq N_1(f_p)$, qed.

One should notice the difference between (L 12) and the elementary theorems on term-by-term integration of normally convergent series (n° 4): convergence (almost everywhere!) of the series $\sum |f_n(x)|$ is a *consequence*, and no longer a *cause*, of the relation $\sum m(|f_n|) < +\infty$.

We can now establish Cauchy's criterion for the convergence in mean.

(L 13) (Riesz-Fischer Theorem[81]) *If a sequence $(f_n)$ of integrable functions satisfies Cauchy's criterion for convergence in mean then it converges*

---

[81] This result shows that the *the normed vector space of classes of integrable functions is complete*, i.e. is a Banach space (Appendix to Chap. III, n° 5). Historically, a large part of the theory of Banach spaces has been motivated by integration theory.

*in mean to an integrable function f and one can extract a subsequence dominated by an integrable function and converging almost everywhere to f.*

Suppose that for every $r > 0$

$$N_1(f_s - f_t) < r \text{ for } s \text{ and } t \text{ large.}$$

We are to prove the existence of an integrable function $f$ such that $\lim N_1(f - f_n) = 0$. As in every metric space it is enough to show that from the given Cauchy sequence we can extract a subsequence which converges in mean: see the first proof of the usual Cauchy criterion in Chap. III, n° 10, Theorem 13.

For every $s \in \mathbb{N}$ denote by $n_s$ the *least* integer such that

$$k \geq n_s \ \& \ h \geq n_s \Longrightarrow N_1(f_k - f_h) \leq 1/2^s.$$

It is clear that $n_s \leq n_{s+1}$. So

$$(21) \qquad\qquad N_1(f_{n_{s+1}} - f_{n_s}) \leq 1/2^s \text{ for every } s.$$

For the differences

$$(22) \qquad\qquad g_s = f_{n_{s+1}} - f_{n_s},$$

which are integrable, by (L 8), we therefore have $\sum N_1(g_s) < +\infty$. By (L 12), the series $\sum g_n$ converges *absolutely* almost everywhere and its sum $g$ is also the limit *in mean* of its partial sums. But (22) shows that

$$(23) \qquad\qquad g_1(x) + \ldots + g_s(x) = f_{n_{s+1}}(x) - f_{n_1}(x).$$

Since the left hand side tends to $g(x)$ a.e., we deduce that

$$\lim f_{n_s}(x) = g(x) + f_{n_1}(x)$$

exists almost everywhere; if we denote by $f$ the function on the right hand side – its values at the points where the limit does not exist can be chosen arbitrarily –, we have

$$(24) \qquad\qquad g - g_1 - \ldots - g_s = f - f_{n_{s+1}} \text{ a.e.}$$

by (23), so

$$\lim N_1(f - f_{n_{s+1}}) = 0,$$

by (L 11). We have thus extracted from $f_n$ a subsequence which converges to $f$ both almost everywhere and in mean. It remains to show the existence of an integrable function $p \geq 0$ such that

$$|f_{n_s}(x)| \leq p(x)$$

for every $s$ and every $x$. But (24) shows that

$$|f_{n_{s+1}}| \leq |f| + |g| + \sum |g_n|$$

and we know that $f$, $g$ and the sum of the series $\sum |g_n|$ are integrable, qed.

The statement (L 13) applies in particular to every sequence $(f_n)$ which converges in mean since such a sequence trivially satisfies Cauchy's criterion. *It does not follow* that it converges almost everywhere: we know only that we can extract a subsequence converging a.e. But if, for some other reason, one also knows that $\lim f_n(x) = g(x)$ exists a.e. then $g$ is necessarily the limit in mean of $f_n$. Indeed, the limit in mean $f$ is actually the limit a.e. of a sequence $g_n$ extracted from the given sequence; if $\lim f_n(x) = g(x)$ outside a negligible set $N$ and $\lim g_n(x) = f(x)$ outside of negligible set $N'$ then $f(x) = g(x)$ outside $N \bigcup N'$. In other words:

(L 14) *If a sequence of integrable functions $f_n$ converges in mean to a function $f$ and almost everywhere to a function $g$, then $f = g$ almost everywhere.*

Since every integrable function $f$ is, by definition, the limit in mean of functions belonging to $L(X)$, the Riesz-Fischer theorem shows the existence of a sequence $(f_n)$ of *continuous* functions *of compact support* such that $f(x) = \lim f_n(x)$ almost everywhere; one can even assume the $f_n$ dominated by an integrable function.

*Exercise.* Let $f$ be an integrable real function. (i) Show that there are $f_n \in L(X)$ such that

$$\sum N_1(f_n) < +\infty, \sum f_n(x) = f(x) \text{ a.e.}$$

(ii) Put $\varphi'(x) = \sum f_n^+(x)$ and $\varphi''(x) = \sum f_n^-(x)$. Show that $\varphi'$ and $\varphi''$ are lsc, integrable, and that $f = \varphi' - \varphi''$ *almost everywhere* (compare with note 23 of n° 11).

(vii) *Lebesgue's grand theorem.* Theorem (L 19) below, without doubt the most useful of the whole theory, is the definitive version of the "dominated convergence" theorem of which we gave a pale unproven glimpse in n° 4.

(L 15) *Let $(f_n)$ be an increasing sequence of integrable real functions. For $f = \sup f_n = \lim f_n$ to be integrable it is necessary and sufficient that $\sup \mu(f_n) < +\infty$. If so*

$$\lim N_1(f - f_n) = 0, \quad \mu(f) = \lim \mu(f_n).$$

The necessity of the condition is clear since $f_n \leq f$ for every $n$. If it is satisfied the relation

$$N_1(f_q - f_p) = \mu(f_q - f_p) = \mu(f_q) - \mu(f_p),$$

valid for $p < q$ by (23), shows that $(f_n)$ is a Cauchy sequence for convergence in mean. It therefore converges in mean, necessarily to the function $f(x) = \lim f_n(x)$ by (L 14), qed.

(L 16) *Let $(f_n)$ be a countable family of integrable real functions; for the upper envelope $f(x) = \sup f_n(x)$ to be integrable it is necessary and sufficient that the $f_n$ be dominated by a function with finite upper integral.*

If $f$ is integrable, it is clear that the condition is satisfied. If conversely we have $f_n \leq p$ where $\mu^*(p) < +\infty$, the partial upper envelopes $g_n = \sup(f_1, ..., f_n)$, which are integrable by (L 10), form an increasing sequence whose integrals are all $\leq \mu^*(p)$; since $\sup(f_n) = \sup(g_n)$, we need only apply (L 15).

(L 17) *Every decreasing sequence $(f_n)$ of positive integrable functions converges in mean to the integrable function $f(x) = \lim f_n(x)$.*

Recall first that in $\mathbb{R}$ every decreasing sequence with positive terms converges (even when all the terms are equal to $+\infty$) and therefore satisfies Cauchy's criterion if its limit is finite. This said, for $p < q$ we have

$$N_1(f_q - f_p) = \mu(|f_q - f_p|) = \mu(f_q - f_p) = \mu(f_q) - \mu(f_p)$$

by (L 7) and (L 8). Now the sequence $\mu(f_n) \in \mathbb{R}$ is decreasing and has positive terms, hence converges; so $N_1(f_q - f_p) \leq r$ for $p$ and $q$ large. The function $f(x) = \lim f_n(x)$ is almost everywhere finite since $\mu^*(f) \leq \mu^*(f_n)$ for every $n$; it therefore converges almost everywhere to $f$; by (L 14) this is the limit in mean of $f_n$, qed.

(L 18) *The lower envelope of a countable family of positive integrable functions is integrable.*

Apply (L 17) to the functions $\inf(f_1, ..., f_n)$.

(L 19) (Lebesgue's dominated convergence theorem) *Let $(f_n)$ be a sequence of integrable functions which converges almost everywhere to a function $f$. Assume that there exists a positive function $p$ such that*

$$\mu^*(p) < +\infty \quad \& \quad |f_n(x)| \leq p(x) \text{ a.e. for every } n.$$

*Then $f$ is integrable and*

$$f(x) = \text{l.i.m. } f_n(x), \quad \int f(x)d\mu(x) = \lim \int f_n(x)d\mu(x).$$

We first show how to write the usual Cauchy criterion in a form adapted to the following proof.

If $(u_n)$ is a sequence of complex numbers, this says that, for every $r > 0$, there exists an integer $n$ such that $|u_i - u_j| \leq r$ for any $i, j \geq n$. By the definition of an upper bound this is equivalent to the relation

$$\sup_{i,j \geq n} |u_i - u_j| \leq r.$$

If then one puts

$$\sup_{i,j \geq n} |u_i - u_j| = v_n$$

for every $n$, one obtains a decreasing sequence of positive numbers and Cauchy's criterion is equivalent to the relation

$$\lim v_n = 0.$$

This established, let us prove (L 19). By (L 14) it is enough to show that $(f_n)$ is a Cauchy sequence for convergence in mean. The sequence of functions

$$g_n = \sup_{i,j \geq n} (|f_i - f_j|) \leq 2p$$

is decreasing. Since $N_1(|f_i - f_j|) \leq N_1(g_n)$ for any $i, j \geq n$, the proof reduces to showing that the $g_n$ converge in mean to 0. By Cauchy's criterion in $\mathbb{R}$, $g_n(x)$ converges to 0 at every $x$ where $\lim f_n(x)$ exists, so almost everywhere. It is then enough, by (L 14), to show that the $g_n$ converge in mean, which (L 17) guarantees so long as the $g_n$ are integrable. Now $g_n$ is the upper envelope of the countable family of the $|f_i - f_j|$ $(i, j \geq n)$, and these integrable functions are dominated by the function $2p$, whence the result by (L 16), qed.

(viii) *Integrable sets.* A set $A \subset X$ is said to be *integrable* if its characteristic function is integrable; then one puts $\mu(A) = \mu(\chi_A)$. If $A$ and $B \subset A$ are integrable then so is $A - B$, by (L 8), and

$$\mu(A - B) = \mu(A) - \mu(B).$$

If $A$ and $B$ are integrable so are $A \cap B$ and $A \cup B$, by (L 10). If $A$ and $B$ are equal up to a negligible set, and if one of them is integrable, then so is the other.

(L 20) *Every open or closed set $A$ such that $\mu^*(A) < +\infty$ is integrable.*

The first case follows from (L 11). The second follows from the first if one shows that there exists an open integrable $U \supset A$, since then $A = U - (U - A)$ where $U$ and $U - A$ are open and have finite outer measure, so are integrable. In fact, more generally,

$$\mu^*(A) = \inf_{A \subset U} \mu^*(U)$$

for *every* set $A$. This is clear if the left hand side is infinite. If it is finite, there exists for every $r > 0$ an lsc function $\varphi \geq \chi_A$ such that $\mu^*(A) \leq \mu^*(\varphi) \leq \mu^*(A) + r$; the open sets $U_n = \varphi(x) > 1 - 1/n$ contain $A$ for every $n$ and $\mu^*(\varphi) \geq (1 - 1/n)\mu^*(U_n)$, whence

$$(1 - 1/n)\mu^*(U_n) \leq \mu^*(A) + r$$

and therefore $\mu^*(U_n) \leq \mu^*(A) + 2r$ for $n$ large, qed.

In particular, the set $X$ itself is integrable if and only if $\mu^*(X) < +\infty$; since the function 1 is the upper envelope of $f \in L(X)$ such that $0 \leq f \leq 1$, and since $|f| \leq ||f||_1$ for every $f \in L(X)$, we then have

$$|\mu(f)| \leq \mu(X)||f||_X$$

for every $f \in L(X)$; the measure $\mu$ is therefore *bounded* or *of finite total mass*. If, conversely, we have a bound $|\mu(f)| \leq M||f||_X$ then it is clear that $\mu(1) \leq M$.

We also see (in the general case) that *every compact set is integrable*.

(L 21) *The intersection $A = \bigcap A_n$ of every countable family of integrable sets is integrable.*

The characteristic function $\chi_A$ of $A$ is of course the lower envelope of the functions $\chi_n$ of the $A_n$, whence the result by (L 18). If the sequence $(A_n)$ is decreasing, then, by (L 17),

$$\mu\left(\bigcap A_n\right) = \lim \mu(A_n) = \inf \mu(A_n).$$

(L 22) *For the union $A = \bigcup A_n$ of a countable family of integrable sets to be integrable it is necessary and sufficient that there exists a set $B$ such that*

(25)    $$\mu^*(B) < +\infty \quad \text{and} \quad A_n \subset B \text{ for every } n.$$

Necessity is clear: choose $B = \bigcup A_n$. If it is satisfied the characteristic functions of the $A_n$ are dominated by the function $\chi_B$, whence the result by (L 16).

Further,

$$\mu\left(\bigcup A_n\right) \leq \sum m(A_n)$$

by (3'), with equality if the $A_n$ are pairwise disjoint, by (L 12).

When the sequence $A_n$ is *increasing* (L 15) allows one to replace the condition (25) by $\sup \mu(A_n) < +\infty$; then

$$\mu(A) = \lim \mu(A_n) = \sup \mu(A_n).$$

Let us show for example that the Cantor set $C$ is of measure zero (for Lebesgue measure on $\mathbb{R}$). This set is constructed by removing from $[0,1]$ its middle interval $]1/3, 2/3[$, then from each of the two remaining intervals their middle interval, then from each of the four remaining intervals their middle interval, and and so on indefinitely. The total sum of the lengths of the excluded intervals is equal to

$$1/3 + 2/3^2 + 2^2/3^3 + ... = 1,$$

and since they are pairwise disjoint, we have $m([0,1] - C) = 1$, whence $m(C) = 0$.

(viii) *Measurable sets.* We say that a set $A \subset X$ is *measurable* if $A \cap K$ is integrable for every compact set $K \subset X$. It is clear that every integrable set is measurable, as well as $X$, or the complement of any measurable set, or the union and intersection of a countable family of measurable sets.

Every open or closed set $M$ is measurable; if $M$ is closed, $M \cap K$ is compact and so integrable for every compact $K \subset X$, so that $M$ is measurable. If $M$ is open then $X - M$ is closed, so measurable, hence $M$ is too.

These results allow us, starting from open sets, closed sets and from sets of measure zero, to construct extraordinarily complicated measurable sets: countable intersections of countable unions of countable intersections of open sets, for example. In fact, the difficulty is rather to construct *non*measurable sets explicitly, a practically impossible task without using transfinite induction as in Chap. I in some form or another. *In current practice* one has no chance of meeting nonmeasurable sets; even so this is no excuse for evading proofs of measurability ...

(L 23) *For a measurable set $A$ to be integrable it is necessary and sufficient that $\mu^*(A) < +\infty$.*

Assume we have proved that $X$ *is the union of a countable family of compact $K_n$*; since the $A \cap K_n$ are by hypothesis integrable, the set

$$A = \bigcup A \cap K_n$$

is then integrable by (L 22). It remains to prove the existence of $K_n$. It is obvious if $X$ is an interval in $\mathbb{R}$. If $X$ is an open subset of $\mathbb{C}$ let $D$ be the set, countable, of $x \in X$ with rational coordinates. For every $x \in X$ there exists an $n$ such that the closed disc $B(x, 1/n)$ is contained in $U$, then a $d \in D$ such that $|x - d| \le 1/2n$; the closed disc $D(d, 1/2n)$ is then contained in $B(x, 1/n) \subset X$ and contains $x$. This shows that $X$ is the union of a countable family of compact discs of the form $B(d, 1/p)$, whence the result. If finally $X = U \cap F$ with $U$ open and $F$ closed, and if $U = \bigcup K_n$, then $X = \bigcup K_n \cap F$, qed.

The preceding result shows that the notion of measurable set does not differ from that of an integrable set except when $\mu(X) = +\infty$.

To conclude, let us give a characterisation of integrable real functions which will lead us back to Lebesgue's original point of view:

(L 24) *Let $f$ be a real function such that $N_1(f) < +\infty$. For $f$ to be integrable it is necessary and sufficient that, for any $a$ and $b$ the set $\{a \le f(x) \le b\}$ should be measurable.*

To establish the necessity of the condition one chooses a sequence of functions $f_n \in L(X)$ and a negligible set $N$ such that $f(x) = \lim f_n(x)$ for every $x \in X - N$. Since

$$[a, b] = \bigcap ]a - 1/p, b + 1/p[,$$

the relation $a \le f(x) \le b$ means that for every $p$ one has

(26)    $a - 1/p < f_n(x) < b + 1/p$   for every sufficiently large $n$.

For $n$ and $p$ given (26) defines an open $U_{p,n}$ and the $x \in X - N$ such that (26) is satisfied for every $n \geq q$ are the points of the measurable set

$$U_{p,q} \cap U_{p,q+1} \cap \ldots = A_{p,q}.$$

For $p$ given the $x$ satisfying (26) are thus the elements of the measurable set

$$A_{p,1} \cup A_{p,2} \cup \ldots = B_p.$$

Finally, to say that (26) holds for every $p$ means that $x$ belongs to the measurable set $B = \bigcap B_p$. Since the set defined by the condition $a \leq f(x) \leq b$ is equal to $B$ to within a negligible set, it is measurable.

Suppose conversely that the set $\{a \leq f(x) \leq b\}$ is measurable for any $a$ and $b$, and assume first that $f \geq 0$. For $n, p \geq 1$ let us put

$$A_{n,p} = \left\{ \frac{p}{n} \leq f(x) < \frac{p+1}{n} \right\}.$$

Since

$$[a, b[ = \bigcup [a, b - 1/q]$$

the set $A_{n,p}$ is the union of a countable family of sets of the form $\{u \leq f(x) \leq v\}$, so is measurable. If $\chi_{n,p}$ is the characteristic function of $A_{n,p}$ then $\chi_{n,p} \leq nf(x)/p$; since we have assumed $N_1(f)$ finite the function $\chi_{n,p}$ is integrable by (L 23), and so also is the function

$$f_n(x) = \sum_{1 \leq p \leq n^2} \frac{p}{n} \chi_{n,p}.$$

We have $f_n(x) \leq f(x)$ for any $n$ and $x$, as well as

$$f(x) - f_n(x) \leq 1/n \quad \text{if } f(x) \leq n$$

as a figure, that of n° 30 for example, will show better than a calculation. It follows that $f(x) = \lim f_n(x)$ for every $x$, whence the integrability of $f^+$, by the dominated convergence theorem. Finally, if $f$ is not positive, apply (L 24) to $f^+$ and $f^-$, qed.

If one accepts that every reasonable set is measurable, it follows that in practice all the functions one meets in classical analysis are integrable, so long, of course, as $N_1(f) < +\infty$.

*Exercise.* Assume that $X$ is an interval $\mathbb{R}$, choose an $a \in X$, and, for every $x \in X$, define

$$\mu(x) = \begin{cases} \mu([a, x]) & \text{if } x \geq a \\ -\mu(]a, x[) & \text{if } x < a. \end{cases}$$

Show that $\mu(x)$ is increasing, right continuous, and that, for every $f \in L(X)$, $\mu(f)$ is the Stieltjes integral of $f$ with respect to the function $\mu(x)$.

# VI – Asymptotic Analysis

§ *1. Truncated expansions* – § *2. Summation formulae*

## § 1. Truncated expansions

### 1 – Comparison relations

Recall that in Chap. II, n° 3 and 4, we introduced relations which, given scalar functions defined on a set $X \subset \mathbb{R}$, allowed us to compare their "orders of magnitude" on a neighbourhood of a point $a$ adherent to $X$, the case $a = +\infty$ or $-\infty$ not excluded, quite the contrary. These are the following:

$$(1.1) \qquad f(x) = O(g(x)) \qquad \text{when } x \to a,$$

which is equivalent to the existence of a constant $M \geq 0$ such that $|f(x)| \leq M|g(x)|$ for every $x \in X$ near to $a$;

$$(1.2) \qquad f(x) \asymp g(x) \qquad \text{when } x \to a,$$

which means that simultaneously $f(x) = O(g(x))$ and $g(x) = O(f(x))$;

$$(1.3) \qquad f(x) \sim g(x) \qquad \text{when } x \to a,$$

equivalent to

$$\lim_{x \to a} f(x)/g(x) = 1;$$

and finally

$$(1.4) \qquad f(x) = o(g(x)) \qquad \text{when } x \to a,$$

which is equivalent to $\lim f(x)/g(x) = 0$.

We also saw that (3) can be expressed as

$$f(x) = g(x) + o(g(x))$$

since $|f(x)/g(x) - 1| \leq \varepsilon$ can be written as $|f(x) - g(x)| \leq \varepsilon|g(x)|$.

Chapter IV provided us with formulae of this type applicable to the power, exponential and logarithmic functions:

$$
\begin{array}{llll}
(1.5) & x^b & = & O(x^a), \ x \to 0 \quad \Longleftrightarrow \mathrm{Re}(b) \geq \mathrm{Re}(a), \\
(1.6) & x^b & = & o(x^a), \ x \to 0 \quad \Longleftrightarrow \mathrm{Re}(b) > \mathrm{Re}(a), \\
(1.7) & x^b & = & O(x^a), \ x \to +\infty \Longleftrightarrow \mathrm{Re}(b) \leq \mathrm{Re}(a), \\
(1.8) & x^b & = & o(x^a), \ x \to +\infty \Longleftrightarrow \mathrm{Re}(b) < \mathrm{Re}(a), \\
(1.9) & x^s & = & o(a^x), \ x \to +\infty \quad \text{if } a > 1, \ s \in \mathbb{C}, \\
(1.10) & \log x & = & o(x^s), \ x \to +\infty \quad \text{if } \mathrm{Re}(s) > 0, \\
(1.11) & \log x & = & o(1/x^s), \ x \to 0 \quad \text{if } \mathrm{Re}(s) > 0.
\end{array}
$$

It is useful to remember the three last formulae as:

$$
\begin{array}{lll}
\lim a^{-x} x^s & = & 0 \quad \text{when } x \to +\infty \text{ if } a > 1, s \in \mathbb{C}, \\
\lim x^s \log x & = & 0 \quad \text{when } x \to +\infty \text{ if } \mathrm{Re}(s) < 0, \\
\lim x^s \log x & = & 0 \quad \text{when } x \to 0 \text{ if } \mathrm{Re}(s) > 0.
\end{array}
$$

The theory of power series and Taylor's formula provide other general results. For a power series

$$ f(x) = a_m x^m + \ldots + a_p x^p + a_r x^r + \ldots $$

where we write only the nonzero terms, we have

$$(1.12) \qquad f(x) - (a_m x^m + \ldots + a_p x^p) \sim a_r x^r \qquad \text{when } x \to 0$$

since the left hand side can be written $a_r x^r(1 + ?x + \ldots)$ with a series which tends to 1; this leads to formulae such as

$$
\begin{array}{lll}
e^x & = & 1 + x + x^2/2 + x^3/6 + o(x^3), \\
\log(1 + x) & = & x - x^2/2 + x^3/3 - x^4/4 + O(x^5),
\end{array}
$$

etc. when $x \to 0$. As for Taylor's Formula, this shows that if a function $f$ is of class $C^{n+1}$ on a neighbourhood of a point $a \in \mathbb{R}$, then

$$(1.13) \quad f(a+h) = f(a) + f'(a)h + \ldots + f^{(n)}(a)h^n/n! + O\left(h^{n+1}\right)$$

as $h \to 0$ and even

$$(1.14) \qquad f(a+h) - \left[f(a) + f'(a)h + \ldots + f^{(n)}(a)h^n/n!\right] \sim$$
$$\sim f^{(n+1)}(a)h^{n+1}/(n+1)!$$

if $f^{(n+1)}(a) \neq 0$ [Chap. V, eqn. (18.11) and (18.14)].

## 2 – Rules of calculation

The symbols $O$ and $o$ obey rules of calculation which are easy to remember and also to prove, with the exception of those which apply to the quotients of asymptotic relations. We shall restrict ourselves to stating them in telegraphic style – there is little point in discussing them at length when we have already used them on several occasions in the preceding chapters – with minimal indications of their proofs.

$$(2.1) \qquad f = O(g) \ \& \ g = O(h) \Longrightarrow f = O(h).$$

For if $|f(x)| \leq A|g(x)|$ and if $|g(x)| \leq B|h(x)|$, then $|f(x)| \leq AB|h(x)|$.

$$(2.2) \qquad f = O(g) \ \& \ g = o(h) \Longrightarrow f = o(h).$$

For if $|f(x)| \leq A|g(x)|$ and if $|g(x)| \leq r|h(x)|$, then $|f(x)| \leq \varepsilon|h(x)|$ provided $r \leq \varepsilon/A$.

$$(2.3) \qquad f \ = \ O(h) \ \& \ g = O(h) \Longrightarrow f + g = O(h).$$
$$(2.4) \qquad f \ = \ o(h) \ \& \ g = o(h) \Longrightarrow f + g = o(h).$$
$$(2.5) \qquad f' \ = \ O(g') \ \& \ f'' = O(g'') \Longrightarrow f'f'' = O(g'g'').$$
$$(2.6) \qquad f' \ = \ O(g') \ \& \ f'' = o(g'') \Longrightarrow f'f'' = o(g'g'').$$

One might also write some of these rules in the following way, remembering the fact that a symbol such as $O(g)$ denotes any function $f$ such that $f = O(g)$:

$$O(h) + O(h) = O(h), \quad o(h) + o(h) = o(h),$$
$$O(o(h)) = o(O(h)) = o(h),$$
$$O(g)O(h) = O(gh), \quad O(g)o(h) = o(gh).$$

*Example 1.* Let us multiply term-by-term the relations

$$e^x = 1 + x + x^2/2 + O(x^3), \qquad \sin x = x - x^3/6 + O(x^5)$$

valid for $x \to 0$; calculating *à la* Newton one finds

$$\begin{aligned} e^x \sin x \ &= \ (1 + x + x^2/2)(x - x^3/6) + (1 + x + x^2/2)O(x^5) + \\ &\quad + (x - x^3/6)O(x^3) + O(x^3)O(x^5) = \\ &= \ x + x^2 + x^3/3 - x^4/6 - x^5/12 + O(x^4) + O(x^5) + \ldots + O(x^8); \end{aligned}$$

but as $x^n = O(x^4)$ for $n \geq 4$ we have

$$e^x \sin x = x + x^2 + x^3/3 + O(x^4);$$

one cannot derive anything more precise starting from these relations.

*Example 2.* When $x \to 0$,

$$\left(x^4 + x^2\right)^{1/3} = x^{2/3} \left(1 + x^2\right)^{1/3} = x^{2/3} \left[1 + x^2/3 - x^4/9 + O(x^6)\right]$$

by the binomial series, whence

$$\left(x^4 + x^2\right)^{1/3} = x^{2/3} + x^{8/3}/3 - x^{14/3}/9 + O(x^{20/3}).$$

In these calculations we have used the fact that $x^a O(x^b) = O(x^{a+b})$, a particular case of (5).

There are also rules concerning the relation $f \sim g$.

$$(2.7) \qquad\qquad f \sim g \ \& \ g \sim h \Longrightarrow f \sim h.$$

For $f = g + o(g) = h + o(h) + o(h + o(h)) = o(h) + o(O(h)) = h + o(h).$

$$(2.8) \qquad f' \sim g' \ \& \ f'' \sim g'' \Longrightarrow f'f'' \sim g'g'' \ \text{and} \ f'/f'' \sim g'/g''.$$

For $f'f''/g'g'' = (f'/g')(f''/g'')$, the product of two ratios tending to 1, etc. Or simply multiply the relations $f' = g' + o(g')$ and $f'' = g'' + o(g'')$.

*Example 3.* Consider the ratio

$$\frac{x^2 - x + \log x}{x^2 - (\log x)^2}$$

as $x$ tends to $+\infty$. In the numerator, $x$ and $\log x$ are $o(x^2)$, so it is $\sim x^2$. In the denominator, $\log x$ is $o(x)$, so $(\log x)^2$ is $o(x^2)$, so that the denominator also is $\sim x^2$. The fraction we are considering therefore tends to 1 as $x \to +\infty$.

As we have already noted elsewhere, at infinity a polynomial is equivalent to its term of highest degree; a rational fraction is therefore equivalent to the quotient of the terms of highest degree in its numerator and denominator.

## 3 – Truncated expansions

The preceding examples – and more so those which follow – show that to study the behaviour of a function on a neighbourhood of a point $a$, it is useful to compare it to functions as simple as possible. If for example the function is represented by a convergent power series in $x - a$, one compares it to the partial sums of the latter, i.e. to polynomials. In general, and even in the most elementary situations, it is necessary to choose comparison functions a little less simple.

Assume that one wishes to study the behaviour of a function $f$ near $x = 0$, or only when $x \to 0+$. It may happen that there are a constant $a \neq 0$ and a real exponent $s$ such that $f(x) \sim ax^s$. Then – by definition – $f(x) = ax^s + o(x^s)$, which encourages us to consider the difference $f(x) - ax^s$. It may happen that there are a constant $b \neq 0$ and a real exponent $t$ such that $f(x) - ax^s \sim bx^t$; necessarily $t > s$. It may then happen that there are a constant $c \neq 0$ and a real exponent $u > t$ such that $f(x) - ax^s - bx^t \sim cx^u$, and so on.

In this context we shall call a *generalised polynomial* any function of the form

(3.1) $$p(x) = a_1 x^{s_1} + \ldots + a_n x^{s_n}$$

where the $a_k$ are *nonzero* constants and where the real exponents $s_k$ satisfy

$$s_1 < s_2 < \ldots < s_n;$$

we shall then say that $f$ admits a *truncated expansion of order $s$* at the origin if there exists a generalised polynomial $p$ such that

(3.2) $$f(x) = p(x) + o(x^s) \qquad \text{when } x \to 0.$$

Since $x^t = o(x^s)$ for $t > s$, we may assume that, in (1),

(3.3) $$s_1 < s_2 < \ldots < s_n \leq s;$$

we then say that $p$ is the *principal part of order $s$* of $f$ at the origin.

It is unique, for (2) clearly implies

$$f(x) = a_1 x^{s_1} + o\left(x^{s_1}\right) \sim a_1 x^{s_1}$$

and so

$$a_1 = \lim f(x)/x^{s_1},$$

which determines $a_1$; the higher coefficients are obtained similarly from the relations

$$f(x) - a_1 x^{s_1} - \ldots - a_k x^{s_k} \sim a_{k+1} x^{s_{k+1}}.$$

It is clear that if one has two expansions $f(x) = p(x) + o(x^s)$ and $g(x) = q(x) + o(x^s)$ of the same order, then adding gives a truncated expansion of $f + g$. If the orders are different, naturally it is the smallest which is valid for the sum: from the relations

$$e^x = 1 + x + x^2/2 + x^3/6 + o(x^3), \quad \cos x = 1 - x^2/2 + x^4/24 + o(x^5)$$

one can deduce no more than

$$e^x + \cos x = 2 + x + x^3/6 + o(x^3)$$

since it *might be* that $e^x$ has a truncated expansion of order 5 containing nonzero terms in $x^4$ and $x^5 \ldots$

It is also easy to multiply truncated expansions term-by-term. An example will suffice to indicate the method. One writes

(3.4) $e^x \cos x = \left[1 + x + x^2/2 + x^3/6 + o(x^3)\right] \cdot \left[1 - x^2/2 + x^4/24 + o(x^5)\right]$

and multiplies mentally term-by-term; first one sees the terms of the form $ax^s$, then the terms of the form $ax^s o(x^t) = o(x^{s+t})$, and finally a term $o(x^s)o(x^t) = o(x^{s+t})$. Among the terms of the form $o(x^u)$ only that or those having the smallest exponent $u$ are retained since all the others are themselves $o(x^u)$; and among the terms of the form $ax^s$, only those of exponent $s \leq u$ are to be retained, for the same reason. In the case (4), obviously $u = 3$ because of the product $o(x^3).1$, so that

$$e^x \cos x = 1 + x - x^3/3 + o(x^3)$$

without needing to calculate any more terms. The fact that the exponents may be neither integers nor positive does not change the method at all, since it rests on the fact that $x^a = o(x^b)$ for $a > b$ when $x$ tends to 0.

## 4 – Truncated expansion of a quotient

Suppose we are given two truncated expansions

$$f(x) = p(x) + o(x^s), \qquad g(x) = q(x) + o(x^t)$$

on a neighbourhood of $x = 0$ and seek to deduce the most precise truncated expansion possible of $h(x) = f(x)/g(x)$. Let $bx^\alpha$ be the term of *lowest* degree in the generalised polynomial $q(x)$; then

$$h(x) = f_1(x)/g_1(x)$$

where the truncated expansion of $g_1(x) = b^{-1}x^{-\alpha}g(x)$ begins with the monomial 1 and is of order $t - \alpha$, and that of $f_1(x) = b^{-1}x^{-\alpha}f(x)$ is now of order $s - \alpha$. So we reduce to finding a truncated expansion of $1/g_1$ in the case where $g_1$ is of the form

(4.1)    $g_1(x) = 1 - r(x) + o(x^u), \qquad r(x) = b_2 x^{u_2} + \ldots + b_n x^{u_n},$

with nonzero coefficients and exponents satisfying

(4.2)                                $0 < u_2 < \ldots < u_n \leq u.$

On putting $g_2(x) = r(x) + o(x^u)$, whence $g_1 = 1 - g_2$, one has

(4.3)    $1/g_1(x) = 1 + g_2(x) + \ldots + g_2(x)^{N-1} + g_2(x)^N/g_1(x)$

for every integer $N > 0$. The function $g_2(x)^k$ is a sum of monomials whose degrees are of the form $k_2 u_2 + \ldots + k_n u_n$ with $\sum k_i = k$, $k_i \geq 0$, and of similar monomials where at least one of the factors is $o(x^u)$, so are themselves $o(x^u)$. Since $o(x^u)$ already figures in the second term $g_2(x)$ of (3) one cannot hope to deduce a truncated expansion of the left hand side of order $> u$ from (3). The last term of (3) is equivalent to its numerator since $g_1(x) \sim 1$; in the numerator, $g_2(x)$ is equivalent to the term of lowest degree of $r(x)$, whence

$$(4.4) \qquad g_2(x)^N / g_1(x) \sim b_2^N x^{Nu_2} = b_2^N x^{Nu_2} + o\left(x^{Nu_2}\right).$$

(3) thus yields a relation of the form

$$1/g_1(x) = 1 + c_2 x^{u_2} + \ldots + o(x^u) + b_2^N x^{Nu_2} + o\left(x^{Nu_2}\right),$$

where the inexplicit terms are of degrees $> u_2$.

If $Nu_2 < u$, the term $o(x^u)$ is negligible with respect to the second $o$ and we have only obtained a truncated expansion of order $Nu_2 < u$; if on the contrary $Nu_2 \geq u$, the last term of (4) is itself $o(x^u)$. Since we have no wish to spend our energy in vain, nor to lose information, we must choose for $N$ the *smallest integer* $\geq u/u_2$: going further adds only terms all negligible with respect to $x^u$ arising from the powers of $g_2(x)$, while not going so far diminishes the order of the expansion we obtain.

The method is general, but, to gain an understanding, it will be better to remember the principle and apply it to examples.

*Example 1.* We seek a truncated expansion of order 1 at $x = 0$ of the function $h(x) = e^x/x^2 \sin x$. Here $h(x) \sim x^{-3}$, so that a relation of the form $h(x) = p(x) + o(x)$ can be written as $x^3 h(x) = q(x) + o(x^4)$. We have to find a truncated expansion of order 4 for

$$x^3 h(x) = e^x/(\sin x/x),$$

so for

$$(4.5) \qquad e^x = 1 + x + x^2/2 + x^3/6 + x^4/24 + o(x^4),$$

for

$$(4.6) \qquad \sin x/x = 1 - x^2/3! + x^4/5! + o(x^4) = 1 - r(x) + o(x^4)$$

and for the reciprocal of (6). Since $r(x)$ is $\sim x^2$ up to a constant factor, its square is $\sim x^4$ and its cube $\sim x^6 = o(x^4)$. One may simply write

$$\begin{aligned} x/\sin x &= 1 + \left(x^2/3! - x^4/5!\right) + \left(x^2/3! - x^4/5!\right)^2 + o(x^4) = \\ &= 1 + x^2/3! + \left[(1/3!)^2 - 1/5!\right] x^4 + o(x^4) = \\ &= 1 + x^2/6 + 7x^4/360 + o(x^4). \end{aligned}$$

It remains to multiply by (5) and by $x^{-3}$, which gives

$$e^x/x^2 \sin x =$$
$$= x^{-3}\left(1 + x + x^2/2 + x^3/6 + x^4/24\right)\left(1 + x^2/6 + 7x^4/360\right) + o(x) =$$
$$(4.7) = 1/x^3 + 1/x^2 + 2/3x + 1/3 + x/5 + o(x).$$

This formula provides very precise information on the behaviour of $f(x) = e^x/x^2 \sin x$ as $x$ tends to 0. To a first approximation, $f(x) \sim 1/x^3$, which means that the *ratio* between the two members tends to 1. But their *difference* increases indefinitely, and in a precise way is $\sim 1/x^2$, which does not prevent the difference $f(x) - 1/x^3 - 1/x^2$ from growing indefinitely and, in fact from being $\sim 2/3x$; this time, the difference $f(x) - 1/x^3 - 1/x^2 - 2/3x$ tends to $1/3$, etc.

We remark, to conclude these generalities, that in practice one does not confine oneself to using the power functions $x^s$; frequently, particularly when examining the behaviour of a function "at infinity", one has to compare a given function with the functions $e^{sx}$ or $\log x$, $\log \log x$, etc, and more generally with functions of the form $e^{sx} x^t \log^n x$, mainly when having to determine the convergence of an integral on an interval unbounded to the right (Chap. V, n° 22). The idea is always to order these monomials by order of decreasing magnitude, so that, in a sum of monomials, each term is negligible with respect to the preceding term.

## 5 – Gauss' convergence criterion

The $u_{n+1}/u_n$ criterion allows us to determine whether many simple series converge, but, as we know, it is inconclusive if the ratio tends to 1. In his research on the hypergeometric series, C. F. Gauss obtained a very useful result for this case; the proof rests on simple, but ingenious, asymptotic evaluations.

**Gauss' convergence criterion.** *Let $\sum u_n$ be a series with positive terms and suppose that there exists a number $s$ such that*

$$(5.1) \qquad u_{n+1}/u_n = 1 - s/n + O(1/n^2).$$

*Then the series converges if $s > 1$ and diverges if $s \leq 1$.*

Note that in this case the d'Alembert ratio tends to 1. It is easy to remember the criterion: remember that for the series to converge it is preferable for the ratio not to tend to 1 too rapidly, in other words that $s$ it should be *greater* than a certain limit, namely 1.

First let us present two examples – we need them in the proof – of series for which we have a relation (1). For the series $v_n = 1/n^\alpha$

$$(5.2) \qquad v_{n+1}/v_n = (1 + 1/n)^{-\alpha} = 1 - \alpha/n + O(1/n^2)$$

by Newton's binomial formula. For $w_n = 1/n.\log n$ we have

$$w_{n+1}/w_n = [1 - 1/(n+1)] \log(n)/\log(n+1);$$

then

$$1 - \frac{1}{n+1} = 1 - \frac{1}{n}\frac{1}{1+1/n} = 1 - \frac{1}{n}(1 - 1/n + \ldots) = 1 - 1/n + O(1/n^2),$$

$$\frac{\log n}{\log(n+1)} = \frac{\log n}{\log n + \log(1+1/n)} = \frac{1}{1 + \log(1+1/n)/\log n} =$$

$$= 1 - \frac{\log(1+1/n)}{\log n} + O\left(\frac{\log(1+1/n)}{\log n}\right)^2 =$$

$$= 1 - \frac{1/n + O(1/n^2)}{\log n} + O\left(\frac{1/n + O(1/n^2)}{\log n}\right)^2 =$$

$$= 1 - 1/n.\log n + O(1/n^2),$$

whence

(5.3)        $$w_{n+1}/w_n = 1 - 1/n - 1/n.\log n + O(1/n^2).$$

Note that the term $1/n.\log n$ is $o(1/n)$ but not $O(1/n^2)$.

To establish Gauss' criterion we also need a

**Lemma.** *Let* $\sum u_n$ *and* $\sum v_n$ *be two series with positive terms; if*

(5.4)        $$u_{n+1}/u_n \le v_{n+1}/v_n \qquad \text{for } n \text{ large}$$

*and if the series* $\sum v_n$ *converges, then so does the series* $\sum u_n$.

We may assume that (4) holds for any $n$. On multiplying the first $n$ relations we obtain $u_n/u_1 \le v_n/v_1$, whence $u_n = O(v_n)$, qed.

Now we come to Gauss' criterion. If $s > 1$ there is an $\alpha$ such that $1 < \alpha < s$ and on comparing (1) and (2) we see that

$$v_{n+1}/v_n - u_{n+1}/u_n = (s-\alpha)/n + O(1/n^2) \sim (s-\alpha)/n,$$

so that the left hand side is $> 0$ for $n$ large. Since the series $v_n$ converges for $\alpha > 1$, so does the series $u_n$.

For $s < 1$, one chooses $\alpha$ between $s$ and $1$. The results are the opposite, so that if the series $\sum u_n$ were convergent, so would be the series $\sum v_n$, absurd for $\alpha < 1$.

If $s = 1$, the above comparison is unusable because one does not know the sign of an expression such as $O(1/n^2)$. But using (3) one has

$$u_{n+1}/u_n - w_{n+1}/w_n = 1/n.\log n + O(1/n^2) \sim 1/n.\log n,$$

a result $> 0$ for $n$ large. Since $\sum 1/n.\log n$ diverges (Chap. II, n° 12), so does $\sum u_n$, qed.

*Exercise.* Show that, for $a, b$ real, $b$ not a negative integer, the series

$$u_n = \frac{a^2(a+1)^2 \ldots (a+n)^2}{(n+1)^{1/2}b^2(b+1)^2 \ldots (b+n)^2}$$

converges if and only if $b - a > 1/4$.

## 6 – The hypergeometric series

The series

$$
\begin{aligned}
F(a,b,c;z) &= \\
(6.1) \quad &= 1 + \sum_{n=1}^{\infty} \frac{a(a+1)\ldots(a+n-1)\,b(b+1)\ldots(b+n-1)}{c(c+1)\ldots(c+n-1)} \frac{z^n}{n!} = \\
&= \sum a_n z^n,
\end{aligned}
$$

already found in Euler, plays a much more important rôle in mechanics, astronomy, physics, etc. than in mathematics proper, for the majority of the users' "special functions" are particular cases of it. Moreover, it is probably the first series whose convergence was studied correctly – by Gauss who, in 1813, showed that it converges for $|z| < 1$, diverges for $|z| > 1$ and, much less easy, examined what happens for $|z| = 1$.

First

$$
(6.2) \qquad |u_{n+1}/u_n| = |z|.|(a+n)(b+n)/(c+n)(n+1)|,
$$

an expression which tends to $|z|$, whence the radius of convergence. We shall therefore suppose $|z| = 1$ in what follows, and also that $a$, $b$, $c$ are *real*, to reduce the difficulties. Clearly we have to eliminate the case where one of these parameters is a negative integer since then the series reduces to a polynomial or is meaningless.

(i) We have

$$
\begin{aligned}
a_{n+1}/a_n &= \\
&= (a+n)(b+n)/(c+n)(n+1) = \frac{(1+a/n)(1+b/n)}{(1+c/n)(1+1/n)} = \\
&= \left(1 + \frac{a+b}{n} + \frac{ab}{n^2}\right)\left[1 - c/n + c^2/n^2 + O(1/n^3)\right] \cdot \\
&\quad \cdot \left[1 - 1/n + 1/n^2 + O(1/n^3)\right] = \\
(6.3) \quad &= 1 - \frac{c+1-a-b}{n} + \frac{ab - (c+1)(a+b) + c^2 + c + 1}{n^2} + O(1/n^3).
\end{aligned}
$$

Since $|u_{n+1}/u_n| = 1 - s/n + O(1/n^2)$ with $s = c+1-a-b$, we obtain a first result from Gauss' criterion:

$$
(6.4) \qquad c > a+b \quad \Longleftrightarrow \quad \text{absolute convergence for } |z| = 1.
$$

(ii) Suppose $s < 0$; then $1 - s/n > 1$ and since

$$
|u_{n+1}/u_n| \sim 1 - s/n,
$$

so is the left hand side for $n$ large, so the $u_n$ increase. Whence

(6.5) $\qquad\qquad c < a + b - 1 \implies$ diverges for $|z| = 1$.

(iii) It remains to examine the interval $a + b - 1 \le c \le a + b$, on which $0 \le s \le 1$ and where the series does not converge absolutely. First assume $s > 0$ and write

$$a_{n+1}/a_n = 1 - v_n \qquad \text{with } v_n \sim s/n,$$

whence

$$a_{n+1} = a_p(1 - v_p)\ldots(1 - v_n)$$

for $n > p$. For $n$ large, $v_n$ is $> 0$ like $s$ and tends to 0, so is $< 1$, so that for $p$ well chosen, the product $(1 - v_p)\ldots(1 - v_n)$ is positive, decreases when $n$ increases and so tends to a limit. Since $\log(1 - v_n) \sim -v_n \sim -s/n$, the series with negative terms $\sum \log(1 - v_n)$ diverges, which shows that $\lim(1 - v_p)\ldots(1 - v_n) = 0$ (see the similar arguments on infinite products in Chap. IV, n° 17). Consequently, $a_n$ tends to 0, decreasing, up to a factor $a_p$.

For $s > 0$, the hypergeometric series $\sum a_n z^n$ is then decidable by Dirichlet's theorem (Chap. III, n° 11, Theorem 15 or Corollary 1). Consequently

(6.6) $\quad a + b - 1 < c \le a + b \implies \begin{cases} \text{convergence for } |z| = 1,\ z \neq 1, \\ \text{divergence for } z = 1. \end{cases}$

Divergence for $z = 1$ follows from the fact that the series $\sum a_n$ has negative terms for $n$ large, so we may apply Gauss' criterion here with $s \le 1$.

(iv) If $s = 0$, then, by (3),

(6.7) $\qquad\qquad a_{n+1}/a_n = 1 + k/n^2 + O(1/n^3),$

with

$$k = ab - (c + 1)(a + b) + c^2 + c + 1 = (a - 1)(b - 1)$$

since $c = a + b - 1$. We have to distinguish three cases.

If $k > 0$, the left hand side of (7), which is $\sim 1 + k/n^2$, is $> 1$ for $n$ large. Consequently, the series $\sum a_n z^n$ diverges for $|z| = 1$.

If $k = 0$, i.e. if $a = 1$ (in which case $c = b$) or if $b = 1$ (in which case $c = a$), it is clear that the series reduces to $\sum z^n$, so diverges for $|z| = 1$.

If $k < 0$ we have $|u_{n+1}/u_n| = 1 + v_n$ where $v_n \sim k/n^2$ is the general term of a convergent series all of whose terms are negative for $n$ large; the infinite product of the $1 + v_n$ is then absolutely convergent, from which it follows that $|u_n|$ tends to a *nonzero* limit (Chap. IV, n° 17, Theorem 13), which prevents the series from converging.

To sum up, for $a$, $b$, $c$ real, we obtain the following table, which the reader is not asked to memorise:

$$a + b < c \qquad \text{absolute convergence for } |z| = 1$$
$$a + b - 1 < c \le a + b \qquad \text{convergence for } |z| = 1,\ z \neq 1$$
$$c \le a + b - 1 \qquad \text{divergence for } |z| = 1.$$

We said above that the hypergeometric series includes many important series as particular cases. The first is the binomial series

$$(1+z)^s \;=\; \sum s(s-1)\ldots(s-n+1)\,z^{[n]} =$$
$$=\; \sum -s(-s+1)\ldots(-s+n-1)(-z)^{[n]} =$$
$$=\; F(-s,1,1;-z).$$

For $s$ *real*, we have complete results as to the behaviour of the series for $|z| = 1$:

$$
\begin{aligned}
s > 0 \qquad & \text{absolute convergence on } |z| = 1, \\
-1 < s \le 0 \qquad & \text{convergence for } |z| = 1,\; z \ne 1, \\
s \le -1 \qquad & \text{divergence for } |z| = 1.
\end{aligned}
$$

## 7 – Asymptotic study of the equation $xe^x = t$

In this n° we shall detail a most ingenious exercise[1] whose principal interest, at our level, is to make full use of the $O$ and $o$ techniques; as we said above, *in this domain* it is much less useful to learn the general theorems than to perform practical work.

The problem is to study the behaviour as $t \to +\infty$ of the root $x$ of the equation

(7.1) $$xe^x = t.$$

The method consists of first obtaining a very crude estimate of the order of magnitude of $x$ as a function of $t$, then to substitute the result in (1) to derive a second more precise estimate, then to substitute the second result in (1) to derive a third estimate more precise than the second, and so on.

First we show that *for every $t \ge 0$, the equation (1) has a unique solution $x \ge 0$ and that this is a continuous function of $t$.* For $x \ge 0$, the map $x \mapsto xe^x$ is continuous and strictly increasing, zero for $x = 0$ and it increases indefinitely with $x$: it therefore maps $\mathbb{R}_+$ onto $\mathbb{R}_+$ and has a continuous inverse map, whence the existence, uniqueness and continuity of $x$ as a function of $t$.

Since $x = 1$ for $t = e$, we see that

$$t > e \Longrightarrow x > 1 \Longrightarrow e^x < xe^x = t \Longrightarrow x < \log t,$$

whence, for $t$ large ($t > e$), $0 < \log x < \log\log t$ and thus

$$\log x = O(\log\log t).$$

---

[1] taken from N.G. of Bruijn, *Asymptotic Methods in Analysis* (Gröningen, Nordhoof, 1960). We also advise reading Chap. III of the *Calcul infinitésimal* of J. Dieudonné (Paris, Hermann, 1968), mainly n° 8.

But $xe^x = t$ implies $x = \log t - \log x$ whence

(7.2) $$x = \log t + O(\log \log t).$$

Since $\log \log t = o(\log t)$, we have $x \sim \log t$ at infinity.

Putting $\log t = y$, we have $x = y + O(\log y)$, a result that we can substitute into the equation (1), put in the form $x = \log t - \log x$. To do this we have to evaluate

$$\log x = \log[y + O(\log y)] = \log\{y[1 + O(\log y/y)]\} = \log y + \log[1 + O(\log y/y)]$$

and since in general $\log(1 + z) \sim z = O(z)$ as $z \to 0$, we have

$$\log x = \log y + O(\log y/y).$$

Since $x = \log t - \log x$ and $y = \log t$, we now obtain

(7.3) $$x = \log t - \log \log t + O(\log \log t/ \log t),$$

a more precise result than (2).

Now we substitute (3) in $x = \log t - \log x$, all the work being to deduce information on $\log x$ from (3). Again putting $y = \log t$ we have $x = y - \log y + O(\log y/y)$, i.e.

$$x = y(1 + z) \qquad \text{with } z = -\log y/y + O(\log y/y^2).$$

Since $z$ tends to 0 we have

$$
\begin{aligned}
\log x &= \log y + z - z^2/2 + O(z^3) = \\
&= \log y - \log y/y + O\left(\log y/y^2\right) - \\
&\quad - \frac{1}{2}\left[-\log y/y + O\left(\log y/y^2\right)\right]^2 + O(z^3) = \\
&= \log y - \log y/y + O\left(y^{-2}\log y\right) - \\
&\quad - \frac{1}{2}y^{-2}\log^2 y + O\left(y^{-3}\log^2 y\right) + O\left(y^{-4}\log^2 y\right) + O(z^3).
\end{aligned}
$$

Since $z \sim -y^{-1}\log y$ we have $O(z^3) = O\left(y^{-3}\log^3 y\right) = y^{-2}\log y . O\left(y^{-1}\log^2 y\right)$, a term negligible with respect to the term $O\left(y^{-2}\log y\right)$ figuring in the result like the other terms in $O$. Thus in actual fact we have

$$\log x = \log y - \log y/y - \frac{1}{2}y^{-2}\log^2 y + O\left(y^{-2}\log y\right),$$

a relation in which each term is negligible with respect to the preceding. The relation $x = \log t - \log x$ therefore leads to

(7.4) $$
\begin{aligned}
x &= \log t - \log \log t + \log \log t/ \log t + \\
&\quad + \frac{1}{2}\left(\log \log t/ \log t\right)^2 + O\left(\log \log t/ \log^2 t\right),
\end{aligned}
$$

an estimate again more precise than (3).

If he continues one stage further the courageous reader will find that

$$
\begin{aligned}
x = {} & \log t - \log\log t + \frac{\log\log t}{\log t} + \frac{1}{2}\left(\frac{\log\log t}{\log t}\right)^2 - \\
& - \frac{\log\log t}{\log^2 t} + \frac{1}{3}\left(\frac{\log\log t}{\log t}\right)^3 - \frac{3}{2}\frac{(\log\log t)^2}{\log^3 t} + O\left(\frac{\log\log t}{\log^3 t}\right).
\end{aligned}
$$

## 8 – Asymptotics of the roots of $\sin x . \log x = 1$

(de Bruijn, p. 33, exercise 1). On examining the graphs of the functions $\sin x$ and $1/\log x$ one sees immediately that for every $n > 1$ the equation has two roots between $2n\pi$ and $(2n + 1)\pi$; one of them, $x_n$, lies between $2n\pi$ and $2n\pi + \pi/2$, the other, $y_n$, between $2n\pi + \pi/2$ and $2n\pi + \pi$. Let us examine, for example, the behaviour of $x_n$.

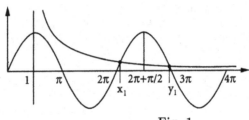

Fig. 1.

Since it is geometrically clear that $x_n \sim 2\pi n$, let us put

(8.1)                         $x_n = 2\pi n + u_n = 2\pi n(1 + v_n)$

with $0 < u_n < \pi/2$, $0 < v_n < 1$ and

(8.2)   $\log x_n = \log(2\pi n) + \log(1 + v_n)$,         $\sin x_n = \sin u_n = 1/\log x_n$.

Now

$1/\log(2\pi n). \sin u_n = \log(x_n)/\log(2\pi n) = 1 + \log(1+v_n)/\log(2\pi n) = 1 + w_n$.

Since $1 + v_n < 2$, the third member tends to 1, thus also the first, which shows that

(8.3)                         $u_n \sim \sin u_n \sim 1/\log(2\pi n)$,

then that $u_n \sim 1/\log(2\pi n)$ tends to 0 and that

(8.4)                         $v_n = u_n/2\pi n \sim 1/2\pi n. \log(2\pi n)$.

Hence

$$w_n = \log(1 + v_n)/\log(2\pi n) = \left[v_n + O\left(v_n^2\right)\right]/\log(2\pi n).$$

Since $w_n$ tends to 0 we have

$$\log(2\pi n).\sin u_n =$$
$$= 1/(1 + w_n) = 1 - w_n + O\left(w_n^2\right)$$
$$= 1 - v_n/\log(2\pi n) + O\left(v_n^2\right)/\log(2\pi n) + O\left(v_n^2\right)/\log^2(2\pi n) =$$
$$= 1 - v_n/\log(2\pi n) + O\left(v_n^2\right)/\log(2\pi n),$$

whence

(8.5)     $\sin u_n = 1/\log(2\pi n) - v_n/\log^2(2\pi n) + O\left(v_n^2\right)/\log^2(2\pi n).$

But $\sin u_n = u_n + O\left(u_n^3\right) = u_n + O\left(1/\log^3(2\pi n)\right)$ by (3), whence

$$\begin{aligned}
u_n &= 1/\log(2\pi n) - v_n/\log^2(2\pi n) + \\
&\quad + O\left(v_n^2\right)/\log^2(2\pi n) + O\left(1/\log^3(2\pi n)\right) = \\
&= 1/\log(2\pi n) + O\left(1/n\log^3(2\pi n)\right) + \\
&\quad + O\left(1/n^2\log^4(2\pi n)\right) + O\left(1/\log^3(2\pi n)\right).
\end{aligned}$$

On the right hand side the second and the third terms are negligible with respect to last, which therefore dominates. At the point where we are now we cannot say anything more precise than

(8.6)          $u_n = 1/\log(2\pi n) + O\left(1/\log^3(2\pi n)\right),$

which nevertheless improves on (3).

To get further we write that $\sin u_n = u_n - u_n^3/6 + O\left(u_n^5\right)$. Using (6) gives

$$\begin{aligned}
\sin u_n &= u_n - \left[1/\log(2\pi n) + O\left(1/\log^3(2\pi n)\right)\right]^3/6 + O\left(1/\log^5(2\pi n)\right) = \\
&= u_n - \left[1/\log^3(2\pi n) + O\left(1/\log^5(2\pi n)\right)\right]/6 + O\left(1/\log^5(2\pi n)\right),
\end{aligned}$$

whence, by (5),

$$\begin{aligned}
u_n &= 1/\log(2\pi n) - v_n/\log^2(2\pi n) + O\left(v_n^2\right)/\log^2(2\pi n) + \\
&\quad + 1/6\log^3(2\pi n) + O\left(1/\log^5(2\pi n)\right) + O\left(1/\log^5(2\pi n)\right);
\end{aligned}$$

since $v_n = O\left(1/n.\log(2\pi n)\right)$ by (4), the terms containing $v_n$ are negligible with respect to $O\left(1/\log^5(2\pi n)\right)$ and there remains

(8.7)     $u_n = 1/\log(2\pi n) + 1/6\log^3(2\pi n) + O\left(1/\log^5(2\pi n)\right),$

which improves (6). Here, as in the preceding n°, the reader can continue the calculations and/or examine the behaviour of the other series of roots $y_n$.

## 9 – Kepler's equation

We saw in Chap. II, at the end of n° 16, that Kepler's equation $u - e.\sin u = \omega t$ has one and only one root $u$ provided that the eccentricity $e$ of the elliptical motion is $< 1$. To simplify the notation a little, and to avoid confusing $e$ with $2,71828\ldots$, let us write it as

$$(9.1) \qquad\qquad u = \varphi + \varepsilon \sin u.$$

Laplace (see the 130-page notice by Gillispie in the supplement to DSB), the author of a treatise on celestial mechanics much more advanced than Newton's *Principia*, proved (?) that $u$ is the sum of a power series in $\varepsilon$ whose coefficients, depending on $\varphi$, are determined by an extraordinarily simple formula:

$$(9.2) \qquad u = \varphi + \varepsilon(\sin \varphi)/1! + \varepsilon^2 (\sin^2 \varphi)'/2! + \varepsilon^3 (\sin^3 \varphi)''/3! + \ldots.$$

One can make (2) plausible by putting $u = \varphi + v$, so that $v = \varepsilon \sin u$ tends to 0 with $\varepsilon$, and seeking an asymptotic evaluation of $v$; this does not replace a power series, but one works with the tools at one's disposal.

Clearly $v = O(\varepsilon)$ and indeed

$$\begin{aligned} v &= \varepsilon(\sin \varphi.\cos v + \cos \varphi.\sin v) = \\ &= \varepsilon \sin \varphi \left(1 + O(\varepsilon^2)\right) + \varepsilon \cos \varphi.O(\varepsilon) = \varepsilon \sin \varphi + O(\varepsilon^2), \end{aligned}$$

whence

$$u = \varphi + \varepsilon \sin \varphi + O(\varepsilon^2) = \varphi + \varepsilon \sin \varphi + \varepsilon^2 w \quad \text{with } w = O(1).$$

Then by (1)

$$\begin{aligned} \varepsilon \sin \varphi + \varepsilon^2 w &= \varepsilon \sin \varphi \left(\varphi + \varepsilon \sin \varphi + \varepsilon^2 w\right) = \\ &= \varepsilon \sin \varphi.\cos \left(\varepsilon \sin \varphi + \varepsilon^2 w\right) + \varepsilon \cos \varphi.\sin \left(\varepsilon \sin \varphi + \varepsilon^2 w\right) = \\ &= \varepsilon \sin \varphi. \left[1 - \left(\varepsilon \sin \varphi + \varepsilon^2 w\right)^2 /2 + O(\varepsilon^4)\right] + \\ &\quad + \varepsilon \cos \varphi.\left[\varepsilon \sin \varphi + \varepsilon^2 w + O(\varepsilon^3)\right], \end{aligned}$$

whence

$$\varepsilon^2 w = \varepsilon^2 \sin \varphi \cos \varphi + \varepsilon^3 \left(w \cos \varphi - \sin^3 \varphi/2\right) + O(\varepsilon^4)$$

i.e.

$$w = \sin \varphi \cos \varphi + \varepsilon \left(w \cos \varphi - \sin^3 \varphi/2\right) + O(\varepsilon^2).$$

Since $w = O(1)$, i.e. is bounded, one infers immediately that $w = \sin \varphi \cos \varphi + O(\varepsilon)$, whence, substituting in the right hand side of the preceding relation,

$$w = \sin \varphi \cos \varphi + \varepsilon \left(\sin \varphi.\cos^2 \varphi - \sin^3 \varphi/2\right) + O(\varepsilon^2).$$

Thus

$$u = \varphi + \varepsilon \sin \varphi + \varepsilon^2 \sin \varphi \cos \varphi + \varepsilon^3 \left( \sin \varphi. \cos^2 \varphi - \sin^3 \varphi/2 \right) + O(\varepsilon^4),$$

which yields the first terms of formula (2). The calculation becomes more and more painful as one pushes further and further.

The result was then extended by Lagrange to much more general equations, namely, in his notation,

$$(9.3) \qquad z = x + yf(z).$$

Since, for these Gentlemen, all functions arising in Nature, or even in mathematics, are analytic outside isolated points – as it happens, they were right to believe so if $f$ is, but then to prove this ... –, Lagrange set himself to calculate the expansion of $z = \sum a_n y^n$ as a power series, where the coefficients $a_n$ of course depend on $x$. The direct way would be, for $x$ given, to differentiate (3) indefinitely with respect to $y$, and to deduce the relations

$$z' = f(z) + yf'(z)z', \qquad z'' = 2f'(z)z' + y \left[ f''(z)z'^2 + f'(z)z'' \right],$$

$$z''' = 2f''(z)z'^2 + 2f'(z)z'' + f''(z)z'^2 + f'(z)z'' + \\ + y \left[ f'''(z)z'^3 + 2f''(z)z'z'' + f''(z)z'z'' + f'(z)z''' \right],$$

etc. and to find successively their values for $y = 0$:

$$z(0) = x, \qquad z'(0) = f(x), \qquad z''(0) = 2f'(x)f(x) = \left[ f(x)^2 \right]',$$

$$z'''(0) = 3f''(x)f(x)^2 + 6f'(x)^2 f(x) = \left[ f(x)^3 \right]'',$$

etc. This is, in an other form, what we have done above for Kepler's equation. The first results suggest the formula

$$(9.4) \qquad z = x + \sum y^n \left[ f(x)^n \right]^{(n-1)} /n!$$

using Maclaurin. But one falls rapidly, as above, into impossible calculations. Lagrange's method of establishing (4), at least formally, is considerably more ingenious.

His idea was to consider $z$ as a function of $y$ *and of $x$* and to differentiate (3) with respect to each of these two variables in order to calculate the coefficients $D_2^n z(x, 0)$ of the Maclaurin series of $z$ with respect to $y$.

To start with, thanks to the Chain Rule (Chap. III, n° 21), one finds

$$(9.5) \qquad D_1 z = 1 + yf'(z)D_1 z, \qquad D_2 z = f(z) + yf'(z)D_2 z,$$

whence $(D_1 z - 1)D_2 z = [D_2 z - f(z)] D_1 z$, and consequently

$$(9.6) \qquad D_2 z = f(z)D_1 z.$$

For $y = 0$, (5) gives

$$(9.7) \qquad D_1 z(x,0) = 1, \qquad D_2 z(x,0) = f(x)$$

since then $z = x$. Differentiating (6) with respect to $y$,

$$
\begin{aligned}
D_2^2 z &= f'(z) D_2 z D_1 z + f(z) D_2 D_1 z = f'(z) D_2 z D_1 z + f(z) D_1 D_2 z = \\
&= D_1 \left[ D_2 z . f(z) \right] = D_1 \left[ D_1 z . f(z)^2 \right]
\end{aligned}
$$

by (6) and $D_1 D_2 = D_2 D_1$ (Chap. III, n° 23). Whence $D_2^2 z(x,0) = \left[ f(x)^2 \right]'$ since, for $y = 0$, we have $D_2 z . f(z) = f(x)^2$ by (7). Suppose we have proved that

$$(9.8) \qquad D_2^n z = D_1^{n-1} \left[ D_1 z . f(z)^n \right]$$

for any $x$ and $y$, and differentiate. Using (6) again, and $D_1 D_2 = D_2 D_1$, we get

$$
\begin{aligned}
D_2^{n+1} z &= D_1^{n-1} D_2 \left[ D_1 z . f(z)^n \right] = \\
&= D_1^{n-1} \left[ D_2 D_1 z . f(z)^n + D_1 z . n f(z)^{n-1} f'(z) D_2 z \right] = \\
&= D_1^{n-1} \left[ D_1 D_2 z . f(z)^n + D_2 z . n f(z)^{n-1} f'(z) D_1 z \right] = \\
&= D_1^n \left[ D_2 z . f(z)^n \right] = D_1^n \left[ D_1 z . f(z)^{n+1} \right],
\end{aligned}
$$

which is (8) for $n + 1$.

The relation (8) is therefore valid for any $n$ and yields the formula

$$D_2^n z(x,0) = \left[ f(x)^n \right]^{(n-1)}$$

which justifies (4), at least formally.

In fact, Lagrange went even further; instead of just expanding $z$ he expanded an "arbitrary" function of $z$, say $u = \varphi(z)$. Since $D_1 u = \varphi'(z) D_1 z$ and $D_2 u = \varphi'(z) D_2 z$, the relation (6) becomes

$$D_2 u = f(z) D_1 u,$$

which allows one to calculate as we have just done, this time with

$$(9.9) \qquad D_2^n u(x,0) = \left[ \varphi'(x) f(x)^n \right]^{(n-1)}$$

and "thus"

$$\varphi(z) = \varphi(x) + \sum y^n \left[ \varphi'(x) f(x)^n \right]^{(n-1)} / n!.$$

The proof consists of establishing the relation

$$D_2^n u = D_1^{n-1} \left[ D_1 u . f(z)^n \right],$$

which replaces (8), by induction as above; for $y = 0$ we have $D_1 u = \varphi'(z) D_1 z = \varphi'(x)$, whence (9).

**10 – Asymptotics of the Bessel functions**

Consider the differential equation

$$(10.1) \qquad x'' + (1 - c/t^2)x = 0,$$

where $x$ is an unknown function of the real variable $t \neq 0$ and $c$ is a nonzero constant. We set ourselves to study the asymptotic behaviour of its solutions for $t$ large. We shall divide this relatively difficult but highly instructive exercise into several parts.

*Passage to an integral equation*

Since (1) can be written

$$(10.1') \qquad x'' + x = cx/t^2,$$

one may assume that at infinity its solutions resemble those of the much simpler equation $y'' + y = 0$, which has as its solutions at least (and, we shall see, at most) the functions

$$y(t) = ae^{it} + be^{-it}, \qquad \text{whence } y'(t) = iae^{it} - ibe^{-it},$$

where $a$ and $b$ are arbitrary constants. In the general case one puts

$$(10.2) \qquad x(t) = a(t)e^{it} + b(t)e^{-it}, \qquad x'(t) = ia(t)e^{it} - ib(t)e^{-it}$$

where $a(t)$ and $b(t)$ are now functions that one may easily calculate from $x$ and $x'$, by multiplying the relations (2) by $e^{it}$ or $e^{-it}$. This is the *method of variation of constants* (Johann Bernoulli, end of the XVII[th] century, for equations of the first order, Lagrange in the general case) which applies to all differential equations in which the unknown function and its derivatives occur linearly, but which one applies here in a nonclassical way since one is making believe that the function $cx(t)t^{-2}$ occurring on the right hand side of (1') is known [if it were the method would provide all the solutions of (1') in terms of integrals involving the right hand side].

The second relation (2), which seems to contradict the Chain Rule grossly, is in fact equivalent to

$$(10.3) \qquad a'(t)e^{it} + b'(t)e^{-it} = 0.$$

It then follows that

$$x'' = -ae^{it} - be^{-it} + ia'e^{it} - ib'e^{-it} = -x + ia'e^{it} - ib'e^{-it}$$

by the relations (2). Equation (1) can therefore be written as

$$(10.4) \qquad ia'(t)e^{it} - ib'(t)e^{-it} = cx(t)/t^2.$$

One deduces from (3) and (4) that

(10.5)        $2ia'(t) = ct^{-2}x(t)e^{-it}$,        $2ib'(t) = -ct^{-2}x(t)e^{it}$,

whence, using the FT,

$$2ia(t) - 2ia(t_0) = c\int_{t_0}^t x(u)e^{-iu}u^{-2}du,$$

$$2ib(t) - 2ib(t_0) = -c\int_{t_0}^t x(u)e^{iu}u^{-2}du.$$

We must assume $t_0 \neq 0$ and $t$ of the same sign as $t_0$ because of the factor $u^{-2}$, not integrable on a neighbourhood of 0. We shall assume them $> 0$ in all that follows, the opposite case being treated similarly. Substituting in the first relation (2), one finds

(10.6)        $$x(t) = p_0(t) + c\int_{t_0}^t x(u)\sin(t - u)u^{-2}du$$

with $p_0(t) = a_0e^{it} + b_0e^{-it}$, where $a_0 = a(t_0)$, $b_0 = b(t_0)$. Instead of, like (1), involving the function $x$ and its derivatives, (6) involves $x$ and an integral featuring the function $x$ itself; this is an *integral equation*. It does not even assume $x$ to be differentiable: the continuity of $x$ is enough for (6) to make sense. It is (6) which will allow us to examine the behaviour of $x$ at infinity.

One may conversely verify that every continuous solution of (6) is in fact $C^\infty$ and satisfies (1). The theorem on differentiation under the $\int$ sign with variable limits (Chap. V, n° 12, Theorem 13) in fact shows that the right hand side is differentiable and that

(10.7)        $$x'(t) = p_0'(t) + c\int_{t_0}^t x(u)\cos(t - u)u^{-2}du,$$

since the function integrated in (6) is zero for $u = t$. This relation in its turn shows that $x'$ is differentiable and that

(10.8)      $$x''(t) = p_0''(t) - c\int_{t_0}^t x(u)\sin(t - u)u^{-2}du + cx(t)t^{-2}$$

since $x(u)\cos(t - u)u^{-2} = x(t)t^{-2}$ for $u = t$. Since $p_0 + p_0'' = 0$, one finds (1) again, on adding the result to (6). The fact that $x$ is $C^\infty$ is then obvious, either because the function integrated in (6) has continuous derivatives of arbitrary order with respect to $t$, or because the differential equation shows that if $x$ is $C^p$, then it is automatically $C^{p+2}$ away from the origin.

*First bound for the solutions*

Let us agree provisionally that there is a solution of (6) defined for $t > 0$ and then show it is bounded at infinity. Indeed, let $M(t)$ be the upper bound of $|x(u)|$ on the interval $[t_0, t]$ and $M_0$ that of $|p_0(u)|$ in $\mathbb{R}$, clearly finite. (6) shows that, for $t_0 \le t' \le t$,

$$|x(t')| \le M_0 + |c|M(t) \int_{t_0}^{t'} u^{-2}du \le M_0 + |c|M(t)/t_0,$$

whence, passing to the sup,

(10.9) $$M(t) \le M_0 + |c|M(t)/t_0.$$

If we have chosen $t_0$ large enough that $|c|/t_0 \le \frac{1}{2}$ we may deduce that $M(t) \le 2M_0$, qed.

Now let us show that there exist constants $a_1$, $b_1$ such that

(10.10) $$x(t) = a_1 e^{it} + b_1 e^{-it} + O(1/t) = p_1(t) + O(1/t).$$

Since $x(t)$ is indeed bounded, the integral in (6), taken from $t_0$ to $+\infty$, is absolutely convergent like that of the function $1/u^2$. Thus

(10.11) $$x(t) = p_0(t) + c \int_{t_0}^{+\infty} x(u) \sin(t - u) u^{-2} du -$$
$$- c \int_{t}^{+\infty} x(u) \sin(t - u) u^{-2} du.$$

Expressing $\sin(t - u)$ in terms of complex exponentials, we see that the first integral is, like $p_0(t)$, a linear combination of $e^{it}$ and $e^{-it}$ with coefficients independent of $t$, whence

(10.12) $$x(t) = p_1(t) - c \int_{t}^{+\infty} x(u) \sin(t - u) u^{-2} du$$

where $p_1(t)$ is a linear combination of $e^{it}$ and $e^{-it}$ with constant coefficients. Since the function $x(u) \sin(t - u)$ is bounded for $u \ge t_0 > 0$, the integral is, up to a constant factor, majorised by that of $u^{-2}$, i.e. by $1/t$, qed.

The relation (12) allows us to complete the existence theorem for solutions – we shall prove it below – with a uniqueness theorem: *there exists only one solution of (12) for $p_1$ given*. Since $p_1$ depends on two arbitrary constants, this means that the set of solutions is a vector space of dimension 2 over $\mathbb{C}$.

On subtraction we reduce to proving that, if $p_1$ is zero, then so likewise is $x$. But denote now by $M(t)$ the upper bound of $|x(u)|$ for $u \ge t$. For $t' \ge t$ we clearly have

$$|x(t')| \le |c|M(t)/t$$

since the integral of $u^{-2}$ between $t'$ and $+\infty$, equal to $1/t'$, is $\le 1/t$. Whence, on passing to the sup, $M(t) \le |c|M(t)/t$. Substituting this result in the integral equation, we now find

$$|x(t')| \leq |c|^2 M(t) \int_{t'}^{+\infty} u^{-3} du,$$

whence $M(t) \leq |c|^2 M(t)/2t^2$. Substituting again in the equation, we will find $M(t) \leq |c|^3 M(t)/3! t^3$, etc. In short, $M(t) \leq M(t)|c/t|^n/n!$ for any $n$. Since, for $t \neq 0$ given, the right hand side tends to 0 when $n \to +\infty$, we find $M(t) = 0$ for any $t > 0$, whence $x(t) = 0$, qed.

*Existence of solutions*

To go further than (10) in studying $x(t)$ at infinity, one might, as always, iterate the calculation, i.e. substitute (10) in (12) and so on indefinitely. We are going to adopt a slightly different method which will at the same time show the existence of the solutions; this is the *method of successive approximations*, which consists of extending the method of constructing the roots of an equation $x = f(x)$ expounded in Chap. II, n° 16, Theorem 12 and in Chap. III, n° 24 (implicit functions) to integral equations; it can be used to show the existence, at least locally, of the solutions of almost all reasonable differential or integral equations. Since all the integrals which now appear are extended over $[t, +\infty]$, we shall adopt the simplified notation

$$\int = \int_t^{+\infty}$$

up to the end of this n°, clearly not confusing this with an indefinite integral *à la* Leibniz. In this notation

$$\int u^{-n-1} du = t^{-n}/n$$

for $n \geq 1$ by the FT.

The method of successive approximations consists of starting from the function $p_1(t)$ in (12), to which $x(t)$ is equal up to the addition of a $O(1/t)$ term, constructing a sequence of functions $x_n(t)$ on $t > 0$ by putting $x_1 = p_1$ and

$$(10.13) \qquad x_{n+1}(t) = p_1(t) - c \int x_n(u) \sin(t-u) u^{-2} du,$$

and showing that the $x_n$ converge to a solution of (6).

For $n = 1$

$$|x_2(t) - x_1(t)| \leq M|c| \int u^{-2} du = M|c/t|$$

where $M = \|p_1\|_{\mathbb{R}} < +\infty$. It follows that

$$
\begin{aligned}
|x_3(t) - x_2(t)| &= |c|. \left| \int [x_2(u) - x_1(u)] \sin(t-u) u^{-2} du \right| \leq \\
&\leq M|c|^2 \int u^{-3} du = M|c/t|^2/2!.
\end{aligned}
$$

If one has proved that

(10.14) $$|x_n(t) - x_{n-1}(t)| \le M|c/t|^{n-1}/(n-1)!$$

one finds

$$|x_{n+1}(t) - x_n(t)| \;=\; |c| \cdot \left| \int [x_n(u) - x_{n-1}(u)] \sin(t-u) u^{-2} du \right| \le$$

$$\le M|c|^n/(n-1)! \int u^{-n-1} du = M|c/t|^n/n!,$$

which shows in passing that

(10.15) $$x_{n+1}(t) = x_n(t) + O(t^{-n})$$

at infinity. Since the series $\sum [x_{n+1}(t) - x_n(t)]$ is, by (14), dominated by the series $\exp(M|c|/t)$, it converges normally on every interval $t \ge t_0 > 0$, so that $x_n(t)$ converges to a limit $x(t)$ for every $t > 0$, and does so uniformly on every interval $[t_0, +\infty[$. One may then pass to the limit under the $\int$ sign in (13) because of the presence of the *integrable*[2] factor $u^{-2}$. It is then clear that $x(t)$ satisfies (6).

Further, by (14),

$$|x(t) - x_n(t)| \;\le\; \sum_{p \ge 0} |x_{n+p+1}(t) - x_{n+p}(t)| \le$$

$$\le M \sum_{p \ge 0} |c/t|^{n+p}/(n+p)! \le M \sum_{p \ge 0} |c/t|^{n+p}/n!p! =$$

$$= M \cdot \exp(|c/t|)|c/t|^n/n!,$$

a result which implies

(10.16) $$x(t) = x_n(t) + O(t^{-n})$$

at infinity since the factor $\exp(|c/t|)$ tends to 1.

*Exercise.* Let $I$ be a compact interval, $p$ a continuous function on $I$ and $K(t, u)$ a continuous function on $I \times I$. Put

$$M = \sup_{t \in I} \int |K(t, u)| du.$$

Show that, if $M < 1$, the integral equation

---

[2] Let $I$ be an arbitrary interval, $\mu(x)$ an absolutely integrable function on $I$, and $(f_n)$ a sequence of bounded functions which converges uniformly on $I$ to a limit $f$, clearly bounded. The functions $f_n\mu$ and $f\mu$, majorised up to constant factors by $\mu$, are then integrable and one has $\left| \int [f_n(u) - f(u)] \mu(u) du \right| \le \|f_n - f\| \cdot \int |\mu(u)| du$, whence $\lim \int f_n(u)\mu(u) du = \int f(u)\mu(u) du$. Cf. Chap. V, n° 31, Example 1.

$$x(t) = p(t) + \int K(t,u)x(u)du$$

has one and only one solution (the integrals are over $I$). Analogy with a system of linear equations?

### Asymptotics of the solutions: general form

It is clear that in order to obtain asymptotic evaluations of $x(t)$ one should seek them for the $x_n(t)$. To simplify the calculations we shall assume that we are in the case where the "trigonometric binomial" $p_1(t)$ in (10) reduces to $e^{it}$; the case where it is equal to $e^{-it}$ is treated in the same way (the two solutions are even complex conjugates if $c \in \mathbb{R}$), and in the general case it is clear that $x$ is a linear combination of the functions corresponding to these two particular cases.

Since $x_1(t) = e^{it}$, the relation (13) shows that

$$
(10.17) \qquad
\begin{aligned}
2ix_2(t) &= 2ie^{it} - 2ic \int e^{iu}\sin(t-u)u^{-2}du = \\
&= 2ie^{it} - c\int \left(e^{it} - e^{2iu-it}\right)u^{-2}du = \\
&= 2ie^{it} - ce^{it}/t + ce^{-it}\int e^{2iu}u^{-2}du.
\end{aligned}
$$

Here we meet, and we will meet again, an integral of the form $\int e^{2iu}u^{-p}du$ extended over $[t,+\infty[$. By repeatedly integrating by parts it is easy to find a truncated expansion of arbitrarily high order when $t$ tends to infinity. Generally, if $\mathrm{Re}(\alpha) \leq 0$ to ensure the convergence of the integrals, one has, using *exceptionally* the $\int$ sign *à la* Leibniz,

$$
\begin{aligned}
\int e^{\alpha u}u^{-p}du &= e^{\alpha u}/\alpha u^p + \frac{p}{\alpha}\int e^{\alpha u}u^{-p-1}du = \\
&= e^{\alpha u}/\alpha u^p + pe^{\alpha u}/\alpha^2 u^{p+1} + \frac{p(p+1)}{\alpha^2}\int e^{\alpha u}u^{-p-2}du
\end{aligned}
$$

etc[3]. When one integrates from $t$ to $+\infty$, the integrated-out parts, zero at infinity, yield the product of $e^{\alpha t}/t^p$ by a polynomial in $1/t$; the integral of $e^{\alpha u}u^{-N-2}$, majorised by that of $u^{-N-2}$, is $O\left(t^{-N-1}\right)$. One deduces from this that, for any $N > p$, there is a relation

---

[3] Note that instead of trying to pass from an integral in $u^{-p}$ to an integral in $u^{-p+1}$ as was done in Chap. V, n° 15, Example 2 in the illusory hope of calculating a primitive explicitly, here one passes from an integral in $u^{-p}$ to integrals in $u^{-p-1}$, $u^{-p-2}$, etc. whose order of magnitude one may evaluate, even if unable to calculate them explicitly.

$$(10.18) \quad e^{-it} \int_t^{+\infty} e^{2iu} u^{-p} du \;=\; -e^{it} \left(?/t^p + ?/t^{p+1} + \ldots + ?/t^N\right) +$$
$$+ O\left(t^{-N-1}\right)$$

with coefficients ? which depend on $p$, but not on $N$; the reader may calculate them: we don't need them now. Returning to (17), the case where $p = 2$, we finally find

$$(10.19) \quad 2ix_2(t) = 2ie^{it} + e^{it} \left(?/t + ?/t^2 + \ldots + ?/t^N\right) + O\left(t^{-N-1}\right)$$

for any $N$.

It is the same for $x_n(t)$ for any $n$. One shows this by induction using (13):

$$2ix_{n+1}(t) =$$

$$= 2ie^{it} - 2ic \int x_n(u) \sin(t-u) u^{-2} du =$$

$$= 2ie^{it} - c \int e^{iu} \left[? - ?/u - \ldots - ?/u^N + O\left(u^{-N-1}\right)\right] 2i \sin(t-u) u^{-2} du =$$

$$= 2ie^{it} + c \sum_{0 \le p \le N} ? \int \left(e^{it} - e^{2iu-it}\right) u^{-p-2} du + \int O\left(u^{-N-3}\right) du.$$

The integrals $\int e^{it} u^{-p-2} du$ yield the product of $e^{it}$ by a polynomial in $1/t$ without constant term. The integrals $\int e^{2iu-it} u^{-p-2} du$ likewise by (18) have truncated expansions of arbitrarily high order. Finally, the integral $\int O\left(u^{-N-3}\right) du$ is $O\left(t^{-N-2}\right)$. So for every $n$ and every $N$ there is a relation of the form

$$(10.20) \quad x_n(t) = e^{it} \left(1 + ?/t + ?/t^2 + \ldots + ?/t^N\right) + O\left(t^{-N-1}\right).$$

In view of (16) one obtains an expansion

$$(10.21) \quad x(t) = e^{it} \left(1 + a_1/t + a_2/t^2 + \ldots + a_N/t^N\right) + O\left(t^{-N-1}\right)$$

for $x(t)$, whose coefficients do not depend on $N$; for if one has this for $e^{-it}x(t)$ or for every other function with truncated expansions of order 12 and 15, the second, with its terms of degree $> 12$ removed, yields a truncated expansion of order 12; now a given function can have only one truncated expansion of given order, as we saw in n° 3; the two expansions must therefore have the same terms of degree $\le 12$.

One sometimes expresses this fact by writing (21) in the form of an *asymptotic series*

$$(10.22) \quad e^{-it}x(t) \approx \sum_{n=0}^{\infty} a_n/t^n;$$

this way of writing by no means states that the series on the right hand side represents the function considered: in almost every case of this kind, including the one which now occupies us, as we shall see when all will be calculated, the series is *divergent*. The form (22) is, *by definition*, equivalent to the fact that the relation (21) is valid for any $N$. In other words, the difference between the left hand side of (22) and the $N$-th partial sum of the second member, instead of tending to 0 *for t given when N increases*, tends to 0 *for N given when t increases*, and as rapidly as the first term neglected; a nuance not to be forgot ...

This is for example what happens on a neighbourhood of $t = 0$ when one writes the Maclaurin formula for a function $x(t)$ which is not analytic but is of class $C^\infty$. For any $N$, one has

$$x(t) = x(0) + x'(0)t + \ldots + x^{(N)}(0)t^N/N! + O\left(t^{N+1}\right),$$

in other words

$$x(t) \approx \sum x^{(n)}(0)t^n/n!,$$

but the series has no reason to represent the function if it converges – the case of $\exp(-1/t^2)$ – and even less if it diverges, which is the general case since the derivatives at the origin can be chosen arbitrarily (Chap. V, n° 29).

### Term-by-term differentiation of asymptotic expansions

The problem now arises of calculating the coefficients $a_n$ in the expansion (22) explicitly, preferably without drowning oneself in calculation. In doing this in a more explicit way than we have done above we could find the recurrence relations allowing us to calculate the $a_n$. A more elegant[4] and above all more instructive, method, consists of showing that on differentiating (22) term-by-term, one obtains the analogous asymptotic expansions for $x'$ and $x''$; on substituting into the differential equation (1) one will find the needed recurrence relations immediately.

It is not at all obvious, and it is generally false, that one can deduce an asymptotic series for the derivative from the asymptotic series of a given function by differentiating term-by-term. The derivative of a function $O(t^r)$ at infinity ($r \in \mathbb{R}$) has no reason to be $O\left(t^{r-1}\right)$: the function $x(t) = \sin(t^2)/t$ is $O(1/t)$ at infinity, but its derivative $x'(t) = 2\cos(t^2) - \sin(t^2)/t^2$ is $O(1)$ and not $O(1/t^2)$ at infinity. This is the problem we have already met in connection with differentiating term-by-term the sum of a series of differentiable functions: one may, thanks to FT, majorise a function starting from a majoration of its derivative, but the inverse operation is impossible.

---

[4] One of the Goncourt brothers, famous literary critics of the XIX[th] century, relates in his *Journal* that during the reception of a new immortal, X, into the Académie française, the academician Y charged with delivering the eulogy on X had the regrettable idea of describing the oratorical style of X as elegant. The latter, furious, stood up and replied: Elegant yourself, Sir! (I quote from memory).

The reality is that here, as in the case of a convergent series or sequence, it is the existence of an asymptotic series *for the derivative* which enables us to obtain one for the function itself. For assume that the derivative of a function $f(t)$ has an expansion

(10.23)     $f'(t) = a_0 + a_1/t + \ldots + a_{N+1}/t^{N+1} + O\left(1/t^{N+2}\right)$

at infinity. One cannot integrate it from $t$ to $+\infty$ because of the first two terms, but one reduces to the case where they are zero on replacing $f(t)$ by $g(t) = f(t) - a_0 t - a_1 \log t$. Then $g'(t) = O(t^{-2})$, the derivative is integrable from $t$ to $+\infty$, the function $g$ tends to a finite limit $g(+\infty)$ when $t \to +\infty$ and, by the FT extended to the interval $[t, +\infty[$ (by passage to the limit),

$$g(+\infty) - g(t) = \int g'(u)du = a_2/t + a_3/2t^2 + \ldots + a_{N+1}/Nt^N + O\left(1/t^{N+1}\right)$$

since the integral from $t$ to $+\infty$ of an $O(u^{-k})$ function is majorised up to a constant factor by that of $u^{-k}$. Returning to the original function $f(t)$, the relation (23) implies

(10.24)     $f(t) \quad = \quad a_0 t + a_1 \log t + b - a_2/t - a_3/2t^2 - \ldots -$
$\qquad\qquad\qquad - a_{N+1}/Nt^N + O\left(1/t^{N+1}\right)$

for any $N$, with an inevitable constant $b$, since knowing $f'$ determines $f$ only up to a constant. It is then clear, on comparing (23) and (24), that the asymptotic expansion of $f'$ is obtained by differentiating that of $f$ term-by-term.

Returning to the function $x(t)$ which concerns us, we must show directly that $x'(t)$ and $x''(t)$ have asymptotic expansions analogous to (22). To do this, let us again consider the integral equation (10.12)

$$x(t) = e^{it} - c \int_t^{+\infty} x(u) \sin(t - u)u^{-2}du$$

and apply to it the formula of differentiation under the $\int$ sign with variable limits in the case of an infinite interval, namely

$$\frac{d}{dt} \int_{\varphi(t)}^{+\infty} f(t, u)du = \int_{\varphi(t)}^{+\infty} D_1 f(t, u)du - f[t, \varphi(t)]\varphi'(t)$$

(Chap. V, n° 12, Theorem 13, which extends immediately to the case of an infinite interval using[5] n° 25, Theorem 24, of the same Chap. V). The latter assumes that, when $t$ remains in a compact set the function $D_1 f(t, u)$ is majorised by a fixed integrable function of $u$. Now, in the case of (12),

---

[5] One writes that the integral of $\varphi(t)$ to $+\infty$ is the difference between the integrals from $a$ to $+\infty$ and from $a$ to $\varphi(t)$ for a fixed $a$.

$$|D_1 f(t, u)| = |x(u)\cos(t - u)u^{-2}| \le M u^{-2}$$

since the function $x(u)$ is bounded on every interval $t \ge t_0 > 0$; there is no problem. The function $f(t, u)$ which we are integrating here from $t$ to $+\infty$ vanishes for $u = \varphi(t)$, the wholly integrated part of the differentiation formula disappears, and there remains

$$(10.25) \qquad x'(t) = ie^{it} - c \int x(u)\cos(t - u)u^{-2} du$$

where one integrates from $t$ to $+\infty$; compare to (7) and (8).

*Exercise.* By differentiating the recurrence relation between the $x_n(t)$ show that

$$x''_{n+1}(t) + x_{n+1}(t) = a x_n(t) t^{-2}$$

and that, for every $r$, the derivatives $x_n^{(r)}(t)$ of the $x_n$ converge to $x^{(r)}(t)$ uniformly on $t \ge t_0 > 0$.

On substituting (21) in (25), one has

$$x'(t) =$$

$$= ie^{it} - c \int \left[ 1 + a_1/t + a_2/t^2 + \ldots + a_N/t^N + O\left(t^{-N-1}\right) \right] e^{iu}\cos(t - u)u^{-2} du$$

$$= ie^{it} - c \int \left( 1 + a_1/t + a_2/t^2 + \ldots + a_N/t^N \right) e^{iu}\cos(t - u)u^{-2} du + O\left(t^{-N-2}\right).$$

Arguing as above – replacing the sinus by a cosinus clearly changes the method not at all –, one obtains an asymptotic series

$$(10.26) \qquad e^{-it} x'(t) \approx \sum b_n/t^n$$

similar to (22). As for $x''(t) = -(1 - c/t^2)x(t)$, one obtains an asymptotic series for it directly starting from that for $x(t)$.

*Coefficients of the asymptotic expansion*

Let us now put $e^{-it}x(t) = y(t)$. We have $y(t) \approx \sum a_n t^{-n}$ by (22), and on the other hand we know that the derivatives

$$(10.27) \quad y'(t) = e^{-it}\left[ x'(t) - ix(t) \right], \quad y''(t) = e^{-it}[x''(t) - 2ix'(t) - x(t)]$$

also have asymptotic expansions of the same type. They too can be derived from the expansion of $y(t)$ by differentiating the latter term-by-term as for a power series in $1/t$. [This means that the expansion of $x''(t)$ too can be derived from that of $x(t)$ by differentiating term-by-term, not forgetting to differentiate the factors $e^{it}$]. The expansions of $y'$ and $y''$ must thus be

$$y'(t) \approx \sum -na_n t^{-n-1}, \qquad y''(t) \approx \sum n(n+1)a_n t^{-n-2}.$$

Now let us exploit the differential equation we started from. Since we put $x(t) = e^{it}y(t)$ we have $x''(t) + x(t) = e^{it}[y''(t) + 2iy'(t)]$, whence $ct^{-2}y = y'' + 2iy'$. Thus

$$c\left(a_0 t^{-2} + a_1 t^{-3} + a_2 t^{-4} + \ldots\right)$$
$$\approx \left(2.1a_1 t^{-3} + 3.2a_2 t^{-4} + \ldots\right) - 2i\left(a_1 t^{-2} + 2a_2 t^{-3} + 3a_3 t^{-4} + \ldots\right).$$

Because the asymptotic series of a given function is unique it is legitimate to calculate as for a formal series. Since $a_0 = 1$ we find

$$-2ia_1 = c, \quad -2ia_2 = (c - 1.2)a_1/2, \quad -2ia_3 = (c - 2.3)a_2/3$$

and generally

$$a_n/a_{n-1} = i[c - n(n-1)]/2n;$$

one deduces $a_n$ by multiplying together the first $n$ relations.

Note that the ratio $|a_{n+1}/a_n|$ tends to $+\infty$. The radius of convergence of the power series $\sum a_n z^n$ is therefore zero, which confirms that the expansion as an *asymptotic* series $x(t) \approx e^{it} \sum a_n t^{-n}$ is the exact opposite of an expansion as a *convergent* series.

*Exercise.* Show that the differential equation (1) is satisfied by convergent series of the form $t^a \sum_{n>0} a_n t^n$, with a non integer exponent $a$ and coefficients to be determined.

The method used here in the case of the Bessel equation has given rise to an ocean of literature concerning either other special functions, or general linear differential equations; Chapters XIV and XV of Dieudonné, *Calcul infinitésimal*, give a faint glimpse of the general case and of the theory of the Bessel functions, about which voluminous treatises have been written.

The best classical reference on these and the other "special functions" is the "Bateman Project", *Higher Transcendental Functions* (McGraw Hill, 1953–1955, 3 vols). To understand the subject, it would be better to read N. Vilenkin, *Special functions and the theory of group representations* (American Mathematical Society, 1968) (translated from original Russian edition (Moscow, 1965)), which is based on ideas which are totally foreign to the "experts" on the classical theory and of much more general scope than these (harmonic analysis on non commutative Lie groups). Since they are well above the level of this book, there is no point in citing more recent and inaccessible references.

# § 2. Summation formulae

## 11 – Cavalieri and the sums $1^k + 2^k + \ldots + n^k$

In Chapter II, n° 11 we gave a very effective direct method for integrating the function $x^k$ when $k$ is a positive integer: one divides the interval of integration $[a, b]$ by points $aq^k$ forming a *geometric* progression, with $q = (b/a)^{1/n}$, and lets $n$ tend to infinity. But the first mathematicians to perform this calculation proceeded in another way: like Archimedes in the case $k = 2$, they used a subdivision of $[a, b]$ by the points of an *arithmetic* progression. In the simple case where the interval of integration is of the form $[0, a]$ one puts $q = a/n$ and uses the points $q, 2q, \ldots nq$; the integral sought is then clearly the limit of the sums

$$(11.1) \qquad \sigma_n = \left[ q^k + (2q)^k + \ldots + (nq)^k \right] a/n = \\ = \left( 1^k + 2^k + \ldots + n^k \right) a^{k+1}/n^{k+1}.$$

For $k = 1$ it had been known for a long time that

$$(11.2) \qquad 1 + 2 + \ldots + n = n(n+1)/2 = n^2/2 + n/2,$$

whence $\sigma_1 = a^2 n(n+1)/2n^2$, an expression which tends to $a^2/2$. For $n = 2$, the case treated by Archimedes, who already knew that

$$(11.3) \qquad 1^2 + \ldots + n^2 = n(n+1)(2n+1)/6 = n^3/3 + n^2/2 + n/6,$$

one has $\sigma_2 = a^3 \left( 1/3 + 1/2n + 1/6n^2 \right)$, which tends to $a^3/3$.

The Italian Cavalieri studied the case where $k = 4$ around 1630, using the formula

$$(11.4) \qquad 1^3 + \ldots + n^3 = n^2 (n+1)^2 /4 = n^4/4 + n^3/2 + n^2/4,$$

which gave him the value $a^4/4$ for the integral. Around 1646 he extended the calculations up to $k = 9$, with the help of the formula

$$(11.5) \quad 1^9 + \ldots + n^9 = n^{10}/10 + n^9/2 + 3n^8/4 - 7n^6/10 + n^4/2 - 3n^2/20.$$

These calculations are all the more praiseworthy than the modern mechanism of algebra, with its condensed notation was then strongly in flux. John Wallis set out the method in his *Arithmetica Infinitorum* of 1656, but no one was yet able to find the formula (5) corresponding to an arbitrary value of the exponent $k$.

Fermat, who did not publish, took up the problem around 1636 – it was his idea to use a geometric progression –, but instead of trying to calculate $1^k + \ldots + n^k$ exactly he was content to find an approximate value adequate to solve the problem, namely

(11.6)    $1^k + \ldots + (n-1)^k \quad < \quad n^{k+1}/(k+1) <$
$$< \quad 1^k + \ldots + n^k < (n+1)^{k+1}/(k+1).$$

This shows that up to a factor $a^{k+1}$ the Riemann sum (1) lies between the products of $1/(k+1)$ by 1 and $(n+1)^{k+1}/n^{k+1}$, which tend to 1.

(6) is proved by induction on $n > 2$. The case $n = 2$ is obvious. If (6) has been proved for an integer $n$ it follows that

$$1^k + \ldots + n^k < \frac{n^{k+1}}{k+1} + n^k \quad \text{and} \quad \frac{n^{k+1}}{k+1} + (n+1)^k < 1^k + \ldots + (n+1)^k.$$

It is therefore enough to show that

$$\frac{n^{k+1}}{k+1} + n^k < \frac{(n+1)^{k+1}}{k+1} < \frac{n^{k+1}}{k+1} + (n+1)^k,$$

and then, putting $x = 1/n$, that

$$1 + (k+1)x < (1+x)^{k+1} < 1 + (k+1)(1+x)^k.$$

Since $x > 0$ the binomial formula proves the first inequality. The second can be written as

$$1 + \sum_{p=0}^{k} \binom{k+1}{p+1} x^{p+1} < 1 + (k+1) \sum_{p=0}^{k} \binom{k}{p} x^{k+1}$$

and reduces to the inequality

$$\binom{k+1}{p+1} = \frac{k+1}{p+1}\binom{k}{p} < (k+1)\binom{k}{p}$$

between binomial coefficients.

*Exercise*[6]. (a) Prove the equalities

$$S_n^1 \quad := \quad 1 + 2 + \ldots + n = \frac{1}{2}n(n+1)$$
$$S_n^2 \quad := \quad 1^2 + 2^2 + \ldots + n^2 = n(n+1)(2n+1)/6$$
$$S_n^3 \quad := \quad 1^3 + 2^3 + \ldots + n^3 = \left[\frac{1}{2}n(n+1)\right]^2 = (1 + 2 + \ldots + n)^2.$$

(b) For
$$S_n^p := 1^p + 2^p + \ldots + n^p,$$

establish the identity

---

[6] Walter, *Analysis I*, p. 36. The signs := mean that the expression which follows the sign = is the definition of that which precedes the sign :.

$$(p+1)S_n^p + \binom{p+1}{2}S_n^{p-1} + \ldots + S_n^0 = (n+1)^{p+1} - 1$$

discovered by Pascal in 1654.

(c) Show that for every $p > 1$ there exist $p$ real numbers $c_1, \ldots, c_p$ such that

$$S_n^p = n^{p+1}/(p+1) + n^p/2 + c_1 n^{p-1} + \ldots + c_{p-1}n + c_p.$$

Hints:

$$(x+1)^{p+1} - x^{p+1} = \binom{p+1}{1}x^p + \binom{p+1}{2}x^{p-1} + \ldots + 1 \quad (p, n \in \mathbb{N}, \ n \geq 1).$$

Add these equations term-by-term for $x = 1, 2, \ldots, n$. The assertions (a) and (c) can be proved by induction or from Pascal's identity.

## 12 – Jakob Bernoulli

In 1713, in his *Ars Conjectandi*, the most famous, if not the first, of the treatises on the calculus of probabilities, Jakob Bernoulli published – rather, it was published for him, for he died in 1705 before finishing his book – the general method which allows one, for $k \in \mathbb{N}$, to express $1^k + \ldots + n^k$ as a polynomial of degree $k+1$ in $n$. He calculated the first sums afresh and noted in passing that he had been able to calculate "in less than half a quarter hour" that

$$1^{10} + \ldots + 1000^{10} = 91 \ 409 \ 924 \ 241 \ 424 \ 243 \ 424 \ 241 \ 924 \ 242 \ 500.$$

*Exercise.* If the human species had had thirteen fingers instead of ten, Bernoulli would have had to calculate the sum $1^{13} + \ldots + 2197^{13}$. Find the result in less than half an hour using numeration to base 13.

His general method[7] was to start from the relation

(12.1) $$\binom{n}{k} = \sum_{p=0}^{n-1} \binom{p}{k-1} = \sum_{p=1}^{n} \binom{p-1}{k-1}$$

between the binomial coefficients (which he wrote explicitly, as everyone did then); one may prove this easily by induction on $n$, writing that

$$\binom{n}{k} = \binom{n-1}{k-1} + \binom{n-1}{k} = \binom{n-1}{k-1} + \sum_{p=0}^{n-2}\binom{p}{k-1}.$$

For $k = 3$, one thus finds

---

[7] See Vol. III of Moritz Cantor, pp. 343–347.

(12.2)  $\quad n(n-1)(n-2)/3! \;=\; \sum (p-1)(p-2)/2! =$

$$= \sum \left( p^2/2 - 3p/2 + 1 \right),$$

which, using the formulae for the exponents $k = 0$ and $1$, yields the formula for $k = 2$. The formula (1) for $k = 4$ then allows one to calculate the $\sum p^3$ from the formulae already obtained, and so on.

But Bernoulli went much further. He stated that

(12.3)  $\quad \displaystyle\sum_1^n p^k = n^{k+1}/(k+1) + n^k/2 + \frac{k}{2} A n^{k-1} +$

$$+ \frac{k(k-1)(k-2)}{2.3.4} B n^{k-3} + \frac{k(k-1)\ldots(k-4)}{6!} C n^{k-5} + \ldots$$

with coefficients $A$, $B$, $C$, ... *independent of* $k$, and exponents $k-1$, $k-3$, $k-5, \ldots$ The two first terms were obvious because he knew the explicit formulae for $k \leq 10$, but no one knows how he divined the relation (3) from them, which he merely stated after a list of explicit formulae. Of course, if one accepts (3), the first formulae easily give the values

$$A = 1/6, \quad B = -1/30, \quad C = 1/42, \quad D = -1/30, \quad E = 5/66, \quad \text{etc.}$$

A much less magical method is to postulate that, in conformity with the first formulae, one has[8]

(12.4)  $\quad\quad\quad\quad\quad 1^k + \ldots + n^k = A_{k+1}(n)$

with $A_1(x) = x$ for $k = 0$ and, for $k \geq 1$, a polynomial of degree $k + 1$ *without constant term*, then to establish those properties of these conjectured polynomials which allow one to calculate them "without calculations" and, to finish, to verify that they satisfy (4).

To start with, the relation

$$n^k = \left( 1^k + \ldots + n^k \right) - \left( 1^k + \ldots + (n-1)^k \right)$$

implies the polynomial identity

(12.5)  $\quad\quad\quad\quad\quad A_{k+1}(x) - A_{k+1}(x-1) = x^k$

since the difference of the two sides is a polynomial which vanishes at every $x \in \mathbb{N}$. This relation already determines the $A_k$ up to additive constants, for the difference between two solutions is a polynomial of period 1, so constant;

---

[8] Bernoulli uses the notation $S\,n^k = 1^k + \ldots + n^k$, which, once again, violates all the tabus concerning phantom and free variables. See Hairer and Wanner, p. 15, for a photographic reproduction of Bernoulli's table of the first ten formulae.

if one then assumes $A_k(0) = 0$ for $k \geq 1$ then the $A_k$ are entirely determined by (5). Now in deriving (5) one sees that $A'_{k+1}(x)/k$ satisfies it for $k - 1$. If one writes $a_k$ for the constant term of $A'_{k+1}(x)/k$, then

(12.6).    $A'_2(x) = x + a_1, \quad A'_{k+1}(x)/k = A_k(x) + a_k \quad$ for $k \geq 2$.

The $A_k$ having zero constant term, one obtains step-by-step, by straightforward calculation of the successive primitives,

$$
\begin{aligned}
A_1(x) &= x, \\
A_2(x) &= x^2/2 + a_1 x, \\
A_3(x) &= x^3/3 + a_1 x^2 + 2a_2 x, \\
A_4(x) &= x^4/4 + a_1 x^3 + 3a_2 x^2 + 3a_3 x, \\
A_5(x) &= x^5/5 + a_1 x^4 + 4a_2 x^3 + 6a_3 x^2 + 4a_4 x;
\end{aligned}
$$

etc. The general formula

(12.7)    $$A_k(x) = x^k/k + \sum_{p=1}^{k-1} \binom{k-1}{p-1} a_p x^{k-p},$$

now obvious, is just (3) with

$$a_1 = 1/2, \quad a_2 = A/2, \quad a_3 = 0, \quad a_4 = B/4, \quad a_5 = 0, \quad a_6 = C/6,$$

etc.

Though not immediately providing the numerical values of the $a_p$, at least (6) proves the existence of a relation (7) with the same coefficients $a_p$ for all the formulae. This, Moritz Cantor calls it Jakob Bernoulli's "idea of genius", seems relatively humdrum to me, even for the period; for if there was anything they knew how to do, it was to calculate the derivatives or primitives of polynomials in $x \ldots$

To obtain the numerical values of the coefficients one uses a remark which, here again, was surely within the scope of the genial inventor: by (5) one must have

(12.8)    $A_k(0) = A_k(-1) \quad$ for $k \geq 2$

and so $A_k(-1) = 0$. Whence, by (7), a relation[9]

(12.9)    $$1/k - a_1 + \binom{k-1}{1} a_2 - \binom{k-1}{2} a_3 + \ldots +$$

$$+ (-1)^{k-1} \binom{k-1}{k-2} a_{k-1} = 0 \quad (k \geq 2)$$

---

[9] Bernoulli clearly knew this, for he wrote, without proof, that to calculate the coefficients in his first ten formulae one uses the fact that $A_k(1) = 1$; he details the calculation for $A_8$. See the text in Walter, *Analysis I*, pp. 162–163.

which enables one to calculate the coefficients step-by-step.

It remains to show that with this choice of the $a_k$ the $A_k$ do indeed satisfy (4). Since

$$(12.10) \quad A_{k+1}(n) = [A_{k+1}(n) - A_{k+1}(n-1)] + \\ + [A_{k+1}(n-1) - A_{k+1}(n-2)] + \ldots + \\ + [A_{k+1}(1) - A_{k+1}(0)] + A_{k+1}(0)$$

and since $A_{k+1}(0) = 0$, (4) will in fact be a consequence of (5), clearly true for $k = 0$. We shall prove (5) by induction on $k$.

First, by the simplest of the relations between binomial coefficients, (7) *defines polynomials satisfying* (6) *for any* $a_p$. If one has already verified that $A_k(x) - A_k(x-1) = x^{k-1}$ then the formula (6) shows that $A'_{k+1}(x) - A'_{k+1}(x-1) = kx^{k-1}$, whence $A_{k+1}(x) - A_{k+1}(x-1) = x^k$ up to an additive constant. This must be zero for $k = 0$ since $A_1(x) = x$. It is zero for $k \geq 1$ because the choice (9) of the $a_p$ is equivalent to $A_k(0) = A_k(-1)$ and shows that (5) is valid for $x = 0$. Hence (5), and consequently (4) for any $k$.

Posterity has preferred, for reasons which will appear later, to use the polynomials $B_k(x)$, $k \geq 0$, of degree $k$, possibly with nonzero constant terms

$$(12.11) \quad B_k(0) = b_k,$$

and chosen so as to replace (6) by

$$(12.12) \quad B'_k(x) = kB_{k-1}(x), \quad k \geq 1,$$

and (8) by

$$(12.13) \quad B_k(1) = B_k(0), \quad k \geq 2,$$

for every $k \geq 0$. One chooses

$$B_0(x) = 1 = b_0$$

to simplify the formulae as much as possible. Again calculating straightforwardly one obtains

$$\begin{aligned} B_1(x) &= b_0 x + b_1, \\ B_2(x) &= b_0 x^2 + 2b_1 x + b_2, \\ B_3(x) &= b_0 x^3 + 3b_1 x^2 + 3b_2 x + b_3, \\ B_4(x) &= b_0 x^4 + 4b_1 x^3 + 6b_2 x^2 + 4b_3 x + b_4, \\ B_5(x) &= b_0 x^5 + 5b_1 x^4 + 10b_2 x^3 + 10b_3 x^2 + 5b_4 x + b_5 \end{aligned}$$

and more generally

$$(12.14) \qquad B_k(x) = \sum_{p=0}^{k} \binom{k}{p} b_p x^{k-p},$$

a formula which, here again, implies (12) for any choice of the $b_k$. It depends only on (12) and is not sufficient to determine the $b_p$; but (13) can be written

$$(12.15) \qquad b_0 + \binom{k}{1} b_1 + \ldots + \binom{k}{k-1} b_{k-1} = 0 \qquad \text{for } k \geq 2,$$

i.e.

$$1 + 2b_1 = 0,$$
$$1 + 3b_1 + 3b_2 = 0,$$
$$1 + 4b_1 + 6b_2 + 4b_3 = 0,$$

etc., which allows one to calculate the *Bernoulli numbers* $b_p$ afresh, step-by-step. Euler, who discovered them in another way as we shall see, and must certainly have sought an explicit "formula" for the solution, had come to the conclusion that there probably was none; posterity has confirmed this, and has even quasi-proved it, by observing that the $b_p$ increase with a speed too prodigiously fast to be expressible by algebraic, exponential and other functions. You will find a little later their values for $p \leq 30$, calculated by Euler; it seems quite implausible, considering the taste of the Bernoullis for numerical calculations, that Jakob had not pushed the calculations beyond $b_{10} = 5/66$, but he did not publish them.

Let us now show that instead of (5) one has

$$(12.16) \qquad B_k(x+1) - B_k(x) = kx^{k-1}.$$

This is clear for $k = 0$. If (16) holds for $k-1$ the relation (12) shows that it is true up to an additive constant. But by (13) it is correct without an additive constant for $x = 0$. Whence (16).

Finally, (16) shows that

$$B_k(n+1) = [B_k(n+1) - B_k(n)] + [B_k(n) - B_k(n-1)] + \ldots + $$
$$+ [B_k(2) - B_k(1)] + B_k(1) = k \left( n^{k-1} + \ldots + 1^{k-1} \right) + b_k,$$

whence

$$(12.17) \qquad 1^{k-1} + \ldots + n^{k-1} = [B_k(n+1) - b_k]/k.$$

A comparison with (4) shows that

$$B_k(x+1) = kA_k(x) + b_k$$

or, by (16),

$$B_k(x) = kA_k(x) - kx^{k-1} + b_k$$

$$(12.18) \qquad = x^k + \sum_{p=1}^{k-1} k\binom{k-1}{p-1} a_p x^{k-p} + b_k.$$

(14) then shows that $kb_1 = ka_1 - k$, whence

$$a_1 = b_1 + 1 = 1/2$$

and, for $p > 2$,

$$k\binom{k-1}{p-1} a_p = \binom{k}{p} b_p,$$

whence

$$a_p = b_p/p.$$

In view of Bernoulli's relations between the $a_p$ and the coefficients $A$, $B$, $C$, etc. we see that they are just $b_2$, $b_4$, $b_6$, etc.

We still have to show that, according to the first formulae,

$$(12.19) \qquad b_3 = b_5 = \ldots = 0.$$

Since $b_k = B_k(0) = B_k(1)$ this will follow from the relation

$$(12.20) \qquad B_k(1 - x) = (-1)^k B_k(x).$$

To establish this one puts $C_k(x) = (-1)^k B_k(1 - x)$ and confirms by a one-line calculation that the $C_k$ satisfy the conditions (12) and (13) as well as $C_0(x) = 1$. Now these conditions determine the $B_k$ fully.

Here, to conclude, are the values of the Bernoulli numbers as calculated by Euler:

$$b_0 = 1, \quad b_1 = -1/2, \quad b_2 = 1/6, \quad b_4 = -1/30, \quad b_6 = 1/42,$$
$$b_8 = -1/30, \quad b_{10} = 5/66, \quad b_{12} = -691/2730, \quad b_{14} = 7/6,$$
$$b_{16} = -3617/510, \quad b_{18} = 43867/798, \quad b_{20} = -174611/330,$$
$$b_{22} = 854513/123, \quad b_{24} = -236364091/2730, \quad b_{26} = 8553103/6,$$
$$b_{28} = -23749461029/870, \quad b_{30} = 8615841276005/14322,$$
$$b_{32} = -7709321041217/510, \quad b_{34} = 2577687858367/6.$$

## 13 – The power series for cot $z$

By the definition of the binomial coefficients the recurrence relation

$$b_0 + \binom{n+1}{1} b_1 + \ldots + \binom{n+1}{n} b_n = 0 \qquad \text{for } n > 0$$

can be rewritten as

$$\sum_{0\leq p\leq n} \frac{b_p}{p!} \frac{1}{(n+1-p)!} = \begin{cases} 1 & \text{if} \quad n=0 \\ 0 & \text{if} \quad n\geq 1 \end{cases}$$

and, in this form, evokes the formula for the multiplication of formal power series; to be precise, it is equivalent to the identity

$$(13.1) \qquad \sum_0^\infty b_p X^p/p! \sum_0^\infty X^q/(q+1)! = 1,$$

or, multiplying by $X$, to

$$[\exp(X) - 1] . \sum b_p X^{[p]} = X$$

where, we recall, we have put $X^{[p]} = X^p/p!$. We do not know the radius of convergence of the series $b_p z^{[p]}$ *a priori*, but we know that the power series

$$(13.2) \qquad z^{-1}(e^z - 1) = 1 + z/2! + z^2/3! + \ldots$$

converges for any $z$. By the general theorems on analytic functions (Chap. II, n° 22, particular case of Theorem 17), we know that the reciprocal of the function (2) admits an expansion in a power series on a neighbourhood of $z = 0$; by (1), this series must be $\sum b_p z^{[p]}$. This shows on the one hand that the radius of convergence $R$ of this series is not zero – a nonobvious result since for the moment we do not know the order of magnitude of the $b_n$ – and on the other hand that

$$(13.3) \quad z/(e^z - 1) = \sum b_n z^{[n]} = 1 - z/2 + z^2/12 - z^4/720 + \ldots$$

for $|z|$ small enough. In fact, and as we shall see with the help of general theorems on analytic functions, the relation (3) is valid in the *largest* disc of centre 0 where the left hand side is analytic or holomorphic, i.e. where $e^z - 1$ does not vanish, whence

$$R = 2\pi,$$

a result which we shall find again a little later, without recourse to Cauchy or Weierstrass.

In the formula (1), let us replace the constants $b_p$ by the *Bernoulli polynomials*

$$B_p(t) = \sum \binom{p}{k} b_{p-k} t^k = p! \sum_{m+n=p} b_m t^n/m!n!;$$

it follows that

$$\begin{aligned} \sum B_p(t) X^p/p! &= \sum b_m t^n X^{m+n}/m!n! = \sum b_m X^{[m]}(tX)^{[n]} = \\ &= \exp(tX) \sum b_m X^{[m]}, \end{aligned}$$

whence, in view of of (3),

$$\text{(13.4)} \qquad \sum B_p(t) z^{[p]} = z e^{tz} / (e^z - 1),$$

a relation valid, here again and for the same reasons as above, for $|z| < 2\pi$; on the contrary, $t$ can be an arbitrary complex number.

From this one may deduce the power series expansions of the functions $\coth z$ and $\cot z$. For the first, observe that, by (3),

$$z.\coth z = z.\frac{e^z + e^{-z}}{e^z - e^{-z}} = z.\frac{e^{2z} + 1}{e^{2z} - 1} = z + \frac{2z}{e^{2z} - 1} = z + \sum b_n (2z)^n / n!$$

whence

$$\text{(13.5)} \qquad \begin{aligned} z.\coth z \;=\; & 1 + z^2/3 - z^4/45 + 2z^6/945 - \\ & - z^8/4725 + 2z^{10}/18711 - \ldots. \end{aligned}$$

For $z.\cot z = iz.\coth iz$ one then obtains

$$\text{(13.6)} \quad \begin{aligned} z.\cot z \;=\; & 1 - z^2/3 - z^4/45 - 2z^6/945 - z^8/4725 - \ldots, \\ =\; & 1 - \sum |b_{2n}| \, (2z)^{[2n]}. \end{aligned}$$

Now we saw at the end of n° 22 of Chap. II that if one puts

$$\cot x = 1/x - c_1 x - c_3 x^3 - \ldots,$$

then

$$\pi^{2p} c_{2p-1} = 2 \sum 1/n^{2p} = 2\zeta(2p).$$

Comparing with (6), we see that $c_{2p-1} = |b_{2p}| \, 2^{2p} / (2p)!$, whence

$$\text{(13.7)} \qquad |b_{2p}| = \frac{2(2p)!}{(2\pi)^{2p}} \zeta(2p),$$

which reduces the calculation of the sums $\sum 1/n^{2p}$ to that of the Bernoulli numbers. Stirling's formula, which we shall establish in a little while, will show that $b_{2p}$ increases very rapidly when $p \to +\infty$, as the first numerical values have already suggested.

The formula (7) enables one to calculate the radius of convergence $R = 2\pi$ of the power series $\sum b_n z^n / n!$ directly; indeed,

$$\frac{1}{2} \sum_{n \geq 2} |b_n z^n| / n! = \sum_{n=2}^{\infty} \zeta(2n) \, (|z|/2\pi)^{2n} = \sum_n \sum_p (|z|/2\pi p)^{2n},$$

and since this is a series with positive terms, the convergence of the left hand side is equivalent to the unconditonal convergence of the double series obtained (Chap. II, n° 18, Theorem 13) so presupposes, in particular, that of the partial series obtained by summing over $n$ for given $p$; this requires $|z|/2\pi p < 1$ for every $p \geq 1$ and thus $|z| < 2\pi$. Convergence for $|z| < 2\pi$ is then obtained by interchanging the summations with respect to $n$ and $p$ and recognising the convergence of the series $\sum |z|^2 / \left(4\pi^2 p^2 - |z|^2\right)$.

An even quicker method is to remark that, for $s > 1$,

$$\frac{1}{s-1} = \int_1^{+\infty} x^{-s} dx < \zeta(s) < 1 + \int_1^{+\infty} x^{-s} dx = \frac{s}{s-1}$$

(Chap. V, (24.1)), so that $\zeta(2p)$ lies between 1 and 2 for any $p \geq 1$, whence $b_{2p} z^{2p}/(2p)! \asymp (z/2\pi)^{2p}$ and

$$b_{2p} \asymp (2p)!/(2\pi)^{2p}.$$

## 14 – Euler and the power series for arctan $x$

The sums of powers and the Bernoulli numbers reappear *chez* Euler in 1739 when he calculates the integral

$$\arctan x = \int_0^x \frac{dt}{1+t^2}$$

by the method of Cavalieri and others, i.e. as the limit of the Riemann sums $s_n = \sum nx/\left(n^2 + p^2 x^2\right)$ corresponding to the subdivisions of $[0,x]$ into intervals of length $x/n$, the sum being extended over the $p \in [1,n]$. Since

$$\frac{nx}{n^2 + p^2 x^2} = \frac{x/n}{1 + p^2 x^2/n^2} = \frac{x}{n} \sum_{k \geq 0} (-1)^k \frac{p^{2k} x^{2k}}{n^{2k}},$$

the Riemann sum considered can be written

(14.1)     $$s_n = \sum_{k=0}^{\infty} (-1)^k \left(1^{2k} + 2^{2k} + \ldots + n^{2k}\right) x^{2k+1}/n^{2k+1},$$

which reintroduces the sums of powers, here the even powers, of the first $n$ integers. One remarks in passing that the first series converges only if $|px/n| < 1$, i.e. $|x| < 1$ since $p \in [1,n]$, but this is a detail.

Without referring explicitly to the Bernoulli formulae, Euler uses them to write that

$$\begin{aligned} s_n &= nx/n - \left(n^3/3 + n^2/2 + n/6\right) x^3/n^3 + \\ &\quad + \left(n^5/5 + n^4/2 + n^3/3 - n/30\right) x^5/n^5 + \ldots \\ &= x - \left(1/3 + 1/2n + 1/6n^2\right) x^3 + \end{aligned}$$

$$+ \left(1/5 + 1/2n + 1/3n^2 - 1/30n^4\right) x^5 - \ldots$$
$$= \left(x - x^3/3 + x^5/5 - \ldots\right) - \left(x - x^3 + x^5 - x^7 + \ldots\right) x^2/2n -$$
$$- \left(x - 2x^3 + 3x^5 - 4x^7 + \ldots\right) x^2/6n^2 -$$
$$- \left(x - 5x^3 + 14x^5 - 30x^7 + \ldots\right) x^4/30n^4 -$$
$$- \left(x - 28x^3/3 + 42x^5 - 132x^7 + \ldots\right) x^6/42n^6 + \&c.$$

The expressions between ( ) may seem bizarre to you, but for Euler it is obvious that the coefficient of $x^m/?n^m$ $(m = 2, 4, \ldots)$ is the series

$$v_m(x) = x - \frac{(m+1)(m+2)}{2.3}x^3 + \frac{(m+1)(m+2)(m+3)(m+4)}{2.3.4.5}x^5 - \ldots,$$

so obvious that he does not prove it, and for good reason: he would have to use (12.14) and (12.17), which he does not write. Moritz Cantor, though, who has seen many other displays of acrobatics, tells us (p. 673) "its infinite form does not please Euler and he launches into a stunning [verblüffende] transformation" of his formulae.

Indeed, using the binomial series for a negative integral exponent,

$$
\begin{aligned}
mv_m(x) &= mx - m(m+1)(m+2)x^3/3! + \\
&\quad + m(m+1)(m+2)(m+3)(m+4)x^4/4! - \ldots = \\
&= \left[(1-ix)^{-m} - (1+ix)^{-m}\right]/2i = \\
&= \left[(1+ix)^m - (1-ix)^m\right]/2i \left(1+x^2\right)^m = \\
&= \left[mx - m(m-1)(m-2)x^3/3! + \right. \\
&\quad \left. + m(m-1)\ldots(m-4)x^4/4! + \ldots\right]/\left(1+x^2\right)^m
\end{aligned}
$$

by the binomial theorem. Finally one finds easily that

$$
(14.2) \qquad s_n = \left(x - x^3/3 + x^5/5 - x^7/7 + \ldots\right) -
$$
$$
- \frac{x^3}{2n(1+x^2)} - \frac{x^2}{2.6n^2(1+x^2)^2}\cdot\frac{2x}{1} -
$$
$$
- \frac{x^4}{4.30n^4(1+x^2)^4}\left(\frac{4x}{1} - \frac{4.3.2}{1.2.3}x^3\right) - \ldots.
$$

For $x = 1$ for example, in which case the first term of (3) equals $\pi/4$, one finds

$$
\begin{aligned}
\pi &= \frac{4n}{n^2+1} + \frac{4n}{n^2+4} + \frac{4n}{n^2+9} + \ldots + \frac{4n}{n^2+n^2} \\
&\quad + \frac{1}{6}\cdot\frac{1}{1n^2} - \frac{1}{42}\cdot\frac{1}{2^3.3n^6} + \frac{5}{66}\cdot\frac{1}{5n^{10}} - \ldots,
\end{aligned}
$$

a formula "correspondingly more exact as $n$ is large" according to Euler who immediately adds that despite appearances, the series (2) converges only "up to a certain rank", "whatever that means" after which its terms again start to increase ...

Recall that if one expands $1/(1+t^2)$ as a geometric progression the integral for $\arctan x$ immediately gives

$$\arctan x = x - x^3/3 + x^5/5 - \dots$$

for $|x| < 1$, which is the first term of (2). I do not know what Euler had in mind in publishing his "stunning" calculations, but one has to admit that his introduction of the Bernoulli numbers into the machine leads, as always with him, to mathematical pyrotechnics.

The situation and the calculations would in fact be more lucid if instead of starting from the function $1/(1 + x^2)$ one started from an "arbitrary" function $f$. For let us write

(14.3) $$\int_0^1 f(t)dt = \lim \frac{1}{n} \sum_{p=0}^{n-1} f(p/n) = \lim \mu_n(f)$$

and use the Maclaurin series

(14.4) $$f(x) = \sum f^{(k)}(0)x^k/k!$$

which replaces the geometric series $1/(1+x^2) = 1 - x^2 + x^4 - \dots$ Calculating formally – Euler never did otherwise –, we find, using (12.14) and (12.17),

$$
\begin{aligned}
\mu_n(f) &= \sum_{\substack{k \geq 0 \\ p < n}} \frac{f^{(k)}(0)}{n^{k+1}k!} p^k = \sum_{k \geq 0} \frac{f^{(k)}(0)}{n^{k+1}k!(k+1)} \left[ B_{k+1}(n) - b_{k+1} \right] = \\
&= \sum_{\substack{k \geq 0 \\ p \leq k}} \frac{f^{(k)}(0)}{n^{k+1}(k+1)!} \binom{k+1}{p} b_p n^{k+1-p} = \\
&= \sum_{p \geq 0} \frac{b_p}{n^p} \sum_{k=p}^{\infty} \binom{k+1}{p} f^{(k)}(0)/(k+1)!
\end{aligned}
$$

or, putting $k = p + h$,

$$
\begin{aligned}
\mu_n(f) &= \sum_{p=0}^{\infty} n^{-p} b^p \sum_{h=0}^{\infty} \binom{h+p+1}{p} f^{(h+p)}(0)/(h+p+1)! \\
&= \sum_{p=0}^{\infty} \frac{b_p}{p! n^p} \sum_{h=0}^{\infty} f^{(h+p)}(0)/(h+1)!.
\end{aligned}
$$

The series

(14.5) $\qquad f^{(p)}(0)/1! + f^{(p+1)}(0)/2! + f^{(p+2)}(0)/3! + \ldots,$

involving $h$ is the Maclaurin series of $f^{(p-1)}(1)$ without its first term $f^{(p-1)}(0)$. Thus one finds

(14.6) $\qquad \mu_n(f) = \displaystyle\sum_{p=0}^{\infty} \frac{b_p}{p!n^p} \left[ f^{(p-1)}(1) - f^{(p-1)}(0) \right].$

For $p = 0$, one has $b_0 = 1$ and there remains $f^{(-1)}(1) - f^{(-1)}(0)$, where $f^{(-1)}$ is in reality a primitive $F$ of $f$ as one sees on putting $p = 0$ in (5). The term $p = 0$ in (6) is precisely the integral of $f$ over $[0, 1]$ that we are calculating, so that (6) actually expresses the difference between the latter and the sum $\mu_n(f)$. For $p = 1$, one has $b_1 = -\frac{1}{2}$ and one finds $[f(0) - f(1)]/2n$. For $p \geq 2$, the odd $p$ do not feature. By the definition of $\mu_n(f)$, multiplying the two sides by $n$ and adding $f(1)$ to the two sides, one thus finds in the final analysis the formula

(14.7) $\qquad f(0) + f(1/n) + \ldots + f(n/n) =$

$$= n \int_0^1 f(t)dt + \frac{1}{2}[f(0) + f(1)] +$$

$$+ \sum_{p=1}^{\infty} \frac{b_{2p}}{(2p)!n^{2p-1}} \left[ f^{(2p-1)}(1) - f^{(2p-1)}(0) \right].$$

One would find Euler's results again – apart of course from the "stunning" transformation which is very specific to the function $1/(1 + t^2)$ – replacing the function $t \mapsto f(t)$ by $t \mapsto f(tx)$, whose derivatives are the functions $f^{(k)}(tx)x^k$; this transforms the integral (3) of $f$ over $[0, 1]$ into its integral over $[0, x]$.

If on the other hand one applies this formula to $t \mapsto f(nt)$, which replaces $f^{(k)}(x)$ by $n^k f^{(k)}(nx)$, one obtains

(14.8) $\qquad f(0) + f(1) + \ldots + f(n) =$

$$= \int_0^n f(t)dt + \frac{1}{2}[f(n) + f(0)] +$$

$$+ \sum_{p=1}^{\infty} \frac{b_{2p}}{(2p)!} \left[ f^{(2p-1)}(n) - f^{(2p-1)}(0) \right].$$

It goes without saying that these purely formal calculations are in general *meaningless* apart from the case where $f$ is a polynomial and where the Maclaurin series reduces to a finite sum. (*Exercise.* Verify the formula for $f(x) = x^k$.) Even if the function $f$ is represented everywhere by a convergent Maclaurin series, it is not clear that these permutations and groupings of terms are legitimate, and in fact the result (8) is almost always a divergent series. If on the other hand you apply (8) to a function of period 1, all the terms of the right hand side are zero apart from the first two, and you will find, for $n = 1$ for example, the fanciful formula

$$\frac{1}{2}[f(0) + f(1)] = \int_0^1 f(t)dt \dots$$

But this is a beautiful exercise in calculation, and we shall see later that one can, as one does in replacing the Taylor *series* by a finite sum with a controllable "remainder", obtain a result which yields a very precise asymptotic evaluation of the left hand side of (8).

## 15 – Euler, Maclaurin and their summation formula

The relation (14.8), which is the formal version of the *Euler-Maclaurin summation formula*, had in fact already been published by Euler in 1736 in the *Commentarii Academiae Petropolitanae* and would appear again in Maclaurin's *Treatise of Fluxions* of 1741; their methods are almost identical, and there is every reason to believe that Maclaurin had not seen Euler's memoir before sending his manuscript to the printer. In both cases we have formal calculations. Let us, for example, set out the heroic Scot's method, who, at this late date, still militates on Newton's side.

Starting (in modern notation) from the formula

$$(15.1) \qquad \int_0^1 f^{(p)}(t)dt = \sum_{n=0}^{\infty} f^{(p+n)}(0)/(n+1)!$$

which one obtains by integrating the Taylor (or, on this occasion, Maclaurin) series of $f^{(p)}(t)$ or, for $p = 0$, of a primitive of $f$ as in (14.5), Maclaurin tries to express $f(0)$ as an (infinite ...) linear combination

$$(15.2) \qquad f(0) = \sum_{p=0}^{\infty} a_p \int_0^1 f^{(p)}(t)dt$$

of the left hand sides, with *universal* constants $a_p$, i.e. valid for every function $f$. On substituting the expressions (1) in (2), he finds the identity

$$(15.3) \qquad \sum a_p f^{(p+n)}(0)/(n+1)! = f(0)$$

summing over all pairs of integers $p, n \geq 0$. Since the derivatives can be chosen arbitrarily, as Emile Borel proved a little later, it is necessary (or it suffices) that the terms containing the derivatives of order $\geq 1$ disappear, i.e. that for every $k \geq 1$ the total coefficient of $f^{(k)}(0)$ corresponding to the pairs $(n,p)$ such that $n + p = k$ should be zero. This can be written

$$(15.4) \quad a_0/(k+1)! + a_1/k! + \dots + a_k/1! = \sum a_p/(k-p+1)! = 0;$$

now clearly $a_0 = 1$ since $f(0)$ occurs in (3) only for the pair $(0, 0)$. Maclaurin and Euler then deduced the numerical values of the $a_p$, and if one puts $a_p = b_p/p!$, though they did not, one again finds that the coefficients satisfy

(15.5)
$$b_0 = 1, \qquad \sum_{0 \leq p \leq k} \binom{k+1}{p} b_p = 0$$

since the binomial coefficient equals $(k+1)!/p!(k-p+1)!$. Miracle: the $b_p$ are the Bernoulli numbers!

This done, (2) can be written

(15.6)
$$f(0) = \int_0^1 f(t)dt - \frac{1}{2}[f(1) - f(0)] +$$
$$+ \sum \frac{b_{2p}}{(2p)!} \left[ f^{(2p-1)}(1) - f^{(2p-1)}(0) \right]$$

as these Gentlemen clearly affirm, after calculating the first $b_p$, that they vanish for $p = 3, 5$, &c. However, like everyone else at the time, they provided only the first terms of the series.

On replacing $t \mapsto f(t)$ by $t \mapsto f(t+x)$ one obtains

(15.7)
$$f(x) = \int_x^{x+1} f(t)dt - \frac{1}{2}[f(x+1) - f(x)] +$$
$$+ \sum \frac{b_{2p}}{(2p)!} \left[ f^{(2p-1)}(x+1) - f^{(2p-1)}(x) \right] ;$$

on replacing $x$ by $p$ and adding from 0 to $n-1$ one recovers (14.8).

## 16 – The Euler-Maclaurin formula with remainder

Following these excursions into the history of the subject, let us move on to the correct methods, due to Jacobi (1834) for the expression of the remainder, and to H. Wirtinger (1902) for the method of integration by parts, as Hairer and Wanner tell us (p. 162). This is exactly the method we explained for obtaining Taylor's formula (Chap. V, n° 18), except that instead of choosing polynomials $P_k$ satisfying $P_0 = 1$, $P_k' = P_{k-1}$ and vanishing at the right end of the interval of integration, one chooses polynomials taking the same value at its two end-points. If these are 0 and 1, we must then assume $P_k = B_k$ and the method expounded in Chap. V leads, under the same hypotheses, to the relation

$$f(1) - f(0) = \sum_{p=1}^r \frac{(-1)^{p-1}}{p!} f^{(p)}(x) B_p(x) \bigg|_0^1 + \frac{(-1)^r}{r!} \int_0^1 f^{(r+1)}(x) B_r(x)dx.$$

Since $B_1(x) = x - \frac{1}{2}$ and $B_p(0) = B_p(1) = b_p$ for $p \geq 2$, it follows that

$$f(1) - f(0) \;=\; \frac{1}{2}[f'(0) + f'(1)] + \sum_{p=2}^{r}(-1)^{p-1}b_p \left[f^{(p)}(1) - f^{(p)}(0)\right]/p! +$$

$$+ \frac{(-1)^r}{r!} \int_0^1 f^{(r+1)}(x)B_r(x)dx.$$

Since $b_3 = b_5 = \ldots = 0$ one can replace $(-1)^{p-1}$ by $-1$ in the $\sum$; by applying the result to a primitive of $f$, which transforms $f(1) - f(0)$ into the integral of $f$ over $[0,1]$, and $f'(0) + f'(1)$ into $f(0) + f(1)$, one finally finds

$$(16.1) \quad \frac{1}{2}[f(0) + f(1)] \;=\; \int_0^1 f(x)dx + \sum_{p=2}^{p=r} b_p \left[f^{(p-1)}(1) - f^{(p-1)}(0)\right]/p!$$

$$- \frac{(-1)^r}{r!} \int_0^1 f^{(r)}(x)B_r(x)dx.$$

To obtain the Euler-Maclaurin formula one considers a function $f$ defined and of class $C^r$ on an interval $[0,n]$, applies (1) to each function $f(x + k)$, and adds the relations so obtained. On the left hand side one finds

$$\frac{1}{2}[f(0)+f(1)]+\ldots+\frac{1}{2}[f(n-1)+f(n)] = f(0)+\ldots+f(n)-\frac{1}{2}[f(0)+f(n)].$$

On the right hand side the sum of the integrals in $f$ yields that of $f$ over $[0,n]$. In the $\sum$ on the right hand side all the terms cancel in pairs, except for the values of the derivatives at $n$ and $0$. Finally, to write the sum of the integrals conveniently in terms of $B_r$, one introduces the function $B_r^*(x)$ of period 1 equal to $B_r(x)$ on $[0,1]$, clearly given by

$$(16.2) \qquad\qquad B_r^*(x) = B_r(x - [x])$$

where $[x]$ is the integer part of $x$; then

$$(16.3) \qquad \int_0^1 f^{(r)}(x+k)B_r(x)dx = \int_k^{k+1} f^{(r)}(x)B_r^*(x)dx,$$

which, by addition, yields the integral of the same function over $[0,n]$. For $f$ of class $C^{2r}$ one then has the final result, namely

$$(16.4) \quad f(0) + \ldots + f(n) \;=\; \int_0^n f(x)dx + \frac{1}{2}[f(0) + f(n)] +$$

$$+ \sum_{p=1}^{p=r} \frac{b_{2p}}{(2p)!} \left[f^{(2p-1)}(n) - f^{(2p-1)}(0)\right] -$$

$$- \frac{1}{(2r)!} \int_0^n f^{(2r)}(x)B_{2r}^*(x)dx.$$

For $r = 3$, for example,

$$f(0) + \ldots + f(n) =$$

$$= \int_0^n f(x)dx + [f(0) + f(n)]/2 + [f'(n) - f'(0)]/12 -$$

$$- [f'''(n) - f'''(0)]/720 + [f'''''(n) - f'''''(0)]/30240 -$$

$$- \frac{1}{6!} \int_0^n f^{(6)}(x)B_6^*(x)dx.$$

*Exercise.* Let $f$ be a function of class $C^{2r}$ on $\mathbb{R}$. Show that

$$\sum_{n \in \mathbb{Z}} f(n) = \int_{-\infty}^{\infty} f(x)dx - \frac{1}{(2r)!} \int_{-\infty}^{\infty} f^{(2r)}(x)B_{2r}^*(x)dx$$

subject to hypotheses to be found.

## 17 – Calculating an integral by the trapezoidal rule

If, in (16.4), one transfers the term $\frac{1}{2}[f(0) + f(n)]$ to the left hand side, it becomes

$$\frac{1}{2}[f(0) + f(1)] + \ldots + \frac{1}{2}[f(n-1) + f(n)]$$

and is simply the sum of the areas of the trapezia constructed on the verticals joining the integer points of the $x$ axis to the corresponding points of the curve. If $f$ is a function of class $C^{2r}$ on $[0, 1]$ and if one applies the preceding results to the function $f(x/n)$, defined between 0 and $n$, which replaces $f^{(k)}(x)$ by $n^{-k}f^{(k)}(x/n)$, one immediately finds the relation

$$\int_0^1 f(x)dx \; = \; [f(0) + f(1/n)]/2n + \ldots + [f(1 - 1/n) + f(1)]/2n -$$

$$\text{(17.1)} \qquad - [f'(1) - f'(0)]/12n^2 + [f'''(1) - f'''(0)]/720n^4 - \ldots -$$

$$- b_{2r}\left[f^{(2r-1)}(1) - f^{(2r-1)}(0)\right]/(2r)!n^{2r} +$$

$$+ \frac{1}{(2r)!n^{2r+1}} \int_0^1 f^{(2r)}(x)B_{2r}^*(nx)dx.$$

The left hand side represents the "curvilinear" area $m(f)$ bounded by the graph of $f$, the $x$ axis and the verticals $x = 0$ and $x = 1$. On the right hand side one then has the sum $T_n(f)$ of the areas of the trapezia inscribed in the graph of $f$ and having as vertical sides the lines $x = k/n$. If generally one puts

$$c_p(f) = b_{2p}\left[f^{(2p-1)}(1) - f^{(2p-1)}(0)\right]/(2p)!,$$

one then finds

$$\text{(17.2)} \quad T_n(f) = m(f) + c_1(f)/n^2 + \ldots + c_r(f)/n^{2r} + (\ldots)/n^{2r+1}$$

where

(17.3) $$(\dots) = -\frac{1}{(2r)!} \int_0^1 f^{(2r)}(x) B_{2r}^*(nx) dx.$$

This expression remains bounded as $n$ increases indefinitely, for the functions $B^*$ are of period 1 and are polynomials on $[0,1]$, so bounded in $\mathbb{R}$. The relation (2) then can be written

(17.4)    $T_n(f) = m(f) + c_1(f)/n^2 + \dots + c_r(f)/n^{2r} + O\left(1/n^{2r+1}\right)$

and shows that, if $f$ is $C^\infty$, the difference $T_n(f) - m(f)$ is represented by the asymptotic series $\sum c_p(f)/n^p$ in the sense of n° 10. This also means that

$$T_n(f) - m(f) \sim c_1(f)/n^2, \qquad T_n(f) - m(f) - c_1(f)/n^2 \sim c_2(f)/n^4,$$

etc.

The situation becomes curious if $f$ is the restriction to $[0,1]$ of a periodic function that is indefinitely differentiable on $\mathbb{R}$ and not only on $[0,1]$. Then $f^{(k)}(1) = f^{(k)}(0)$ for any $k$, so (4) reduces to

$$m(f) = T_n(f) + O(1/n^k) \qquad \text{for any } k.$$

**18 – The sum $1 + 1/2 + \dots + 1/n$, the infinite product for the $\Gamma$ function, and Stirling's formula**

By simple arguments one may prove the existence of a constant $C$, or $\gamma$, *Euler's constant*, such that

(18.1)          $\lim(1 + 1/2 + \dots + 1/n - \log n) = C = \gamma,$

a result which provides an excellent order of magnitude for $1 + \dots + 1/n$ for $n$ large. But the Euler-Maclaurin formula provides a complete asymptotic expansion for it.

First of all, consider again the general formula (16.4) and assume that in it the derivative $f^{(2r)}(x)$ is absolutely integrable on the interval $[0, +\infty]$. This is then true for $f^{(2r)}(x) B_{2r}^*(x)$ too, since the functions $B^*$ are bounded. The integral from 0 to $n$ is then the difference between the integrals from 0 to $+\infty$ and from $n$ to $+\infty$. Putting

(18.2)        $C(f) = \dfrac{1}{2}f(0) - \displaystyle\sum_{p=1}^{r} b_{2p} f^{(2p-1)}(0)/(2p)! -$

$$-\frac{1}{(2r)!} \int_0^{+\infty} f^{(2r)}(x) B_{2r}^*(x) dx,$$

it follows that

$$(18.3) \quad f(0) + \ldots + f(n) = \int_0^n f(x)dx + C(f) + \frac{1}{2}f(n) +$$

$$+ \sum_{p=1}^{r} b_{2p}f^{(2p-1)}(n)/(2p)! + \rho_r(n)$$

with a "remainder" $\rho_r(n)$ given by

$$(18.4) \qquad \rho_r(n) = \frac{1}{(2r)!} \int_n^{+\infty} f^{(2r)}(x)B_{2r}^*(x)dx.$$

If $f$ and its successive derivatives tend to 0 at infinity then

$$C(f) = \lim \left[ f(0) + \ldots + f(n) - \int_0^n f(x)dx \right]$$

for every $r$: the "remainder" $\rho_r(n)$ tends to 0 since the function under the $\int$ sign is by hypothesis absolutely integrable at infinity. This shows that the constant $C(f)$ does not depend on the number $r$ chosen. One might call it "Euler's constant for $f$" because he had already exhibited it (notation $C$ or $\gamma$) in the case where $f(x) = 1/x$.

In this particular case, and in other similar cases of functions which are defined for $x > 0$ but infinite at $x = 0$, one has to modify the formulae, i.e. consider the sum $f(1) + \ldots + f(n)$. This comes down to applying the initial formula to the function $f(x+1)$ or, equivalently, to replacing the limit 0 by 1 in the derivatives and integrals. For $f(x) = 1/x$ the derivatives at $x = n$ are easily calculated and the remainder is $O\left(n^{-2r}\right)$ since the function integrated is $O\left(x^{-2r-1}\right)$. On replacing $r$ by $r+1$ the formula (3) can in this case be written

$$(18.5) \quad 1 + 1/2 + \ldots + 1/n =$$
$$= \log n + C + 1/2n - 1/12n^2 + 1/120n^4 -$$
$$- 1/252n^6 + 1/240n^8 - 1/132n^{10} + 691/32760n^{12} -$$
$$- 1/12n^{14} + \ldots - b_{2r}/2r.n^{2r} + O\left(1/n^{2r+2}\right).$$

Thus one sees that the sum $1 + 1/2 + \ldots + 1/n = s_n$ is approximately equal to $\log n$, the error being approximately equal to Euler's constant

$$C = \gamma = 0,577\ 215\ 664\ldots.$$

But (5) is much more precise. For example, in the simplest formula

$$(18.6) \qquad s_n = \log n + C + 1/2n + \int_n^{+\infty} x^{-2}B_1^*(x)dx,$$

one has $|B_1^*(x)| \leq \frac{1}{2}$ since $B_1^*(x) = B_1(x) = x - \frac{1}{2}$ between 0 and 1. The integral in (6) therefore lies between $-1/2n$ and $1/2n$, so that, on adding the term $1/2n$ to the formula, one obtains a result between $0$ and $1/n$. In other words,

(18.7)                    $s_n = \log n + C + \theta_n/n$      with $0 \le \theta_n \le 1$.

For $n = 10^6$ one thus finds $s_n = 6.\log 10 + C$ to within $10^{-6}$; since $10$ lies between $e^2$ and $e^3$ its log lies between $2$ and $3$, which shows that $s_n$ lies between $12$ and $19$; certainly a not very exact result, but obtained in probably less time than it would take a machine to calculate a million terms of the harmonic series to a dozen decimal places so as to obtain the result to within $10^{-6}$.

To improve this rough estimate one needs to know that

$$\log 10 = 2,302\ 585\ 092,$$

a result generously provided, among many others, by the Founders, whence one deduces $s_n = 14,392\ 726\ldots$ The same argument shows that on calculating the sum of the first $10^{100}$ terms of the harmonic series one finds a result equal, to within $1$, to $100.\log 10 \sim 230$. One finds in Hairer and Wanner, II.10, apart from the very precise numerical results, a reproduction p. 167 of a letter from Euler to Johann Bernoulli, dating from 1740, in Latin, and in an impeccable script, where the former informs the latter of his numerical results.

From this one may deduce an expansion of the function $\Gamma$ as an *infinite product*. We have already seen [Chap. V, eqn. (23.6)] that

(18.8)                    $\Gamma(s) = \lim n! n^s / s(s+1)\ldots(s+n)$

for $\mathrm{Re}(s) > 0$. The reciprocal of the right hand side can again be rewritten as

(18.9)                    $s.\lim(1+s)(1+s/2)\ldots(1+s/n)n^{-s};$

now $n^{-s} = e^{-s.\log n}$ and $\log n = (1 + 1/2 + \ldots + 1/n) - C + o(1)$ by (6); so

$$n^{-s} \sim e^{-s(1+1/2+\ldots+1/n-C)} = e^{Cs}e^{-s}e^{-s/2}\ldots e^{-s/n},$$

whence

$$(9) = se^{Cs}.\lim \prod_{p=1}^{n}(1+s/p)e^{-s/p}.$$

But, for $p$ large,

$$(1+s/p)e^{-s/p} = (1+s/p)\left(1 - s/p + O\left(1/p^2\right)\right) = 1 + O\left(1/p^2\right)$$

is, for $\mathrm{Re}(s) > 0$ and even for every $s \in \mathbb{C}$, the general term of an absolutely convergent infinite product (Chap. IV, n° 17, Theorem 13), a product whose value is $\neq 0$ for every $s \neq -1, -2, \ldots$ Returning to (8), we conclude that, for $\mathrm{Re}(s) > 0$, the $\Gamma$ function is everywhere $\neq 0$, and is given by

$$(18.10) \qquad 1/\Gamma(s) = se^{Cs} \prod_{1}^{\infty} (1 + s/n)e^{-s/n}$$

where $C = \gamma$ is Euler's constant, a famous result due to the latter.

This formula is in fact valid for any $s \in \mathbb{C}$. First, it is clear that on retracing the calculations which brought us from (10) to (8), we have

$$(18.11) \qquad se^{Cs} \prod_{1}^{\infty} (1 + s/n)e^{-s/n} = \lim s(s + 1) \ldots (s + n)/n^s n!$$

for any $s \in \mathbb{C}$; the limit exists, like the infinite product, on all $\mathbb{C}$ and not only for $\mathrm{Re}(s) > 0$. But if one denotes the right hand side of (11) by $f(s)$, one has, for any $s \in \mathbb{C}$,

$$sf(s + 1) = \lim s(s + 1) \ldots (s + n + 1)/n^{s+1} n! = f(s)$$

since $n^{s+1} n! \sim (n + 1)^s (n + 1)!$ as one sees immediately. Now we know (Chap. V, n° 22, Example 1) that $\Gamma(s + 1) = s\Gamma(s)$ for $\mathrm{Re}(s) > 0$. The two members of this formula being holomorphic in $\mathbb{C}$, the negative integers removed, (Chap. V, n° 25, Example 5), *and therefore analytic* – in mathematics, one may stoop to swindles so long as one warns the victims in advance – the equality is valid without restriction (principle of analytic continuation: Chap. II, n° 20). Now consider the product $g(s) = f(s)\Gamma(s)$. We know that $g(s) = 1$ for $\mathrm{Re}(s) > 0$ by (10), and that $g(s + 1) = g(s)$ for any nonnegative integer $s$. It follows clearly that $g(s) = 1$ everywhere, qed.

Combining (10) and (11), one also finds that

$$(18.12) \qquad 1/\Gamma(s) = \lim s(s + 1) \ldots (s + n)/n^s n!$$

for any $s \in \mathbb{C}$. Consequently,

$$
\begin{aligned}
1/\Gamma(s)\Gamma(1 - s) &= \\
&= \lim s(s + 1)(s + 2) \ldots (s + n)(1 - s)(2 - s) \ldots (n + 1 - s)/n(n!)^2 = \\
&= s.\lim \left(1^2 - s^2\right)\left(2^2 - s^2\right) \ldots \left(n^2 - s^2\right)(n + 1 - s)/n(n!)^2 = \\
&= s.\lim \left(1 - s^2\right)\left(1 - s^2/2^2\right) \ldots \left(1 - s^2/n^2\right)
\end{aligned}
$$

since $(n + 1 - s)/n$ tends to 1. Whence

$$1/\Gamma(s)\Gamma(1 - s) = s\prod \left(1 - s^2/n^2\right) = \frac{1}{\pi}\sin \pi s$$

[Chap. IV, eqn. (18.16)], a formula due to Euler and which one also writes

$$(18.13) \qquad \Gamma(s)\Gamma(1 - s) = \pi/\sin \pi s.$$

There are all sorts of other ways to establish these properties of the Gamma function, among many others.

Among the functions whose derivatives are integrable at infinity there appears $f(x) = \log x$, for which

$$f^{(r)}(x) = (-1)^{r-1}(r-1)! x^{-r}.$$

Formula (18.3) clearly applies for $r > 1$ and immediately gives

$$(18.14) \qquad \log(n!) \;=\; n \log n - n + 1 + \frac{1}{2}\log n + C(f) + $$
$$+ \sum b_{2p}/2p(2p-1)n^{2p-1} + \rho_r(n);$$

the first three terms come from calculating the integral of $\log x$ from 1 to $n$ (primitive: $x.\log x - x$) and then $\rho_r(n) = O\left(1/n^{2r-1}\right)$ since $f^{(2r)}(x) = O\left(x^{-2r}\right)$ at infinity; this assumes that $r \geq 1$ for otherwise the integral for the remainder would be divergent.

Rather than going over the expansion again, let us just deduce *Stirling's formula* from it, for $r = 2$. In this case we obtain

$$(18.15) \qquad \log(n!) - n\log n + n - \frac{1}{2}\log n - c = 1/12n + O\left(1/n^3\right)$$

where $c = 1 + C(f)$. The left hand side is the log of

$$u_n = n! e^{n-c}/n^{n+\frac{1}{2}}$$

and since the right hand side tends to 0, we see that $u_n$ tends to 1, whence

$$(18.16) \qquad\qquad n! \sim e^c n^n e^{-n} \sqrt{n}.$$

While we have no information on Euler's constant $\gamma$ for the harmonic series – one does not even know whether it is algebraic or transcendental –, we can, here, calculate

$$(18.17) \qquad c = \log\sqrt{2\pi}, \qquad \text{whence } n! \sim \sqrt{2\pi n}(n/e)^n,$$

but the method is not particularly transparent. We start from Wallis' formula (Chap. V, n° 17)

$$\pi/2 \;=\; \lim \frac{2^2 4^2 \ldots (2n)^2}{1^2 3^2 \ldots (2n-1)^2(2n+1)} =$$
$$= \lim \frac{2^4 4^4 \ldots (2n)^4}{1^2 2^2 3^2 \ldots (2n)^2(2n+1)} = \lim \frac{2^{4n}(n!)^4}{\left((2n)!\right)^2 (2n+1)}$$

and write $(2n!)^2 \sim (2n)^{2n+\frac{1}{2}} e^{-2n+c}$ by (16). It follows that

$$\pi/2 \sim \frac{2^{4n} n^{4n+2} e^{-4n+4c}}{(2n)^{4n+1}e^{-4n+2c}(2n+1)} = \frac{n e^{2c}}{2(2n+1)} \sim e^{2c}/4,$$

whence $e^{2c} = 2\pi$ and Stirling's formula (17).

## 19 – Analytic continuation of the zeta function

In the Euler-Maclaurin formula let us choose $f(x) = 1/x^s$ with $\mathrm{Re}(s) > 1$, so that the series

$$(19.1) \qquad \zeta(s) = \sum 1/n^s = \sum f(n)$$

converges. Since here

$$(19.2) \qquad f^{(r)}(x) = (-1)^r s(s+1)\ldots(s+r-1)/x^{s+r},$$

(16.4) can be written

$$(19.3) \quad f(1) + \ldots + f(n) = \int_1^n x^{-s}dx + \frac{1}{2}(1 + n^{-s}) +$$

$$+ \sum_{p=1}^r b_{2p}s(s+1)\ldots(s+2p-2)\left(n^{-s-2p+1} - 1\right)/(2p)! + \rho_r(n)$$

with a remainder

$$\rho_r(n) = \frac{s(s+1)\ldots(s+2r-1)}{(2r)!} \int_1^n B_{2r}^*(x)x^{-s-2r}dx.$$

When $n$ increases indefinitely, the left hand side tends to $\zeta(s)$, the first integral on the right hand side tends to $1/(s-1)$ since $\mathrm{Re}(s) > 1$, the terms containing a power of $n$ tend to 0, and the integral in the remainder converges. Multiplying by $s - 1$, one finds, in the limit,

$$\zeta(s) \;=\; \frac{1}{s-1} + \frac{1}{2} -$$

$$- \sum_{p=1}^r b_{2p}(s+1)s\ldots(s+2p-2)/(2p)! + \sigma_r(s)$$

or, writing out the first terms,

$$(19.4) \qquad \zeta(s) \;=\; \frac{1}{s-1} + \frac{1}{2} + \frac{s}{6.2!} - \frac{s(s+1)(s+2)}{30.4!}$$

$$+ \frac{s(s+1)(s+2)(s+3)(s+4)}{42.6!} + \ldots$$

$$+ b_{2r}\frac{s(s+1)\ldots(s+2r-2)}{(2r)!} + \sigma_r(s),$$

where we have put

$$(19.5) \qquad \sigma_r(s) = \frac{s(s+1)\ldots(s+2r-1)}{(2r)!} \int_1^{+\infty} B_{2r}^*(x)x^{-s-r}dx.$$

These formulae assume $\mathrm{Re}(s) > 1$, but the integral (5) converges for $\mathrm{Re}(s) > 1 - r$, and the other terms of (4) are polynomials in $s$. We may therefore use (4) to *define* $\zeta(s)$ in the half plane $\mathrm{Re}(s) > 1 - r$, apart from the point $s = 1$; and since $r$ is an arbitrary integer $> 0$, in this way we obtain a definition of the zeta function valid in the whole complex plane, the point $s = 1$ deleted.

The point of these calculations is that they furnish a *holomorphic* function on $\mathbb{C} - \{1\}$, equal to $\zeta(s)$ on the half plane $\mathrm{Re}(s) > 1$ where the series converges; to see this it suffices to argue from the integral (5) as we did in Chap. V, n° 25 for the function $\Gamma(s)$: since Bernoulli's function is bounded on $\mathbb{R}$, the function which one integrates, holomorphic in $s$, is, on the whole half plane $\mathrm{Re}(s) + r > 1 + \varepsilon$, dominated up to a constant factor by the function $x^{-1-\varepsilon}$, integrable on $[1, +\infty]$; Theorem 24 bis of Chap. V, n° 25 then yields the result.

In fact, the function $\zeta$ is even (sic) analytic. Not yet having the general Cauchy-Weierstrass theory at our disposal we have to use a workaday method to prove it. We write

$$x^{-s} = \exp(-s.\log x) = \sum (-1)^n s^n \log^n x/n!,$$

substitute this result in the integral (5), and integrate it term-by-term, leaving the justification of this operation until later. We obtain the series

$$(19.6) \quad \sum a_n s^n/n! \quad \text{where} \quad a_n = (-1)^n \int_1^{+\infty} B_{2r}^*(x) \log^n x.x^{-r} dx.$$

Putting $M = \sup |B_{2r}^*(x)|$, we then have

$$(19.7) \qquad\qquad |a_n| \leq M \int_1^{+\infty} \log^n x.x^{-r} dx$$

and since, at the least, we have to satisfy ourselves that the radius of convergence of the power series (6) does not reduce to 0, we need to evaluate the integral (7). Convergence is obvious for $r > 1$ since $\log^n x$ is $O(x^\alpha)$ at infinity, for every $\alpha > 0$. The change of variable $x = e^u$ reduces this integral to $\int u^n e^{(1-r)u} du$ where one integrates now from 0 to $+\infty$. A second change of variable $(r - 1)u = v$ reduces us to

$$(r-1)^{-n-1} \int_0^{+\infty} v^n e^{-v} dv = (r-1)^{-n-1} \Gamma(n+1) = (r-1)^{-n-1} n!$$

by Chap. V, n° 22, Example 1. The inequality (7) then becomes

$$|a_n| \leq Mn!/(r-1)^{n+1}.$$

The series (6) is therefore majorised up to a constant factor by the series with general term $|s|^n/(r-1)^n$, which converges for $|s| < r - 1$.

The term-by-term integration used to obtain (6) is justified by Theorem 20 of Chap. V, n° 23. On the one side, the series to be integrated, with general term

$$u_n(x) = (-1)^n B_{2r}^*(x) \log^n x . x^{-r} s^n / n!,$$

converges normally on every compact subset of $[1, +\infty[$ i.e. on every interval $[1, b]$ with $b < +\infty$, for, putting $M = \sup |B_{2r}^*(x)|$, one has, on this interval,

$$|u_n(x)| \leq M \log^n b . |s|^n / n!,$$

the general term of a convergent series independent of $x \in [1, b]$. On the other hand,

$$\sum |u_n(x)| \leq M . \exp(|s| . \log x) x^{-r} = M x^{|s|-r} = p(x)$$

is a function integrable on $[1, +\infty[$ since, to make the power series (6) converge, we have already had to assume $|s| < r - 1$ and thus $|s| - r < -1$. The formal calculation above is therefore justified.

The integral (5) is therefore an analytic function of $s$ in the disc $|s| < r-1$. So likewise by (4) is the function $(s - 1)\zeta(s)$. But since $r$ is an integer that may be chosen arbitrarily large, it follows that $(s - 1)\zeta(s)$ is analytic on all of $\mathbb{C}$, qed.

We have thus shown that *the function $(s - 1)\zeta(s)$ is the restriction to the half plane* $\mathrm{Re}(s) > 1$ *of an analytic function on all of* $\mathbb{C}$. The latter is unique by the principle of analytic continuation of Chap. II, n° 20. Later we shall see that there is a simple relation between $\zeta(s)$ and $\zeta(1 - s)$.

Formula (19.4), valid for every $s \neq -1$, applies mainly when $s$ is an integer $\leq 0$. The remainder (5) is then zero if one chooses $r$ suitably, when one finds a *rational* value for $\zeta(s)$. One can calculate it for the small values of $r$:

$$
\begin{aligned}
\zeta(0) &= -1/2, & (r = 1) \\
\zeta(-1) &= -1/2 + 1/2 - 1/6.2! = -1/12, & (r = 1) \\
\zeta(-2) &= -1/3 + 1/2 - 2/6.2! = 0, & (r = 2) \\
\zeta(-3) &= -1/4 + 1/2 - 3/6.2! + 3.2/30.4! = 1/120, & (r = 2)
\end{aligned}
$$

etc. In fact,

$$\zeta(1 - 2r) = -b_{2r}/2r, \qquad \zeta(-2r) = 0$$

for any $r \geq 1$, as we shall see in Chapter XII, using other methods.

# VII – Harmonic Analysis
# and Holomorphic Functions

§ 1. Analysis on the unit circle – § 2. Elementary theorems on Fourier series – § 3. Dirichlet's method – § 4. Analytic and holomorphic functions – § 5. Harmonic functions and Fourier series – § 6. From Fourier series to integrals

## 1 – Cauchy's integral formula for a circle

It is not the tradition to treat Fourier series and the theory of analytic functions together. Nevertheless the two theories are closely related. If

$$f(z) = \sum a_n z^n$$

is a power series of radius of convergence $R > 0$ the function

$$(1.1) \qquad f\left(re^{2\pi it}\right) = \sum a_n r^n e^{2\pi int}$$

which, for $0 \leq r < R$, represents $f$ on the circle $|z| = r$ is an absolutely convergent trigonometric series of period 1. It follows [Chap. V, eqn. (5.13)] that

$$(1.2) \qquad \int_0^1 f\left(re^{2\pi it}\right) e^{-2\pi int} dt = \begin{cases} a_n r^n & \text{for } n \geq 0, \\ 0 & \text{for } n < 0. \end{cases}$$

The integral (2) is zero for $n < 0$ since only positive powers $n$ appear in the series (2); this shows, in passing, that the function $t \mapsto f\left(re^{2\pi it}\right)$ is very far from being the most general periodic function.

As we have seen in Chap. V, it follows from this that for $|z| < r$

$$(1.3) \qquad f(z) = \int_0^1 \frac{re^{2\pi it}}{re^{2\pi it} - z} f\left(re^{2\pi it}\right) dt.$$

If we perform the change of variable $\zeta = re^{2\pi it}$ in (3) (a priori forbidden since the values are complex) and if we calculate à la Leibniz, we have $d\zeta = 2\pi i re^{2\pi it} dt$, which allows us to write (3) in Cauchy's form

(1.4)
$$\int_{|\zeta|=r} f(\zeta)\frac{d\zeta}{\zeta - z} = \begin{cases} 2\pi i f(z) & \text{for } |z| < r \\ 0 & \text{for } |z| > r \end{cases}$$

where we integrate along of the circumference $|\zeta| = r$ oriented traditionally; we are in fact dealing with a particular case of a much more general formula – one may integrate along arbitrary closed curves[1] –, obtained by Cauchy much later than (4), and which cannot be obtained by calculations of the preceding type. But (4) nevertheless shows that, in the disc $|z| < r < R$, *one may calculate f from its values on the circumference $|z| = r$* through an explicit formula of the simplest kind.

Why does one find 0 for $|z| > r$ in (4)? Because, putting $u = e^{2\pi i t}$, one may write

(1.5)
$$\frac{ru}{ru - z} = -\frac{ru/z}{1 - ru/z} = -\sum (ru/z)^{n+1}$$

and obtain a convergent series which can be integrated term-by-term in (3). Therefore

(1.6) $$\int_0^1 \frac{re^{2\pi i t}}{re^{2\pi i t} - z} f\left(re^{2\pi i t}\right) dt = -\sum (r/z)^{n+1} \int_0^1 f\left(re^{2\pi i t}\right) e^{2\pi(n+1)it} dt,$$

which causes the coefficients of index $< 0$ in the Fourier series representing $f\left(re^{2\pi i t}\right)$ to appear; but by (2) they vanish.

If for example $f(z) = z^n$ with $n \in \mathbb{N}$, one finds

(1.7) $$2\pi i z^n = \int_{|\zeta|=r} \frac{\zeta^n d\zeta}{\zeta - r} \quad \text{for } |z| < r$$

(and 0 for $|z| > r$) or, putting $a = z/r$,

(1.8) $$a^n = \int_0^1 \frac{e^{2\pi i(n+1)t}}{e^{2\pi i t} - a} \, dt \quad \text{for } |a| < 1, \ n \in \mathbb{N}.$$

---

[1] Let $t \mapsto \gamma(t) = (x(t), y(t))$ be a differentiable map of a compact interval $I$ into $\mathbb{C}$, whence a "curve", the trajectory of the point $\gamma(t)$. If $f(z)$ is a continuous function of $z$ defined on an open set containing the curve we put

$$\int_\gamma f(z)dz = \int_I f[\gamma(t)]\gamma'(t)dt$$

and more generally

$$\int_\gamma u(x,y)dx + v(x,y)dy = \int_I \{u[\gamma(t)]x'(t) + v[\gamma(t)]y'(t)\} dt$$

if $u$ and $v$ are continuous on a neighbourhood of $\gamma(I)$. Formula (4) corresponds to the case where $\gamma(t) = re^{2\pi i t}$. If $s = \theta(t)$ is a map of class $C^1$ of $I$ onto an interval $J$, the integral $\int f(z)dz$ does not change if one replaces $t \mapsto \gamma(t)$ by $s \mapsto \gamma[\theta(s)]$, by the Chain Rule: we have $\int \varphi[\theta(t)]\theta'(t)dt = \int \varphi(s)ds$, where the integrals are taken over $I$ and $J$ respectively. See Vol. III, Chap. VIII, n° 2.

One may go further and also calculate the derivatives

$$(1.9) \qquad f^{(k)}(z) = \sum n(n-1)\ldots(n-k+1)a_n z^{n-k}$$

of $f$. One proceeds in the same way, but this time using the relation

$$(1.10) \qquad \sum n(n-1)\ldots(n-k+1)q^{n-k} = k!/(1-q)^{k+1},$$

which follows by differentiation (Chap. II, eqn. (19.14)) in place of the formula $\sum q^n = 1/(1-q)$; one substitutes the $a_n$ given by (2) in (9), whence

$$f^{(k)}(z) =$$

$$= \sum n(n-1)\ldots(n-k+1)z^{n-k}r^{-n} \int f\left(re^{2\pi it}\right) e^{-2\pi int} dt =$$

$$= \sum n(n-1)\ldots(n-k+1) \int \left(z/re^{2\pi it}\right)^{n-k} \left(re^{2\pi it}\right)^{-k} f\left(re^{2\pi it}\right) dt$$

where one integrates over $[0,1]$; one verifies, as for $k = 0$, that one may integrate the series term-by-term, whence, using (10),

$$f^{(k)}(z) = k! \int \left(1 - z/re^{2\pi it}\right)^{-k-1} \left(re^{2\pi it}\right)^{-k} f\left(re^{2\pi it}\right) dt,$$

i.e.

$$(1.11) \qquad f^{(k)}(z) = k! \int_0^1 \frac{re^{2\pi it}}{\left(re^{2\pi it} - z\right)^{k+1}} f\left(re^{2\pi it}\right) dt$$

or again, *à la* Leibniz,

$$(1.11') \qquad 2\pi i f^{(k)}(z) = k! \int_{|\zeta|=r} f(\zeta) \frac{d\zeta}{(\zeta - z)^{k+1}} \quad \text{for } |z| < r.$$

(11') can be deduced formally from (4) by differentiating the factor $1/(\zeta - z)$ appearing in the integral (4) $k$ times with respect to $z$. In fact, Theorem 9 of Chap. V, n° 9 (differentiation under the $\int$ sign) would allow us to justify this operation *a priori*, starting from (6) without intermediate calculations, since to differentiate an analytic function with respect to $z$ is the same as differentiating it with respect to $x = \text{Re}(z)$.

The preceding assumes that the function $f$ is *analytic*; what happens if it is only *holomorphic*, i.e. $C^1$ in the real sense on a disc $|z| < R$ and a solution of the Cauchy equation

$$(1.12) \qquad D_2 f = i D_1 f \ ?$$

As we want to show that $f$ is in fact analytic we are forced to reverse the procedure, i.e. to introduce the Fourier coefficients

$$(1.13) \qquad a_n(r) = \int_0^1 f\left(re^{2\pi it}\right) e^{-2\pi int} dt$$

for $r < R$, and to show that
   (i) they are of the form

$$(1.14) \qquad a_n(r) = a_n r^n$$

with numerical coefficients $a_n$ independent of $r$,
   (ii) we have $a_n = 0$ for $n < 0$,
   (iii) and

$$(1.15) \qquad f\left(re^{2\pi it}\right) = \sum a_n(r) e^{2\pi int}$$

for any $t \in \mathbb{R}$ and $r < \mathbb{R}$. On substituting (14) in (15) we will find an expansion of $f(z)$ as an *entire* series, by (ii).

We might establish points (i) and (ii) as of now, the first using (12), the second by observing that, by (13), the function $a_n(r)$ must remain bounded as $r$ tends to 0. Point (iii), on the other hand, assumes known the fact that the Fourier series of a function of class $C^1$ is absolutely convergent and represents the given function. These points will be justified later.

All this shows that the foundations of the theory of the analytic or holomorphic functions rests on Fourier series or can be deduced therefrom. We shall see that conversely one may use the Cauchy formula to obtain the first theorems on Fourier series.

In this chapter you will find only those properties of holomorphic functions which can be derived from the theory of Fourier series. Everything that depends on integrals over arbitrary curves (the Cauchy theory) will be expounded in Volume III.

# § 1. Analysis on the unit circle

## 2 – Functions and measures on the unit circle

The purpose of this § is to present some definitions and notations which we shall use constantly, and to clarify a number of preliminary questions.

We shall adopt the notation $\mathbb{T}$ (the one-dimensional "torus") to denote the set of the complex numbers $u$ such that $|u| = 1$; some other authors denote it by $\mathbb{U}$ (the "unitary" group in one variable), not to speak of those who prefer to write $\mathbb{R}/\mathbb{Z}$... The aim of the theory of Fourier series is to expand "arbitrary" functions defined on $\mathbb{T}$ in series whose general term is a multiple of $e^{2\pi i n t} = u^n$, putting $u = e^{2\pi i t}$. Note that if one puts

$$(2.1) \qquad \chi(u) = u^n \quad \text{for every } u \in \mathbb{T}$$

for $n \in \mathbb{Z}$ one obtains a continuous function on $\mathbb{T}$ such that

$$(2.2) \qquad \chi(uv) = \chi(u)\chi(v) \quad \text{for any } u, v \in \mathbb{T}.$$

We shall see a little later that this equation has no continuous solutions other than the functions $u \mapsto u^n$, and it is this remark which is at the origin of the contemporary generalisations of the theory.

(i) *How to eliminate the factors* $2\pi$

As far as possible I intend neither to bore the reader with the factors $2\pi$ and the exponentials which uselessly encumber this kind of mathematics, nor to inflict them on my two typists. I will therefore use the notation

$$(2.3) \qquad \mathbf{e}(t) = e^{2\pi i t}, \qquad \mathbf{e}_n(t) = e^{2\pi i n t} = \mathbf{e}(nt) = \mathbf{e}(t)^n$$

where the factors $2\pi$, relegated to the exponents, are invisible and can be absorbed into "macros" that can be typed globally; this convention is already to be found in Hardy and Wright.

I earnestly advise the reader to reread n° 14 of Chap. IV on the imaginary exponentials, since it will be used constantly. The exponentials (1) have period 1 and in the sequel we shall consider functions of period 1 alone: a function $f$ of period $T$ is transformed into a function of period 1 on considering $t \mapsto f(Tt)$ instead of $f(t)$. Users may have excellent reasons to drag cohorts of functions $\cos(2\pi n t/T)$ after themselves, but we have none here.

(ii) *Functions on* $\mathbb{T}$ *and periodic functions*

To every function $f$ on $\mathbb{T}$ there corresponds, on $\mathbb{R}$, a function $t \mapsto f[\mathbf{e}(t)]$ of period 1. Conversely, every function $f(t)$ of period 1 on $\mathbb{R}$ can be considered as a function on the unit circle $\mathbb{T}$: one puts

$$(2.4) \qquad f(u) = f(t) \quad \text{if } u = \mathbf{e}(t), \qquad \text{whence } f(t) = f[\mathbf{e}(t)];$$

since $f(t + 1) = f(t)$ there is no ambiguity, $t$ being determined modulo an integer when one knows $u$. This is an abuse of notation since a function on $\mathbb{R}$ is not, strictly speaking, a function on $\mathbb{T}$; but it is indispensable to be able to adopt both of these two points of view; and to use different notations for the two functions which correspond "canonically" according to (4) would make the text unreadable.

This correspondence between functions defined on these different sets preserves continuity. Since the map $t \mapsto \mathbf{e}(t)$ of $\mathbb{R}$ on $\mathbb{T}$ is continuous, the continuity of $f$ at a point of $\mathbb{T}$ trivially implies its continuity at the corresponding points of $\mathbb{R}$. On the other hand, even though the map $t \mapsto \mathbf{e}(t)$ is not *globally* bijective, it maps every compact interval $I \subset \mathbb{R}$ of length *strictly* $< 1$ bijectively onto a closed arc $K \subset \mathbb{T}$ of the circle. Restricted to $I$, the map $t \mapsto \mathbf{e}(t)$ therefore has an inverse map $K \to I$ which is also continuous (Chap. III, n° 9). The map $t \mapsto \mathbf{e}(t)$ thus inversely transforms every continuous function on $I$ into a continuous function on $K$. This is equivalent to saying that a function defined on the circle $|u| = 1$ is continuous if and only if it is a continuous function of the polar angle, or argument, of the point $u$, as is obvious from a sketch. § 4 of Chap. IV on the uniform branches of the "function" $\mathcal{A}rg\, z$ also shows that on a neighbourhood of each point $u_0 \in \mathbb{T}$ or even on $\mathbb{T}$ *with a point removed*, for example on $\mathbb{T} - \{1\}$, but not on the whole of $\mathbb{T}$, one may choose $t$ so that it is a continuous function of $u = \mathbf{e}(t)$.

(iii) *Characterisation of the exponentials*

The correspondence (4) allows us to show that the functional equation (2) has no continuous solutions apart from the functions (1). To start with, note that for every continuous solution of (2)

$$|\chi(u)| = 1 \quad \text{for every } u \in \mathbb{T};$$

this relates to the fact that, endowed with the usual multiplication and topology of the complex numbers, $\mathbb{T}$ is a "compact group": since the continuous function $\chi$ is bounded on the compact set $\mathbb{T}$, it follows that for every $u \in \mathbb{T}$, the family of the numbers $\chi(u^n) = \chi(u)^n$ for $n \in \mathbb{Z}$ is likewise bounded; it remains then to show that the only complex numbers $z$ such that

$$\sup_{n \in \mathbb{Z}} |z^n| < +\infty$$

are those of modulus 1, which is clear. Since (2) implies $\chi\left(u^{-1}\right) = \chi(u)^{-1}$, it follows that every continuous solution of (2) also satisfies

$$(2.5) \qquad\qquad \chi\left(u^{-1}\right) = \overline{\chi(u)}$$

and more generally

$$(2.6) \qquad\qquad \chi\left(uv^{-1}\right) = \chi(u)\overline{\chi(v)}$$

for any $u, v \in \mathbb{T}$.

On the other hand, the continuous function $\chi(x) = \chi(\mathbf{e}(x)) = \chi\left(e^{2\pi i x}\right)$ on $\mathbb{R}$ satisfies the functional equation

$$\chi(x+y) = \chi(x)\chi(y)$$

of Chap. IV, n° 13. We have shown there that every solution of the latter is of the form

(2.7) $$\chi(x) = \exp(cx)$$

with a constant $c \in \mathbb{C}$, *on condition* that we know that the function $\chi$ has a derivative at the origin.

But, in fact, *continuity suffices*, and implies much more than differentiability at the origin. To see this, one chooses on $\mathbb{R}$ a function $\varphi \in \mathcal{D}(\mathbb{R})$ and, as in Chap. V, n° 27, one regularises $\chi$ by means of the convolution product

(2.8) $$\chi \star \varphi(x) = \int_{\mathbb{R}} \chi(x-y)\varphi(y)dy = \int_{\mathbb{R}} \chi(x)\chi(y)^{-1}\varphi(y)dy = c.\chi(x);$$

the constant $c = \int \chi(y)^{-1}\varphi(y)dy$ may be assumed to be nonzero, since, if $\chi \star \varphi$ were zero for every $\varphi \in \mathcal{D}(\mathbb{R})$, then so would be the function $\chi$ (Chap. V, n° 27, Theorem 26). Now the function $\chi \star \varphi$ is $C^\infty$ for every $\varphi \in \mathcal{D}(\mathbb{R})$. So likewise is $\chi$.

We may now apply the result of Chap. IV, n° 13 and write (7). But in the present case $|\chi(x)| = 1$ for every $x \in \mathbb{R}$. Putting $c = a + ib$ with $a$ and $b$ real, we have $|\exp(cx)| = \exp(ax)$ for $x \in \mathbb{R}$, whence $a = 0$ and $\chi(x) = \exp(ibx)$ with $b$ real. For the result to be of period 1 it is necessary and sufficient that $\exp(ib) = 1$, i.e. that $b = 2\pi n$ with an $n \in \mathbb{Z}$. Finally we find $\chi(x) = \exp(2\pi i n x)$, whence $\chi(u) = u^n$, qed.

We shall later call every function of the form $u \mapsto u^n = \chi(u)$ a *character* of $\mathbb{T}$. This terminology comes from the theory of commutative groups (Chap. XI), where one considers the solutions of the functional equation (2) systematically on the given group $G$. If one assumes $G$ commutative and finite – the simplest case, involving only algebra –, every function on $G$ is, and in a unique way, a linear combination of characters of $G$, of which there are $\mathrm{Card}(G)$; this is the simplest version of the Fourier transform, though dating from very much later than Fourier himself (even though Dirichlet had used this idea for the group $\mathbb{Z}/n\mathbb{Z}$ in proving his theorem on arithmetic progressions).

(iv) *Mean value of a function on a circle*

In the theory of analytic functions one often considers the mean value of a function over a circle, and, in that of the periodic functions, over a period interval. There is no difference between these two ways of integrating.

First, on a circle $|z| = r$, a function of $z$, analytic or not, can, as we have seen, be transformed into a function of period 1 by putting

(2.9) $$z = r\mathbf{e}(t) = re^{2\pi it}$$

or into a function of period $2\pi$ on putting $z = re^{it}$. Its *mean value* around the circle $|z| = r$ is, by definition, the number which we denote by

(2.10) $$\int_{\mathbb{T}} f(ru)dm(u) = \int_0^1 f(r\mathbf{e}(t))dt = \frac{1}{2\pi}\int_0^{2\pi} f(re^{it})dt$$

where $\mathbb{T}$, we recall, denotes the unit circle $|u| = 1$ in the complex plane. More generally we put

(2.11) $$m(f) = \int_{\mathbb{T}} f(u)dm(u) = \int_0^1 f(\mathbf{e}(t))dt = \frac{1}{2\pi}\int_0^{2\pi} f(e^{it})dt$$

for every "reasonable", for example regulated[2], function $f$ on $\mathbb{T}$. We will use the norms

$$\|f\| = \|f\|_{\mathbb{T}} = \sup|f(u)|, \qquad \|f\|_1 = \int_{\mathbb{T}} |f(u)|dm(u),$$

$$\|f\|_2 = \left(\int_{\mathbb{T}} |f(u)|^2\, dm(u)\right)^{1/2}$$

as in $\mathbb{R}$.

We may in fact, in (11), integrate over any interval of length 1, since

(2.12) $$\int_0^1 f(t)dt = \int_a^{a+1} f(t)dt = \int_a^{a+1} f(b+t)dt$$

for any $a, b \in \mathbb{R}$ for every function $f$ of period 1 on $\mathbb{R}$ (Chap. V, end of n° 2).

Besides, even in the case of arbitrary functions of period $T$ on $\mathbb{R}$, the formulae only mention the mean values of the functions over a period interval; it is convenient to write, here again,

(2.13) $$m(f) = \int_0^1 f(t)dt = \frac{1}{2\pi}\int_0^{2\pi} = \frac{1}{T}\int_0^T = \oint,$$

as we study functions of period $1, 2\pi$ or $T$; the sign $\oint$ dispenses us from writing the limits of integration since, by definition, it denotes the mean

---

[2] This means, as one prefers, that the corresponding periodic function is regulated, i.e. has left and right limit values at every point of $\mathbb{R}$, or that the function given on $\mathbb{T}$ enjoys the same property, the limit values at a point of $\mathbb{T}$ being defined in the obvious way. The BL theorem being valid for the compact set $\mathbb{T}$, it comes to the same to require that for every $r > 0$ there exists a partition of $\mathbb{T}$ into a finite number of arcs of circles, of any kind, on each of which the given function is constant to within $r$. One may also define, using such partitions, the notion of step function on $\mathbb{T}$, and, generally, transpose to $\mathbb{T}$ the arguments of Chap. V, n° 7.

value over a period (or even over several periods) of the integrand. In fact, and without express mention to the contrary, all the integrals in $dt$ will be, in the rest of this chapter, apart from § 6, extended over any interval of length 1.

As we have seen in Chap. V, n° 5, which the reader is invited to read again, the essential formulae in the theory of absolutely convergent Fourier series (calculation of the coefficients and of the scalar product) stem from the "orthogonality relations" of the exponentials, the relation (5.2) of Chap. V. With the notation just introduced, they can be written

$$\int \mathbf{e}_p(u)\overline{\mathbf{e}_q(u)}dm(u) = \begin{cases} 1 & \text{if} \quad p = q, \\ 0 & \text{if} \quad p \neq q. \end{cases}$$

The scalar product $(f \,|\, g)$ of two periodic functions introduced in Chap. V, eqn. (5.4), will now be written

$$(f \,|\, g) = \int f(u)\overline{g(u)}dm(u) = m(f\bar{g}).$$

The orthogonality relations thus signify that if $\chi$ and $\chi'$ are two characters of $\mathbb{T}$, then $(\chi \,|\, \chi') = 1$ or $0$ according to whether $\chi$ and $\chi'$ are equal or different.

(v) *Measures on* $\mathbb{T}$

The notation $dm(u)$ in (11) indicates that we are integrating with respect to a measure on $\mathbb{T}$. We have defined this notion in Chap. V, n° 30 in the case of a compact set $X \subset \mathbb{C}$: one considers the vector space $C^0(X)$ of scalar functions defined and continuous on $X$, endowed with the norm

(2.14)                $$\|f\|_X = \sup_{x \in X} |f(x)|$$

of uniform convergence on $X$; a measure on $X$ is then, by definition, a map $\mu$ of $C^0(X)$ into $\mathbb{C}$ which is *linear* and *continuous*, i.e. satisfies an inequality

(2.15)                $$|\mu(f)| \leq M(\mu) \|f\|_X$$

which allows one to pass to the limit under the $\int$ sign when integrating a uniformly convergent sequence of functions $f_n \in C^0(X)$ with respect to $\mu$. A measure is said to be *positive* if $\mu(f) \geq 0$ for every function $f \geq 0$. It is clear that formula (11) defines such a measure on $\mathbb{T}$.

In the case of the measure $m$, all that we have said in Chap. V transposes immediately to integration on $\mathbb{T}$, starting with the notion of integrable function (Chap. V, n° 2); it will be the same for an arbitrary measure $\mu$ once we have defined the integrable functions in this case. The proper theory of Fourier series uses the Lebesgue integral – for $m$ or any other measure on $\mathbb{T}$ – and has, historically, constituted one of the principal justifications or motivations for it. Since we cannot yet do this here, we shall confine ourselves, in this chapter, without exceptions, to considering regulated, mostly continuous, functions, when we integrate with respect to an arbitrary measure: there is no benefit in complicating one's existence in exploiting to the full the possibilities of the Riemann integral by circuitous and complicated methods when one can obtain much more complete results more easily using the Lebesgue integral (a principle of Dieudonné's).

We may also define distributions in the sense of Schwartz on $\mathbb{T}$, as we shall see in n° 9.

(vi) *Invariance of the measure m on* $\mathbb{T}$

For the measure $m$ defined on $\mathbb{T}$ by (11), we have

(2.16)    $$\int f(au)dm(u) = \int f(u)dm(u) \quad \text{for every } a \in \mathbb{T};$$

this is the analogue of the translation invariance

(2.17)    $$\int f(x + a)dx = \int f(x)dx$$

of Lebesgue measure on $\mathbb{R}$ and of the analogous property

$$\iint f(x + a, y + b)dxdy = \iint f(x, y)dxdy$$

on $\mathbb{R}^2$: the maps $u \mapsto au$, where $|a| = 1$ (geometrically: rotations about the origin), play the same rôle in $\mathbb{T}$ as the translations $x \mapsto a + x$ in $\mathbb{R}$ or $(x, y) \mapsto (x + a, y + b)$ in $\mathbb{R}^2$. The reader who knows what a group is (additive in the case of $\mathbb{R}$ or $\mathbb{R}^2$, multiplicative in the case of $\mathbb{T}$) will understand. On the multiplicative group $\mathbb{R}^*$ of nonzero real numbers the invariant measure is $dx/|x|$ as one sees on making the change of variable $x \mapsto ax$ with $a \in \mathbb{R}^*$.

To prove (16), it is enough to reduce to (12) by putting $u = \mathbf{e}(t)$ and $a = \mathbf{e}(\alpha)$ with $\alpha \in \mathbb{R}$.

One can show that, among the measures on $\mathbb{T}$, the measure $m$ is the only one that satisfies (16) and attributes the value 1 to the integral of the constant function 1. For this reason, one calls $m$ the *invariant measure* on $\mathbb{T}$. The relation (17) likewise characterises Lebesgue measure up to a constant factor. The measure $m$ is also invariant under symmetries, i.e. satisfies

(2.18)    $$\int f\left(u^{-1}\right)dm(u) = \int f(u)dm(u)$$

for any $f$. This follows from the corresponding property of Lebesgue measure on $\mathbb{R}$: the change of variable $t \mapsto -t$ replaces the integral of $f(t)$ on a period interval by the integral of $f(-t)$ on the symmetric interval, so again on a period interval. Also $\int f(-t)dt = \int f(t)dt$ when one integrates over all of $\mathbb{R}$.

Finally, we shall need double integrals, for example à *propos* the convolution product on $\mathbb{T}$. In Chap. V, n° 30, we showed that if $I$ and $J$ are compact intervals in $\mathbb{R}$ and $f(x, y)$ is a function defined and continuous on the rectangle $I \times J$, then

$$(2.19) \qquad \int d\mu(x) \int f(x, y) d\nu(y) = \int d\nu(y) \int f(x, y) d\mu(x)$$

for any measures $\mu$ and $\nu$ on $I$ and $J$, the common value of the two sides being by definition the double integral $\iint f(x, y) d\mu(x) d\nu(y)$ over $I \times J$. Since, in the case of the invariant measure, integration on $\mathbb{T}$ reduces to an integration over $[0, 1]$, it is clear that, for every function $f(u, v)$ defined and continuous[3] on $\mathbb{T} \times \mathbb{T}$ we will have

$$(2.19') \qquad \int_{\mathbb{T}} dm(u) \int_{\mathbb{T}} f(u, v) dm(v) = \int_{\mathbb{T}} dm(v) \int_{\mathbb{T}} f(u, v) dm(u),$$

the common value being denoted by $\iint f(u, v) dm(u) dm(v)$. In the general case[4] of two arbitrary measures, it is necessary, to establish (19) on $\mathbb{T}$, to use as in Chap. V, n° 30 partitions of unity on $\mathbb{T}$ in order to show that *every continuous function on $\mathbb{T} \times \mathbb{T}$ is the uniform limit of finite sums of functions of the type $g(u)h(v)$*, with $g$ and $h$ continuous on $\mathbb{T}$; the proofs are the same as in Chap. V: one replaces the intervals of $\mathbb{R}$ by arcs of the circle. You may even, if you think it worthwhile, use the diagram in Chap. V, n° 30, so long as you do not forget that, when working on $\mathbb{T}$, the graph of a real function is drawn on the cartesian product $\mathbb{T} \times \mathbb{R}$, i.e. on the surface of the vertical cylinder in $\mathbb{R}^3$ having $\mathbb{T}$ as base.

## 3 – Fourier coefficients

The *Fourier coefficients* of a regulated function $f(t)$ of period 1 will be denoted

$$(3.1) \qquad \hat{f}(n) = \oint f(t)\overline{\mathbf{e}_n(t)} dt = \int_a^{a+1} f(t) e^{-2\pi i n t} dt;$$

---

[3] A function $f$ defined on $\mathbb{T} \times \mathbb{T}$ is continuous at a point $(a, b)$ of $\mathbb{T} \times \mathbb{T}$ if for any $r > 0$, there exists an $r' > 0$ such that

$$\{|u - a| < r' \ \& \ |b - v| < r'\} \Longrightarrow |f(a, b) - f(u, v)| < r.$$

This is the general notion of continuity in a metric space if one defines the distance of two elements of $\mathbb{T} \times \mathbb{T}$ by $d[(u', v'), (u'', v'')] = |u' - u''| + |v' - v''|$ as in the Appendix to Chap. III. The definition amounts to continuity on $\mathbb{R} \times \mathbb{R}$ on considering the function $f[\mathbf{e}(s), \mathbf{e}(t)]$, which is periodic in $s$ and $t$.

[4] Despite appearances, a measure $\mu$ on $\mathbb{T}$ is not a measure on $I = [0, 1]$, for $C^0(\mathbb{T})$ is identified with the vector subspace of $C^0(I)$ formed by the functions such that $f(0) = f(1)$. But every continuous function $f$ on $I$ can be written $f(t) = f_0(t) + c(f)t$, with $f_0$ "periodic" and $c(f) = f(1) - f(0)$. If $\mu$ is a continuous linear form on the periodic functions, one may then extend it to $C^0(I)$ by putting $\mu(f) = \mu(f_0) + \gamma[f(1) - f(0)]$, where $\gamma$ is a constant. One might remove the ambiguity by agreeing to choose $\gamma = 0$, but this is a little artificial.

it is sometimes useful to use a notation such as $a_n(f)$. In the "functions on $\mathbb{T}$" version the formula becomes

$$(3.1') \qquad \hat{f}(n) = \int_{\mathbb{T}} f(u)u^{-n}dm(u) \qquad (n \in \mathbb{Z})$$

where $m$ is the invariant measure defined above. If one uses the notation $\chi(u)$ to denote a character $u \mapsto u^n$ of $\mathbb{T}$, one may put

$$(3.1'') \qquad \hat{f}(\chi) = \int \overline{\chi(u)}f(u)dm(u) = (f \mid \chi),$$

the scalar product of the functions $f$ and $\chi$.

The first relation to establish is the trivial but useful inequality

$$(3.2) \qquad |\hat{f}(n)| \le \int |f(u)|dm(u) = \|f\|_1 \le \|f\|_{\mathbb{T}}.$$

In fact, we shall soon prove much more: the series $\sum |\hat{f}(n)|^2$ converges, so that $\hat{f}(n)$ tends to 0 when $|n|$ increases indefinitely (n° 7).

More generally one may define the Fourier coefficients of an arbitrary measure $\mu$ on $\mathbb{T}$ by

$$(3.1''') \qquad \hat{\mu}(n) = \int u^{-n}d\mu(u) \quad \text{or} \quad \hat{\mu}(\chi) = \int \overline{\chi(u)}d\mu(u).$$

Compatibility with (1') is obtained by associating with every regulated function[5] $f$ the measure $f(u)dm(u)$ of density $f$ with respect to the invariant measure $m$.

If for example $\mu$ is the Dirac measure at the point[6] $u = 1$ of $\mathbb{T}$, given by $\mu(f) = f(1)$, then we have

$$(3.3) \qquad \hat{\mu}(n) = 1 \quad \text{for every } n \in \mathbb{Z},$$

which shows that, in contrast to those of a function, the Fourier coefficients of a measure need not tend to 0 at infinity. In this case, one may only say that the function $\hat{\mu}$ is bounded on $\mathbb{Z}$ since the existence of a bound[7] $|\mu(f)| \le M.\|f\|$ clearly implies $|\hat{\mu}(n)| \le M$ for every $n$.

The notation[8] $\hat{f}(n)$ is intended to display the fact that the theory of Fourier series consists of associating to every function $f$ on the multiplicative

---

[5] In fact it would be enough for $f$ to be absolutely integrable on $\mathbb{T}$ (i.e. on $[0,1]$ for example) in the sense of Chap. V, n° 22, which allows us to extend the definition (1) of the Fourier coefficients to this case.

[6] Do not confuse this with the Dirac measure on $\mathbb{R}$. The latter is a linear form on $C^0(\mathbb{R})$ while here we are concerned only with linear forms on $C^0(\mathbb{T})$. For this reason we resist the temptation of again writing $\delta$ for the Dirac measure at the point $u = 1$ of $\mathbb{T}$.

[7] One almost always writes $\|f\|$ instead of $\|f\|_{\mathbb{T}}$.

[8] It was introduced by André Weil, *L'integration dans les groupes topologiques et ses applications* (Hermann, 1940) in the framework of the most general version of

compact group $\mathbb{T}$ or, equivalently, to every periodic function on $\mathbb{R}$, a function $\hat{f}$ on the discrete additive group $\mathbb{Z}$, its *Fourier transform*[9]. Conversely, one may associate to every function $g$ on $\mathbb{Z}$ which tends rapidly enough to 0 at infinity a function $\hat{g}$ on $\mathbb{T}$, namely the Fourier series

$$(3.4) \qquad \hat{g}(u) = \sum g(n)u^n$$

whose coefficients are the values of the given function on $\mathbb{Z}$; this is the essence of the subject as its contemporary generalisations have shown. On the additive group $\mathbb{R}$, the Fourier transform associates to every absolutely integrable (or even Lebesgue integrable) function $f$ a function

$$(3.5) \qquad \hat{f}(y) = \int_{\mathbb{R}} f(x)\overline{\mathbf{e}(xy)}dx = \int_{\mathbb{R}} e^{-2\pi ixy} f(x)dx$$

defined on the same additive group $\mathbb{R}$; this also appears in the same general framework, also in the Fourier transform in $\mathbb{R}^n$ or the theory of multiple Fourier series for periodic functions of several real variables, etc.

The first fundamental problem of the theory is to decide whether every "reasonable" function $f$ on $\mathbb{T}$, or periodic on $\mathbb{R}$, is represented by its *Fourier series*, i.e. if one has

$$(3.6) \qquad f(u) = \sum \hat{f}(n)u^n \quad \text{or} \quad f(t) = \sum \hat{f}(n)\mathbf{e}_n(t).$$

This is the case, we shall see, if $f$ is $C^1$. When one does not know if the Fourier series of a function $f$ converges and represents $f$ it is prudent to confine oneself to writing something like

$$(3.6') \qquad f(u) \approx \sum \hat{f}(n)u^n \quad \text{or} \quad f(t) \approx \sum \hat{f}(n)\mathbf{e}_n(t)$$

to avoid confusion (no connection with asymptotic expansions!).

Note that, in (4), one adds over all the rational integers and not over $\mathbb{N}$. Since

$$(3.7) \qquad \mathbf{e}_n(t) = \cos(2\pi nt) + i\sin(2\pi nt),$$

---

harmonic analysis on commutative topological groups, invented independently at the same period, with better methods, by the Soviet school (D. A. Raïkov) and, in other way by H. Cartan and R. Godement, *Théorie de la dualité et analyse harmonique dans les groupes abéliens* [= commutative] *localement compacts* (Ann. Ecole Norm. Sup., 64 (1947)), which expounds the whole topic in twenty pages. See Chap. XI, §7.

[9] In the classical theory, one speaks of the "Fourier transform" only in the case of $\mathbb{R}$. But this notion applies to any commutative locally compact group (or even non commutative group, but this is far more complicated).

the series (4) can always be put in the traditional trigonometric form

$$(3.8) \qquad a_0 + \sum_1^\infty b_n \cos(2\pi nt) + c_n \sin(2\pi nt)$$

(take the terms with indices $n$ and $-n$ together), or, if the $b_n$ and $c_n$ are real,

$$\sum I_n \cos[2\pi n(t - \omega_n)]$$

with "phase lags" $\omega_n$ and "intensities" $I_n \geq 0$ for $n > 0$, but using this form complicates the calculations; the formulae

$$(3.9) \qquad \mathbf{e}_n(x + y) = \mathbf{e}_n(x)\mathbf{e}_n(y), \quad \overline{\mathbf{e}_n(t)} = \mathbf{e}_n(t)^{-1} = \mathbf{e}_{-n}(t),$$
$$\mathbf{e}_p(t)\mathbf{e}_q(t) = \mathbf{e}_{p+q}(t)$$

are simpler than the analogous trigonometric formulae and lend themselves to the generalisations to group theory[10].

As we have already said elsewhere, one has always to observe that the series (4) being extended over $\mathbb{Z}$ and not over $\mathbb{N}$, it has a meaning only if it converges *unconditionally*, i.e. if

$$(3.10) \qquad \sum \left| \hat{f}(n) \right| < +\infty.$$

This is the case for the functions $f \in C^1(\mathbb{T})$ as we shall see in n° 8, but (10) is very likely to be false when one attempts to study more general functions. In this case, one gives a sense to the series by putting, by definition,

$$(3.11) \qquad \sum_{\mathbb{Z}} \hat{f}(n)\mathbf{e}_n(t) = \lim_{N \to +\infty} \sum_{-N}^N \hat{f}(n)\mathbf{e}_n(t) = \lim f_N(t),$$

which considerably increases the chances of convergence and amounts, in fact, to considering the traditional series (8) and its usual partial sums $f_N(t)$.

---

[10] If $G$ is a commutative group endowed with a locally compact topology, one calls a character of $G$ any continuous map $\chi : G \longrightarrow \mathbb{T}$ such that $\chi(uv) = \chi(u)\chi(v)$. It is clear that if $\chi'$ and $\chi''$ are two characters, then so likewise is the product function $\chi(u) = \chi'(u)\chi''(u)$; endowed with this multiplication, the set of the characters of $G$ becomes a group; on endowing this with the topology of compact convergence one obtains a new commutative locally compact group, the "dual" $\hat{G}$ of $G$. Since there always exists a positive measure $dm(u)$ on $G$ invariant under the translations $u \mapsto uv$, one may associate a "Fourier transform" $\hat{f}(\chi) = \int \overline{\chi(u)} f(u) dm(u)$ to every function $f$ on $G$ decreasing rapidly enough at infinity. One may then choose a positive invariant measure $dm(\chi)$ on $\hat{G}$ so that conversely $f(u) = \int \chi(u)\hat{f}(\chi)dm(\chi)$ under reasonable hypotheses on $f$. In the case where $G = \mathbb{T}$, the characters are the $\mathbf{e}_n(t)$, whence $\hat{G} = \mathbb{Z}$, and the last formula (9) shows that the "multiplication" on $\hat{G}$ is precisely the addition on $\mathbb{Z}$.

These are trigonometric polynomials since only a finite number of nonzero coefficients feature in $f_N$. If one knew that for every continuous function $f$ on $\mathbb{T}$, the $f_N$ converged uniformly on $\mathbb{T}$ to $f$, one would obtain Weierstrass' approximation theorem of Chap. V, n° 28 for periodic functions. Unfortunately not, even if one demands only simple convergence; to obtain uniform convergence when $f$ is continuous it is enough to substitute for the $f_N$ their arithmetic means $(f_1 + \ldots + f_N)/N$ as we shall see (Fejér's theorem), which will provide a proof – there are others – of the approximation theorem.

In a general way, and as we have already said elsewhere, we have to warn the reader to exercise the most extreme prudence once he steps outside the framework of the $C^1$ functions: most of the statements that one might believe obvious are false and, when they are correct, they are never obvious. This is one of the charms of the theory for those whom it attracts, and the reason why it played such large rôle in the development of analysis during all the XIX$^{\text{th}}$ century and a large part of the following one: when one does not understand one tries to understand, and this often brings one much further than one had imagined. One of the first traps of the theory is to believe that if a *trigonometric series*

$$a_0 + \sum b_n \cos 2\pi n t + c_n \sin 2\pi n t,$$

with arbitrarily given coefficients converges for any $t$, than it must be the Fourier series of its sum. False: though a simple limit of continuous functions and so "measurable" in the sense of Lebesgue, the sum of the series can fail to be integrable. The first theorems proved by Cantor in 1870 say that, for a trigonometric series,

(i)  the coefficients $b_n$ and $c_n$ tend to 0 if the general term $b_n \cos(2\pi n t) + c_n \sin(2\pi n t)$ tends to 0 at every point of an interval $I$ of nonzero length, and therefore if the series converges on $I$,

(ii)  if the series converges to 0 for every $t \in \mathbb{R}$, then all the coefficients are zero ("obvious", but try to prove it ...).

It was in trying to weaken the hypothesis of the statement (ii), i.e. in trying to characterise the sets $E \subset [0, 1]$ such that

$$f(t) = 0 \quad \text{for every } t \in E \Longrightarrow a_n = b_n = 0$$

("sets of uniqueness"), that Cantor was led to construct more and more baroque sets in $\mathbb{R}$, then to his theory of transfinite numbers. Do not confuse this, as we have already said, with the naïve trivialities of Chap. I, which would not have led him to the edge of sanity if he had not been already predisposed. This kind of question continues to be the object of much research[11]; most mathematicians and *a fortiori* users are happy with much less subtle results of universal use.

---

[11] See for example J-P. Kahane and R. Salem, *Ensembles parfaits and séries trigonometriques* (Paris, Hermann, nouvelle éd. 1987) and J.-P. Kahane and P. G. Lemarié-Rieusset, *Séries de Fourier et ondelettes* (Paris, Cassini, 1997).

## 4 – Convolution product on $\mathbb{T}$

The invariance of the measure $m$ under translation leads to useful formulae. For example, the left hand side of formula (2.19') does not change if in it one replaces $f(u,v)$ by $f(u,av)$ for an $a \in \mathbb{T}$ independent of $v$; in particular, one can, for each $u$, replace $v$ by $uv$, or $u^{-1}v$, since $u$ is a variable independent of $v$, whence the formulae

$$(*) \qquad \iint f(u,v)dm(u)dm(v) \;=\; \iint f\left(u, u^{-1}v\right) dm(u)dm(v),$$

$$(**) \qquad \iint f(u,v)dm(u)dm(v) \;=\; \iint f\left(uv, v^{-1}\right) dm(u)dm(v) :$$

one integrates first with respect to $u$, makes the change of variable $u \mapsto uv^{-1}$, then integrates with respect to $v$, finally one replaces $v$ by $v^{-1}$. Similar result in $\mathbb{R}$: if $f(x,y)$ is, for simplicity, continuous and of compact support in $\mathbb{R}^2$, then

$$\iint f(x,y)dxdy = \iint f(x+y,-y)dxdy$$

for

$$\iint f(x,y)dxdy =$$
$$= \int dy \int f(x,y)dx = \int dy \int f(x-y,y)dx \qquad \text{(using } x \longmapsto x-y)$$
$$= \int dx \int f(x-y,y)dy = \int dx \int f(x+y,-y)dy \qquad \text{(using } y \longmapsto -y)$$
$$= \iint f(x+y,-y)dxdy.$$

The invariant measure allows us, as in $\mathbb{R}$ (Chap. V, n° 27), to define the *convolution product*[12]

$$(4.1) f \star g(u) = \int f\left(uv^{-1}\right) g(v)dm(v) = \int f(w)g\left(uw^{-1}\right) dm(w) = g \star f(u)$$

of two regulated functions on $\mathbb{T}$ or, in the "periodic functions on $\mathbb{R}$" version,

$$(4.1') \qquad f \star g(t) = \oint f(t-s)g(s)ds = \oint g(t-s)f(s)ds,$$

integrating over a period. The equality of the two integrals in (1) is obtained by means of the change of variable $v \mapsto uv^{-1} = w$ (or $s \mapsto t-s$), the composition of a translation $v \mapsto uv$ followed by a symmetry $v \longrightarrow v^{-1}$;

---

[12] A symbol such as $f \star g(u)$ denotes the value at the point $u$ of the function $f \star g$ and replaces the expression $(f \star g)(u)$.

see (∗). A convenient way to simplify the theoretical calculations on Fourier series is to remark that if $\chi$ is a character of $\mathbb{T}$, the convolution product

$$(4.2) \qquad f \star \chi(u) \;=\; \int \chi\left(uv^{-1}\right) f(v)dm(v) =$$
$$= \;\chi(u) \int \overline{\chi(v)} f(v)dm(v) = \hat{f}(\chi)\chi(u)$$

is the general term of the Fourier series of $f$. The relation (3.6) can then, when it is true, be written

$$(4.3) \qquad f(t) = \sum f \star e_n(t) \quad \text{or} \quad f(u) = \sum f \star \chi(u)$$

where, in the second case, one sums over all the characters of $\mathbb{T}$. Symbolically, one may also write it as $f = \sum f \star e_n = \sum f \star \chi$, which has the advantage of not presupposing the mode of convergence that one chooses: simple convergence, uniform convergence, convergence in mean, etc. It is precisely the choice of the mode of convergence to make (3) correct that is the whole theory of Fourier series; (3) is always correct *in the sense of distributions* as we shall see, but the convergence of a sequence or series of distributions is, in practice, the weakest invented from Newton to nowadays. (Paradoxically, this is in fact the interest of distributions: everything which converges in a reasonable sense, or does not converge, converges in the sense of distributions).

The convolution product has properties similar to those[13] obtained in Chap. II, n° 18, Example 3, for functions defined on $\mathbb{Z}$: but the proofs are less easy. We shall restrict ourselves to functions which are *regulated* and so bounded; going further requires recourse to the Lebesgue integral and will be expounded in Chap. XI, n° 25 in the general framework of group theory – for this is a matter of group theory as the case of the convolution product on $\mathbb{Z}$ has already shown.

First of all the inequality $|f(uv^{-1})g(v)| \leq \|f\| \, |g(v)|$, valid for all $u$ and $v$, shows that, always,

$$\|f \star g\| \leq \|f\| \, \|g\|_1 \leq \|f\| \, \|g\| \, \cdot$$

We shall deduce from this that *the function $f \star g$ is continuous*. If $f$ is continuous this follows directly from Chap. V, n° 9 (Theorem 9, (i)) since we are integrating the continuous function $f(uv^{-1})$ with respect to the measure $g(v)dm(v)$. In the general case there exists a sequence $(f_n)$ of continuous functions (see the lemma in n° 8 below) such that $\lim \|f - f_n\|_1 = 0$; then, by (4),

$$\|f \star g - f_n \star g\| \leq \|g\| \, \|f - f_n\|_1 \, ,$$

---

[13] apart from the existence of a unit element: this would be a function $e(u)$ such that one had $\int f\left(uv^{-1}\right) e(v)dv = f(u)$ for any $f$; the only candidate is the Dirac "function", which is a measure and not a function.

from which $f * g$ is the uniform limit of the continuous functions $f_n * g$, whence the result.

Now let us establish the relations

$$(4.5) \qquad f \star (g + h) = f \star g + f \star h, \qquad (f \star g) \star h = f \star (g \star h)$$

for $f$, $g$ and $h$ regulated. The first is obvious. To obtain the associativity formula let us first consider an integral of the form

$$(4.6) \qquad \int \int \varphi(uv) f(u) g(v) dm(u) dm(v)$$

where $\varphi$ is continuous and $f$ and $g$ are regulated. Theorem 10 of Chap. V, n° 9, and the invariance of the measure show that it is equal to

$$\int g(v) dm(v) \int \varphi(uv) f(u) dm(u)$$

$$= \int g(v) dm(v) \int \varphi(z) f(zv^{-1}) dm(z)$$

$$= \int \varphi(z) dm(z) \int f(zv^{-1}) g(v) dm(v);$$

whence the relation

$$(4.7) \qquad \int \int \varphi(uv) f(u) g(v) dm(u) dm(v) = \int \varphi(z). f \star g(z) dm(z).$$

This done, let us consider the triple integral

$$(4.8) \qquad I(\varphi) = \int \int \int \varphi(uvw) f(u) g(v) h(w) dm(u) dm(v) dm(w)$$

where $f$, $g$ and $h$ are regulated. Theorem 10 of Chap. V, n° 9, which is clearly valid for multiple integrals, shows us that on the one hand

$$\begin{aligned}
I(\varphi) &= \int h(w) dm(w) \int \int \varphi(uvw) f(u) g(v) dm(u) dm(v) \\
&= \int h(w) dm(w) \int \varphi(xw) f \star g(x) dm(x) \qquad \text{by (7)} \\
&= \int \int \varphi(xw) \cdot f \star g(x) \cdot h(w) dm(x) dm(w) \\
&= \int \varphi(z). (f \star g) \star h(z) dm(z)
\end{aligned}$$

applying (7) again, to the functions $f \star g$ and $h$.

But one can also calculate $I(\varphi)$ alternatively as

$$I(\varphi) = \int f(u)dm(u) \int\int \varphi(uvw)g(v)h(w)dm(v)dm(w)$$

$$= \int f(u)dm(u) \int \varphi(ux).g \star h(x)dm(x) \qquad \text{by (7)}$$

$$= \int\int \varphi(ux)f(u)g \star h(x)dm(u)dm(x)$$

$$= \int \varphi(z).f \star (g \star h)(z)dm(z)$$

by applying (7) now to $f$ and $g \star h$. Comparing these results we see that the function $F = f \star (g \star h) - (f \star g) \star h$ satisfies $\int \varphi(z)F(z)dm(z) = 0$ for every continuous $\varphi$ on $\mathbb{T}$. Now $F$ is itself continuous. One may therefore choose $\varphi$ to be the conjugate of $F$, whence $\int |F(z)|^2 dm(z) = 0$ and $F = 0$ (Chap. V, n° 7, Theorem 7), which proves associativity for regulated functions, if not yet for all the integrable functions of the Appendix to Chap. V.

Along with (4) for the uniform norm, we also have

(4.9) $$\|f \star g\|_1 \leq \|f\|_1.\|g\|_1.$$

Replacing $f$ and $g$ by their absolute values does not change the right hand side, but increases the left hand side, since

$$|f \star g(u)| \leq \int |f(uv^{-1})|.|g(v)|dm(v) = |f| \star |g|(u);$$

so it is enough to prove (8) for *positive* $f$ and $g$. Relation (7) with $\varphi = 1$ shows that

$$\|f \star g\|_1 = \|f\|_1 \|g\|_1$$

in this case, qed.

The Fourier series of a convolution product is calculated very simply from the formula

(4.10) $$\widehat{f \star g}(n) = \hat{f}(n)\hat{g}(n).$$

To see this, use the associativity of the convolution product:

$$\widehat{f \star g}(n)\mathbf{e}_n = (f \star g) \star \mathbf{e}_n = f \star (g \star \mathbf{e}_n) =$$
$$= f \star (\hat{g}(n)\mathbf{e}_n) = \hat{g}(n)f \star \mathbf{e}_n = \hat{g}(n)\hat{f}(n)\mathbf{e}_n.$$

For the inverse Fourier transform, which starts from a function $f(n)$ in $L^1(\mathbb{Z})$, i.e. such that $\sum |f(n)| < +\infty$, and leads to a function

(4.11) $$\hat{f}(u) = \sum f(n)u^n,$$

one has likewise, for the convolution product on $\mathbb{Z}$ (Chap. II, n° 18, Example 3),

$$\widehat{f \star g}(u) \;=\; \sum f \star g(n)u^n = \sum f(p)g(n-p)u^n = \sum f(p)g(q)u^{p+q} =$$
$$=\; \sum f(p)g(q)u^p u^q = \sum f(p)u^p \sum g(q)u^q = \hat{f}(u)\hat{g}(u).$$

*The Fourier transform thus interchanges convolution products and ordinary products.*

This last result is particularly obvious in the framework of measures. Let $\mu$ and $\nu$ be two measures on $\mathbb{T}$ and let us calculates the product

$$(4.12) \qquad \hat{\mu}(\chi)\hat{\nu}(\chi) \;=\; \int \overline{\chi(u)}d\mu(u) \int \overline{\chi(v)}d\nu(v) =$$
$$=\; \int\int \overline{\chi(u)\chi(v)}d\mu(u)d\nu(v) =$$
$$=\; \int\int \overline{\chi(uv)}d\mu(u)d\nu(v)$$

of their Fourier transforms. This leads us to consider more generally the map

$$(4.13) \qquad \lambda : f \longmapsto \int\int f(uv)d\mu(u)d\nu(v)$$

of $C^0(\mathbb{T})$ into $\mathbb{C}$; this is clearly a linear form on $C^0(\mathbb{T})$, and it is continuous:

$$|\lambda(f)| \le \left| \int d\mu(v) \right| \int f(uv)d\nu(u) \right\| \le M(\mu)M(\nu)\|f\|.$$

Consequently, $\lambda$ is again a measure which one calls the *convolution product of the measures* $\mu$ and $\nu$, notation $\lambda = \mu \star \nu$. Clearly $\lambda$ is positive if $\mu$ and $\nu$ are. With these conventions the formula (12) can be written

$$(4.14) \qquad \hat{\mu}(\chi)\hat{\nu}(\chi) = \hat{\lambda}(\chi) \qquad \text{where } \lambda = \mu \star \nu.$$

One finds (4) again on considering the measures $d\mu(u) = f(u)dm(u)$ and $d\nu(u) = g(u)dm(u)$, as the reader can easily verify.

There is no simple formula analogous to (1) for defining or calculating the convolution product of two measures; the simplicity of the definition (13) shows once more the advantage in defining measures as linear forms on the continuous functions and not starting from a function of sets.

## 5 – Dirac sequences in $\mathbb{T}$

As in $\mathbb{R}$, the convolution product is linked to *Dirac sequences* on $\mathbb{T}$, formed by regulated functions $\varphi_n(u)$ such that

$$(5.1) \qquad f(1) = \lim \int f(u)\varphi_n(u)dm(u)$$

for every regulated function $f$ continuous at the point $u = 1$.

The conditions to impose on them are the same as in Chap. V, n° 27. The first is that

(D 1) $$\int \varphi_n(u)dm(u) = 1 \quad \text{for every } n;$$

and then $f(1) = \int f(1)\varphi_n(u)dm(u)$, whence

(5.2) $$\left| \int f(u)\varphi_n(u)dm(u) - f(1) \right| \leq \int |f(u) - f(1)|.|\varphi_n(u)|dm(u).$$

Let us take an $r > 0$, and, on the right hand side of (2), distinguish the contributions of the arcs $|u-1| < \delta$ and $|u-1| > \delta$ of $\mathbb{T}$ for some $\delta > 0$ yet to be decided. Since $f$ is continuous at the origin, one can, for $r$ given, choose $\delta$ so that

(5.3) $$|u - 1| < \delta \Longrightarrow |f(u) - f(1)| < r.$$

If one assumes that

(D 2) $$\sup \int |\varphi_n(u)| \, dm(u) = M < +\infty,$$

the contribution of this "small" arc to the total integral is $\leq Mr$. On the "large" arc $|u - 1| > \delta$ we have $|f(u) - f(1)| \leq 2\|f\|$ since $f$ is bounded; the corresponding integral is thus, up to a factor $2\|f\|$, bounded by that of $|\varphi_n(u)|$. Assume now that, for any $r$ and $\delta$, there exists an integer $N(\delta, r)$ such that

$$n > N(\delta, r) \Longrightarrow \int_{|u-1|>\delta} |\varphi_n(u)| \, dm(u) < r,$$

in other words that

(D 3) $$\lim_{n \to \infty} \int_{|u-1|>\delta} |\varphi_n(u)| \, dm(u) = 0 \quad \text{for every } \delta > 0.$$

The preceding arguments now show that for every $r$ and every $\delta$ satisfying (3) we will have

(5.4) $$\int |f(u) - f(1)|.|\varphi_n(u)| \, dm(u) \leq (M + 2\|f\|)r$$

for every $n > N(\delta, r)$, whence (1).

The conditions (D 1), (D 2) and (D 3) may therefore be taken as the *definition* of the Dirac sequences on $\mathbb{T}$.

The most frequent case is that where the $\varphi_n$ are all positive; (D 1) then implies (D 2) with $M = 1$. To achieve (D 3) it is simplest to assume that for every $\delta > 0$ the functions $\varphi_n$ converge *uniformly* to 0 on the arc $|u - 1| \geq \delta$ of $\mathbb{T}$, in other words that

(5.5) $$\lim \varphi_n(u) = 0 \quad \text{uniformly on every compact } K \subset \mathbb{T} - \{1\}$$

since the points of a compact subset of $\mathbb{T}$ not containing the point $u = 1$ remain "standing off" from it.

If one applies (1) to the function $u \mapsto f\left(vu^{-1}\right)$ for a given $v \in \mathbb{T}$, one obtains more generally the formula

$$(5.6) \qquad f(v) = \lim \int f\left(vu^{-1}\right) \varphi_n(u) dm(u) = \lim f \star \varphi_n(v),$$

so long as $f$ is assumed continuous at the point $v$.

In practice, one needs a more precise result.

**Lemma.** *If $f$ is continuous on an* open *arc $J$ of $\mathbb{T}$ then*

$$(5.7) \qquad\qquad f(v) = \lim f \star \varphi_n(v)$$

*uniformly on every* compact *$K \subset J$, for every Dirac sequence on $\mathbb{T}$.*

Applied to the function $u \mapsto f\left(vu^{-1}\right)$, the relation (4) shows that, if $f$ is continuous at the point $v$,

$$(5.8) \qquad\qquad |f \star \varphi_n(v) - f(v)| \le (M + 2\|f\|)r$$

for $n$ large; but to obtain uniform convergence on $K$ one has to find an integer $N$ such that (8) will be valid for $n > N$ for all the $v \in K$ simultaneously. Now, for $r$ and $v$ given, the integer $N$ depends only, as we have seen, on the choice of a $\delta$ such that

$$(5.9) \qquad\qquad \left|f\left(vu^{-1}\right) - f(v)\right| < r \quad \text{for } |u - 1| < \delta.$$

So it all reduces to showing that, for any $r$, there exists a $\delta$ satisfying (9) for all $v \in K$ simultaneously. Since $vu^{-1}$ is "close" to $v$ for $u$ "close" to 1, we are manifestly dealing with a uniform continuity property of $f$.

Assume now that $f$ continuous on the open arc $J$ of $\mathbb{T}$ and let $K$ be a compact arc contained in $J$ (figure 1). Since $\mathbb{T} - J$ and $K$ are compact and disjoint, the distance $d(\mathbb{T} - J, K) = d$ is strictly positive. Since

$$\left|vu^{-1} - v\right| = |v - vu| = |1 - u|$$

for every $u \in \mathbb{T}$, we see that

$$(v \in K) \ \& \ (|u - 1| < d) \Longrightarrow vu^{-1} \in J.$$

For every $\delta < d$, the set $K(\delta) \supset K$ of points $vu^{-1}$ with $v \in K$ and $|u-1| \le \delta$ is thus contained in $J$; moreover it is compact like $K$ and the arc $|u-1| \le \delta$ of $\mathbb{T}$ (use BW or, more elementarily, define the arcs of $\mathbb{T}$ by inequalities between the polar angles of their points). But since $f$ is continuous in $J$ it is uniformly

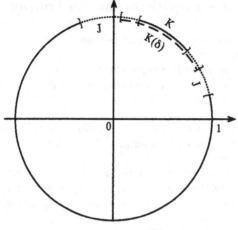

**Fig. 1.**

continuous on $K(\delta)$. So we see that for any given $r > 0$ there exists a $\delta > 0$ such that (8) holds for all $v \in K$, qed.

We leave to the reader the task of verifying, as in Chap. V, n° 27, that if the $\varphi_n$ are indefinitely differentiable, then so likewise are the functions $f \star \varphi_n$. This follows in the usual way from the standard theorem on differentiation under the $\int$ sign (Chap. V, n° 9).

# § 2. Elementary theorems on Fourier series

### 6 – Absolutely convergent Fourier series

Almost all the simultaneously *simple and important* results in the theory of Fourier series, especially those which can be generalised, can be deduced from one fundamental statement:

**Theorem 1 (Weierstrass).** *Every continuous periodic function is the uniform limit of trigonometric polynomials.*

Instead of proving this now, we shall, in this §, show how one may use it; we shall present an "elementary" proof later (n° 12, Theorem 8 and n° 23), more complicated than the general "abstract" Stone-Weierstrass theorem of Chap. V, n° 28.

The most immediate consequence of Theorem 1 is the following:

**Theorem 2.** *If $f$ is a continuous function on $\mathbb{T}$ such that $\sum |\hat{f}(n)| < +\infty$, then*

$$f(u) = \sum \hat{f}(n)u^n \quad \text{for every } u \in \mathbb{T}.$$

Let us denote the right hand side by $g(u)$. This is the sum of an absolutely convergent Fourier series, whence (Chap. V, n° 5) $\hat{g}(n) = \hat{f}(n)$ for every $n$. Putting $f = g + h$, we see that all the Fourier coefficients of the function $h$ vanish.

The relation $\hat{h}(n) = 0$ for every $n$ means that, with respect to the standard scalar product

$$(f \mid g) = \int f(u)\overline{g(u)}dm(u)$$

of two functions on $\mathbb{T}$, the function $h$ is "orthogonal" to all the characters $u \mapsto u^n$ of $\mathbb{T} : (h \mid \chi) = 0$. It is therefore also orthogonal to every linear combination of a finite number of these functions, i.e. to every trigonometric polynomial $p$.

Now $h$ is continuous like $f$ (by hypothesis) and $g$ (the sum of a normally convergent series of continuous functions). By Theorem 1, there therefore exists a sequence $(p_n)$ of trigonometric polynomials which converges to $h$ uniformly on $\mathbb{T}$. Since the functions $h(u)\overline{p_n(u)}$ convergent to $|h(u)|^2$ uniformly on $\mathbb{T}$ we deduce that

$$\int |h(u)|^2 dm(u) = \lim \int h(u)\overline{p_n(u)}dm(u) = \lim(h \mid p_n) = 0.$$

The function $|h(u)|^2$ being continuous and positive, we have $h = 0$ (Chap. V, n° 2), whence $f = g$, qed.

Here is an easy consequence of Theorem 2: *for a continuous function $f$ to be represented by an absolutely convergent Fourier series, it is necessary* *and sufficient that* $\sum |\hat{f}(n)| < +\infty$. The condition is sufficient by Theorem 2. We know on the other hand (Chap. V, n° 5) that if a function $f$, necessarily continuous, is the sum of an absolutely convergent Fourier series, then the coefficients of the latter must be the numbers $\hat{f}(n)$; the one and only series which represents $f$ is then *the* Fourier series of $f$.

The theorem of Cantor mentioned above shows much more: two distinct trigonometric series (i.e. not having the same coefficients) and everywhere convergent (absolutely *or not*) cannot have the same sum.

## 7 – Hilbertian calculations

Let us denote by $\mathcal{H}$ the complex vector space (of infinite dimension) of regulated functions on $\mathbb{T}$ and let us endow it with the usual scalar product

$$(7.1) \qquad (f \mid g) = \int f(u)\overline{g(u)}dm(u).$$

It has the same properties as in Chap. V, n° 3: it is a linear function of $f$ for $g$ given, we also have

$$(7.2) \qquad (g, f) = \overline{(f \mid g)}$$

and finally

$$(7.3) \qquad (f \mid f) = \int |f(u)|^2 dm(u) \geq 0$$

for any $f$. As in Chap. V and, more generally, as in every pre-Hilbert space (Appendix to Chap. III), it therefore satisfies the Cauchy-Schwarz inequality

$$(7.4) \qquad |(f \mid g)|^2 \leq (f \mid f)(g \mid g).$$

We deduce that the expression

$$(7.5) \qquad \|f\|_2 = (f \mid f)^{1/2} = \left( \int |f(u)|^2 dm(u) \right)^{1/2}$$

has the properties

$$(7.6) \qquad \|\lambda f\|_2 = |\lambda|.\|f\|_2, \qquad \|f + g\|_2 \leq \|f\|_2 + \|g\|_2,$$

of a "norm" on the vector space $\mathcal{H}$ of regulated functions on $\mathbb{T}$, except that the relation $\|f\|_2 = 0$ shows only that the set $\{f(u) \neq 0\}$ is countable (Chap. V, n° 7, Theorem 7) and not that $f = 0$. This is not important since two regulated functions which are equal outside a countable set have the same integrals and so the same Fourier series.

As in the case of the norm of uniform convergence, the second relation (6) shows that the expression

$$d_2(f,g) = \|f - g\|_2 = \left( \int |f(u) - g(u)|^2 \, dm(u) \right)^{1/2}$$

is a "distance" between $f$ and $g$ (distance in quadratic mean).

This done, one says that two functions $f$ and $g$ are *orthogonal* if $(f \mid g) = 0$, a concept we have already used above. We now have the Pythagoras relation

$$(7.7) \qquad \|f + g\|_2^2 = \|f\|_2^2 + \|g\|_2^2$$

since $(f + g \mid f + g) = (f \mid f) + (f \mid g) + (g \mid f) + (g \mid g)$. This extends to a finite sum of pairwise orthogonal functions $f_i$: indeed

$$\left( \sum f_i \mid \sum f_i \right) = \sum (f_i \mid f_j) = \sum (f_i \mid f_i) \quad \text{if } (f_i \mid f_j) = 0 \text{ for } i \neq j.$$

In particular, $(\mathbf{e}_p \mid \mathbf{e}_q) = 0$ or 1, whence

$$(7.8) \qquad \left( \sum a_p \mathbf{e}_p \mid \sum b_q \mathbf{e}_q \right) = \sum a_p \overline{b_p}$$

at least when dealing with finite sums, i.e. trigonometric polynomials.

With these definitions, the Fourier coefficients of a function $f$ are, as we have already seen, given by $\hat{f}(n) = (f \mid \mathbf{e}_n)$ where $\mathbf{e}_n$ is the exponential function $\mathbf{e}_n(t)$, in version $\mathbb{R}$, or $u^n$, in version $\mathbb{T}$. If we consider the partial sum

$$(7.9) \qquad f_N = \sum_{|n| \leq N} \hat{f}(n) \mathbf{e}_n$$

of the Fourier series of $f$, we have $(f_N \mid \mathbf{e}_n) = \hat{f}(n) = (f \mid \mathbf{e}_n)$ for $|n| \leq N$ by (8), and so $(f - f_N \mid \mathbf{e}_n) = 0$. The function $f - f_N$ being orthogonal to the exponentials $\mathbf{e}_n$ such that $|n| \leq N$ it is also orthogonal to every linear combination of the these, and in particular to the function $f_N$ itself. Since $f = (f - f_N) + f_N$, Pythagoras' theorem shows that

$$(7.10) \qquad (f \mid f) = (f_N \mid f_N) + (f - f_N \mid f - f_N) \geq (f_N \mid f_N).$$

But by (8)

$$(7.11) \qquad (f_N \mid f_N) = \sum_{|n| \leq N} \left| \hat{f}(n) \right|^2.$$

The partial sums of the series with positive terms $\sum \left| \hat{f}(n) \right|^2$ are therefore bounded above by $(f \mid f)$; they consequently converge, and we have

$$(7.12) \qquad \sum \left| \hat{f}(n) \right|^2 \leq (f \mid f) = \int |f(u)|^2 dm(u) = \oint |f(t)|^2 dt$$

for every regulated function on $\mathbb{T}$.

These calculations, which generalise the traditional ones in $\mathbb{R}^3$ starting from the "unit vectors" of a system of rectangular coordinates, are valid in every pre-Hilbert space. If for example you have a sequence of continuous functions $P_n(t)$ on a compact interval $I \subset \mathbb{R}$ – these are often, in practice, polynomials or the solutions of differential equations – satisfying $\int P_k(t)\overline{P_h(t)}dt = 0$ or 1 according to whether $k \neq h$ or $k = h$, and if for every continuous function $f$ in $I$ you put

$$(7.13) \qquad c_n(f) = \int_I f(t)\overline{P_n(t)}dt,$$

then you obtain the inequality $\sum |c_n(f)|^2 \leq \int |f(t)|^2 dt$. In the good cases, one hopes – while there is life there is hope – to obtain not only an equality but even an expansion

$$(7.14) \qquad f(t) = \sum c_n(f)P_n(t)$$

in a convergent series. The Fourier series were, historically, the first case to present themselves, and have, of course, inspired the many later generalisations of which we have just described the simplest.

## 8 – The Parseval-Bessel equality

The inequality (7.12) is in reality an *equality* as we have seen in Chap. V, n° 5 in the simple case of absolutely convergent Fourier series. We shall now prove this for every regulated function, using Weierstrass' approximation theorem.

By (7.10) and (7.11), it reduces to proving that $\|f - f_N\|_2$ tends to 0, i.e. that

$$(8.1) \qquad \lim_{N \to \infty} \int_0^1 \left| f(t) - \sum_{|n| \leq N} \hat{f}(n)\mathbf{e}_n(t) \right|^2 dt = 0,$$

but to write integrals of this kind explicitly would be the best method of not understanding the proof, and to avail oneself of Knuth's software in vain.

In the complex vector space $\mathcal{H}$ of the preceding n°, let $\mathcal{H}_N$ be the set of trigonometric polynomials involving only the $\mathbf{e}_n$, $|n| \leq N$, in other words, the vector subspace generated by these $2N + 1$ functions; it contains $f_N$. Since $f - f_N$ is orthogonal to the $\mathbf{e}_n \in \mathcal{H}_N$, it is orthogonal to every $p \in \mathcal{H}_N$ as we saw above. On writing $f - p = (f - f_N) + (f_N - p)$ and observing that $f_N - p \in \mathcal{H}_N$ we then have

$$(8.2) \qquad \begin{aligned} (f - p \mid f - p) &= (f - f_N \mid f - f_N) + (f_N - p \mid f_N - p) \\ &\geq (f - f_N \mid f - f_N) \end{aligned}$$

for any $p \in \mathcal{H}_N$. In other words, $f_N$ *is the point of the vector subspace* $\mathcal{H}_N$ *lying at the minimum distance from* $f$, which is plausible (figure 2) since it the "orthogonal projection" of $f$ onto $\mathcal{H}_N$.

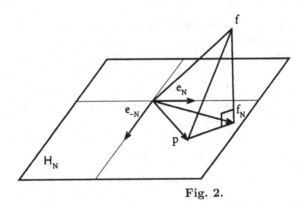

**Fig. 2.**

It follows that, to establish (1), it is enough to prove that, for every $r > 0$, there exists *a* (indefinite article) trigonometric polynomial $p$ such that

$$(8.3) \qquad (f - p \mid f - p) < r^2;$$

such a polynomial $p$ belongs of course to every $\mathcal{H}_N$ of sufficiently large index, so that, by (7.10),

$$(8.4) \quad 0 \le (f \mid f) - (f_N \mid f_N) = (f - f_N \mid f - f_N) \le (f - p \mid f - p) < r^2$$

for $N$ large, which will establish the Parseval-Bessel equality.

Relation (3) is a theorem on approximation by trigonometric polynomials; but instead of measuring the "distance" between two functions $f$ and $g$ by the uniform convergence norm – which is doomed to failure if $f$ is not continuous –, one measures it by the function $\|f - g\|_2$ which leads to *convergence in quadratic mean*, while on using the distance

$$\|f - g\|_1 = \int |f(u) - g(u)| dm(u)$$

one obtains *convergence in mean* (Chap. V, end of n° 4), much less easy to manipulate than the preceding in this context.

Let us return to the proof of (3). There is no problem if $f$ is continuous: Weierstrass provides a trigonometric polynomial $p$ such that $|f(u) - p(u)| < r$ for any $u$, which is incomparably better than (3). In the general case, it reduces to showing that $f$ can be approximated in quadratic mean by *continuous* functions $g$, for if $\|f - g\|_2 < r$ and $\|g - p\|_2 < r$, it follows that $\|f - p\|_2 < 2r$. To do this it is enough to have a general result which could also well be obtained from the definition of the integrable functions at Chap. V:

**Lemma.** *Let $f$ be a regulated function on a compact interval $I \subset \mathbb{R}$ (resp. on $\mathbb{T}$). Then, for every $r > 0$, there exists a continuous function $g$ on $I$ (resp. $\mathbb{T}$) such that $\|f - g\|_2 < r$, or $\|f - g\|_1 < r$.*

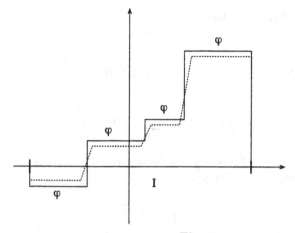

**Fig. 3.**

One may assume $I = [0, 1]$. There exists a step function $\varphi$ on $I$ such that

$$|f(t) - \varphi(t)| \leq r$$

for any $t \in I$, whence

$$\|f - \varphi\|_1 \leq r$$

and the same inequality for the other norm. It therefore suffices to establish the lemma for the step function $\varphi$. Figure 3 indicates the method: one replaces $\varphi$ by a continuous piecewise linear function equal to $\varphi$ except on neighbourhoods of the discontinuities of $\varphi$; if one puts $M = \sup |\varphi(t)|$ and if $\varphi$ has $n$ discontinuities in $[0, 1]$, one may choose the $n$ intervals on which one modifies it so that their lengths are less than $r/Mn$; the contribution of such an interval to the integral of $|f - \varphi|$ is then less than the length $r/Mn$ of the latter multiplied by the maximum of $|f - \varphi|$ on this interval, so to $Mr/Mn = r/n$, whence, for the $n$ intervals which actually contribute to the integral of $|f - \varphi|$, a total contribution of $\leq r$. For approximation in quadratic mean, one chooses intervals of length $< r^2/M^2n^2$. The case of periodic functions is treated similarly, arguing on $\mathbb{T}$ instead of on $I$.

[Artificial proof of too limited a result. In the Lebesgue theory (Bourbaki model), an integrable (resp. square integrable) function is, almost by definition, a limit in mean (resp. in quadratic mean) of continuous functions. There is nothing else to prove, except perhaps the integrability of the regulated functions, a "result" whose proof takes three lines and which, at this level, is totally uninteresting.]

However it may be, these roundabout procedures provide a result, which, fundamental though it is, is still too limited:

**Theorem 3 (Parseval-Bessel[14]).** *Let $f$ be a regulated periodic function. Then the series $\sum |\hat{f}(n)|^2$ is convergent and*

$$(8.5) \qquad \sum |\hat{f}(n)|^2 = \|f\|_2^2 = \int |f(u)|^2 dm(u) = \oint |f(t)|^2 dt,$$

$$(8.5') \qquad \lim_{N \to \infty} \int_0^1 \left| f(t) - \sum_{|n| \le N} \hat{f}(n)\mathbf{e}_n(t) \right|^2 dt = 0.$$

**Corollary.** *Let $f$ and $g$ be two regulated periodic functions. Then the series $\sum \hat{f}(n)\overline{\hat{g}(n)}$ converges absolutely and*

$$(8.6) \qquad \sum \hat{f}(n)\overline{\hat{g}(n)} = (f \mid g) = \int f(u)\overline{g(u)}dm(u) = \oint f(t)\overline{g(t)}dt.$$

The proof consists of using the algebraic identity

$$(8.7) \qquad \begin{aligned} 4(f \mid g) \;=\; & (f + g \mid f + g) - (f - g \mid f - g) + \\ & + i(f + ig \mid f + ig) - i(f - ig \mid f - ig) \end{aligned}$$

which follows formally – calculate mechanically by expanding the squares without writing any integrals – from the fact that the scalar product $(f \mid g)$ is, for $g$ given, a linear function of $f$ and satisfies $(g \mid f) = \overline{(f \mid g)}$; the relation (7) generalises the identity

$$(8.8) \qquad 4u\bar{v} = |u + v|^2 - |u - v|^2 + i|u + iv|^2 - i|u - iv|^2$$

between complex numbers. Having done this, one applies Parseval-Bessel to the functions $f + g$, $f - g$, $f + ig$ and $f - ig$ that appear in (7), and applies (8) to $u = \hat{f}(n)$ and $v = \hat{g}(n)$. Next one remarks that the series $\sum \hat{f}(n)\overline{\hat{g}(n)}$ is a linear combination of four absolutely convergent series, so converges absolutely, and that its sum is the scalar product $(f \mid g)$.

---

[14] This equality was published by M.-A. Parseval (1755–1836) in 1805 in the Mémoires de l'Académie des Sciences; Parseval considered two series of the form $P(t) = \sum a_n t^n$ and $Q(t) = \sum b_n t^{-n}$ (summing over $\mathbb{N}$), and remarked, up to notation, that $P(t)Q(t) = \sum c_n t^n$ (summing over $\mathbb{Z}$) with $c_0 = \sum a_n b_n$, and then considered it obvious that

$$\sum a_n b_n = \frac{1}{\pi} \int_0^\pi \left[ P\left(e^{iu}\right) Q\left(e^{iu}\right) + P\left(e^{-iu}\right) Q\left(e^{-iu}\right) \right] du;$$

just a simple formal calculation. The astronomer Friedrich Wilhelm Bessel (1784–1846), ultrafamous for his work in celestial mechanics, published the inequality $\sum \left|\hat{f}(n)\right|^2 < \|f\|^2$ in a memoir of 1828 on periodic phenomena, where he used the expansion in Fourier series without reference to its proof or to problems of convergence. I. Grattan-Guinness, *Joseph Fourier 1768–1830* (MIT Press, 1972), pp. 240 and 376. It was at the end of the century, with the appearance of the first works on "functional" Hilbert spaces, that the theorem would be proved correctly and its importance highlighted.

Another corollary: assume that the Fourier series of a *regulated* function $f$ converges absolutely, i.e. that

$$\sum \left| \hat{f}(n) \right| < +\infty$$

and let $g$ be the sum of the latter. By n° 5 of Chap. V, we must have $\hat{g}(n) = \hat{f}(n)$, which suggests that $g = f$ "more or less"; this is precisely what we proved in Theorem 2 assuming $f$ *continuous*.

In the general case, let us again consider the function $h = f - g$. It is regulated and its Fourier coefficients $\hat{h}(n) = \hat{f}(n) - \hat{g}(n)$ are all zero. Parseval-Bessel then shows that $\int |h(t)|^2 \, dt = 0$ and thus that $h(t) = 0$ except maybe on a countable set $D$ of points (Chap. V, n° 7, Theorem 7). We thus see that

$$f(t) = \sum \hat{f}(n) e_n(t)$$

for every $t \notin D$.

*Exercise* – Consider a sequence of polynomials $P_n(t)$ on a compact interval $I$, satisfying $\int P_k(t) \overline{P_h(t)} dt = 0$ or $1$ and such that $d°(P_n) = n$ for every $n$. Show that every continuous function $f$ on $I$ is the uniform limit of (finite) linear combinations of the $P_n$ and deduce that $\int |f(t)|^2 \, dt = \sum |c_n(f)|^2$ [notation (7.13)].

The preceding results show that if one associates to each $f$ its Fourier transform $\hat{f} : n \mapsto \hat{f}(n)$, one obtains *a linear map of $\mathcal{H}$ into $L^2(\mathbb{Z})$ which preserves scalar products*. The reader who would like to understand the difference in effectiveness between the integrals of Riemann and those of Lebesgue may ask himself the question of whether this map is surjective. Negative response *chez* Riemann, positive *chez* Lebesgue, whose theory there found one of its first great successes.

To understand the problem, let us start from a function $c(n)$ in $L^2(\mathbb{Z})$; we are to find an $f \in \mathcal{H}$ such that $\hat{f}(n) = c(n)$ for every $n$. If $f$ exists we must have

(8.9)     $$f_N(u) = \sum_{|n| \leq N} c(n) u^n \quad \text{and} \quad \lim \| f - f_N \|_2 = 0.$$

Now the $f_N$ form a *Cauchy sequence* in $\mathcal{H}$, since for $p < q$ one has, by (7.8),

(8.10)     $$\| f_p - f_q \|_2^2 = \sum_{p < |n| \leq q} |c(n)|^2,$$

a result arbitrarily small for $p$ large since $\sum |c(n)|^2 < +\infty$. The question asked is thus to decide if the convergence in quadratic mean is, in $\mathcal{H}$, guaranteed by Cauchy's criterion, in other words: is $\mathcal{H}$ a *complete* space in the sense of the Appendix to Chap. III? Negative response in Riemann theory, positive (Riesz-Fischer theorem) in the Lebesgue theory where one considers the much more general "square integrable" functions. This is one of the many reasons which show that one can probably never surpass the present theory of integration – that of Lebesgue – or, to be more prudent, that it is not worth seeking to do this if one is not interested in the ultra fine and ultra specialised mathematics of Baire's successors.

For lack of great "modern"results, i.e. not more than a century old, one may always turn back to Euler and Fourier.

*Example 1 (Fourier).* Consider the periodic function equal to $t$ for $|t| < \frac{1}{2}$; the values at the end-points are immaterial. Integrating by parts, we have, for $n \neq 0$,

$$\hat{f}(n) = \int_{-\frac{1}{2}}^{\frac{1}{2}} t \overline{e_n(t)} dt = \left. \frac{t \overline{e_n(t)}}{-2\pi i n} \right|_{-\frac{1}{2}}^{\frac{1}{2}} + \frac{1}{2\pi i n} \int_{-\frac{1}{2}}^{\frac{1}{2}} \overline{e_n(t)} dt;$$

the last integral is zero since $e_n$ is orthogonal to $e_0$; since $e_n(t) = (-1)^n$ for $t = \pm\frac{1}{2}$, it follows that

$$\hat{f}(0) = 0, \quad \hat{f}(n) = (-1)^{n+1}/2\pi i n.$$

The integral of $t^2$ being equal to $1/12$, one finds the relation

$$\sum 1/4\pi^2 n^2 = 1/12$$

where the sum is taken over all nonzero $n \in \mathbb{Z}$. Whence again the relation $\sum 1/n^2 = \pi^2/6$.

*Example 2.* Consider the function of period 1 such that

$$f(t) = e^{2\pi i z t} \qquad \text{for } |t| < \frac{1}{2},$$

where $z$ is a complex number, not an integer, since otherwise the interest of the problem evaporates. We have

$$\hat{f}(n) = \int_{-\frac{1}{2}}^{\frac{1}{2}} e^{2\pi i(z-n)t} dt = \left. \frac{e^{2\pi i(z-n)t}}{2\pi i(z-n)} \right|_{-\frac{1}{2}}^{\frac{1}{2}}$$

since, for every $\lambda \in \mathbb{C}$, the derivative of $e^{\lambda t}$ is $\lambda e^{\lambda t}$ (Chap. IV, n° 10, obvious since $e^{\lambda t} = \sum \lambda^n t^n/n!$). Whence

(8.11)     $$\hat{f}(n) = (-1)^n \sin \pi z/\pi(z-n).$$

Considering now the function $g(t) = \overline{f(-t)}$, on passing to the $\mathbb{T}$ interpretation and bearing in mind the symmetry of the invariant measure we have

$$(8.12) \qquad \hat{g}(n) = \int \overline{f(u^{-1})} u^{-n} dm(u) = \int \overline{f(u)} u^n dm(u) =$$

$$= \int \overline{f(u) u^{-n}} dm(u) = \overline{\hat{f}(n)}.$$

The last corollary then shows – a general result of course – that

(8.13) $$\sum \hat{f}(n)^2 = \oint f(t)f(-t)dt.$$

In the present case we have $f(t)f(-t) = 1$ for any $t$, whence, by (11), the identity

$$1 = \sum \frac{\sin^2 \pi z}{\pi^2(z-n)^2},$$

or, replacing $z$ by $z/\pi$,

(8.14) $$\frac{1}{\sin^2 z} = \sum_{\mathbb{Z}} \frac{1}{(z-n\pi)^2}.$$

We have already obtained this formula in Chap. II, n° 21, by formally differentiating the expansion of $\cot z$ in series of rational fractions, then in Chap. III, n° 17, Example 4 by more orthodox arguments .

In his memoir on the propagation of heat, Fourier calculated similar expansions; he considered for example the function of period $\pi$ (and not $2\pi$) equal to $\sin x$ (or to $\cos x$) between 0 and $\pi$ and expanded it as a function of period $2\pi$. He also considered the function of period $2\pi$ equal to $\cos x$ between $-\pi/2$ and $\pi/2$ and zero between $\pi/2$ and $3\pi/2$, etc. These examples were particularly bold at the time, since they yield a series of the form $\sum a_n \cos nx$ whose sum is equal to $\cos x$ in the first interval and to 0 in the second, i.e. a series of analytic functions whose sum is not analytic[15]; Lagrange, who had all the same met "Fourier" series in 1759 *à propos* the vibrating string problem but rejected them because of their periodicity and who attempted to found all of analysis on power series, criticised Fourier's memoir briskly. The reader will easily find the coefficients in these formulae and may be interested to trace the graphs of these bizarre functions, as Fourier himself did.

### 9 – Fourier series of differentiable functions

In the sequel we shall write $C^p(\mathbb{T})$ for the set of periodic functions of class $C^p$; and $\mathcal{D}(\mathbb{T}) = C^\infty(\mathbb{T})$ as in $\mathbb{R}$. As we shall see, Theorem 2 always applies to the functions of $C^1(\mathbb{T})$. Let us first make several remarks on the formula for integration by parts.

This is particularly simple in the case of two periodic functions $f$ and $g$ of class $C^1$ in $\mathbb{R}$. When one integrates $fg' + f'g$ over an interval $[a, a+1]$ in

---

[15] although the problem of vibrating strings had already suggested this kind of phenomenon to d'Alembert, Euler and Daniel Bernoulli, who did not pursue it (and argued about this subject). One may find Fourier's text, explanations and a biography of the prefect of the Isère, a position he occupied while writing his memoir, in I. Grattan-Guinness, *Joseph Fourier 1768–1830* (MIT Press, 1972).

$\mathbb{R}$, the term $f(a+1)g(a+1) - f(a)g(a)$ cancels to zero because $f$ and $g$ are periodic. Writing generally $f'(u)$ for the function which, on $\mathbb{T}$, corresponds through $f'(e(t)) = f'(t)$ to the periodic function[16] $f'(t)$ on $\mathbb{R}$, we then have

$$(9.1) \qquad \int f'(u)g(u)dm(u) = -\int f(u)g'(u)dm(u)$$

or, in the $\mathbb{R}$ version,

$$(9.2) \qquad \oint f'(t)g(t)dt = -\oint f(t)g'(t)dt.$$

This result extends to the periodic functions which are primitives of regulated functions, but this requires some explanations. Periodic or not, a function $f$ is, on $\mathbb{R}$, a primitive of a regulated function $f'$ if (i) $f$ is *continuous*, (ii) $f$ admits right and left derivatives at each point $t \in \mathbb{R}$, equal to the limits $f'(t+)$ and $f'(t-)$ of $f'$; the derivative $f'$ is thus periodic if $f$ is. Theorem 12 bis of Chap. V, n° 13, i.e. the FT, being valid for the primitives of regulated functions, one has

$$f(1) - f(0) = \oint f'(t)dt,$$

so that *the mean value of $f'$ is zero if $f$ is periodic*. Formula (2) remains valid since, if $f$ and $g$ are *periodic* primitives of regulated functions $f'$ and $g'$, necessarily periodic, the function $fg$ is manifestly a periodic primitive of $f'g + fg'$; since $fg$ is periodic, the integral of $f'g + fg'$ over a period is zero as we have just seen, whence (2).

One must not believe that a regulated periodic function $f$ always admits a *periodic* primitive. If indeed – the only possibility up to an additive constant – one puts

$$F(t) = \int_0^t f(x)dx$$

as in Chap. V, n° 13, it is clear that $F$ is periodic if and only if the mean value of $f$ is zero: write that $F(1) = F(0)$. One could adopt the preceding formula for $t \in [0, 1[$ and define $F$ on $\mathbb{R}$ by periodicity, but then one would have

$$F(1-) - F(1+) = F(1-) - F(0+) = \lim[F(1-\varepsilon) - F(\varepsilon)] = \int_0^1 f(t)dt,$$

whence a discontinuity for $t = 1$ and more generally for $t \in \mathbb{Z}$; not being continuous, $F$ cannot be a primitive of $f$. In a case of this kind, one has to add to the right hand side of (2.20') a term equal to the difference

---

[16] We have $f'(u) = 2\pi i. \lim[f(uv) - f(u)]/(v-1)$ as $v \in \mathbb{T}$ tends to 1.

$f(1-)g(1-) - f(0+)g(0+)$; a notorious source of errors[17] in calculation ...

This done, let us consider a function $f \in C^1(\mathbb{T})$ and write $Df = f'$ for its derivative, a periodic continuous function. An integration by parts then shows, by (2), that

$$\int_0^1 Df(t)e^{-2\pi int}dt = 2\pi in \int_0^1 f(t)e^{-2\pi int}dt$$

i.e.

(9.3) $$\widehat{Df}(n) = 2\pi in\hat{f}(n).$$

This calculation is again valid if $f$ is a periodic primitive of a regulated periodic function, as we have seen above. It does not apply to Example 1 of the preceding n°, for the periodic function equal to $t$ on $[-\frac{1}{2}, \frac{1}{2}[$, not being everywhere continuous, is not a primitive on $\mathbb{R}$.

Now we know that the series $\sum \left|\widehat{Df}(n)\right|^2$ converges since $Df$ is regulated; so likewise is the series $\sum 1/n^2$. The series $\sum \left|\widehat{Df}(n)/n\right|$ is therefore absolutely convergent (Cauchy-Schwarz inequality for series). But $\widehat{Df}(n)/n = \hat{f}(n)$ up to a constant factor. Consequently:

**Theorem 4.** *Let $f$ be a periodic continuous function, the primitive of a regulated function on $\mathbb{R}$ (for example, a periodic function of class $C^1$ on $\mathbb{R}$); then the Fourier series of $f$ is absolutely convergent and*

$$f(t) = \sum \hat{f}(n)e_n(t)$$

*for any $t \in \mathbb{R}$.*

If $f$ is of class $C^2$ one may iterate (3) and obtain

$$\widehat{D^2f}(n) = (2\pi in)^2 \hat{f}(n),$$

and so on.

The Parseval-Bessel inequality now shows that

(9.4) $$\sum \left|n^p\hat{f}(n)\right|^2 < +\infty$$

if $f \in C^p(\mathbb{T})$, and *a fortiori*

---

[17] Pay attention to the fact that the $f \in C^p(\mathbb{T})$ must be of class $C^p$ on $\mathbb{R}$ and not only on a period interval such as $[0, 1]$, since this last property is compatible with the existence of discontinuities at 0 and 1 of the derivatives of the periodic function considered. For a function $f$ of class $C^p$ on $[0, 1]$ to be extendable to a periodic function of class $C^p$ on $\mathbb{R}$ it is necessary and sufficient that $f^{(k)}(0) = f^{(k)}(1)$ for every $k \leq p$.

(9.5) $$\hat{f}(n) = o(1/n^p).$$

One may wonder whether, conversely, these properties characterise the Fourier coefficients of functions of class $C^p$; the answer is negative: for $p = 1$, the relation (4) is satisfied by every primitive of a regulated function, which allows many discontinuities in the derivative. But it is worth looking closer.

First, (3) shows that if $f$ is a periodic primitive of a regulated function $Df$, then the Fourier series

(9.6) $$\sum \widehat{Df}(n)\mathbf{e}_n(t) \approx \sum 2\pi i n \hat{f}(n)\mathbf{e}_n(t)$$

of $Df$ is obtained *as if* one may differentiate that of $f$ term-by-term, even though the general theorem on term-by-term differentiation (Chap. III, n° 17, Theorem 19 and Example 2) does not necessarily apply here: the discontinuities of $Df$ can prevent its Fourier series from converging uniformly or even simply (whence the $\approx$ sign).

If, however, the right hand side of (6) converges absolutely for a given regulated periodic function $f$, i.e. if $\sum |n\hat{f}(n)| < +\infty$, a more restrictive condition than $\hat{f}(n) = o(1/n)$, then a fortiori $\sum |\hat{f}(n)| < +\infty$; one may then, as we have seen in n° 8, assume that

$$f(t) = \sum \hat{f}(n)\mathbf{e}_n(t)$$

everywhere by modifying $f$ on a countable set; since the derived series converges uniformly, we conclude that $f$ is differentiable and that $Df(t) = \sum 2\pi i n \hat{f}(n)\mathbf{e}_n(t)$ is a continuous function: the function $f$ is thus of class $C^1$. More generally, if the Fourier coefficients of a function $f$ satisfy $\sum |n^p \hat{f}(n)| < +\infty$, the function is of class $C^p$: iterate the argument.

The case where $p = \infty$, i.e. of the functions $f \in \mathcal{D}(\mathbb{T})$, is simpler. If $f$ is $C^\infty$, in which case (5) applies for any $p$ to the Fourier coefficients of all the successive derivatives of $f$, one may differentiate the Fourier series of $f$ term-by-term *ad libitum* and obtain series which are all normally convergent and represent the successive derivatives of $f$; note that, except for that of $f$, they have no constant term. If, conversely, one takes coefficients $c(n)$ satisfying (5) for any $p$ and if one puts $f(t) = \sum c(n)\mathbf{e}_n(t)$, an absolutely convergent series and so the Fourier series of $f$, it is clear that the products $n^r c(n)$ again satisfy (5) for any $r$ and that the series obtained on differentiating the series $\sum c(n)\mathbf{e}_n(t)$ formally $r$ times will converge normally; the standard theorem on term-by-term differentiation (Chap. III, n° 17, Theorem 19) then applies to the series $\sum c(n)\mathbf{e}_n(t)$: $f$ is a $C^\infty$ function of which $c(n)$ are the Fourier coefficients. In conclusion:

**Theorem 5.** *Let $c(n)$ be a scalar function on $\mathbb{Z}$. For there to be a function $f \in C^\infty(\mathbb{T})$ such that $\hat{f}(n) = c(n)$ for every $n$ it is necessary and sufficient that $c(n) = O(1/n^p)$ for every $p \in \mathbb{N}$. One may then differentiate the Fourier series of $f$ term-by-term any number of times.*

A function $c$ on $\mathbb{Z}$ satisfying (5) for *every $p$* is said to be *of rapid decrease.*

## 10 – Distributions on $\mathbb{T}$

The identification of the functions on $\mathbb{T}$ with the functions of period 1 on $\mathbb{R}$ has allowed us to define the spaces $C^p(\mathbb{T})$ in an obvious way for every $p \in \mathbb{N}$, and also $\mathcal{D}(\mathbb{T}) = C^\infty(\mathbb{T})$. As on the Schwartz space $\mathcal{D}(\mathbb{R})$ (Chap. V, n° 34) one may define the norms

$$(10.1) \qquad \|\varphi\|^{(k)} = \|\varphi\| + \|D\varphi\| + \ldots + \|D^k\varphi\|$$

on $\mathcal{D}(\mathbb{T})$, and the distances

$$(10.1') \qquad d_k(\varphi, \psi) = \|\varphi - \psi\|^{(k)},$$

where $\|\varphi\| = \sup |\varphi(u)|$ denotes the norm of uniform convergence on $\mathbb{T}$ (or, in terms of periodic functions, on $\mathbb{R}$) and where the $D^r\varphi = \varphi^{(r)}$ are the successive derivatives, again periodic, of the function $\varphi$. A concept of convergence is associated with these norms: a sequence $\varphi_n \in \mathcal{D}(\mathbb{T})$ converges to a $\varphi \in \mathcal{D}(\mathbb{T})$ if $\lim d_k(\varphi, \varphi_n) = 0$ for every $k$, in other words, if for every $k > 0$ one has $\lim D^k\varphi_n = D^k\varphi$ uniformly on $\mathbb{T}$. This is the mode of convergence which allows us to differentiate the given sequence term-by-term *ad libitum*, to calculate the derivatives of the limit.

This said, a *distribution* on $\mathbb{T}$ is, as on $\mathbb{R}$, a linear map $T : \mathcal{D}(\mathbb{T}) \to \mathbb{C}$ which is continuous in the following sense: there exist a $k \in \mathbb{N}$ and a constant $M \geq 0$ such that

$$(10.2) \qquad |T(\varphi)| \leq M. \|\varphi\|^{(k)} \qquad \text{for every } \varphi \in \mathcal{D}(\mathbb{T}),$$

i.e. $|T(\varphi) - T(\psi)| \leq M. \|\varphi - \psi\|^{(k)}$. The smallest integer possible $k$ is called the *order* of $T$. Then $\lim T(\varphi_n) = T(\varphi)$ if the $\varphi_n \in \mathcal{D}(\mathbb{T})$ converge uniformly to a $\varphi \in \mathcal{D}(\mathbb{T})$ as do all their successive derivatives of order $\leq k$: the others are not involved.

The examples given in Chap. V, n° 34 in the case of $\mathbb{R}$ transpose easily to here, so long as one does not try to integrate the periodic functions on all of $\mathbb{R}$, an integral of this kind clearly being divergent. In particular, every integrable function $f$ on $\mathbb{T}$ defines a distribution $T_f : \varphi \mapsto \int \varphi(u) f(u) dm(u)$, and every measure $\mu$ on $\mathbb{T}$ a distribution $T_\mu : \varphi \mapsto \int \varphi(u) d\mu(u)$. These distributions are of order 0. A distribution such as $\varphi \mapsto \int \varphi^{(r)}(u) f(u) dm(u)$ is of order $r$; we shall see later that up to an additive constant[18], every distribution on $\mathbb{T}$ is of this type.

It would be convenient to use the Leibniz notation $T(\varphi) = \int \varphi(u) dT(u)$ for distributions; the definition of the *derivative*

$$(10.3) \qquad T'(\varphi) = -T(\varphi')$$

---

[18] A constant $c$ is also the constant *function* $u \mapsto c$, so is also a *distribution*, namely $\varphi \mapsto c \int \varphi(u) dm(u) = c.m(\varphi)$.

of a distribution would then be written

$$(10.4) \qquad \int \varphi(u) dT'(u) = - \int \varphi'(u) dT(u)$$

as in the formula for integration by parts (9.2) from which it is directly derived. Frowned on by Schwartz, this notation has not gained currency; but we shall use it on occasion.

As in the case of functions and of measures on $\mathbb{T}$, one may associate *Fourier coefficients*

$$(10.5) \qquad \hat{T}(n) = \int u^{-n} dT(u) = T(\mathbf{e}_{-n})$$

to every distribution $T$ on the torus. Now the fact that the Fourier series of a function $\varphi \in \mathcal{D}(\mathbb{T})$ converges uniformly together with all its derived series clearly means that *the series $\varphi = \sum \hat{\varphi}(n) \mathbf{e}_n$ converges in the sense of the space $\mathcal{D}(\mathbb{T})$*: we have

$$(10.6) \qquad \lim_{N \to \infty} \|\varphi - \varphi_N\|^{(k)} = 0 \quad \text{for every } k \in \mathbb{N}$$

where, as always, the $\varphi_N$ are the partial sums of the Fourier series of $\varphi$.

One may thus "integrate" the Fourier series of $\varphi$ term-by-term with respect to any distribution $T$ on the torus. Since the value of $T$ on the function $\mathbf{e}_n$ is just, by definition, the Fourier coefficient $\hat{T}(-n)$ of $T$, one finds

$$(10.7) \qquad T(\varphi) = \sum \hat{T}(-n) \hat{\varphi}(n) \quad \text{for every } \varphi \in \mathcal{D}(\mathbb{T}).$$

This relation resembles Parseval-Bessel more if one writes it in the form

$$T(\bar{\varphi}) = \int \overline{\varphi(u)} dT(u) = \sum \overline{\hat{\varphi}(n)} \hat{T}(n).$$

One may again interpret it as an *expansion of $T$ in Fourier series*. Let us associate a distribution $T_f$ to every reasonable function $f$ on $\mathbb{T}$ by putting $T_f(\varphi) = \int \varphi(u) f(u) dm(u)$. In particular, write $\mathbb{E}_n$ for the distribution associated to the function $t \mapsto \mathbf{e}_n(t)$ or $u \mapsto u^n$, whence

$$\mathbb{E}_n(\varphi) = \hat{\varphi}(-n) \quad \text{for every } \varphi \in \mathcal{D}(\mathbb{T}).$$

Formula (7) can then be written

$$(10.8) \qquad T(\varphi) = \sum \hat{T}(n) \mathbb{E}_n(\varphi) \quad \text{for every } \varphi \in \mathcal{D}(\mathbb{T})$$

or, symbolically, in the form

$$T = \sum \hat{T}(n) \mathbb{E}_n;$$

this manner of writing has a sense if one defines the sum of a series $\sum T_n$ of distributions as the distribution $T$ such that

$$T(\varphi) = \sum T_n(\varphi)$$

for every $\varphi \in \mathcal{D}(\mathbb{T})$, which assumes, at the least, (and, in fact, precisely[19]) that the right hand side converges for every $\varphi \in \mathcal{D}(\mathbb{T})$.

For example let us choose $T = T_f$ where $f$ is a regulated function on $\mathbb{T}$, whence $\hat{T}(n) = \hat{f}(n)$. For every $\varphi \in \mathcal{D}(\mathbb{T})$, by Parseval-Bessel,

$$
\begin{aligned}
T_f(\bar{\varphi}) &= \int f(u)\overline{\varphi(u)}dm(u) = \sum \hat{f}(n)\overline{\hat{\varphi}(n)} = \sum \int \hat{f}(n)u^n\overline{\varphi(u)}dm(u) = \\
&= \lim_{N \to \infty} \int \sum_{|n|<N} \ldots = \lim \int f_N(u)\overline{\varphi(u)}dm(u)
\end{aligned}
$$

where the $f_N$ are the partial sums of the Fourier series of $f$. From the distribution point of view this may be written

(10.9)        $T_f(\varphi) = \lim T_{f_N}(\varphi)$   i.e.   $T_f = \lim T_{f_N}$;

in other words, *qua distribution*, the function $f$ is the limit of the partial sums of its Fourier series. This does not mean that the latter converges to $f$ in the usual sense! This is the one of the sleights of hand allowed by the theory of distributions ...

On the other hand we note that the derivative $T' = DT$ of a distribution $T$ has for Fourier coefficients the numbers

(10.10)   $\widehat{DT}(n) = DT(e_{-n}) = -T(e'_{-n}) = -T(-2\pi in e_{-n}) = 2\pi in\hat{T}(n);$

in other terms, another trick, *the formula (9.3) is valid for every distribution on* $\mathbb{T}$.

Can one characterise the functions $n \mapsto c(n)$ on $\mathbb{Z}$ which are the Fourier coefficients of a distribution? If $T$ is a distribution, by definition one has an inequality of the form

(10.11)                  $|T(\varphi)| \leq M. \|\varphi\|^{(k)},$

valid for every $\varphi \in \mathcal{D}(\mathbb{T})$. But if $\varphi(t) = e_n(t)$, we have, up to the factor $2\pi i$, that $D\varphi(t) = ne_n(t)$, $D_2\varphi(t) = n^2 e_n(t)$, etc. and so

$$\|e_n\|^{(k)} = 1 + |2\pi n| + |2\pi n|^2 + \ldots + |2\pi n|^k,$$

---

[19] If a series $\sum T_n(\varphi)$ converges for any $\varphi \in \mathcal{D}(\mathbb{T})$, then $T(\varphi) = \sum T_n(\varphi)$ is again a distribution, i.e. satisfies an estimate of the form $|T(\varphi)| \leq M. \|\varphi\|^{(k)}$. The proof is obtained without any calculation from the general theorems of functional analysis.

an expression $\sim |2\pi n|^k$ for $|n|$ large (order of growth of a polynomial at infinity). One concludes from (2) that there exists an integer $k$ such that

$$(10.12) \qquad \hat{T}(n) = O(|n|^k) \qquad \text{for } |n| \text{ large.}$$

Conversely, every function $c(n)$ satisfying $c(n) = O(n^k)$ for *one* integer $k \in \mathbb{N}$ defines a distribution by the formula

$$(10.13) \qquad T(\varphi) = \sum c(-n)\hat{\varphi}(n).$$

First, the series converges since the product of a function "of slow increase" by a function of rapid decrease is clearly of rapid decrease. One has $T(e_n) = c(n)$ since the Fourier coefficients of $e_n$ are all zero apart from the $n$-th (orthogonality relations). It remains to establish the continuity, in the sense of $\mathcal{D}(\mathbb{T})$, of the linear form $\varphi \mapsto T(\varphi)$.

First,

$$(10.14) \qquad \widehat{D^r\varphi}(n) = (2\pi i n)^r \hat{\varphi}(n)$$

for any $r$ for every $\varphi \in \mathcal{D}(\mathbb{T})$ and so

$$(10.15) \qquad \sum |(2\pi i n)^r \hat{\varphi}(n)|^2 = \int |D^r\varphi(u)|^2 \, dm(u) \leq \|D^r\varphi\|^2$$

since the mean value of a function is bounded by its uniform norm. Now we write (13) in the form

$$(10.16) \qquad T(\varphi) = c(0)\hat{\varphi}(0) + \sum \frac{c(-n)}{(2\pi i n)^r}(2\pi i n)^r \hat{\varphi}(n)$$

with $r = k + 1$ and put

$$u_n = c(-n)/(2\pi i n)^r, \qquad v_n = (2\pi i n)^r \hat{\varphi}(n).$$

By (12), we have $u_n = O(1/n)$ and therefore $\sum |u_n|^2 < +\infty$. The relation (14) and Parseval-Bessel show that also $\sum |v_n|^2 = \|D^r\varphi\|_2^2 < +\infty$. The Cauchy-Schwarz inequality then shows that

$$\left| \sum u_n \bar{v}_n \right| \leq M.\|D^r\varphi\|_2 \leq M.\|D^r\varphi\|$$

where $M^2 = \sum |u_n|^2$ depends only on $T$. Since $r = k + 1$, we finally have a majoration

$$(10.17) \qquad |T(\varphi)| \leq |c(0)|.|\hat{\varphi}(0)| + M\|D^{k+1}\varphi\|,$$

which shows that $T$ truly is a distribution. In conclusion:

**Theorem 6.** *Let* $n \mapsto c(n)$ *be a scalar function on* $\mathbb{Z}$. *For there to be a distribution* $T$ *on* $\mathbb{T}$ *such that* $\hat{T}(n) = c(n)$ *for every* $n$, *it is necessary and sufficient that there exists a* $k \in \mathbb{N}$ *such that* $c(n) = O(n^k)$.

One says then that the function $c(n)$ is *of slow increase* or is *tempered*.

*Example.* Consider with Fourier the series

$$\sin t - \sin(2t)/2 + \sin(3t)/3 - \ldots;$$

Fourier calculates its partial sums by differentiating, as was done in Chap. V, n° 16 for square waves, which, for $|t| < \pi$, puts them in the form

$$\frac{t}{2} - \frac{1}{2} \int_0^t \frac{\cos(N + \frac{1}{2})x}{\cos x/2} \, dx;$$

an integration by parts shows that the integral tends to 0, whence

$$t/2 = \sin t - \sin(2t)/2 + \sin(3t)/3 - \ldots \qquad \text{for } |t| < \pi.$$

When Fourier presented his first manuscript to the Académie, Lagrange had objections; for example, he wrote the preceding formula in the form

$$\frac{1}{2}(\pi - t) = \sin t + \sin(2t)/2 + \sin(3t)/3 + \ldots,$$

and differentiated to obtain

$$(*) \qquad\qquad -\frac{1}{2} = \cos t + \cos 2t + \cos 3t + \ldots,$$

then integrated the result between 0 and $t$, whence

$$-t/2 = \sin t + \sin(2t)/2 + \sin(3t)/3 + \ldots$$

and a superb contradiction! Fourier replied that the formula from which Lagrange started is valid only for $0 < t < 2\pi$ and that he consequently had no right to integrate the derived series[20] from $t = 0$.

He might have started by observing that it is not very catholic to differentiate the initial series term-by-term since the series $\sum \cos nt$ is clearly divergent for any $t$; but since he himself did so constantly, Fourier did not use this argument ...

In fact, formula $(*)$ makes sense (but is wrong) *in the sense of distributions*. Using Euler's relations it can be written as

$$\sum_{n \in \mathbb{Z}} \mathbf{e}_n = 0,$$

---

[20] See Grattan-Guinness, *Joseph Fourier 1768–1830*, p. 172.

and if we interpret the left hand side as a series of distributions, (\*) means that

$$\sum_{\mathbb{Z}} \hat{\varphi}(n) = 0 \qquad \text{for every } \varphi \in \mathcal{D}(\mathbb{T}).$$

But the result should be $\varphi(0)$ or $\varphi(1)$ depending on whether you are in $\mathbb{R}$ or $\mathbb{T}$. Now $\varphi(0) = \delta(\varphi)$ where $\delta$ is the Dirac measure at the origin 1 on $\mathbb{T}$. The correct formula is therefore

$$\sum_{n \in \mathbb{Z}} \mathbb{E}_n = \delta$$

an identity between *distributions* equivalent to the obvious formula $\hat{\delta}(n) = 1$. One should therefore replace (\*) by

$$\sum_{1}^{\infty} \oint \varphi(t) \cos nt\, dt = \frac{1}{2}\varphi(0) - \frac{1}{2} \oint \varphi(t) dt$$

a formula equivalent to

$$\varphi(0) = \sum_{\mathbb{Z}} \hat{\varphi}(n).$$

The presence of the additional term $\frac{1}{2}\varphi(0)$ is easy to explain; the series (\*) was indeed obtained by differentiating a series whose sum, equal to $\frac{1}{2}(\pi - t)$ for $0 < t < 2\pi$, is discontinuous at $t = 0$ (or, in version $\mathbb{T}$, at $u = 1$); the distribution obtained by differentiating it must therefore contain a Dirac measure at the origin as in the case of the function equal to 1 for $t > 0$ and to 0 for $t < 0$ (Chap. V, n° 35, Example 2). Note in passing that if one considered distributions on $\mathbb{R}$ and not on $\mathbb{T}$, the derivative of the function $\sum \sin(nt)/n$ would include a Dirac measure at each multiple of $2\pi$.

A method of stripping all the mystery from the distributions consists of considering their successive primitives. A *primitive S of a distribution T* must, by definition, satisfy the relation $S' = T$, i.e.

$$S(D\varphi) = -T(\varphi)$$

for every $\varphi \in \mathcal{D}(\mathbb{T})$. Then, if $S$ exists, by (10) one has $\hat{T}(0) = 0$ and

(10.18)                    $\hat{S}(n) = \hat{T}(n)/2\pi i n$

for $n \neq 0$. Since the sequence $\hat{T}(n)/n$ is slowly increasing, $S$ will exist if and only if

(10.19)                    $\hat{T}(0) = T(\mathbf{e}_0) = \int dT(u) = 0,$

the "integral" of the constant function 1 with respect to $T$. In this case $S$ is unique up to an additive constant, namely the term $\hat{S}(0)$ of its Fourier series; if one chooses this to be zero one obtains a standard primitive of $T$, which it is natural to denote $D^{-1}T$ or $T^{(-1)}$; then

$$(10.20) \qquad \widehat{D^{-1}T}(0) = 0, \quad \widehat{D^{-1}T}(n) = \hat{T}(n)/2\pi in \quad \text{for } n \neq 0.$$

When $\hat{T}(0) \neq 0$, one may apply the argument to the terms of nonzero index of the Fourier series of $T$, whence a distribution $S$ such that $T = \hat{T}(0) + S'$, i.e. such that

$$(10.21) \qquad T(\varphi) = \hat{T}(0)m(\varphi) - S(D\varphi)$$

for every $\varphi \in \mathcal{D}(\mathbb{T})$; one can, here again, insist that $\hat{S}(0) = 0$ to standardise $S$.

The interest of this operation is that on applying it repeatedly to a distribution $T$ such that $\hat{T}(0) = 0$, i.e. "orthogonal" to the constant functions, one increases the chances of convergence *in the usual sense* of the Fourier series of $T$ since one divides its coefficients by the powers of $n$. Since these coefficients are of slow increase, it is clear that on choosing an integer $r$ sufficiently large, the Fourier coefficients of the primitive of order $r$ of $T$ form an absolutely convergent series, in other words are those of a continuous function $f$. This means that $T$ is the derivative of order $r$ of the function $f$ in the sense of distributions, or again that every distribution on $\mathbb{T}$ is given by a formula

$$(10.22) \qquad T(\varphi) = (-1)^r \int \varphi^{(r)}(u)f(u)dm(u) + c \int \varphi(u)dm(u)$$

where $c = \hat{T}(0)$ is a constant. Despite appearances, the notion of a distribution on the torus is thus hardly more general than that of a function in the usual sense: one integrates its derivatives.

We said[21] in n° 7 that in the modern theory of integration, every function $c \in L^2(\mathbb{Z})$ is the Fourier transform of a "square integrable" function on $\mathbb{T}$. Though unable to prove this now, we remark that, by Theorem 6, there exists a distribution $T$ such that $\hat{T}(n) = c(n)$; it is given by the formula (13). In fact, the latter is meaningful for every regulated function $f$ since then the series $\sum |\hat{f}(n)|^2$ converges, hence also $\sum c(-n)\hat{f}(n)$; if one puts

$$(10.23) \qquad T(f) = \sum c(-n)\hat{f}(n)$$

again in this case, the Cauchy-Schwarz inequality for series shows that

$$|T(f)|^2 = \left|\sum c(-n)\hat{f}(n)\right|^2 \leq M^2\|f\|_2^2$$

where $M^2 = \sum |c(n)|^2$. Hence a bound of the form

---

[21] This paragraph is not important in the sequel.

(10.24) $$|T(f)| \leq M\|f\|_2 \leq M\|f\|$$

for every regulated function on $\mathbb{T}$, and in particular for every continuous function, which shows that the distribution $T$ is a *measure* on $\mathbb{T}$. In fact, $T$ is defined by a measure of the form $g(u)dm(u)$ where $g$ is the square integrable function (*à la* Lebesgue) on $\mathbb{T}$ such that $\hat{g}(n) = c(n)$ for every $n$, and (24) is just the extension to these functions of the Cauchy-Schwarz inequality of Chap. V, n° 2.

## § 3. Dirichlet's method

### 11 – Dirichlet's theorem

When Dirichlet discovered Fourier's work, at the beginning of the 1820s, he tried to justify it by rigorous methods. Fourier having discovered the general formula which we now write

$$(11.1) \qquad \hat{f}(n) = \int f(u) u^{-n} dm(u)$$

after dozens of pages of implausible calculations, and Dirichlet having heard from Cauchy that the sum of a series is the limit of its partial sums, he started by calculating those of a Fourier series (we shall simplify the calculation a little by using convolution products):

$$(11.2) \qquad f_N = \sum_{|n| \leq N} f \star e_n = f \star \left( \sum_{|n| \leq N} e_n \right) = f \star D_N$$

where

$$(11.3) \qquad D_N(u) = \sum_{|n| \leq N} u^n = u^{-N} + u^{-N+1} + \ldots + u^N =$$
$$= \frac{u^{-N} - u^{N+1}}{1 - u} \quad \text{for } u \neq 1.$$

It follows that

$$(11.4) \qquad f_N(u) = f \star D_N(u) = \int_{\mathbb{T}} f\left(uv^{-1}\right) \frac{v^{N+1} - v^{-N}}{v - 1} \, dm(v)$$

On putting $v = e(t)$ we have

$$(11.5) \quad D_N(v) = \frac{e((N+1)t) - e(-Nt)}{e(t) - 1} =$$
$$= \frac{e\left(\left(N + \tfrac{1}{2}\right)t\right) - e\left(-\left(N + \tfrac{1}{2}\right)t\right)}{e(t/2) - e(-t/2)} = \frac{\sin(2N+1)\pi t}{\sin \pi t}$$

as one sees on multiplying the two terms of the fraction by $e(-t/2) = e^{-\pi i t}$ and using Euler's formulae. The calculation obviously assumes that $v \neq 1$, i.e. $t \notin \mathbb{Z}$; the value $D_N(1) = 2N + 1$ follows from definition (3). On passing to the language of periodic functions, the partial sums $f_N$ are again given by

$$(11.6) \quad f_N(s) = \oint f(s - t) D_N(t) dt = \oint f(s - t) \frac{\sin(2N+1)\pi t}{\sin \pi t} \, dt.$$

Since we are dealing with the convolution products (on $\mathbb{T}$) of $f$ by the sequence of functions $D_N$, and since we would like the result to tend to $f(s)$

when $N$ increases indefinitely, it would seem, at first sight, that we should use the method of Dirac sequences expounded in n° 5. The condition

$$\int D_N(u)dm(u) = \oint D_N(s)ds = 1$$

is satisfied, because this mean value is the Fourier coefficient of index 0 of the trigonometric polynomial $D_N$. But the $D_N$ change sign more and more often as $N$ increases; it is neither obvious (nor even correct) that the integral of $|D_N(u)|$ remains bounded as $N$ increases. Finally, if one works on an arc $|u - 1| > \delta$ of $\mathbb{T}$, then $|D_N(u)| \leq |1 - u^{2N+1}|/\delta$ by (3), which is insufficient to make $D_N(u)$ tend to 0. In short, a bad idea.

Moreover, if the $D_N$ did form a Dirac sequence, the Fourier series of every continuous function would converge uniformly to the latter by the lemma of n° 5: this would be Paradise. On Earth, although converging "almost everywhere" in the sense of Lebesgue measure[22] (a famous and very difficult result of Lars Carleson, 1966, valid for "square-integrable" functions in Lebesgue's sense), it can still very well diverge for values of $u$ forming an uncountable set[23]. In other words, the method does not work because if it did it would lead to a false result.

Having lived and died (1805–1859) too early to have heard of Lebesgue, Dirac, Carleson and even of Weierstrass' approximation theorem, Dirichlet did not ask himself these questions and, using (4) – so in reality (5) – calculated the difference

(11.7)     $$f_N(u) - f(u) = \int \left[ f(uv^{-1}) - f(u) \right] D_N(v)dm(v)$$

or, replacing $v$ by $v^{-1}$ since $D_N$ is symmetric,

(11.8)     $$f_N(u) - f(u) = \int \frac{f(uv) - f(u)}{v - 1} \left( v^{N+1} - v^{-N} \right) dm(v)$$

---

[22] In Chap. V, n° 11 we defined the (Lebesgue) measure of an open $U$ contained in a compact interval; n° 31, where we defined the integral of a positive lsc function on $\mathbb{R}$, likewise allowed us to define the measure of any open $U \subset \mathbb{R}$. This being so, a subset $N$ of $\mathbb{R}$ is said to be *of measure zero* if for every $r > 0$ there exists an open $U$ such that $N \subset U$, $m(U) < r$. Granted this, a property – the convergence of a series of functions for example – is said to be *true almost everywhere* if the set of $x$ where it is false is of measure zero. Every countable set is of measure zero, but not conversely. See the Appendix to Chap. V.

[23] The first example was that of the German P. du Bois-Reymond: "Before 1873, it was the general belief, of Lejeune Dirichlet, of Riemann, of Weierstrass, among others, that this series always converges to the limit $f(x)$ when $f(x)$ is continuous. Now, in trying to find a proof of this theorem, I came upon an argument to prove the contrary". Letter of 1883 to the Frenchman G. Halphen (Dugac, p. 62). In 1926 the Soviet mathematician A. N. Kolmogoroff produced an integrable (but not *square* integrable) function in the sense of Lebesgue whose Fourier series *diverges everywhere*. Newton would probably have said that one does not meet such functions in Nature.

The right hand side of (8) resembles the difference between the Fourier coefficients of indices $-N-1$ and $N$ of the function

(11.9)        $$g_u(v) = [f(uv) - f(u)]/(v-1);$$

but this function, as regulated as $f$ for $v \neq 1$, has, *a priori*, no meaning for $v = 1$; its integral may well diverge on a neighbourhood of this point, which prevents one from speaking of its Fourier coefficients; the integral (8) is defined only because it involves the quotient $\left(v^{N+1} - v^{-N}\right)/(v-1)$, an everywhere continuous trigonometric polynomial.

Since $v - 1 = \mathbf{e}(t) - 1 \sim 2\pi i t$ when $t$ tends to 0, i.e. when $v$ tends to 1, one always has

(11.10)        $$\lim[f(uv) - f(u)]/(v-1) = f'(s)/2\pi i$$

*if* this derivative exists at the point $u = \mathbf{e}(s)$ considered. The function $g_u$ then has left and right limit values at *every* point $v \in \mathbb{T}$, so is regulated on all of $\mathbb{T}$. In this case it is legitimate to write that

(11.11)        $$f_N(u) - f(u) = \widehat{g_u}(-N-1) - \hat{g}_u(N);$$

and to show that the left hand side tends to 0, it is enough to know that the Fourier coefficients of a regulated function tend to 0 at infinity, which the Parseval-Bessel *inequality* (7.12) makes obvious without recourse to Weierstrass' theorem. Thus:

**Theorem 7.** *Let $f$ be a regulated periodic function. Then*

(11.12)        $$f(u) = \sum \hat{f}(n)u^n = \lim_{N \to +\infty} \sum_{|n| \leq N} \hat{f}(n)u^n$$

*at every point $u \in \mathbb{T}$ where $f$ is differentiable.*

**Corollary (Riemann).** *The behaviour on an open interval of the Fourier series of a regulated periodic function $f$ depends only on the behaviour of $f$ on this interval.*

If in fact $f = g$ on an open interval $U$ then the function $f - g$ has a derivative at every point of $U$. Its Fourier series therefore converges to 0 at every $t \in U$. This means that, for every $t \in U$, only two cases are possible: (i) the Fourier series of $f$ and $g$ are simultaneously divergent at $t$, (ii) they are simultaneously convergent and have the same sum. Another translation: if two regulated periodic functions $f$ and $g$ are equal on an interval with centre $t$, then their Fourier series at $t$ are either simultaneously divergent, or simultaneously convergent with the same sum on a neighbourhood of $t$.

Dirichlet in fact went somewhat further than Theorem 1, for the sum of the square wave series, to mention just this one, is not differentiable in the strict sense at the points where it is discontinuous; there it has only left and right derivatives; so one has to modify the preceding calculations. Now the symmetry of the function $D_N$ shows that its integral over $\left[-\frac{1}{2}, 0\right]$ or over $\left[0, \frac{1}{2}\right]$ is equal to $\frac{1}{2}$; this allows us to replace (7), or the $\mathbb{R}$ version, by

$$(11.13)\ f_N(s) - \frac{1}{2}[f(s+) + f(s-)] =$$

$$= \int_0^{\frac{1}{2}} [f(s+t) - f(s+)]D_N(t)dt + \int_0^{\frac{1}{2}} [f(s-t) - f(s-)]D_N(t)dt.$$

The quotient

$$[f(s+t) - f(s+)]/\sin \pi t$$

appears in the first integral. If $f$ has a right derivative at the point $s$ (obvious definition), this quotient tends to a limit when $t > 0$ tends to 0; for $0 \le t \le \frac{1}{2}$, this quotient then has, like $f$, left and right limit values; the first integral is thus, as in (11), the value at $N$ of the Fourier transform of a regulated periodic function that vanishes on $\left]\frac{1}{2}, 1\right[$, so tends to 0 as $N$ increases. Same argument for the second integral. Whence a simple result, which has been refined in many ways (see for example A. Zygmund, *Trigonometrical Series*, Cambridge UP, 1969):

**Theorem 7 bis (Dirichlet, 1829).** *Let $f$ be a regulated periodic function and $f_N$ the partial sum of order $N$ of its Fourier series. Then*

$$(11.14) \qquad \lim f_N(s) = \frac{1}{2}[f(s+) + f(s-)]$$

*at every point where $f$ has left and right derivatives.*

*Exercise* Dirichlet's Theorem is still valid if the function $t \to |f(s+t) - f(s)|/|t|$ is integrable. *Example:* $f(s+t) = f(s) + O(t^\alpha)$ when $t \to 0$, with an $\alpha > 0$, in which case the graph of $f$ at $s$ has a vertical tangent.

*Example 1. Expansion of* $\cot z$ *as a series of rational fractions.* Consider the function of period 1 on $\mathbb{R}$ given by

$$(11.15) \qquad f(t) = \cos 2\pi zt \qquad \text{for } |t| < \frac{1}{2},$$

where $z \in \mathbb{C}$ is not a rational integer, for otherwise there would be no problem. Since $f\left(-\frac{1}{2}\right) = f\left(\frac{1}{2}\right)$, the periodic function which extends $f$ to all of $\mathbb{R}$ is continuous everywhere and it is clear that it satisfies the hypotheses of Theorem 7 bis. We have

$$
\begin{aligned}
\hat{f}(n) &= \int_{-\frac{1}{2}}^{\frac{1}{2}} \cos 2\pi zt . e^{-2\pi int} dt = \frac{1}{2} \int_{-\frac{1}{2}}^{\frac{1}{2}} \left[ e^{2\pi i(z-n)t} + e^{-2\pi i(z+n)t} \right] dt = \\
&= \frac{1}{2} \left. \frac{e^{2\pi i(z-n)t}}{2\pi i(z-n)} + \frac{e^{-2\pi i(z+n)t}}{-2\pi i(z+n)} \right|_{-\frac{1}{2}}^{\frac{1}{2}} = (-1)^n \frac{z . \sin \pi z}{\pi(z^2 - n^2)}
\end{aligned}
$$

as we see using Euler's formulae. Thus

(11.16)     $\pi.\cos 2\pi zt = \sum (-1)^n \dfrac{z.\sin \pi z}{z^2 - n^2}\, e^{2\pi i n t}$   for $|t| \leq \dfrac{1}{2}$,

an absolutely convergent Fourier series. In particular, for $t = \frac{1}{2}$,

(11.17)     $\pi.\cot \pi z = z \sum \dfrac{1}{z^2 - n^2} = \dfrac{1}{z} + 2z \sum_{n=1}^{\infty} \dfrac{1}{z^2 - n^2}.$

This is the formula due to Euler which we have already met several times, and established at Chap. IV, n° 18, using the infinite product for the sine function. The method we have just presented – to be found essentially in Fourier – is surely the simplest proof.

For $t = 0$, (16) yields the expansion

(11.18)     $\dfrac{\pi}{\sin \pi z} = \dfrac{1}{z} + 2z \sum_{n \geq 1} \dfrac{(-1)^n}{z^2 - n^2}.$

*Example 2. The Bernoulli polynomials.* Recall (Chap. VI, n° 12) that the Bernoulli polynomials are defined by the recurrence relations

(11.19)     $B_0(x) = 1, \qquad B_k'(x) = k B_{k-1}(x)$

and by the condition

(11.20)     $B_k(0) = B_k(1)$     for $k \geq 2$.

The inventor was not acquainted with Fourier series, but condition (20) is exactly what one needs to transform the $B_k$, for $k \geq 2$, into continuous periodic functions $B_k^*$, by putting

(11.21)     $B_k^*(t) = B_k(t)$     for $0 \leq t \leq 1$

as we did in Chap. VI *à propos* the Euler-Maclaurin formula. The hypotheses of the Dirichlet theorems are clearly satisfied. Adopting for once the notation $a_n(f) = \hat{f}(n)$, we have, integrating by parts and assuming $k \geq 2$, $n \neq 0$,

$$a_n(B_k^*) = \int_0^1 B_k(t)\mathbf{e}_{-n}(t)dt = \dfrac{1}{2\pi i n} \int_0^1 B_k'(t)\mathbf{e}_{-n}(t)dt;$$

(19) now shows that

(11.22)     $a_n(B_k^*) = k a_n(B_{k-1}^*)/2\pi i n$     $(k \geq 2,\ n \neq 0)$.

On writing this relation for $k - 1, k - 2, \ldots, 2$ we obtain

(11.23)     $a_n(B_k^*) = k! a_n(B_1^*)/(2\pi i n)^{k-1}.$

Since $B_1(t) = t - \frac{1}{2}$ and since the Fourier coefficients of a constant vanish for $n \neq 0$, we have

$$a_n\left(B_1^*\right) = \int_0^1 te_{-n}(t)dt = -\left.\frac{te_{-n}(t)}{2\pi in}\right|_0^1 + \frac{1}{2\pi in}\int_0^1 e_n(t)dt;$$

the last integral is zero and what remains is

(11.24) $\qquad\qquad a_n\left(B_1^*\right) = -1/2\pi in,$

whence finally

(11.25) $\qquad a_n\left(B_k^*\right) = -k!/(2\pi in)^k \quad$ for $k \geq 1,\ n \neq 0.$

For $n = 0$, we have $(k+1)a_0\left(B_k^*\right) = \oint B_{k+1}'(t)dt = 0$ by (20) if $k \geq 1$, and $a_0\left(B_0^*\right) = 1$ trivially.

Formula (25) shows that the Fourier series is absolutely convergent for $k \geq 2$, whence

(11.26) $\qquad \sum_{n\neq 0} e_n(t)/(2\pi in)^k = -B_k(t)/k! \quad$ for $k \geq 2,\ 0 \leq t \leq 1,$

the sum being taken over all nonzero $n \in \mathbb{Z}$. For $k = 2$ for example, one finds

$$\sum_1^\infty \cos(2\pi nt)/\pi^2 n^2 = t^2 - t + 1/6 \qquad (0 \leq t \leq 1).$$

For $t = 0$, the left hand side of (26) reduces to $\sum 1/(2\pi in)^k$, so is zero for odd $k$; for $k = 2p$, $p \geq 1$, on the other hand,

(11.27) $\qquad\qquad \sum 1/n^{2p} = (-1)^{p+1}(2\pi)^{2p}b_{2p}/(2p)!$

where $b_k = B_k(0)$ (Chap. VI, (3.7)). We should not forget that the left hand side is twice the sum of the Riemann series.

For $k = 1$, the function $B_1^*$, equal to $t - \frac{1}{2}$ for $0 < t < 1$, is discontinuous at the points $t \in \mathbb{Z}$. On grouping the terms of index $n$ and $-n$ of its Fourier series, we again have

(11.28) $\qquad \frac{1}{2} - t = \sum_{n=1}^\infty \sin(2\pi nt)/\pi n \quad$ for $0 < t < 1,$

the series being zero for $t = 0$ or 1 as one may check without invoking Dirichlet. For $t = \frac{1}{4}$ one obtains Leibniz' series for $\pi/4$.

## 12 – Fejér's theorem

We observed in the preceding n° that the Dirichlet kernels do not form a Dirac sequence in the sense of n° 5. At the end of the XIX$^{\text{th}}$ century, the Italian Cesàro had the idea of making divergent sequences $(u_n)$ converge by considering their arithmetic means

(12.1)                    $v_n = (u_1 + \ldots + u_n)/n.$

If you apply this to the sequence $1, 0, 1, 0, \ldots$, you will find that it then "converges" to $\frac{1}{2}$. The method does not always work, even if one iterates – every sequence which tends to $+\infty$ is recalcitrant –, but it is reassuring at least to know that if the sequence converges to $u$ in the usual sense, then it also converges to $u$ in the Cesàro sense: if $|u - u_n| < r$ for $n > p$ and if one writes that

$$v_n = (u_1 + \ldots + u_p)/n + (u_{p+1} + \ldots + u_n)/n,$$

the first quotient is, for $p$ given, $< r$ for $n$ large; on replacing each $u_k$ by $u$ in the second, one commits an error bounded by $(n - p)r/n < r$, whence a total error $< 2r$ for $n$ large, qed.

One may also apply this procedure to a series $\sum u_n$, replacing the standard partial sums $s_n = u_1 + \ldots + u_n$ by their means

(12.2)                    $\sigma_n = (s_1 + \ldots + s_n)/n.$

This allows one to make convergent series which are not; one finds again, for example, the formula

$$1 - 1 + 1 - 1 + 1 - \ldots = \frac{1}{2},$$

conforming to the somewhat premature anticipations of Jakob Bernoulli (Chap. II, n° 7). The subject has been the object of much research, but it is rarely used outside of "fine" analysis.

If one goes back to the Dirichlet formula

$$f_N(t) = \oint f(t - x)D_N(x)dx = f \star D_N(t)$$

for the partial sums of the Fourier series of a function $f$ it is clear that their arithmetic means are the functions $f \star F_N$ where the function

(12.3)                    $F_N = (D_0 + \ldots + D_{N-1})/N$

was introduced by L. Fejér (1880–1959).

In contrast to the $D_N$, the Fejér functions form a Dirac sequence on the unit circle $\mathbb{T}$. To see this, one has to calculate them. Putting $q = e^{\pi i t}$, one has, by (10.5),

$$D_k(t) = \left(q^{2k+1} - q^{-2k-1}\right) / \left(q - q^{-1}\right),$$

whence, adding from 0 to $N - 1$,

$$\begin{aligned}
N\left(q - q^{-1}\right) F_N(t) &= \\
&= \left(q + q^3 + \ldots + q^{2N-1}\right) - \left(q^{-1} + q^{-3} + \ldots + q^{-2N+1}\right) = \\
&= q\left(q^{2N} - 1\right) / \left(q^2 - 1\right) - q^{-1}\left(q^{-2N} - 1\right) / \left(q^{-2} - 1\right) = \\
&= \left(q^{2N} - 2 + q^{-2N}\right) / \left(q - q^{-1}\right)
\end{aligned}$$

and finally

(12.4) $$F_N(t) = \frac{\left(q^N - q^{-N}\right)^2}{N\left(q - q^{-1}\right)^2} = \frac{\sin^2 \pi N t}{N \sin^2 \pi t},$$

for $t \neq 0$, with $F_N(0) = N$ by continuity or by (3).

To show that the $F_N$ form a Dirac sequence on $\mathbb{T}$ it then suffices to show that the $F_N$ are positive (obvious), that their integrals on $\mathbb{T}$ are equal to 1 (obvious, since this is so for the $D_k$, hence for their arithmetic means) and finally that, for any $r > 0$ and $\delta > 0$, the contribution of the arc $|u - 1| > \delta$ of $\mathbb{T}$ to the integral of $F_N$ is $< r$ for $N$ large or, equivalently, that

(12.5) $$\int_{\delta \leq |t| \leq 1/2} F_N(t) dt < r \qquad \text{for } N \text{ large.}$$

But on this domain of integration, by (4) one has

(12.6) $$F_N(t) \leq 1/N \sin^2 \pi \delta,$$

so that the $F_N$ converge uniformly to 0 on $\delta \leq |t| \leq \frac{1}{2}$ for any $\delta > 0$, qed.

**Theorem 8 (Fejér).** *For every regulated periodic function $f$ the arithmetic means of the partial sums of the Fourier series of $f$ converge to $\frac{1}{2}[f(t+) + f(t-)]$ for any $t$. If $f$ is continuous in an open interval $J$, the convergence to $f(t)$ is uniform on every compact $K \subset J$.*

The second assertion follows from the lemma of n° 5.

To establish the first one writes, as in (11.13),

(12.7) $$f \star F_N(t) - \frac{1}{2}[f(t+) + f(t-)] =$$

$$= \int [f(t + s) - f(t+)]F_N(s)ds + \int [f(t - s) - f(t-)]F_N(s)ds,$$

the integrals being taken over $(0, \frac{1}{2})$, and then argues as in n° 5.

One may note in passing that assuming $f$ continuous everywhere one obtains a proof of Weierstrass' approximation theorem (without having used it beforehand ...).

**Corollary.** *Let $f$ be a regulated periodic function. Then*

$$(12.8) \qquad \lim_{N \to \infty} \sum_{-N}^{N} \hat{f}(n)\mathbf{e}_n(t) = \frac{1}{2}[f(t+) + f(t-)]$$

*at every point where the Fourier series of $f$ converges.*

For the partial sums $f_N(t)$, if they converge, converge to the same limit as their arithmetic means, which always converge to the right hand side of (8). The corollary does not claim that the relation (8) is true for arbitrary $t$ and $f$.

## 13 – Uniformly convergent Fourier series

Dirichlet's theorem demonstrates the *simple* convergence of the Fourier series of a regulated periodic function at all points where it has left and right derivatives. In the case of the square waves we have shown by *ad hoc* calculations (Chap. III, n° 11) that in fact the series converges *uniformly* on every compact interval not containing discontinuities of $f$. One may refine the proof of Theorem 7 so as to cover this case and many others, for example the series (11.28).

The arguments which follow being somewhat subtle, the reader is invited to consider them more as an exercise.

**Theorem 9.** *Let $f$ be a regulated function on $\mathbb{T}$ and $J$ an open arc on which $f$ is a primitive of a regulated function (is, for example, of class $C^1$). Then the Fourier series of $f$ converges to $f$ uniformly on every compact arc $K \subset J$.*

The proof we are going to set out calls on current techniques in functional analysis and can be divided into several stages.

(i) Consider again the function

$$(*) \qquad g_u(v) = [f(uv) - f(u)]/(v - 1)$$

that we used in proving Dirichlet's theorem. As we saw then, $g_u$ is regulated on $\mathbb{T}$ if $f$ has left and right derivatives at $u$, so, under the hypotheses of Theorem 9, for every $u \in J$. Then

$$f_N(u) - f(u) = \widehat{g_u}(-N - 1) - \widehat{g_u}(N)$$

and the theorem reduces to showing that, as $N \to +\infty$, *the functions*

$$u \mapsto \widehat{g_u}(N) = G_N(u)$$

*converge to 0 uniformly on every compact $K$ of $J$,* i.e. that for every $r > 0$ there exists an $N$ such that

(13.1)                    $(u \in K)$  &  $(|n| > N) \Longrightarrow |\widehat{g_u}(n)| < r.$

(ii) Consider the vector space[24] $L^1(\mathbb{T})$ of regulated functions on $\mathbb{T}$, endowed with the norm $\|f\|_1 = \int |f(v)| dm(v)$. Then $g_u \in L^1(\mathbb{T})$ for every $u \in J$, and the simplest estimate for the Fourier coefficients of an integrable function shows that

(13.2)    $|G_n(u') - G_n(u'')| = |\widehat{g_{u'}}(n) - \widehat{g_{u''}}(n)| \leq \|g_{u'} - g_{u''}\|_1$

for any $u'$ and $u'' \in J$.

Suppose we have shown that the map $u \mapsto g_u$ of $J$ in $L^1(\mathbb{T})$ is continuous, i.e. that for every $u \in J$ and every $r > 0$ there exists an $r' > 0$ such that

(13.3)        $(u' \in J)$  &  $(|u' - u| < r') \Longrightarrow \|g_{u'} - g_u\|_1 < r.$

The relation (2) then shows that

(13.4)  $(u' \in J)$  &  $(|u' - u| < r') \Longrightarrow |G_n(u') - G_n(u)| < r$   for every $n$.

This means precisely that the functions $G_n$ are *equicontinuous* on $J$ (Chap. III, n° 5). The fact that the $G_n(u)$ converge to 0 uniformly on every compact $K \subset J$ will then follow from the following general lemma:

**Lemma.** *If a sequence of functions $f_n$ defined and* equicontinuous *on a compact set $K$ converges simply on $K$, then it converges uniformly on $K$.*

Suppose that $f$ is the limit of the $f_n$ and let us choose an $r > 0$. For every $a \in K$ there exists an open ball $B(a)$ with centre $a$ in $K$ such that

$$x \in B(a) \Longrightarrow |f_n(x) - f_n(a)| \leq r \quad \text{for every } n;$$

this is the definition of equicontinuity. The inequality remains valid for $f$ by passage to the limit, which proves the continuity of $f$; since $|f_n(a) - f(a)| \leq r$ for $n$ large one deduces that, for $n$ large,

$$|f(x) - f_n(x)| \leq 3r$$

for every $x \in B(a)$. But, since $K$ is compact, one may (Borel-Lebesgue) cover it by a finite number of balls $B(a_i)$. The above inequality is then, for $n$ large, valid on all these balls, so on $K$, qed.

(iii) To prove the continuity of the map $u \mapsto g_u$ of $J$ into $L^1(\mathbb{T})$, let us first consider, in this part of the proof, the numerator $f(uv) - f(u)$ of $(*)$. This is the difference between, on the one hand, the function $f_u : v \mapsto f(uv)$ obtained by "translating" the function $f$, and on the other hand the constant

---

[24] The authentic $L^1$ space in Lebesgue theory contains many other functions, but, since it certainly contains the regulated functions, this is what we deal with here.

function $c_u : v \mapsto f(u)$. Since $\|c_{u'} - c_{u''}\|_1 = |f(u') - f(u'')|$, it is clear that $u \mapsto c_u$ is a continuous map of $J$ into $L^1(\mathbb{T})$.

As for $u \mapsto f_u$, this is a continuous map of $\mathbb{T}$ (and not only of $J$) into $L^1(\mathbb{T})$. This is obvious if $f$ is continuous on $\mathbb{T}$, for $f$ being now uniformly continuous on $\mathbb{T}$, we have $|f(u'v) - f(u''v)| \leq r$ for every $v \in \mathbb{T}$, so also $\|f_{u'} - f_{u''}\|_1 \leq r$ so long as $|u' - u''| < r'$. In the general case, we may choose, thanks to the lemma of n° 8, a function $\varphi \in C^0(\mathbb{T})$ such that $\int |f(v) - \varphi(v)| dm(v) = \|f - \varphi\|_1 < r$. Since we are integrating with respect to an *invariant* measure, we again have $\|f_u - \varphi_u\|_1 < r$ for every $u \in \mathbb{T}$. If we now choose functions $\varphi \in C^0(\mathbb{T})$ which converge to $f$ in $L^1(\mathbb{T})$, the corresponding maps $u \mapsto \varphi_u$ of $\mathbb{T}$ into $L^1(\mathbb{T})$ converge to $u \mapsto f_u$ *uniformly* on $\mathbb{T}$. A uniform limit of continuous functions with values in any metric space being again continuous, the required result follows.

So we see that the numerator of the formula $(*)$, considered as a function of $u \in J$ with values in $L^1(\mathbb{T})$, is continuous.

(iv) Next we have to take account of the denominator $v - 1$ and, to do this, use our hypotheses. We shall first give the proof in the case where $f = 0$ on $J$; and show later that the general case reduces to this.

Since the compact sets $K$ and $\mathbb{T} - J$ are disjoint their distance $d$ is $> 0$. Since $|uv - u| = |v - 1|$, we see that

$$(13.5) \qquad (u \in K) \quad \& \quad (|v - 1| < d) \quad \Longrightarrow \quad uv \in J$$
$$\Longrightarrow \quad f(uv) = f(u) = 0.$$

When we restrict to the $u \in K$, the functions of $v$ appearing in the numerator of the formula $(*)$ are thus all zero on the arc $|v - 1| < d$ of $\mathbb{T}$. Let us put

$$(13.6) \qquad h(v) = (v-1)^{-1} \text{ if } |v - 1| > d, \qquad h(v) = 0 \text{ if not.}$$

The formula that defines $g_u$ shows that, for $u \in K$,

$$(13.7) \qquad g_u(v) = h(v)\,[f_u(v) - c_u(v)] \quad \text{for every } v \in \mathbb{T}.$$

This is essentially the definition of $g_u$ on the arc $|v - 1| > d$ and, by (5), reduces to the identity $0 = 0$ on the arc $|v - 1| < d$.

Now we have $|h(v)| < 1/d$ for any $v \in \mathbb{T}$ by (6). The relation (7) showing that $g_u = h\,(f_u - c_u)$ for $u \in K$ (though not for every $u \in \mathbb{T}$) and the map $u \mapsto f_u - c_u$ of $\mathbb{T}$ into $L^1(\mathbb{T})$ being continuous, by point (iii), it remains to show that multiplication by the function $h$, which is *bounded* and independent of $u$, preserves continuity. This is no more difficult than in the framework of complex valued functions: it is enough to write that

$$\|hf' - hf''\|_1 = \int |h(v)|.|f'(v) - f''(v)| dm(v) \leq \|h\|.\|f' - f''\|_1$$

for any $f', f'' \in L^1(\mathbb{T})$, where $\|h\| = \sup |h(v)|$ as always. Since, for $u', u'' \in K$ sufficiently close, the distance from $f' = f_{u'} - c_{u'}$ to $f'' = f_{u''} - c_{u''}$ is arbitrarily small, so likewise is the distance from $g_{u'}$ to $g_{u''}$, which proves the theorem in the case where $f = 0$ in $J$.

(v) It remains to pass to the general case. The arc $K$ being compact, and the arc $J$ open, the distance $d$ of $K$ to the compact $\mathbb{T} - J$ is $> 0$, so that the open arc $J'$ of $\mathbb{T}$ defined by $d(u, K) < d/2$ satisfies $K \subset J' \subset J$. By modifying the graph of $f$ outside $J'$ one may construct a function $g$ which, on all of $\mathbb{T}$, is a primitive of a regulated function and which, on $J'$, coincides with $f$. Since $f - g$ vanishes on $J'$ its Fourier series converges uniformly to 0 on $K$ by section (iv) of the proof. Now the Fourier series of $g$ converges to $g$ uniformly in $\mathbb{T}$ (n° 9, Theorem 4) and so to $f$ uniformly on $K$. The relation $f = (f - g) + g$ then completes the proof.

# § 4. Analytic and holomorphic functions

In Chap. II, n° 19, which the reader is strongly urged to review, we said that a function $f$ defined on an open subset $U$ of $\mathbb{C}$ is *analytic* in $U$ if for every $a \in U$ there exists a power series in $z - a$ which, on a sufficiently small disc of centre $a$, converges to $f(z)$. In fact it represents $f(z)$ in the largest disc $D \subset U$ where it converges, for the sum of this power series is analytic in its disc of convergence (Chap. II, n° 19, Theorem 14) and since it is equal to $f$ on a neighbourhood of the centre of $D$, it is equal to $f$ everywhere in $D$ by virtue of the principle of analytic continuation (Chap. II, n° 20); the same argument shows that the one and only power series representing $f$ on a neighbourhood of $a$ is the Taylor series of $f$ at $a$. We know that it converges, but we still do not know up to where it converges ...

We have also shown that, if the function $f$ is analytic, it has a derivative

$$(*) \qquad f'(a) = \lim[f(a + h) - f(a)]/h$$

in the complex sense at each point $a \in U$; the latter can also be obtained by differentiating the power series representing $f$ term-by-term on a neighbourhood of $a$. The existence of the limit $(*)$ shows that as a function of the real variables $x = \mathrm{Re}(z)$ and $y = \mathrm{Im}(z)$ the function $f$ has partial derivatives satisfying the Cauchy formula

$$(**) \qquad D_2 f = iD_1 f \ (= if').$$

On the other hand we have shown (Chap. III, n° 20, corollary of Theorem 21) that, conversely, every *holomorphic* function, i.e. possessing continuous partial derivatives satisfying $(**)$ in an open set $U$, has a complex derivative $(*)$ and that its differential can be written in the form

$$(***) \qquad df = f'(z)dz = f'(z)(dx + idy).$$

In the following n° we shall show that a holomorphic function is necessarily analytic, by a method that exploits Fourier series, after which the terms "analytic" and "holomorphic" will become synonymous, as we have already announced several times in earlier chapters. Then we shall expound the simplest consequences of this result, without seeking to enter into the detail of a theory to which hundreds of mathematicians have, since Cauchy, added their contribution from their grain of sand to the Empire State Building; Remmert's two volumes, 650 very condensed pages, can scarcely cover the elliptic functions and not at all the modular and automorphic functions, Riemann surfaces, analytic differential equations, special functions, etc., not to speak of the generalisations to several variables.

## 14 – Analyticity of the holomorphic functions

Having recalled these preliminaries let us consider a function $f(z)$ defined and holomorphic on an open disc $D : |z| < R$. We would like to show that it is represented in all this disc by a power series

$$(14.1) \qquad f(z) = \sum a_n z^n.$$

As we have seen in n° 1 of this chapter, or in Chap. V, n° 5, this essentially reduces to showing that the function

$$(14.2) \qquad a_n(r) = \int f(ru)u^{-n} dm(u)$$

is, for every $n \in \mathbb{Z}$, proportional to $r^n$, using only the Cauchy condition or, equivalently, the existence and the continuity of $f'(z)$.

In this direction we write

$$(14.3) \qquad a_n(r) = \int_0^1 f[re(t)]e_{-n}(t)dt$$

and calculate the derivative of $a_n(r)$. We have to perform a differentiation under the $\int$ sign, an operation examined in Chap. V, n° 9, Theorem 9: this is permitted if the function of $r$ and $t$ that one is integrating has a partial derivative with respect to $r$ and if the latter is a continuous function of the pair $(r, t)$. The factor $e_{-n}(t)$ poses no problem. The factor $f[re(t)]$ neither: $f$ is $C^1$ and, for $t$ given, the map $r \mapsto re(t)$ is $C^\infty$. The general relation (21.2) of Chap. III, n° 21, namely that

$$(14.4) \qquad \frac{d}{dr} f[g(r)] = f'[g(r)]g'(r),$$

valid if $f$ is holomorphic and if $g$ is a $C^1$ function of the real variable $r$, then shows that in our case

$$(14.5) \qquad \frac{d}{dr} f[re(t)] = f'[re(t)]\frac{d}{dr}re(t) = f'[re(t)]e(t)$$

is a continuous function of the pair $(r, t)$. Thus

$$(14.6) \qquad \frac{d}{dr} a_n(r) = \int_0^1 f'[re(t)]e(t)e_{-n}(t)dt.$$

Since on the other hand, by the same argument,

$$(14.7) \qquad \frac{d}{dt} f[re(t)] = f'[re(t)]\frac{d}{dt}re(t) = 2\pi i r f'[re(t)]e(t),$$

(6) can again be written

$$2\pi i r \frac{d}{dr} a_n(r) = \int_0^1 \mathbf{e}_{-n}(t) \frac{d}{dt} f[r\mathbf{e}(t)].dt.$$

An integration by parts then gives

$$2\pi i r \frac{d}{dr} a_n(r) = \mathbf{e}_{-n}(t) f[r\mathbf{e}(t)] \Big|_0^1 + 2\pi i n \int_0^1 \mathbf{e}_{-n}(t) f[r\mathbf{e}(t)] dt$$

since $-2\pi i n \mathbf{e}_{-n}(t) = \mathbf{e}'_{-n}(t)$. In the preceding relation the integrated part is zero by periodicity and the integral on the right hand side is just $a_n(r)$. Whence the relation

(14.8) $$r a'_n(r) = n a_n(r)$$

valid for $0 \le r < R$.

Here we have a particularly banal differential equation. Putting $b_n(r) = a_n(r) r^{-n}$ for $r > 0$ and applying the chain rule, one finds that $b'_n(r) = 0$; the function $b_n(r)$ is therefore constant, whence

(14.9) $$a_n(r) = a_n r^n$$

with a coefficient $a_n$ independent of $r$.

For $r \le \rho < R$, one has, by (2),

(14.10) $$|a_n r^n| \le \sup_{|z| \le \rho} |f(z)| = M_f(\rho) < +\infty.$$

For $n < 0$, $r^n$ increases indefinitely when $r$ tends to 0; (10) then shows that

(14.11) $$a_n = 0 \qquad \text{for } n < 0,$$

so that the Fourier series $\sum a_n(r) u^n$ of $f(ru)$ reduces to the *power* series $\sum a_n z^n$ for $z = ru$. Since, on the other hand, the function $u \mapsto f(ru)$ is of class $C^1$ on $\mathbb{T}$, its Fourier series converges absolutely and represents the function in question everywhere.

In particular, the power series $\sum a_n z^n$ converges for $|z| < R$. One may furthermore see this without invoking Theorem 8: choose a $\rho$ such that $|z| < \rho < R$, put $|z| = q\rho$ with $q < 1$, and write

(14.12) $$|a_n z^n| = |a_n \rho^n| q^n \le M_f(\rho) q^n.$$

In conclusion:

**Theorem 10 (Cauchy, 1831).** *Let $f$ be a holomorphic function in an open set $U$ in $\mathbb{C}$. Then $f$ is analytic in $U$ and, for every $a \in U$, the Taylor series of $f$ at $a$ converges and represents $f$ in the largest open disc with centre $a$ contained in $U$.*

It is enough, in the preceding arguments, to replace the disc $|z| < R$ by the largest disc $|z - a| < R$ in question or, if one prefers, to consider the function $f(a + z)$. Now the only power series that can possibly represent $f$ on a neighbourhood of $a$ is the Taylor series of $f$ at $a$ as we know (Chap. II, n° 20). Whence the theorem.

If you believe that Cauchy understood everything immediately, you are in error. He perfectly understood Fourier series and integrals from 1815, and in 1822 had obtained the integral formula for a circle for the holomorphic functions (i.e. satisfying his PDE) by quite another method. Now one needs only a few lines of simple calculations to pass from there to Theorem 10 (see n° 21). Freudenthal, an excellent Dutch mathematician who has seriously examined Cauchy's works, voices the hypothesis, in his notice in the DSB, that he had forgotten his own results. His political, religious and social activities probably occupied too great a place in his life[25] ...

## 15 – The maximum principle

Let $f$ be a holomorphic function in an open $U \subset \mathbb{C}$ and again consider the Cauchy formula (14.2), which, for $n = 0$, can be written as

$$(15.1) \qquad f(a) = \int f(a + ru) dm(u)$$

for every $a \in U$, where one integrates with respect to the invariant measure of $\mathbb{T}$ and where $r$ is small enough for $U$ to contain the closed disc $|z - a| \leq r$. This implies

$$(15.2) \qquad |f(a)| \leq \sup |f(a + ru)|.$$

Assume now that $f$ has a *local maximum* at $a$, i.e. that there exists an $r > 0$ such that

$$(15.3) \qquad |f(z)| \leq |f(a)| \quad \text{for every } z \text{ such that } |z - a| \leq r.$$

---

[25] On Cauchy, see also Bruno Belhoste, *Cauchy, un mathématicien légitimiste au XIX$^{e}$ siècle* (Paris, Belin, 1985) and *Augustin-Louis Cauchy. A Biography* (Springer, 1991), the mathematical information in which does not replace Freudenthal's notice. The book by C. A. Valson, *La vie et les oeuvres du Baron Cauchy* (1868) deserves to be read as a particularly comic example of would-be edifying hagiography, but is difficult to find; it was demolished immediately by Joseph Bertrand (Bull. de la Soc. Math. de France, 1, 1870) who, while insisting on the importance of Cauchy's discoveries, recalled his irresistible need to publish (more than 750 articles), frequently several times, incorrect, incomplete results, such as he had found the same day before breakfast, as we say nowadays. The Cours d'analyse of 1821 has recently been republished in facsimile by Ellipses; reading it could be a very useful exercise (to detect the errors in the argument). On teaching at the Polytechnique, see Bruno Belhoste, Amy Dahan Dalmedico and Antoine Picon, *La formation polytechnicienne 1794–1994* (Dunod, 1994), a collection of articles by twenty or so historians and in the main very interesting.

Applying Parseval-Bessel to the Fourier series

$$f(a + ru) = \sum c_n r^n u^n,$$

one obtains by (3)

$$\sum |c_n|^2 r^{2n} = \int |f(a + ru)|^2 \, dm(u) \leq |f(a)|^2 = |c_0|^2,$$

whence $c_n = 0$ for every $n \geq 1$. The power series for $f$ at the point $a$ then reduces to its constant term, so that there is a disc of centre $a$ on which $f$ is constant.

Now, in Chap. II, n° 20, we proved a *principle of analytic continuation* stating that if, in a *connected* open set $U$, two analytic functions coincide on a neighbourhood of a particular point of $U$, then they coincide in all of $U$. If in particular a holomorphic function in $U$ is constant on a neighbourhood of a particular point of $U$, it is constant in $U$. Conclusion:

**Theorem 11.** *Let $f$ be a holomorphic function in a* connected *open set $U$. Then $f$ is constant if at a point of $U$ it has either a local maximum or a non zero local minimum.*

The case of a local minimum reduces to the preceding case on considering the function $1/f$: this is defined and holomorphic on a neighbourhood of a local minimum of $f$ and has a local maximum there; $1/f$ (and so $f$) is thus constant on a disc, so $f$ is constant on $U$.

The connectedness hypothesis is essential: if $U$ is, for example, the union of two disjoint open discs $D'$ and $D''$, then the behaviour of $f$ on $D''$ has no bearing on its behaviour on $D'$; $f$ might be equal to 1 in $D'$ and to $e^z$ in $D''$.

An open connected set is generally called a *domain*; one most often uses the letter $G$ (in German, domain = Gebiet) to denote connected open sets.

**Corollary 1.** *Let $G$ be a* bounded *domain in $\mathbb{C}$, $K$ its closure, $F = K - G$ its frontier and $f$ a function defined and continuous in $K$ and holomorphic in $G$. Then*

(15.4) $$\|f\|_G = \|f\|_K = \|f\|_F.$$

Since $G$ is bounded, $K$ is bounded and closed, hence compact. The continuous function $|f(z)|$ therefore attains its maximum at a point $a \in K$. If $a \in G$, Theorem 5 shows that $f$ is constant in $G$, hence in $K$, and the corollary is obvious. If $f$ is not constant, the maximum of $|f(z)|$ is thus attained on $F$, whence $\|f\|_K = \|f\|_F$. But since $f$ is continuous in $K$, its value at a point of $F$ is the limit of values taken at points of $G$, whence $\|f\|_F \leq \|f\|_G \leq \|f\|_K$ since $G \subset K$, qed.

**Corollary 2.** *Let $G$ be a bounded domain and $(f_n)$ a sequence of functions defined and continuous on the closure $K$ of $G$ and holomorphic in $G$. Assume that the $f_n$ converge uniformly on the* boundary $F$ *of $G$ to a limit function. Then the $f_n$ converge uniformly on $K$ and the limit function is holomorphic in $G$.*

Consider the functions $f_{pq} = f_p - f_q$. Cauchy's criterion for uniform convergence shows that, for every $r > 0$, one has $\|f_{pq}\|_F \leq r$ for $p$ and $q$ large, and thus (Corollary 1) $\|f_{pq}\|_K \leq r$. The $f_p$ therefore converge uniformly in $K$, so in $G$, and it remains to apply Theorem 17, to be found below (n° 19).

**Corollary 3 ((H. A.) Schwarz' lemma).** *Let $f$ be a function holomorphic and bounded on a disc $|z| < R$ and having a zero of order $p$ at the origin. Then*

$$|f(z)| \leq M |z/R|^p \quad \text{where } M = \sup |f(z)|.$$

The assumption about $f$ implies that $f(z) = z^p g(z)$ where $g$ is, like $f$, the sum of a power series in $|z| < R$. The relation $|z^p g(z)| \leq M$ shows that $|g(z)| \leq M/r^p$ for $|z| = r < R$, so also, by the maximum principle, for $|z| < r$. On letting $r$ tend to $R$, one deduces that

$$|g(z)| \leq M/R^p, \qquad \text{whence } |f(z)| \leq M|z|^p/R^p$$

for every $z$, qed.

Theorem 11 can be extended in part to unbounded domains, but this is more difficult and rather constitutes an exercise:

**Theorem 12.** *Let $G$ be a domain in $\mathbb{C}$ and $f$ a function defined, continuous and* bounded *on the closure of $G$ and holomorphic in $G$. Then*

(15.5)                     $$\|f\|_G = \|f\|_F$$

*where $F = \bar{G} - G$ is the boundary of $G$.*

The case where $G$ is bounded having been treated already, let us assume $G$ unbounded. First consider the simplest case, where $f$ tends to 0 at infinity, i.e. where, for every $\varepsilon > 0$, one has $|f(z)| < \varepsilon$ for every $z \in \bar{G}$ of large enough modulus. Since $f$ is continuous in $\bar{G}$, the inequality $|f(z)| \geq \varepsilon$ defines a closed subset $K$ of $\bar{G}$; since $|f(z)| < \varepsilon$ for $|z|$ large, $K$ is bounded, so compact. There is therefore an $a \in K$ where the function $|f(z)|$ attains its maximum relative to $K$. For every $z \in \bar{G}$ one then has $|f(z)| \leq |f(a)|$, either trivially if $z \in K$, or because $|f(z)| < \varepsilon$ if $z \notin K$. Theorem 11 then shows that $a \in \bar{G} - G$ (qed), unless $f$ is constant, in which case there is nothing to prove.

Now let us pass on to the general case of a function bounded in $G$ but not necessarily tending to 0 at infinity and assume for example that $|f(z)| \leq 1$ on the boundary $F$ of $G$; we are then to show that $|f(z)| \leq 1$ in all $G$ too.

Assume $|f(a)| > 1$ for some $a \in G$ and consider a closed disc $D : |z - a| \leq r$ contained in $G$. Theorem 11 shows that the maximum $M$ of $|f|$ on the boundary of $D$ is $> 1$. Consider now in the domain $H = G - D$ the functions

$$(15.6) \qquad f_n(z) = rf(z)^n/M^n(z - a).$$

Since $f$ is bounded in $G$, the introduction of a denominator $z - a$ shows that the $f_n$ tend to 0 at infinity in $H$. One is therefore in the particular case examined first. Now the boundary of $H$ is clearly the union of the boundary $F$ of $G$ and that of the disc $D$. On $F$, by hypothesis $|f(z)| \leq 1$, and since $M$ and $|(z - a)/r|$ are $>1$, one has $|f_n(z)| \leq 1$ on $F$. The same result holds on the boundary of $D$ since there $|f(z)| \leq M$ and $|z - a| = r$. Thus we see that the function $|f_n(z)|$ is $\leq 1$ on the boundary of $H$, and since it tends to 0 at infinity we conclude that $|f_n(z)| \leq 1$ in $H$. The exponent $n$ in (6) being arbitrary, this forces $|f(z)/M| \leq 1$ in $H$. This relation also being satisfied in $D$, it holds everywhere in $G$, qed.

The hypothesis that the function $f$ is bounded in $G$ and not only on its boundary is essential in the above. All this has been prodigiously refined.

**16 – Functions analytic in an annulus. Singular points. Meromorphic functions**

The arguments of n° 14 in fact apply to a function defined in an annulus $C : R_1 < |z| < R_2$ and in particular on a disc with its centre deleted if $R_1 = 0$. For every circle $|z| = r$ contained in $C$ the Fourier coefficients of the function $f(ru)$ are again of the form $a_n r^n$, but the argument showing that $a_n = 0$ for $n < 0$ no longer applies since, even in the case where $R_1 = 0$, the function, for example $1/z$, has no reason to be bounded on a neighbourhood of 0. What survives is the Fourier series of $f(ru)$, namely

$$(16.1) \qquad \sum a_n r^n u^n = \sum a_n z^n \quad \text{with} \quad a_n r^n = \int f(ru)u^{-n}dm(u),$$

where this time the sum is extended over $\mathbb{Z}$, converges absolutely and represents $f$ in $C$ by Theorem 4 of n° 9. Here again, one may see the convergence directly. Let us choose numbers $r_1$ and $r_2$ such that $R_1 < r_1 < |z| < r_2 < R_2$ (strict inequalities) and let $M$ be the uniform norm of $f$ on the compact annulus $C'$ delimited by the circles of radii $r_1$ and $r_2$. Now $|a_n z^n| \leq M$ by (1), since $|f(ru)| \leq M$, whence

$$(16.2) \qquad |a_n z^n| = \begin{cases} |a_n r_2^n| \cdot |z/r_2|^n < M|z/r_2|^n & \text{for} \quad n \geq 0, \\ |a_n r_1^n| \cdot |z/r_1|^n < M|z/r_1|^n & \text{for} \quad n \leq 0; \end{cases}$$

since $|z/r_2| < 1$, the first inequality proves the absolute convergence of the "positive" part of the series (1), and since $|z/r_1| > 1$, the second proves that of its "negative" part.

The series (1) even converges normally in $C'$ and so on every compact[26] $K \subset C$. Thus, in $C'$,

$$|a_n z^n| \leq \begin{cases} |a_n r_1^n| & \text{if} \quad n \geq 0 \\ |a_n r_2^n| & \text{if} \quad n < 0; \end{cases}$$

now we know that the series $\sum a_n z^n$ converges absolutely in $C$, thus for $z = r_1$ or $r_2$; whence, in $C'$, a bound by a *convergent* series independent of $z$. In conclusion:

**Theorem 13 (Laurent).** *Let $f$ be a holomorphic function in an annulus $C : R_1 < |z| < R_2$. Then we have a series expansion*

$$(16.3) \qquad f(z) = \sum_{n \in \mathbb{Z}} a_n z^n \quad \text{with} \quad a_n = r^{-n} \int_{\mathbb{T}} f(ru) u^{-n} dm(u),$$

*the series converging normally on every compact $K \subset C$.*

An expansion of this type is called a *Laurent series*; it is the sum of a power series in $z$ and of a power series in $1/z$. The first converges at least for $|z| < R_2$ and the second for $|z| > R_1$ since a power series necessarily converges on a disc. This allows us to write that, in $C$, we have a decomposition $f(z) = g(z) + h(z)$ of $f$ into a function $g$ holomorphic for $|z| < R_2$ and a function $h$ holomorphic for $|z| > R_1$.

We may write (16.3) *à la* Leibniz like the Cauchy formula of n° 1. Putting $\zeta = ru$ we have $a_n = \int f(\zeta) \zeta^{-n} dm(u)$; but for $u = \mathbf{e}(t)$ we have $d\zeta = 2\pi i r \mathbf{e}(t) dt = 2\pi i \zeta dm(u)$. Whence

$$2\pi i a_n = \int f(\zeta) \zeta^{-n-1} d\zeta,$$

the "line" integral being taken along any circle $t \mapsto r\mathbf{e}(t)$ contained in $C$. Cauchy's theory will illuminate this point and, in particular, will explain why the result is independent of the circle $|\zeta| = r$ chosen.

Theorem 13 serves mainly to study the behaviour of a holomorphic function on a neighbourhood of an *isolated singular point* $a$, i.e. of a function defined and holomorphic on a neighbourhood of $a$, except at the point $a$ itself. There one has a series expansion $f(z) = \sum c_n (z-a)^n$, whence the distinction between the *poles*, where the series includes only finite number of nonzero terms of degree $< 0$ – the minimal degree, its sign changed, is called the *order* of the pole in question –, and the *essential singular points* where it includes

---

[26] Such a compact subset is contained in $C'$ if one chooses the radii $r_1$ and $r_2$ suitably: the continuous real function $z \mapsto |z|$ attains its minimum and its maximum on $K$, which are *strictly* contained between the radii of $C$ since the limiting circles of $C$, which do not lie in $C$, do not meet $K$.

infinitely many, the case for example of the function $\exp(1/z) = \sum z^{-n}/n!$ at $z = 0$. It is also useful to define the *order* of a *zero* $a$ of $f$, i.e. of a point where $f(a) = 0$: this is the degree of the first nonzero term of the power series of $f$ at $a$.

This leads to the fundamental concept of a function *meromorphic* in an open set $U$: this is a function $f$ defined and analytic in $U - D$, where $D$ is a discrete subset of $U$ (i.e. such that, for every $a \in U$, there exists a disc of centre $a$ containing a *finite* number of points of $D$), and having only polar singularities at the points of $D$. This, for example, is the case of the elliptic functions of Chap. II, n° 23 and, in fact, the "right" definition of the elliptic functions, for there are many, other than the Eisenstein series, is to impose on them just that they should be simultaneously *doubly periodic and meromorphic in* $\mathbb{C}$, as Liouville discovered (n° 18).

We note that, if $D'$ is the set of zeros of a meromorphic function $f$ in $U$, then the union $D \cup D'$ is again a discrete subset of $U$. This results from the fact that, if $a$ is any point in $U$, then $f(z) = (z - a)^p g(z)$, where $g$ is a power series whose constant term is not zero, so such that $g(a) \neq 0$; on a small enough disc of centre $a$ we again have $g(z) \neq 0$ since $g$ is continuous, so that on a neighbourhood of $a$ the function $f$ can have no other zero or pole than the point $a$ itself. On the other hand, the zeros or poles of a function having an essential singular point at $a$ may accumulate at $a$: the zeros of $\sin(1/z)$ converge to 0.

One may perform the usual algebraic operations on the functions meromorphic in a given open set $U$: sum, product, quotient; as we shall see, one again obtains meromorphic functions in $U$.

The case of a sum $f + g$ is obvious: if $f$ and $g$ have poles at the points of two discrete subsets $D$ and $D'$ of $U$, the function $f + g$ is holomorphic outside $D \cup D'$ and it is clear that at a point of $D \cup D'$ it has at most a pole; "at most" because the polar parts of the Laurent series of $f$ and $g$ at a common pole may cancel each other. For $fg$, holomorphic outside $D \cup D'$, one observes that at a point $a \in D \cup D'$ one has the relations

$$f(z) = f_1(z)/(z - a)^p, \qquad g(z) = g_1(z)/(z - a)^q$$

where $f_1$ and $g_1$ are holomorphic on a neighbourhood of $a$ and nonzero at $a$. It is then clear that $fg$ has a pole of order $p + q$ at $a$. Of course it can happen that a pole of $f$ is neutralised by a zero of $g$.

The case of the quotient $f/g$ reduces, as always, to that of the reciprocal $1/g(z)$ of a meromorphic function. On a neighbourhood of a pole $a$ of $g$ one has $g(z) = g_1(z)/(z - a)^q$ where $g_1(z)$ is a power series such that $g_1(a) \neq 0$. The function $g_1$ has a reciprocal $1/g_1(z)$ holomorphic on a neighbourhood of $a$; the formula $1/g(z) = (z - a)^q/g_1(z)$ then shows that the pole $a$ of order $q$ is transformed into a zero of order $q$ of $1/g(z)$. At a point $a$ where $g$ is holomorphic one has $g(z) = (z - a)^q g_1(z)$ where $g_1$ is a power series not vanishing at $a$, whose reciprocal is thus holomorphic on a neighbourhood of $a$, whence clearly a pole of order $q$ for $1/g(z)$ if $g$ has a zero of order $q$ at $a$. Finally we see that $1/g(z)$ is holomorphic outside the zeros of $g$, which are the poles of $1/g$, the poles of $g$ contrariwise providing the zeros of $1/g$. Since the zeros of $g$ form a discrete set in $U$, the function $1/g$ is therefore meromorphic in $U$.

Laurent series can be manipulated like power series. The domain of convergence of a series such as $f(z) = \sum a_n z^n$ is necessarily an annulus $C$ since it is the sum of a power series in $z$ and of a power series in $1/z$ which converge for $|z| < R_2$ and $|1/z| < R_1$ respectively. The multiplication formula

$$\sum a_n z^n \sum b_n z^n = \sum c_n z^n \quad \text{with } c_n = \sum a_p b_{n-p}$$

valid for power series still applies, restricted to an annulus where the two series converge: they then converge absolutely, therefore unconditionally, so on multiplying term-by-term one obtains a double series $\sum a_p b_q z^{p+q}$ which converges unconditionally (Chap. II, n° 22) and in which one may reorder the terms arbitrarily (Chap. II, n° 18, Theorem 13: associativity), for example as a function of the value of $p + q$.

One may also differentiate a Laurent series term-by-term; the simplest way to see this is to write $f(z) = g(z) + h(1/z)$ where $g$ and $h$ are power series, whence

$$f'(z) = g'(z) - h'(1/z)/z^2$$

by the chain rule for analytic functions (Chap. II, n° 22, Theorem 17); as we know, $g'(z)$ is obtained by differentiating the "positive" part of the series $\sum a_n z^n$ term-by-term; since $h(z) = \sum_{n \geq 0} a_{-n} z^n$ and so $h'(z) = \sum_{n \geq 0} n a_{-n} z^{n-1}$, it follows that

$$h'(1/z)/z^2 = \sum_{n \geq 0} n a_{-n} z^{-n+1-2} = \sum_{n \geq 0} n a_{-n} z^{-n-1} = -\sum_{n \leq 0} n a_n z^{n-1}$$

and finally

$$f'(z) = \sum_{n \geq 0} n a_n z^{n-1} + \sum_{n \leq 0} n a_n z^{n-1} = \sum n a_n z^{n-1}$$

as one had hoped, the Laurent series of $f'$ converging in the same annulus $C$ as $f$. Note that there is no term in $1/z$ in the result.

An essential difference from power series will appear when one looks for a *primitive* of $f$, i.e. a holomorphic function $F$ such that $F'(z) = f(z)$ in $C$. The function $F$ is, like $f$, represented by a Laurent series and as we have just seen the derived series contains no term in $1/z$. The problem can therefore have no solution if the series $f(z)$ contains one. The analogous problem for a real variable had, in the XVII[th] century, defied the efforts of several mathematicians before Newton and Mercator, and Newton himself, with his systematic use of Laurent series having a finite number of terms of negative degree, is very reticent on this subject. This "detail", on which the whole of Cauchy's "residue calculus" is founded, apart, the existence of the primitive when $a_{-1} = 0$ is as obvious as for the case of power series: one puts $F(z) = \sum a_n z^{n+1}/(n+1)$. The arguments employed for power series in Chap. II, n° 19 or, equivalently, the fact that in the domain $C$ of convergence of the series, the $a_n$ of positive index and those of negative index are bounded by geometric progressions, show that these operations – differentiation or "integration" term-by-term – lead to series converging in the same annulus as the initial series.

In the case where $f$ contains a term in $1/z$, one may again consider the function $F(z) = \sum a_n z^{n+1}/(n+1)$, where one forgets the term of index $n = -1$; instead of $f = F'$ one obtains the relation

(16.4) $$f(z) = F'(z) + a_{-1}/z.$$

The coefficient $a_{-1}$, the radical obstruction to the existence of a primitive of $f$ in the annulus $C$, is called the residue of $f$; more generally, if one considers a function $f$ holomorphic on a neighbourhood of a point $c$ of $\mathbb{C}$ except at the point itself, which is then an isolated singular point of $f$, the *residue of $f$ at $c$* is by definition the coefficient $a_{-1}$ of $1/(z - c)$ in the expansion of $f$ as a Laurent series $\sum a_n(z - c)^n$ about the point $c$; one writes $\mathrm{Res}_c(f)$ or $\mathrm{Res}(f, c)$. The formula (3) applied for $n = -1$ shows that

(16.5) $$\mathrm{Res}_c(f) = \int f(c + ru) r u \, dm(u) = \int_0^1 f\left(c + r e^{2\pi i t}\right) r e^{2\pi i t} \, dt$$

or, *à la* Leibniz,

(16.6) $$\mathrm{Res}_c(f) = \frac{1}{2\pi i} \int_{|z-c|=r} f(z) dz$$

with an integral taken around the circle $|z - c| = r$ as above; naturally one has to choose $r$ small enough that, except at the point $c$, the function $f$ will be holomorphic in an open set containing the closed disc $|z - c| \le r$.

The preceding arguments show more generally that if $f$ is a meromorphic function in an open set $U$, the existence of a primitive $F$ of $f$ in $U$ presupposes that $\mathrm{Res}_a(f) = 0$ at every pole $a$ of $f$. This necessary condition *is not sufficient, even if $f$ is holomorphic*, except in very particular open sets ("simply connected", i.e. homeomorphic[27] to a disc). To study this question requires the complete Cauchy theory, i.e. the use of the line integrals which we shall develop later in this treatise (Chap. VIII).

To return to the function $1/z$, one might claim that the function

---

[27] Two metric or topological spaces $X$ and $Y$ are said to be *homeomorphic* if there is a continuous bijection $X \longrightarrow Y$ with continuous inverse.

(16.7)     $\mathcal{L}og \, z = \log|z| + i \arg z = \dfrac{1}{2} \log\left(x^2 + y^2\right) + i \arctan y/x$

of Chap. IV, n° 14 and 21 is its primitive; now, in $\mathbb{C}^*$, the latter is all but a function in the strict sense of the term, as we know because of the same problem for $\arg z$. There are open sets $U$ in which the function $1/z$ has a primitive, namely those in which the pseudo-function $\mathcal{L}og \, z$ decomposes into uniform branches: for we know [Chap. IV, §4, section (v)] that, if $L$ is such a branch, one has

$$L(z) = L(a) + \sum_1^\infty (1 - z/a)^n/n$$

on a neighbourhood of every point $a \in U$, so that $L$ is analytic and that $L'(z) = 1/z$ in $U$; the function $L$ is then a primitive of $1/z$ in $U$. If, conversely, $1/z$ has a primitive $f(z)$ in a connected open $U \subset \mathbb{C}^*$, then $\left(e^f\right)' = f'e^f$, so that the function $g = e^f$ satisfies $zg' - g = 0$ or $(g/z)' = 0$; since $U$ is connected one has $g(z) = cz$ for a constant $c$ that we may assume equal to 1 by adding a suitable constant to $f$. Consequently, $f$ is a uniform branch of $\mathcal{L}og$ in $U$. To find a primitive of $1/z$ in $U$ thus reduces exactly to constructing a uniform branch of $\mathcal{L}og$ in $U$. A "punctured" disc (i.e. with its centre removed) of centre 0 is the very type of open set for which the problem has no solution.

An analogous problem arises when one wants to define the non integer powers of a complex number, i.e. the "function" $z \mapsto z^s$ where $s$ is an arbitrary complex number. In view of the formula $a^s = e^{s.\log a}$ of Chap. IV, valid for $a$ real $> 0$, it would seem natural to define

(16.8)                    $z^s = e^{s.\mathcal{L}og \, z}$,

but the ambiguity of $\mathcal{L}og$ then transfers to the left hand side. If, however, one restricts to an open set $U$ on which the multiform correspondence $\mathcal{L}og$ decomposes into uniform branches, the choice of such a branch $L$ yields a holomorphic function $e^{s.L(z)}$ which, in its turn, is a "uniform branch of the multiform function $z^s$"; the latter is unique up to a constant factor of the form $e^{2k\pi is}$. If for example $U = \mathbb{C} - \mathbb{R}_-$, in which case one may choose

(16.9)          $L(z) = \log|z| + i.\arg z$     with $|\arg z| < \pi$

as we have seen in Chap. IV, §4, one finds

(16.10)                    $z^s = |z|^s e^{is.\arg(z)}$

where $|z|^s = \exp(s.\log|z|)$ is the expression defined unambiguously in Chap. IV, n° 14 and where the argument is chosen as above. If for example $s = \frac{1}{2}$, one thus obtains two uniform branches, opposites, of $z^{1/2}$. This type of problem arises frequently in the residue calculus *à la* Cauchy.

## 17 – Periodic holomorphic functions

The method used in n° 14 to obtain the expansion of a holomorphic function in power series applies equally to the Fourier series of periodic holomorphic functions.

A function $f$ defined and holomorphic in an open set $U$ has a period $a \neq 0$ in $U$ if $f(z+a) = f(z)$ for any $z \in U$. This clearly assumes that $z \in U$ implies $z + a \in U$. By considering the function $f(az)$ one reduces to the case where $a = 1$. Then $f(x + 1, y) = f(x, y)$ for any $x + iy = z \in U$, which suggests expanding in a Fourier series with respect to $x$. The only reasonable situation is that where $U$ is a horizontal strip

$$(17.1) \qquad a < \mathrm{Im}(z) < b$$

of finite or infinite height, so that, for every $y \in ]a, b[$, the function $x \mapsto f(x, y) = f(x + iy)$ is defined on all $\mathbb{R}$, is periodic, and $C^{\infty}$ since $f$ is analytic. We then have a much more than absolutely convergent expansion

$$(17.2) \qquad f(x + iy) = \sum a_n(y) e_n(x)$$

with

$$(17.3) \qquad a_n(y) = \oint f(x + iy) e^{-2\pi i n x} dx,$$

the mean value over a period. Whence

$$(17.4) \qquad a_n(y) e^{2\pi n y} = \oint f(z) e^{-2\pi i n z} dx.$$

We shall see that this integral is independent of $y$.

Since the function

$$g(z) = f(z) e^{-2\pi i n z}$$

is as holomorphic and periodic as $f$ is, it is enough to give the proof for $n = 0$, i.e. to show that $a_0(y) = \oint f(x + iy) dx$ is a constant.

The theorem on differentiation under the $\int$ sign (Chap. V, n° 9, Theorem 9) clearly applies to the function $f(x, y)$. Thus, using the Cauchy differential equation,

$$a_0'(y) = \oint D_2 f(x, y) dx = i \oint D_1 f(x, y) dx;$$

by the FT, this integral is the variation of the function $x \mapsto f(x, y)$ over a period interval. Consequently, $a_0'(y) = 0$, qed[28].

---

[28] Most authors employ the Cauchy integral around a rectangular contour to obtain this quasi trivial result; the method adopted here extends to the periodic solutions of many other partial differential equations than $D_1 f = i D_2 f$, and in

Now we have $a_n(y) = a_n e^{-2\pi ny}$ with a constant $a_n$, which puts (3) in the much more pleasant form

$$(17.5) \quad f(z) = \sum a_n e^{2\pi inz} \quad \text{with} \quad a_n = \oint f(x+iy)e^{-2\pi inz}dx.$$

We may differentiate the series term-by-term since to differentiate with respect to $z$ amounts to differentiating with respect to $x$, which is allowed by the theory of Fourier series for $C^\infty$ functions (n° 9, Theorem 5).

The series converges normally on every *closed* strip $c \leq \operatorname{Im}(z) \leq d$ contained in $U$, i.e. such that $a < c \leq d < b$. In such a strip, we have $\left|e^{2\pi inz}\right| = e^{-2\pi ny} \leq e^{-2\pi nc} + e^{-2\pi nd}$ since the monotone function $e^{-2\pi ny}$ lies between its values at $c$ and $d$ on $[c,d]$. The general term of (5) is thus bounded by $|a_n|\, e^{-2\pi nc} + |a_n|\, e^{-2\pi nd}$ on the closed strip in question, but since (5) converges absolutely for $a < \operatorname{Im}(z) < b$ and so for $\operatorname{Im}(z) = c$ or $d$, the two series $\sum |a_n|\, e^{-2\pi nc}$ and $\sum |a_n|\, e^{-2\pi nd}$ converge, so their sum does too. We thus obtain a series independent of $z$ which dominates the series (5) in the closed strip $c \leq \operatorname{Im}(z) \leq d$: whence normal convergence. In conclusion:

**Theorem 14 (Liouville).** *Let $f$ be a holomorphic function of period 1 in an open strip $B : a < \operatorname{Im}(z) < b$. We then have a series expansion*

$$f(z) = \sum a_n e^{2\pi inz}$$

*which converges normally on every closed strip $B' \subset B$. The coefficients $a_n$ are given by the relation*

$$a_n = \oint f(z)e^{-2\pi inz}dx = \int_c^{c+1} f(x+iy)e^{-2\pi in(x+iy)}dx$$

*for arbitrary $y \in \,]a,b[$ and $c \in \mathbb{R}$. We may differentiate the Fourier series of $f$ term-by-term any number of times.*

Conversely, if a *complex Fourier series*, i.e. of the form (5), converges absolutely in a strip $a < \operatorname{Im}(z) < b$, the preceding argument shows that the series converges normally on every closed strip (and so on every compact set) contained in the given open strip. Theorem 17 below will show that the sum of the series is analytic.

## 18 – The theorems of Liouville and of d'Alembert-Gauss

We can now establish the theorem of Liouville to which we alluded *à propos* the differential equation for the function $\wp$ of Weierstrass (Chap. II, end

---

fact Fourier himself, Poisson, Liouville, etc. applied it to the PDEs of physics known at their time – propagation of heat, the wave equation, etc. – whose solutions are not holomorphic functions of $(x,y)$. *Exercise*: find the general form of the periodic solutions in $t$ of the equation $f''_{tt} - f''_{xx} = cf$, where $c$ is a constant.

of n° 23). In its enunciation an *entire function* is, by definition, a holomorphic function on all $\mathbb{C}$: a polynomial, an exponential function, $\exp(\exp(\exp(\sin z)))$, etc.

**Theorem 15 (Liouville).** *Let $f$ be an entire function such that*

$$(18.1) \qquad f(z) = O(z^p) \qquad \text{when} \quad |z| \longrightarrow +\infty,$$

*where $p$ is an integer $\geq 0$. Then $f$ is a polynomial of degree $\leq p$. In particular, a bounded entire function is constant.*

By Theorem 10 one has an expansion $f(z) = \sum a_n z^n$ valid for any $z$. The relation (14.10) then shows that, for every $n$,

$$(18.2) \qquad |a_n| r^n \leq M_f(r)$$

where $M_f(r)$ is the upper bound of $|f(z)|$ on the circumference $|z| = r$, or, equivalently, by the maximum principle, on the disc $|z| \leq r$. So assume that $|f(z)| \leq M|z|^p$ for every $z$ large enough. It follows that $M_f(r) \leq Mr^p$, hence

$$|a_n| \leq Mr^{p-n} \qquad \text{for } r \text{ large},$$

whence $a_n = 0$ for every $n > p$ since then the right hand side tends to 0 at infinity, qed.

One of the most famous and simplest applications of Theorem 13 is a proof (Gauss found four) by the same Liouville of the miraculous[29]

**Theorem 16 (d'Alembert-Gauss).** *Every algebraic equation of degree $\geq 1$ has at least one complex root.*

To see this, consider the function $f(z) = 1/p(z)$ where $p$ is a polynomial not vanishing anywhere. Since $p(z)$ is analytic in $\mathbb{C}$, so also is $f$ (Chap. II, n° 22, Theorem 17 – one may also invoke holomorphy, easier to prove for $1/p$). Now, at infinity, $p(z)$ is equivalent to its term of highest degree, say $az^r$, so that

$$f(z) \sim 1/az^r = O(z^{-r}).$$

Since $r > 0$, $f$ is bounded at infinity, so is constant by Liouville, impossible if $p$ is of degree $> 0$.

Liouville's theorem allow us to complete the proof (Chap. III, n° 23) of the differential equation

---

[29] The complex numbers were invented to calculate the roots of equations of the third degree with the help of formulae involving square roots of negative numbers. The "miraculous" nature of the d'Alembert-Gauss theorem is that it allows us to ascribe roots to equations of any degree and even though, for $n > 4$, no one has ever discovered, or ever will discover, algebraic "formulae", simple or complicated, to calculate the roots of a general equation of degree $n$.

$$\wp'(z)^2 = 4\wp(z)^3 - 20a_2\wp(z) - 28a_4$$

of the Weierstrass function $\wp$. As we said then, it is obvious, if one calculates *à la* Newton, that the difference $f(z)$ between the two sides is a doubly periodic function analytic on a neighbourhood of $z = 0$; it is thus so in all $\mathbb{C}$ because its only singularities must be those of the function $\wp$, namely the periods $\omega$; in other words, $f$ is an entire function. But, just as the function $\sin x$ takes all its values over $\mathbb{R}$ on $[0, 2\pi]$, by its periodicity, likewise a doubly periodic function takes no other values in $\mathbb{C}$ than those on the parallelogram constructed on the fundamental periods $\omega'$ and $\omega''$, i.e. on the *compact* set of the points

$$z = u'\omega' + u''\omega'' \qquad \text{with} \quad u', u'' \in [0, 1].$$

Being continuous, an entire elliptic function is bounded on such a parallelogram, so on all $\mathbb{C}$, so is constant by Liouville. It remains to state that, in the present case, the function $f$ is zero for $z = 0$, as is obvious from the series expansion of the Weierstrass function $\wp$.

In fact, it was *à propos* the elliptic functions that Liouville found his theorem in 1843–44 and it is instructive to follow the evolution of his ideas on this point[30] since they developed following a logic opposite to that, now classical, which we have just expounded.

Liouville first proves, using an idea of Hermite's, that a nonconstant function cannot have two real periods $\alpha$ and $\beta$ whose ratio is nonrational. To do this one writes (in our notation) that $f(t) = \sum a_n \mathbf{e}(nt/\alpha)$ and one checks, on replacing $t$ by $t + \beta$, that $a_n = a_n \mathbf{e}(n\beta/\alpha)$; if $\beta/\alpha \notin \mathbb{Q}$, the exponential is $\neq 1$ for any $n \neq 0$, qed. The result had already been proved in another way by Jacobi, and Liouville reckoned that his proof was equivalent to "looking for difficulties where there were none".

But he then had the idea of using the same method to show *a priori* that if a doubly periodic function, with periods whose ratio is not real, "does not become infinite", i.e. is holomorphic everywhere in $\mathbb{C}$, then it is constant. Despite the 40,000 pages of Liouville's notes – for the most part not published during his life, nor later, and deposited in the Paris Académie des sciences –, we do not know really how he proceeded. All the same, Lützen has recovered a note where Liouville writes that if a "function of $x + \sqrt{-1}y$" admits an imaginary period $\omega = a + \sqrt{-1}b$, then it can be expanded in a Fourier series of the form (in our notation)

(18.3) $$f(z) = \sum a_n \mathbf{e}_n(z/\omega)$$

where $\mathbf{e}_n(z) = \exp(2\pi i n z)$: this is Theorem 14. Assume $\omega = 1$ to simplify; Liouville writes like us that $f(x + iy) = \sum a_n(y)\mathbf{e}_n(x)$; from this he deduces that, for $h \in \mathbb{R}$,

---

[30] See Jesper Lützen, *Joseph Liouville 1809–1882, Master of Pure and Applied Mathematics* (Springer, 1990, 884 p.), chap. XIII.

$$f(x + h + iy) = \sum a_n(y)\mathbf{e}_n(h)\mathbf{e}_n(x),$$

and so

$$a_n(y)\mathbf{e}_n(x) = \oint f(x + h + iy)\mathbf{e}_n(h)dh$$

and consequently that the left hand side "is a function of $x + iy$", so must be of the form $a_n\mathbf{e}_n(x + iy)$ with constant $a_n$, a rather weak argument ...

Liouville made no reference to the holomorphy or analyticity of $f(z)$; for him, in 1844, what mattered was that $f$ should be a "function of $x + \sqrt{-1}y$", which perhaps means that there is an algebraic or analytic expression for $f(z)$ involving only $z$. In fact, one sees him use Cauchy's equation $f'_y = if'_x$ in another note of the same period, and the standard formula

$$a_n(y) = \oint f(x + iy)\mathbf{e}_n(x)dx$$

to establish a differential equation satisfied by the $a_n(y)$ and from them to deduce that they are proportional to $\exp(-2\pi n y)$, which yields (3) directly, as we saw in the preceding $n°$.

Lützen does not tell us how Liouville deduces from (3) that an everywhere holomorphic doubly periodic function is constant, but this was surely as obvious to him as to us: if $f$ has the periods 1 and $\omega$ (nonreal), a case to which one may always reduce, one must have

$$\sum a_n\mathbf{e}_n(z) = \sum a_n\mathbf{e}_n(\omega)\mathbf{e}_n(z)$$

and thus $a_n = a_n\mathbf{e}_n(\omega)$, whence $a_n = 0$ for $n \neq 0$ since the relation $\mathbf{e}_n(\omega) = 1$ requires that $\omega \in \mathbb{Z}$.

To pass to the theorem for arbitrary entire functions Liouville *used* the theory of elliptic functions. If $f$ is a bounded entire function and if $\varphi$ is an elliptic function (actually one of the Jacobi functions), then the composite function $f[\varphi(z)]$ is again elliptic and, being bounded, can have no poles. It is therefore a constant, and consequently so is $f$ (if one knows that $\varphi$ takes all possible complex values[31]). This argument appears in a four line note which Lützen reproduces on p. 543 of his book, the proof being shortened to a "Consequently, etc..".

Liouville announced his ideas on the elliptic functions to the Académie[32] in December 1844 and immediately had to face an offensive from Cauchy at

---

[31] On subtracting a constant, if necessary, it is enough to show that an elliptic function $\varphi$ always has zeros. But if this were not the case, the elliptic function $1/\varphi$ would be holomorphic everywhere, the poles of $\varphi$ included, so constant.

[32] where he entered in 1838 after a battle of which Lützen provides us a particularly edifying summary. The first two hundred pages of his book, which report thoroughly on the social situation of the mathematicians in France at this period, are full of incidents of this kind, and might illustrate the African proverb according to which two (and *a fortiori* fifteen) male crocodiles cannot coexist in the same backwater. The most famous scientists at this period could accrue

the following meeting: the latter recalled how a year previously he had shown how his theory of residues allowed one to reconstruct Jacobi's theory very easily (?); that in 1843 he had announced that if a function $f(z)$ "always has a unique and determined value, if, further, it reduces to a certain constant for all the infinite values of $z$ [which, apparently, means that $f(z)$ tends to a limit when $|z| \to +\infty$], then it reduces to this same constant when the variable $z$ takes an arbitrary finite value "; finally, applying this result to $[f(z) - f(a)]/(z - a)$ for a fixed $a$, a function which tends to 0 at infinity if $f$ is bounded and which, at $z = a$, remains "continuous" since $f$ is differentiable, he deduces the general form of Liouville's theorem: "*If a function $f(z)$ of the real or imaginary variable $z$ always remains continuous* [which, in Cauchy's language, probably means: is everywhere differentiable in the complex sense], *and consequently always finite, it reduces to a simple constant*". We are thus brought back to the quarrels about priority (already discussed, Chap. III, n° 10), an exercise much in vogue in the France of the period, and of which Cauchy was probably the historic champion in all categories. Lützen thinks that Liouville knew the result before Cauchy, but even if so it is nevertheless the date of publication which counts and not the manuscripts which an historian may discover a century and a half later. The situation is not particularly clear ...

This tendency of Liouville's not to publish again led to problems here, *à propos* the elliptic functions (i.e. meromorphic and having two periods with non real ratio). Between 1844 and 1847 he proved theorems which became the starting point for later expositions; essentially they consisted of *characterising the elliptic functions through their poles and zeros*, so avoiding the traditional calculations on elliptic integrals and Jacobi series; it all depends on the nonexistence of everywhere holomorphic elliptic functions. Example: let $f$ and $g$ be two elliptic functions and assume that both of them have simple zeros and simple poles at exactly the same points; then $f/g$ is everywhere holomorphic and elliptic, so constant.

Liouville did not publish these results though he explained them in private to two young Germans, Carl Wilhelm Borchardt and Ferdinand Joachimstahl; on returning to Germany, the first put his notes in order and sent copies to Liouville and to the latter's two friends, Jacobi and Dirichlet. His ideas were then known beyond the Rhine; Borchardt published them in 1880

---

three well remunerated posts (of the order of 6,000 F per annum, while a coach at the X or in another école had to be content with 100 to 150 F per month): Sorbonne, Collège de France, Polytechnique, CNAM, Bureau des Longitudes, etc. Just imagine the competition. The system considerably reduced the chances of the scientists who had not yet acquired high standing obtaining a suitable post, and, because of this, was strongly criticised. Moreover, when the polytechnicien Liouville, after resigning from the Corps des Ponts et Chaussées, found himself in this situation, and had, at the start, to teach nearly forty hours per week in secondary public or private establishments and at the X he claimed he had no more time to do research ... For another very different example, see Maurice Crosland, *Gay-Lussac: Scientist and Bourgeois* (Cambridge UP, 1978).

in the principal German journal when he became its editor. In 1851, having been elected to the Collège de France, Liouville, who clung to his priority against Cauchy, devoted a year to the subject – years are not very long at the Collège ... – before a small audience among whom were Charles Briot and Jean Bouquet, supporters of Cauchy who in 1859 published a *Théorie des fonctions doublement périodiques*, the first overall exposition of the theories of Cauchy and of Liouville; these had been completed elsewhere, since 1844, mainly by Hermite, the first to use Cauchy's ideas. But Liouville had not published, and, after Briot and Bouquet, refused to. In his later years he expressed his resentment of them[33], *"vile thieves but highly dignified Jesuits. Elected as thieves by the* Académie!!!!*"*, underlined in the text. In 1876, when Liouville was elected a foreign member of the Berlin academy, Weierstrass energetically reestablished the truth, recalling that it was all already in Borchardt's notes and that Briot and Bouquet ought to have mentioned that they owed *all* to Liouville. But nobody, Weierstrass included, ever understood why he had not published; he had had fifteen years to do so before Briot and Bouquet.

Nor had Cauchy yet discovered the Laurent series, and although he discovered equation (1.3) of n° 1 of this Chapter in about 1825 concerning holomorphic functions – without using the Fourier series which he was well placed to know –, he seems to have forgotten the result and it was only in 1831–32 and more probably in 1840–41 that he discovered the analyticity of his functions as we said above, with his ideas beginning to clarify about 1850. As Hans Freudenthal has written in his excellent biography in the DSB, *"he would have missed much more if others had cared about matters so general and so simple as those which occupied Cauchy"*. His works are confused, repetitive, with invalid or absurdly complicated proofs, yet nevertheless he produced, at the final count, a formidable branch of analysis and a method of genius for obtaining integrals which nobody had known how to calculate before him. It is curious that they did not attract the attention of his contemporaries, mainly of Germans like Gauss[34] or Jacobi who, at the same time, manipulated analytic functions every day for years (but perhaps without ascribing any importance to analyticity, since they rarely encountered anything else), beginning with elliptic functions; Cauchy, using his own methods, provided in 1846 the first nonmiraculous explanation of their double periodicity, which had been obtained by Abel and Jacobi using

---

[33] Lützen, p. 201. It should be understood that Liouville was a republican and secularist. He was deputy for Toul in the first National Assembly elected after the revolution of 1848. His friend Dirichlet said for his part at the same period that every mathematician had to be a democrat, probably because it is neither necessary nor sufficient to have inherited a title or a fortune to be able to do mathematics. One is not even always to be encouraged to inherit mathematical ability from his father, or, let us be politically correct, from his (or her) mother.

[34] In fact, it seems that Gauss had discovered some of Cauchy's results before the latter, but, as was his wont, had not published them.

complicated addition formulae already established for real variables. It was Riemann (1851) and above all Weierstrass and his pupils who then placed the theory on a solid base. Riemann's works on algebraic functions[35], an extraordinary mixture of topology, algebraic geometry and complex analysis, were so far in advance of the time that it needed a full fifty years before people began to understand and then generalise them, without ever trivialising them. In the meantime, Cauchy's theories, which Briot and Bouquet spread in Germany, and Weierstrass', prospered prodigiously; Remmert notes that the German edition of a book by the Italian G. Vivanti cites 672 titles before 1904. So prodigiously that, in France before 1940 to mention only one case, it monopolised the attention of a large number of mathematicians at the expense of the new branches which were being developed elsewhere, including the much more difficult theory of the holomorphic functions of several variables from which there came the most spectacular progress after 1950, mainly in France (H. Cartan and J.-P. Serre) and in Germany (H. Behnke, H. Grauert, R. Remmert and K. Stein); but this required totally different methods – differentiable varieties, algebraic topology, functional analysis, etc. – and entirely new ideas, the road from one to several complex variables being far too long to reduce to a simple generalisation.

Since we have just spoken of Liouville in a chapter that mainly treats Fourier series, we should mention his discovery, along with the Genevan Charles Sturm, of a formidable generalisation of harmonic analysis; it consists of replacing the exponentials by the "eigenfunctions" of a differential operator satisfying given "boundary conditions".

In Sturm-Liouville theory one considers a differential equation of the form

$$(18.4) \qquad\qquad -x''(t) + q(t)x(t) = 0$$

on the compact interval $I = [0,1]$ where $q$ is a given function, *real* and continuous in $I$. Putting $Lx = -x'' + qx$ (*cf.* the notation $Dx$ for the derived function $x'$), one terms *eigenfunctions* of $L$ the nonzero solutions of the equation

$$(18.5) \qquad\qquad Lx(t) = \lambda x(t)$$

where $\lambda = \mu^2$ is a given constant; *cf.* the eigenvectors of a matrix or of a linear operator in $\mathbb{R}^n$. In the trivial case $Lx = -x''$ one finds the functions $a. \exp(i\mu x) + b. \exp(-i\mu x)$ where $a$ and $b$ are arbitrary constants. The problem then consists of studying the solutions of (5) which satisfy the *boundary conditions*

---

[35] A function $\zeta = f(z)$ is said to be algebraic if one has a relation $P(z, \zeta) = 0$, where $P$ is a given polynomial with complex coefficients. The first difficulty is that, for $z$ given, the equation provides several possible values for $\zeta$. We are not dealing with functions on $\mathbb{C}$ in the strict sense of the term, but with "multiform functions" in the sense of Chap. IV, § 4 or correspondences in the sense of Chap. IV whose graphs are, except in a few respects, the "Riemann surfaces" of Chap. X.

(18.6) $$x'(0) - ux(0) = x'(1) + vx(1) = 0$$

where $u$ and $v$ are given *real* constants. The numbers $\lambda$ for which (5) and (6) have a solution $x \neq 0$ are now called the *eigenvalues* of the "boundary problem". All this came, *chez* Sturm and Liouville, from the PDE of heat propagation from which Fourier had already derived his series.

A first remark (Sturm) is that if $x \neq 0$, the eigenvalue $\lambda$ is real. By calculating the scalar product of $Lx$ and $x$ on $I$, one has, in telegraphic style,

$$
\begin{aligned}
(Lx \,|\, x) &= \int q x \bar{x} - \int x'' \bar{x} = \int q|x|^2 - \left[ x'(1)\overline{x(1)} - x'(0)\overline{x(0)} \right] + \int x'\overline{x'} = \\
&= \int q|x|^2 + \int |x'|^2 + u|x(0)|^2 + v|x'(1)|^2,
\end{aligned}
$$

a real result, since $u$, $v$ and $q(t)$ are real. But since $Lx = \lambda x$, the left hand side reduces to $\lambda \int |x|^2$, whence $\lambda \in \mathbb{R}$ and $\lambda$ is even $> 0$ if the function $q$ has positive values as well as $u$ and $v$ (an hypothesis justified by the physics). These are the same calculations as in algebra to show that the eigenvalues of a hermitian matrix are real.

A second problem is to show that, ignoring the conditions (6), the equation (5) always has solutions, and even a solution for which the initial conditions

(18.7) $$x(0) = a, \qquad x'(0) = b$$

are given. By replacing $q(t)$ by $q(t) - \lambda$ one reduces to the equation $x'' = qx$. Liouville then remarked, as Lützen (Chap. X, p. 447) tells us, that if one considers the differential equations

(18.8) $$x_0'' = 0, \qquad x_1'' = qx_0, \qquad x_2'' = qx_1, \dots,$$

then the function

(18.9) $$x(t) = x_0(t) + x_1(t) + x_2(t) + \dots$$

manifestly satisfies the equation $x'' = qx$: differentiate the series term-by-term; Liouville does not worry himself, at least at the beginning, with justifying this operation: Weierstrass (1815–1897) did not yet reign supreme in the 1830s and one might still, despite Cauchy, or because of Cauchy and his errors, calculate almost as Euler did. If one imposes the conditions

(18.10) $$x_0(t) = a + bt, \qquad x_n(0) = x_n'(0) = 0 \quad \text{for } n > 1,$$

it is clear that the series (9) satisfies (7). But by the FT the conditions (10) impose, for $n \geq 1$, the relation

(18.11) $$x_n(t) = \int_0^t x_n'(t)dt = \int_0^t dt \int_0^t x_n''(t)dt = \int_0^t dt \int_0^t q(t)x_{n-1}(t)dt$$

where, like Liouville, we have violated the ban on denoting both a phantom and a free variable by one and same letter. Whence an impressive formula, written in an entirely explicit way by Liouville,

$$(18.12) \quad x = x_0 + \iint qx_0 + \iint q \iint qx_0 + \iint q \iint q \iint qx_0 + \dots$$

where, for the value $t$ of the variable, the integrals[36] are taken between $0$ and $t$. Of course one has to prove that this series converges, which Liouville did perfectly correctly, by separating (12) into two series corresponding respectively to $x_0(t) = a$ and to $x_0(t) = bt$; modernising his language and putting $M = \sup |q(t)| = \|q\|$, one has[37], for $x_0(t) = a$,

$$|x_1(t)| \leq M \int dt \int |x_0(t)|\, dt = |a|Mt^{[2]},$$

$$|x_2(t)| \leq M \int dt \int |x_1(t)|\, dt \leq |a|M^2 t^{[4]},$$

and more generally $|x_n(t)| \leq |a|M^n t^{[2n]}$, whence convergence; the calculation is similar for $x_0(t) = bt$. We used similar calculations in Chap. VI, n° 10, to demonstrate the existence of solutions of the Bessel equation by means of the *method of successive approximations*; this fits into Liouville's schema, apart from the fact that it takes place on the interval $]0, +\infty[$ with a function $q$ singular at the origin. These calculations of Liouville's show that it is to him, and not to Emile Picard (1890), that one one should attribute the invention of this method as Lützen justifiably notes; Cauchy had another method a little later and would in his turn adopt the method of successive approximations to majorise the solutions, if not to prove their existence.

Having done this, one has to return to the boundary conditions (6), which impose drastic restrictions on the solutions. The first fundamental results are those of Sturm and rely on extremely ingenious arguments; assuming $u$, $v$ and $q > 0$, and then $\lambda > 0$ and $\mu$ real, he shows that the problem (5), (6) has a countable infinity of eigenvalues $\lambda_1 < \lambda_2 < \dots$ [to do this he compares the solutions of $x'' = (q - \lambda)x$ with those, trigonometric, of the equation $x'' = -n(\lambda)^2 x$ where the constant $n(\lambda)$ satisfies $n(\lambda) < \lambda - q(x)$ for every $x$], that to each eigenvalue $\lambda_n$ there corresponds, up to a constant factor, exactly one eigenfunction $u_n(t)$, that one may assume it real, that these are orthogonal on the interval $I$, i.e. that $\int u_p u_q = 0$, and finally that $u_n$ has $n$ zeros which interlace with the $n-1$ zeros of $u_{n-1}$. Liouville, himself,

---

[36] If one denotes by $P$ the operator which associates the primitive which vanishes at 0 to every continuous function on $[0, 1]$, the relation (12) means that

$$x = x_0 + P^2 qx_0 + P^2 qP^2 qx_0 + P^2 qP^2 qP^2 qx_0 + \dots,$$

where $P^2 = P \circ P$ and where each operator $P^2$ applies to all that follows it.

[37] The operator $P$ transforms the function $t^{[n]}$ into $t^{[n+1]}$.

obtained an asymptotic evaluation of the $\lambda_n$ and, above all, showed that the $u_n$ allow one to expand arbitrary functions in "Fourier" series

$$f(t) = \sum c_n(f)u_n(t) \quad \text{with} \quad c_n = (f \mid u_n)/(u_n \mid u_n)^{1/2},$$

the purpose of the denominator being to give the "vector" $u_n$ the length 1 as with the functions $\mathbf{e}_n(t)$ in Fourier's theory.

All this, invented by Sturm and Liouville (and even published ...) in the years 1830–1840, was a half-century ahead of its time. The question was taken up again from 1880–1890; the proofs were corrected; the method was extended to noncompact intervals (example: the Bessel equation of Chap. VI), which is noticeably more difficult and requires expansions in "Fourier" *integrals* involving the eigenfunctions; we enter the framework of the theory of integral equations – they had already appeared *chez* Liouville – then of Hilbert spaces, etc. This theory has given rise to quite remarkable expansions even very recently (*scattering*, the Korteweg-de Vries equation). The Soviet school, particularly B. M. Levitan, has worked enormously at this subject for a full half-century[38].

*Example.* Suppose $q = 0$, a case that seems trivial . Then (5) can be written as $x'' + \mu^2 x = 0$, whence $x(t) = ae^{i\mu t} + be^{-i\mu t}$. The relations (6) can be written

$$i\mu(a - b) - u(a + b) = i\mu \left(ae^{i\mu} - be^{-i\mu}\right) + v \left(ae^{i\mu} + be^{-i\mu}\right) = 0,$$

whence two linear homogeneous equations to determine $a$ and $b$ up to a constant factor. One can have $(a, b) \neq (0, 0)$ only if the determinant

$$\begin{vmatrix} i\mu - u, & -i\mu - u \\ (i\mu + v)e^{i\mu} & (-i\mu + v)e^{-i\mu} \end{vmatrix} = (i\mu - u)(v - i\mu)e^{-i\mu} + (u + i\mu)(v + i\mu)e^{i\mu}$$

is zero; on putting $z = (i\mu - u)(v - i\mu)$ this can be written

$$e^{2i\mu} = z/\bar{z} = z^2/|z|^2 \quad \text{or} \quad e^{i\mu} = \pm z/|z|.$$

This result is of modulus 1, so that $\mu$ is real. On separating the real and imaginary parts we see that $\mu$ must satisfy the relation

$$\cos \mu = \pm \left(\mu^2 - uv\right) / \left(\mu^2 + u^2\right)^{1/2} \left(\mu^2 + v^2\right)^{1/2},$$

---

[38] For a remarkably clear resumé of the state of the question, see the articles on "Sturm-Liouville" in the Soviet encyclopedia of mathematics (*Encyclopaedia of Mathematics*, Reidel, 10 Vol., 1988–1994, the Soviet edition dating from 1977–1985) where one can, more generally, inform oneself on almost every question in mathematics and find a bibliography of the subject (completed by the translators). Amusing detail: the article "cryptology" is entirely the work of the translators; the Soviet editors had omitted it. They also forgot to credit some out-of-favour colleagues ...

a "transcendental" equation, as one said at the time, whose roots are the eigenvalues sought. For $\mu$ large, the right hand side tends increasingly to 1, so that one can obtain an asymptotic expansion of the roots of the equation by the method of Chap. VI, n° 7, an easy exercise; it is more difficult to prove "by hand" that the eigenfunctions allow one to expand arbitrary, say $C^1$, functions in series. For $u = v = 0$, i.e. for the boundary conditions $x'(0) = x'(1) = 0$, one finds $a = b$ and $\cos \mu = \pm 1$, and so $\mu = \pi n$; the eigenfunctions are the functions $\cos \pi n t$ and one recovers Fourier series properly called.

## 19 – Limits of holomorphic functions

One of the most remarkable aspects of the theory of holomorphic functions is that when a sequence of such functions converges in only quite a mild way (convergence in the sense of the theory of distributions suffices) then (i) the limit function is holomorphic, (ii) the derived sequences converge, (iii) the successive derivatives of the limit are the limits of the successive derivatives of the given sequence; Paradise. Since the holomorphic functions are analytic and so $C^\infty$ as functions of $x$ and $y$, Theorem 23 of Chap. III, n° 22 will serve us so long as we prove point (ii) – convergence of the derivatives – which then allows us to apply it, as we have seen, since the Cauchy condition $f'_y = if'_x$ extends trivially to the limit.

The proof rests on a lemma worth isolating:

**Lemma.** *Let $r$ and $R$ be two numbers such that $0 \le r < R < +\infty$ and let $k$ be a positive integer. There exists a constant $M_k(r, R)$ such that, for every function $f$, continuous in $|z| \le R$ and holomorphic in $|z| < R$, one has*

$$(19.1) \qquad \sup_{|z| \le r} \left| f^{(k)}(z) \right| \le M_k(r, R) . \sup_{|z| \le R} |f(z)| .$$

We actually proved in n° 1 [pass to the limit in (1.11) as $r \to R$] that the derivatives of $f$ are given by

$$(19.2) \qquad f^{(k)}(z) = k! \int_{\mathbb{T}} f(Ru) \frac{Ru}{(Ru - z)^{k+1}} dm(u) \quad \text{for } |z| < R .$$

For $|z| \le r$ one has $|Ru - z| \ge R - r$, whence

$$\left| Ru/(Ru - z)^{k+1} \right| \le R/(R - r)^{k+1} .$$

On substituting in (2), one obtains (1) immediately, with $M_k(r, R) = k! R/(R - r)^{k+1}$, qed (One could, in the right hand side of (1), replace the sup extended over $|z| \le R$ by a sup extended over $|z| = R$, but these sup are the same by the maximum principle).

If one writes $D$ and $D'$ for the concentric *closed* discs of arbitrary centre $a$ and radii $R$ and $r < R$, and if one considers a function $f$, continuous in $D$

and holomorphic in the interior, the lemma above, applied to the function $f(a+z)$, shows that for every $k$ there exists a constant $M_k(D', D)$ *independent of $f$* such that

(19.1')
$$\left\| f^{(k)} \right\|_{D'} \leq M_k(D', D) \|f\|_D .$$

This crucial point established, suppose that in an open subset $U$ of $\mathbb{C}$ we have a sequence of analytic functions $f_n(z)$ which converge to a limit $f(z)$ uniformly on every compact $K \subset U$. For every $a \in U$, let us choose the closed discs $D \subset U$ and $D' \subset D$ with centre $a$ as above, and apply (1') to the differences $f_p - f_q$ whose usefulness Cauchy has shown us. His convergence criterion tells us that the right hand side of (1') is $< \varepsilon$ for $p$ and $q$ large enough since the $f_n$ converge uniformly on the compact set $D$. Likewise for the left hand side. Consequently, the $f_n^{(k)}$ converge uniformly in $D'$, i.e. on a neighbourhood of $a$. This means that the $f_n^{(k)}$ converge uniformly on every compact $K \subset U$ since this mode of convergence is a property of local nature (Chap. V, n° 6, Corollary 2 of Borel-Lebesgue).

We may now return to the general arguments of Chap. III, n° 22: since the successive partial derivatives of the $f_n$ are, up to powers of $i$, identical to the derivatives $f_n^{(k)}$ in the complex sense, the limit function is $C^\infty$ and its partial derivatives, being the limits of those of the $f_n$, satisfy the Cauchy condition $D_2 f = i D_1 f$ like them. The limit function $f$ is therefore holomorphic in $U$ and we have $f^{(k)}(z) = \lim f_n^{(k)}(z)$ for every $k$. Whence finally the famous

**Theorem 17 (Weierstrass).** *Let $(f_n)$ be a sequence of holomorphic functions in an open subset $U$ of $\mathbb{C}$. Assume that (i) $\lim f_n(z) = f(z)$ exists for every $z \in U$, (ii) the convergence is uniform on a neighbourhood of every point of $U$. Then the limit function is holomorphic in $U$ and, for every $k \in \mathbb{N}$, the sequence of derivatives $f_n^{(k)}(z)$ converges to $f^{(k)}(z)$ uniformly on every compact $K \subset U$.*

Very clearly, one might state Theorem 15 in terms of series of analytic functions[39]. If in particular such a series converges normally on a neighbourhood of every point of $U$, then its sum is holomorphic, can be differentiated term-by-term, etc.

*Example 1.* This is the case of the Riemann function $\zeta(s) = \sum 1/n^s$ in the open set $\mathrm{Re}(s) > 1$ where the series converges. For every $\sigma > 1$, the series converges normally in the half plane $\mathrm{Re}(s) \geq \sigma$ since there $|1/n^s| \leq 1/n^\sigma$. The function is thus holomorphic in $\mathrm{Re}(s) > \sigma$ for every $\sigma > 1$, so in fact

---

[39] Weierstrass' original proof (1841) in fact concerns series of power series and does not use the Cauchy integral formula of n° 1; it rests on a direct and elementary proof of the inequality (19.1), which allows him to apply his theorem on double series to a convergent series of analytic functions. The present proof is due to Paul Painlevé (1887); see Remmert, *Funktionentheorie 1*, Chap. 8, §§ 3 and 4, who sets out Weierstrass' proof of (19.1) on one page, as simple as it is ingenious.

in $\mathrm{Re}(s) > 1$, and its derivative is given by $\zeta'(s) = -\sum \log n / n^s$. Recall (Chap. VI, n° 19) that in fact the $\zeta$ function can be extended analytically to $\mathbb{C} - \{1\}$, the point $s = 1$ being a simple pole.

*Example 2.* The series

$$\pi . \cot \pi z = 1/z + 2z \sum 1/(z^2 - n^2)$$

converges normally on every compact set $K \subset \mathbb{C} - \mathbb{Z}$ as we have known a long while, so is holomorphic in $\mathbb{C} - \mathbb{Z}$; and the only reason that it does not converge normally on a neighbourhood of a point $n \in \mathbb{Z}$ lies in the term $2z/(z^2 - n^2) = 1/(z - n) + 1/(z + n)$: the series obtained by suppressing the term $1/(z - n)$ converges unproblematically on a neighbourhood of $n$. In other words, on a neighbourhood of each $n \in \mathbb{Z}$, the function is the sum of the pole term $1/(z - n)$ and of a function holomorphic on a neighbourhood of $n$. It therefore has just a simple pole at $n$; it is a meromorphic function on all $\mathbb{C}$, and the series can be differentiated term by term.

*Example 3.* Consider similarly the elliptic functions *à la* Weierstrass of Chap. II, n° 23, the sums of the series $\sum (z - \omega)^{-k}$ extended over the periods $(k > 3)$, or, for $k = 2$, the modified series $\wp(z)$. We have shown that these are analytic by proving by a routine calculation that, in every disc $|z| < R$, they are the sum of a finite number of terms, corresponding to the periods situated in this disc, and of an explicitly calculated power series. The preceding theorem yields the result without any calculation since after subtraction of the exceptional terms in question, the series $\sum (z - \omega)^{-k}$ converges normally in the disc $|z| < R$ as we have already seen. The functions obtained are thus holomorphic in $\mathbb{C}$ with the periods removed. In the neighbourhood of a period these functions are the sum of a holomorphic function on a neighbourhood of $\omega$ and of the term $1/(z - \omega)^k$ of the series, whence a pole of order $k$ at each point of the lattice. The Weierstrass functions are therefore meromorphic in $\mathbb{C}$ and one may differentiate them term-by-term, which confirms, but *without calculation*, the fact that the series $\sum (z - \omega)^{-k}$ are, for $k > 3$, the successive derivatives of $\wp(z)$, up to obvious constant factors.

*Example 4.* The sum of a Fourier series $\sum a_n \mathbf{e}_n(z)$ which converges in a strip $a < \mathrm{Im}(z) < b$, and thus normally in every smaller closed strip, is a holomorphic function.

## 20 – Infinite products of holomorphic functions

Theorem 17 has an analogue, also due to Weierstrass, for infinite products:

**Theorem 18.** *Let* $(u_n(z))$ *be a sequence of functions holomorphic in a domain* $G$ *and suppose the series* $\sum u_n(z)$ *is normally convergent on every compact* $K \subset G$. *Then the function*

$$p(z) = \prod (1 + u_n(z)) = \lim (1 + u_1(z)) \dots (1 + u_n(z))$$

is holomorphic in $G$, its zeros are those of the functions $1 + u_n(z)$, and

$$(20.1) \qquad p'(z)/p(z) = \sum \frac{u_n'(z)}{1 + u_n(z)}$$

at every point where $p(z) \neq 0$.

First we remark that, for every compact $K \subset G$, the series $\sum \|u_n\|_K$ converges, by the definition of normal convergence (Chap. III, n° 8). Thus $\|u_n\|_K < \frac{1}{2}$ for $n$ large, so that the factors $1 + u_n(z)$ which might vanish somewhere in $K$ are finite in number; such a factor can moreover possess only a finite number of zeros in $K$, for otherwise Bolzano-Weierstrass would allow us to construct a sequence of pairwise distinct zeros converging to a point of $K$, thus of $G$, and the principle of isolated zeros (Chap. II, end of n° 19 and n° 20) would show, since $G$ is connected, that $1 + u_n(z)$ is identically zero, a case that one may reasonably exclude in the considerations which follow. Apart from these factors whose product is holomorphic in $G$, the function $p$ is an infinite product all of whose terms are $\neq 0$ for any $z \in K$; since $\sum |u_n(z)| < +\infty$, this product is absolutely convergent and not zero (Chap. IV, n° 17, Theorem 13 whose proof we in any case will have to reproduce).

Let us put $p_n(z) = (1 + u_1(z)) \dots (1 + u_n(z))$ and remain in $K$, forgetting to allow for the factors, finite in number, which vanish in $K$. We have

$$(20.2) \quad \log |p_n(z)| = \log |1 + u_1(z)| + \dots + \log |1 + u_n(z)| \leq$$
$$\leq |u_1(z)| + \dots + |u_n(z)| \leq \|u_1\|_K + \dots + \|u_n\|_K.$$

The series $\sum \|u_n\|_K$ being convergent there exists a finite constant $M(K)$ such that $|p_n(z)| \leq M(K)$ for every $z \in K$. Since we have $p_{n+1} - p_n = p_n u_{n+1}$, it follows that $|p_{n+1}(z) - p_n(z)| \leq M(K) |u_{n+1}(z)|$ for every $z \in K$, whence

$$\|p_{n+1} - p_n\|_K \leq M(K) \|u_{n+1}\|_K.$$

The series $\sum p_{n+1}(z) - p_n(z)$ therefore converges normally in $K$ like the $u_n(z)$. We deduce that the sequence of the $p_n(z)$ converges uniformly on every compact $K \subset G$, and therefore $p(z) = \lim p_n(z)$ is holomorphic in $G$ like the $p_n$, by Theorem 17; the "forgotten" factors in the product do not affect the conclusion since they are holomorphic and finite in number on a neighbourhood of each point of $G$.

It remains to prove (1), a generalisation of the rule

$$(fg)'/fg = f'/f + g'/g$$

for "logarithmic differentiation" of a product. At a point where $p(z) \neq 0$, one has from the latter $p_n'(z)/p_n(z) = \sum_{k \leq n} u_k'(z)/(1 + u_k(z))$. But Theorem 17 assures us that $\lim p'_n(z) = p'(z)$, and since $p_n(z)$ tends to $p(z) \neq 0$, one deduces that $p'(z)/p(z)$ is the limit of the partial sums of the series (1).

In fact, the latter converges normally on every compact $K \subset G$. It suffices, as always, to show this on a neighbourhood of every $a \in G$. To do this, consider a compact disc $D : |z-a| \leq R$ contained in $G$ and a disc $D' : |z-a| \leq r < R$ contained in the interior of the first. The lemma of n° 19 provides an upper estimate $\|u'_n\|_{D'} \leq M \|u_n\|_D$ with a constant $M$ independent of $n$. On the other hand $\|u_n\|_D < \frac{1}{2}$ for $n$ large, as we have seen above, and thus $\|1 + u_n\|_D > \frac{1}{2}$. The uniform norm on $D'$ of the general term of the series (1) is therefore majorised for $n$ large by $2M \|u_n\|_D$, whence the normal convergence of (1) in $D'$, and so on every compact $K \subset U$, qed.

*Example 1.* Consider (Chap. IV, n° 20) Euler's formula

$$P(z) = \prod_{n \geq 1} (1 - z^n)^{-1} = \sum_{n \geq 0} p(n) z^n$$

which appears in the theory of partitions. Theorem 18 shows that the product $\prod (1 - z^n)$ is holomorphic in the disc $D : |z| < 1$ and never vanishes; the left hand side is therefore also holomorphic in $D$, whence the existence of an expansion in power series in $D$. Since we have $p(n) \geq 1$ (this is the least one could say ...), the radius of convergence is equal to 1.

The function $P(z)$ is an example of a curious phenomenon: it is not possible to "extend the function $P$ analytically" outside $D$: if an analytic function defined in a domain $G \supset D$ coincides with $P$ in $D$, then $G = D$. One may understand this by observing that $P(z)$ seems not to tend to any limit when $z$ tends to any root of unity, since then infinitely many factors of the product become infinite, but this is not a proof ...

*Example 2.* Let $q$ be a constant complex such that $|q| < 1$ and let us consider the infinite product

(20.3) $$f(z) = \prod (1 + q^n z),$$

where the product is extended over all $n \geq 1$. Since $\sum |q^n| < +\infty$, Weierstrass' theorem applies, the result being an entire function of $z$. It is clear that $f(z) = (1 + qz)f(qz)$, so that the power series $f(z) = \sum a_n z^n$ which represents $f$ in the whole plane satisfies

$$\sum a_n z^n = (1 + qz) \sum a_n q^n z^n;$$

one deduces that $a_n = q^n a_n + q^n a_{n-1}$, i.e. that

$$(1 - q^n) a_n = q^n a_{n-1}.$$

Since $a_0 = f(0) = 1$, it follows that

$$a_n = q^{1+2+\cdots+n}/(1-q)\ldots(1-q^n),$$

whence the identity

$$(20.4) \quad \prod_{n=1}^{\infty}(1+q^n z) = 1 + \sum_{n=1}^{\infty} \frac{q^{n(n+1)/2}}{(1-q)\ldots(1-q^n)} z^n \qquad (|q|<1,\; z\in\mathbb{C}).$$

*Exercise*: show that

$$(20.5) \qquad \prod_{n=0}^{\infty}(1+q^n z)^{-1} = \sum_{n=0}^{\infty} \frac{(-1)^n z^n}{(1-q)\ldots(1-q^n)}$$

for $|q|<1$, $|z|<1$. What happens for $|z|>1$?

*Example 3.* Consider the infinite product

$$(20.6) \qquad f(z) = z\prod\left(1-z^2/n^2\right)$$

extended over $n \geq 1$. It satisfies the hypotheses of the theorem, with $G = \mathbb{C}$, so represents an entire function having simple zeros at the $n \in \mathbb{Z}$ and is nonzero elsewhere. Theorem 18 shows that

$$(20.7) \quad f'(z)/f(z) = 1/z + \sum 2z/\left(z^2-n^2\right) = \pi.\cot\pi z = (\sin\pi z)'/\sin\pi z.$$

In the connected open set $\mathbb{C}-\mathbb{Z}$ where it is defined, the holomorphic function $f(z)/\sin\pi z$ thus has derivative zero, so is constant; since $f(z)/z$ and $\sin\pi z/z$ tend respectively to 1 and $\pi$ when $z$ tends to 0, this constant is equal to $1/\pi$. Whence

$$(20.8) \qquad \sin\pi z = \pi z\prod\left(1-z^2/n^2\right).$$

This proof of Euler's infinite product manifests the fantastical character of his considerations on "algebraic equations of infinite degree" (Chap. II, end of n° 21). As we have said, the infinite product (6) is an entire function whose only zeros are the $n \in \mathbb{Z}$; on a neighbourhood of such a point, $f(z)$ is the product of $1-z/n$ by an infinite product which no longer vanishes at $n$, so that the $n \in \mathbb{Z}$ are simple zeros of $f$. Since it is clear that $z = n$ is likewise a simple zero of the function $\sin\pi z$ (obvious for $n = 0$, so for any $n$ by periodicity), one deduces that $\sin\pi z = g(z)f(z)$ where $g$ is an everywhere holomorphic function (so analytic) having no zero in $\mathbb{C}$. For such a function, the quotient $g'(z)/g(z)$ is again an entire function, so an everywhere convergent power series, so has in $\mathbb{C}$ a primitive $h(z)$ such that $h'(z) = g'(z)/g(z)$ as we know (Chap. II, n° 19). It follows that

$$\left(ge^{-h}\right)' = g'e^{-h} - gh'e^{-h} = 0,$$

so that $g(z) = ce^{h(z)}$ where $c$ is a constant that one may assume equal to 1 by incorporating a suitable constant into $h$. Euler's argument, corrected, thus proves the existence of a relation of the form

$$(20.9) \qquad \sin \pi z = e^{h(z)} z \prod \left(1 - z^2/n^2\right)$$

with an entire function $h$ about which the preceding argument provides no information whatsoever ...

In fact, Weierstrass invented, and his successors refined, a whole theory that allows one to represent any entire function $f$ by an infinite product that exhibits its zeros, but this is much less simple than Euler's ideas. The first idea is to order the zeros $a_n$ of $f$ in a sequence[40] such that $|a_n| \le |a_{n+1}|$, then to consider the infinite product $\prod (1 - z/a_n)$. This then has exactly the same zeros as $f$, with the same orders of multiplicity, whence $f(z) = g(z) \prod (1 - z/a_n)$ where $g$ is an entire function without zeros, so of the form $e^{h(z)}$. This is Euler's marvellous argument (except that he forgot the factor $g$).

But first one should verify the convergence of the infinite product! Though obvious in the case of the function sinus when one groups the symmetric factors, it can be perfectly *false* in the case of an arbitrary entire function[41].

Weierstrass' idea is now to multiply each factor $1 - z/a_n$ by as simple a function as possible, vanishing nowhere so as not to add parasitic zeros to the product, and making the infinite product convergent. This technique is very shrewd. First, it is clear that, for $z$ given, $z/a_n$ tends to 0, so is in modulus $< 1$ for $n$ large. Consider, generally, $1 - z$ for $|z| < 1$. Then

$$1 - z = \exp \left(-z - z^2/2 - z^3/3 - \ldots\right)$$

[Chap. IV, (13.12)], whence, for every $p$,

$$1 - z = \exp \left(z + \ldots + z^p/p\right)^{-1} \exp \left[-z^{p+1}/(p+1) - \ldots\right].$$

Consider the functions

$$(20.10) \quad E_p(z) = (1 - z) \exp \left(z + \ldots + z^p/p\right) = \exp \left[-z^{p+1}/(p+1) - \ldots\right].$$

The factor $\exp \left(z + \ldots + z^p/p\right)$ never vanishes and tends to 1 when $z$ tends to 0, as does $E_p(z)$. Let us prove that

---

[40] The set of zeros is countable, for, by the principle of isolated zeros, there can be only finitely many in the disc $|z| < p$ for any $p \in \mathbb{N}$. In what follows, we assume that each zero appears as many times among the $a_n$ as its order of multiplicity.

[41] Consider the function $\sin(\pi z^2)$. Its zeros are the $z$ such that $z^2 \in \mathbb{Z}$, so the points of the form $n^{1/2}$ or $in^{1/2}$, and the infinite product is divergent like the series $\sum 1/|n|^{1/2}$.

(20.11) $$|1 - E_p(z)| \leq |z|^{p+1} \qquad \text{for } |z| \leq 1.$$

The function $1 - E_p(z)$ vanishes at the origin and is holomorphic in all $\mathbb{C}$; its derivative

(20.12) $$-E_p'(z) = z^p \exp\left(z + \ldots + z^p/p\right) = z^p \sum_n \left(z + \ldots + z^p/p\right)^{[n]}$$

(exercise!) is an everywhere convergent power series with all its coefficients positive (Theorem 17) and whose term of lowest degree is $z^p$; that of $1 - E_p(z)$ therefore starts with a term in $z^{p+1}$. Schwarz' lemma now shows that $|1 - E_p(z)| \leq M|z|^{p+1}$, where $M$ is the maximum of $|1 - E_p(z)|$ on the circle $|z| = 1$; but since the coefficients of the power series of $1 - E_p(z)$ are, like those of its derivative, all positive, its maximum on $|z| = 1$ is attained for $z = 1$ and so is equal to 1 by (10); whence (11), thanks to Remmert.

This point completed, let us return to the entire function $f(z)$ and to its zeros $a_n$. The function $E_p\left(z/a_n\right) = (1 - z/a_n)\exp(\ldots)$ has a simple zero at $a_n$ and is $\neq 0$ elsewhere. We may therefore try to compare $f(z)$ with the infinite product

(20.13) $$h(z) = \prod E_{p_n}\left(1 - z/a_n\right)$$

where the $p_n$ are chosen to make the product absolutely convergent. Since it can be written as $\prod[1 + u_n(z)]$ with

$$|u_n(z)| \leq |z/a_n|^{p_n+1}$$

by (11), and since, for every compact $K \subset \mathbb{C}$, we have $|z/a_n| \leq \frac{1}{2}$ for every $z \in K$ if $n$ is large enough (the $|a_n|$ increase indefinitely since the zeros of an entire function are isolated in $\mathbb{C}$), we may always choose the $p_n$ to make the series $\sum u_n(z)$ normally convergent on every compact; at the worst, we choose $p_n = n - 1$ for every $n$.

This done, we have, as in the case of the function $\sin \pi z$, a relation

(20.14) $$f(z) = e^{g(z)} \prod E_{p_n}\left(1 - z/a_n\right)$$

with an entire function $g(z)$ about which, a priori, we know nothing.

It goes without saying that the choice $p_n = n - 1$ is not always the best possible, as shown by the case of the function sinus, and that, moreover, il would be useful to determine the function $g(z)$ more precisely. There are theorems that apply to functions that do not increase too fast at infinity. To enter into this difficult subject which has interested (too) many specialists would no doubt exceed the capacity of most of our readers, and, even more certainly, of the author.

*Example 4 (infinite product for the Gamma function)*. Consider Euler's function

$$(20.15) \qquad \Gamma(s) = \int_0^{+\infty} e^{-x} x^{s-1} dx.$$

We already know some of its important properties:

(i) the integral converges absolutely if and only if $\mathrm{Re}(s) > 0$ (Chap. V, n° 22, Example 1), satisfies $\Gamma(s+1) = s\Gamma(s)$, and also

$$(20.16) \qquad s\Gamma(s) = \lim n^s/(1+s)(1+s/2)\dots(1+s/n);$$

(ii) the function $\Gamma$ is holomorphic for $\mathrm{Re}(s) > 0$ and can be continued holomorphically to $G = \mathbb{C} - \{0, -1, -2, \dots\}$ (Chap. V, n° 25, Example 5); in Chap. V, we did not yet know that "holomorphic" and "analytic" are synonymous, but we know this now; this shows in passing that the various methods we have used to continue the function $\Gamma$ analytically to $G$ yield the same function;

(iii) one may (Chap. VI, n° 18) transform (16) into an expansion as an infinite product

$$(20.17) \qquad 1/s\Gamma(s) = e^{\gamma s} \prod (1 + s/n) e^{-s/n}$$

which converges absolutely for every $s \in \mathbb{C}$.

If one only knows that (17) is valid for $\mathrm{Re}(s) > 0$, it is easy to lift this restriction; it amounts to proving that Theorem 18 applies in $\mathbb{C}$: the principle of analytic continuation will do the rest. Remmert, *Funktionentheorie 2*, p. 31, gives what is surely the simplest proof. One starts from the identity

$$1 - (1-w)e^w = w^2 \left[(1 - 1/2!) + (1/2! - 1/3!)w + (1/3! - 1/4!)w^2 + \dots\right]$$

and remarks that the coefficients of the $w^n$ are all $> 0$; for $|w| < 1$ one thus has

$$|1 - (1-w)e^w| < |w|^2 \sum [(1/p! - 1/(p+1)!] = |w|^2.$$

But if one puts the general term of the product (17) in the form $1 - u_n(s)$, one has $u_n(s) = 1 - (1-w)e^w$ for $w = -s/n$; consequently

$$|u_n(s)| \le |s/n|^2 \qquad \text{for } n \ge |s|.$$

In a disc $|s| \le R$, one thus has $|u_n(s)| \le R^2/n^2$ for $n > R$, whence the normal convergence of $\sum u_n(s)$ in $|s| < R$, qed.

Example 5. Let us go back to the theory of elliptic functions with a lattice $L$ of periods (Chap. II, n° 23) and let us consider, with Weierstrass, the infinite product $\prod(1 - z/\omega)$ extended over the periods $\omega \in L$. It clearly does not converge since the convergence of $\sum 1/|\omega|^k$ presupposes that $k \ge 3$ for $k$ an integer (or $k > 2$ for $k$ real). But we have

$$1 - z/\omega = \exp\left(-z/\omega - z^2/2\omega^2 - \ldots\right) =$$
$$= \exp\left(-z/\omega - z^2/2\omega^2\right)\exp\left(-z^3/3\omega^3 - \ldots\right)$$

with $\left|1 - \exp\left(-z^3/3\omega^3 - \ldots\right)\right| \leq M\left|z^3/3\omega^3\right|$ for $|z/\omega| < 1$ by (11); for $|z| \leq R$, this condition holds for $|\omega| > R$, whence a bound by $MR^3/3\,|\omega|^3$ which ensures the normal convergence in the disc considered. We conclude that the infinite product (no connection with the Riemann function $\zeta$)

$$(20.18) \qquad \zeta_L(z) = z\prod_{\omega\neq 0}(1 - z/\omega)e^{z/\omega + z^2/2\omega^2}$$

converges normally on every compact subset of $\mathbb{C} - L$ and even on a neighbourhood of every $\omega \in L$ so long as we isolate the term $1 - z/\omega$. In this way one finds an entire function having simple zeros at the $\omega \in L$ and nonzero elsewhere.

Applying the differentiation formula, one obtains a new bizarre function

$$(20.19)\quad \sigma_L(z) = \zeta_L'(z)/\zeta_L(z) = 1/z + \sum\left[1/(z-\omega) + 1/\omega + z/\omega^2\right]$$

and, differentiating once again,

$$(20.20)\qquad -\sigma_L'(z) = 1/z^2 + \sum\left[1/(z-\omega)^2 - 1/\omega^2\right] = \wp_L(z),$$

the gothic cursive function $\wp$ of this same Weierstrass associated with the lattice $L$. The beauty of these calculations is that they are apparently purely formal; but, in reality, everything converges because Theorem 17, clearly applicable to unconditional convergence, justifies everything once one knows that the infinite product (18) converges.

The relation $\sigma_L'(z) = -\wp_L(z)$ shows that the derivative of the function $\sigma_L$ does not change if we add a period to $z$; hence a relation of the form

$$\sigma_L(z + \omega) = \sigma_L(z) + c(\omega)$$

with a constant $c(\omega)$ clearly satisfying

$$c(\omega' + \omega'') = c(\omega') + c(\omega''),$$

whence $c\left(n_1\omega_1 + n_2\omega_2\right) = n_1c(\omega_1) + n_2c(\omega_2)$, which allows one to calculate it if one knows $c(\omega)$ for two periods forming a basis[42] of $L$. We could continue – the essentials of the theory of elliptic functions can be expounded with hardly any other tools than those of the present § –, but it will be better to delay these explorations for later (Chap. XII).

---

[42] i.e. two periods $\omega_1$ and $\omega_2$ such that every other is a linear combination of them with integer coefficients.

# § 5. Harmonic functions and Fourier series

## 21 – Analytic functions defined by a Cauchy integral

The calculation which, in n° 1, allowed us to represent a power series converging on a disc $|z| < R$ by an integral over a circle with centre 0 and radius $r < R$, can be inverted and generalised: every reasonable function $f$ defined on the circle $|z| = r$ allows one, thanks to Cauchy's integral formula, to define a function $P_f$, its *Poisson transform* (Siméon Denis, 1781–1840, a less brilliant competitor of Fourier and Cauchy, to whom, nevertheless, we owe several important ideas), defined and analytic for $|z| \neq r$. The study of this function, beyond its being an excellent exercise, allows us to prove Weierstrass' approximation theorem again, and, more importantly, to establish the principal properties of the "harmonic" functions, which are, at least locally, the real parts of holomorphic functions, and conversely.

To simplify the formulae, we shall assume that $r = 1$ in what follows. We need only replace $f(u)$ by $f(ru)$ to obtain the general case.

For a regulated periodic function $f$ the function $P_f$ is given by

$$(21.1) \qquad P_f(z) = \frac{1}{2\pi i} \int_{|\zeta|=1} f(\zeta)\frac{d\zeta}{\zeta - z} = \int_{\mathbb{T}} \frac{u}{u - z} f(u)dm(u);$$

we have introduced a factor $2\pi i$ so as to recover as $P_f$ the function $f(z)$ if one chooses for $f(u)$ the restriction to $\mathbb{T}$ of a power series, as in the case of Cauchy's formula (1.4). Recall how, in (1), one passes from the complex Leibniz notation to the integral in $u$: one puts $\zeta = u = \mathbf{e}(t)$, whence $d\zeta = 2\pi i\mathbf{e}(t)dt = 2\pi iudm(u)$.

One may generalise further and replace the expression $f(u)dm(u)$ by a measure $\mu$ on $\mathbb{T}$, whence the Poisson transform

$$(21.2) \qquad P_\mu(z) = \int \frac{u}{u - z}d\mu(u)$$

of $\mu$. If for example $\mu$ is the Dirac measure at the point $u = 1$, one obtains $P_\mu(z) = 1/(1 - z)$. One might use the same formula to define that of a distribution on $\mathbb{T}$ since the function $u \mapsto u/(u-z)$ is indefinitely differentiable on the circle. We shall see that all these functions are analytic for $|z| \neq 1$.

For $|z| < 1$, we have, as in n° 1,

$$u/(u - z) = 1/(1 - u^{-1}z) = \sum z^n/u^n$$

with a series of functions of $u$ which converges normally, so uniformly, on the interval of integration. We may therefore integrate term-by-term by the definition of a measure, i.e. thanks to the estimate

$$(21.3) \qquad \left|\int f(u)d\mu(u)\right| \leq M(\mu)\|f\|$$

valid for every function $f$ defined and continuous on the circle. This done we clearly find

$$(21.4) \quad P_\mu(z) = \sum_{n \geq 0} a_n z^n \quad \text{where} \quad a_n = \int u^{-n} d\mu(u) = \hat{\mu}(n) \quad (|z| < 1)$$

by the definition (3.1"') of the Fourier coefficients of a measure or distribution on $\mathbb{T}$. The series we obtain converges absolutely for $|z| < 1$: since $|u| = 1$ we have $|a_n| \leq M(\mu)$ by (3), and the result follows since $|z| < 1$. In other words, $f$ is analytic in the disc $|z| < 1$.

The reader would be wrong to be overwhelmed by these vast generalisations: we are dealing in trivialities, i.e. assertions following directly from the definitions, not to be confused with theorems which require more or less long and difficult proofs.

*Exercise*: show that, for $z$ given with $|z| < 1$, the series $\sum z^n / u^n$ converges in the space $\mathcal{D}(\mathbb{T})$ or, equivalently, that the series $\sum z^n \mathbf{e}_n(t)$ (sum over the $n \geq 0$), and all those obtained by differentiating term-by-term *ad libitum* with respect to $t$, converge uniformly on $\mathbb{R}$. Deduce that (4) applies to every distribution on $\mathbb{T}$.

For $|z| > 1$, we must, on the contrary, expand in powers of $u/z$, i.e. use the formula

$$u/(u - z) = -u/z \left(1 - uz^{-1}\right) = -\sum u^{n+1}/z^{n+1},$$

whence

$$(21.5) \quad P_\mu(z) = \sum_{n > 0} b_n / z^n \quad \text{where} \quad b_n = -\int u^n d\mu(u) = -\hat{\mu}(-n) \quad (|z| > 1).$$

In this way we find an analytic function of $1/z$, so also of $z$, on the open set $|z| > 1$. It generally has no connection with the function $P_\mu$ obtained for $|z| < 1$; if for example one starts, as in n° 1, from the measure $d\mu(u) = f(u)dm(u)$, where $f$ is a power series converging absolutely on $\mathbb{T}$, then the function (4) is identical with $f$ but (5) is identically zero by the formulae (1.4). No matter, finally we have

$$(21.6) \qquad P_\mu(z) = \begin{cases} \sum_{n \geq 0} \hat{\mu}(n) z^n & \text{for} \quad |z| < 1. \\ -\sum_{n < 0} \hat{\mu}(n) z^n & \text{for} \quad |z| > 1. \end{cases}$$

In the most important case, that of formula (1), the formulae (6) can be written

$$(21.7) \quad P_f(z) = \int_{\mathbb{T}} \frac{u}{u - z} f(u) du = \begin{cases} \sum_{n \geq 0} \hat{f}(n) z^n & \text{for} \quad |z| < 1 \\ -\sum_{n < 0} \hat{f}(n) z^n & \text{for} \quad |z| > 1 \end{cases}$$

since the Fourier coefficients of the measure $d\mu(u) = f(u)dm(u)$ associated with the function $f$ are those of $f$.

## 22 – Poisson's function

Consider a continuous function $f(u)$ on the unit circle $\mathbb{T} : |u| = 1$, and as always put $f(t) = f\left(e^{2\pi it}\right) = f(\mathbf{e}(t))$; let us examine the function

$$(22.1) \qquad P_f(z) = \oint \frac{\mathbf{e}(t)}{\mathbf{e}(t) - z} f(t)dt = \int \frac{u}{u - z} f(u)dm(u).$$

As we have seen above, this formula represents two different analytic functions in the open sets $|z| < 1$ and $|z| > 1$. It is by comparing their behaviour on a neighbourhood of a point $u = \mathbf{e}(t)$ of the limit circle $\mathbb{T}$ that we shall obtain results on the Fourier series of $f$.

To do this we put $z = ru$ with $r \neq 1$, and let $r$ tend to 1 either through values $< 1$, or through values $> 1$.

If $r < 1$, we have, by (21.7),

$$(22.2) \qquad P_f(ru) = \sum_{n \geq 0} \hat{f}(n)r^n u^n.$$

As $r$ tends to 1, we then "clearly" find

$$(22.3) \qquad \lim_{r \to 1,\, r < 1} P_f(ru) = \sum_{n \geq 0} \hat{f}(n)u^n.$$

This passage to the limit is, alas, not always justified; since

$$\sup_{r < 1} \left| \hat{f}(n)r^n u^n \right| = \left| \hat{f}(n) \right|$$

because all the exponents $n$ featuring are positive, the series (2), considered for $u$ fixed, like a series of continuous functions of $r$ in the interval $[0,1]$, will be normally convergent if and only if one assumes that

$$(22.4) \qquad \sum_{n \geq 0} \left| \hat{f}(n) \right| < +\infty.$$

Passage to the limit term-by-term is then allowed by Theorem 9 of Chap. III, $n°$ 8: the sum of the series is a continuous function of $r$ on the *closed* interval $[0,1]$, so that its value $\sum \hat{f}(n)u^n$ for $r = 1$ is the limit of its values when $r < 1$ tends to 1.

For $z = r'u$, with $r' > 1$, we have to start from the formula

$$(22.5) \qquad P_f(r'u) = -\sum_{n < 0} \hat{f}(n)r'^n u^n.$$

If

$$(22.6) \qquad \sum_{n<0} \left| \hat{f}(n) \right| < +\infty,$$

the preceding argument applies again since the fact that $r'$ is $> 1$ is compensated by the presence in (5) of all negative exponents: the series (5) is dominated, on the *closed* interval $r' \geq 1$, by the convergent series (6). Hence we find

$$(22.7) \qquad \lim_{r' \to 1,\ r'>1} P_f(r'u) = -\sum_{n<0} \hat{f}(n)u^n.$$

If the hypotheses (5) and (7) hold, i.e. if

$$(22.8) \qquad \sum_{\mathbb{Z}} |\hat{f}(n)| < +\infty,$$

we then see that the Fourier series of $f$ is given by

$$(22.9) \qquad \sum_{n\in\mathbb{Z}} \hat{f}(n)u^n = \lim_{r\to 1,\ r<1} P_f(ru) - \lim_{r'\to 1,\ r'>1} P_f(r'u).$$

Since we hope that the left hand side has value $f(u) = f(t)$, we have to examine the second more closely. We shall see that, if we choose to let $r$ and $r'$ vary so that $r' = 1/r$, the difference $P_f(ru) - P_f(r'u)$ is then expressed as a very simple integral which tends to $f(u)$ if $f$ is continuous; if the hypothesis (8) holds, we will thus have shown – without using the results of the § 2 – that $f$ is the sum of its Fourier series.

Since $r' = 1/r$, we have $r'^n = r^{-n} = r^{|n|}$ for $n < 0$. By (2) and (5), then

$$(22.10) \qquad P_f(ru) - P_f(u/r) \;=\; \sum \hat{f}(n)r^{|n|}u^n =$$
$$= \sum r^{|n|}u^n \int v^{-n}f(v)dm(v) =$$
$$= \sum \int r^{|n|}u^n v^{-n}f(v)dm(v)$$

by the definition of the Fourier coefficients of $f$; we have written $v$ for the variable of integration to distinguish it from the free variable $u$. Since the function $f$ is regulated – it is unnecessary to assume it continuous for the moment – and since the series $\sum r^{|n|}u^n v^{-n} = \sum r^{|n|}\left(uv^{-1}\right)^n$ is, for $r < 1$ and $u$ given, normally convergent on the circle $|v| = 1$, we may interchange the signs $\int$ and $\sum$ in (10); putting

$$(22.11) \qquad H_f(z) \;=\; \sum \hat{f}(n)r^{|n|}u^n,$$
$$(22.11') \qquad P(z) \;=\; \sum r^{|n|}u^n \quad \text{for } z = ru,\ r < 1,$$

(the series are extended over $\mathbb{Z}$), we thus have

$$(22.12) \qquad H_f(z) = \int P\left(zv^{-1}\right) f(v) dm(v).$$

On putting $u = \mathbf{e}(s)$ and $v = \mathbf{e}(t)$, whence $uv^{-1} = \mathbf{e}(t - s)$, we again obtain

$$(22.13) \qquad H_f(ru) = \oint P[r\mathbf{e}(s - t)]f(t)dt.$$

These changes of notation are a translation exercise, passing from the point of view of "periodic functions on $\mathbb{R}$" to the point of view of "functions on $\mathbb{T}$".

The formula (12) is a convolution product on $\mathbb{T}$, analogous to the one that allowed us to obtain convergence theorems for Fourier series, by Dirichlet's method in n° 11, or in that of Fejér in n° 12. Likewise here: the functions $v \mapsto P(zv)$ allow us to approximate $f$ with the help of (12) or (13) when $r$ tends to 1.

First, let us calculate the function $P$. For $z = ru$, $r < 1$, we have

$$\begin{aligned}
P(z) &= \sum_{\mathbb{Z}} r^{|n|} u^n = 1 + \sum_{n>0} r^{|n|} u^n + \sum_{n>0} r^{|n|} \overline{u^n} = \\
&= 1 + 2\mathrm{Re}\left(\frac{ru}{1 - ru}\right) = 1 + 2\mathrm{Re}\left(\frac{z}{1 - z}\right) = \mathrm{Re}\left(\frac{1 + z}{1 - z}\right),
\end{aligned}$$

or again

$$(22.14) \qquad P(z) = \frac{1 - |z|^2}{|1 - z|^2} = \frac{1 - r^2}{1 - 2r\cos 2\pi s + r^2} \quad \text{for} \quad z = r\mathbf{e}(s).$$

This formula demonstrates the dubious behaviour of $P(z)$ when $z$ tends to 1, and so that of $P(zv^{-1})$ when $z$ tends to $v$. It shows moreover that, for every *real* function $f$ on $\mathbb{T}$, the function

$$(22.15) \quad H_f(z) = \int P\left(zv^{-1}\right) f(v) dm(v) = \int \mathrm{Re}\left(\frac{v + z}{v - z}\right) f(v) dm(v)$$

is the *real part of a holomorphic function* for $|z| < 1$, a trivial result.

## 23 – Applications to Fourier series

Assuming $f$ regulated let us return to the formula

$$(23.1) \qquad H_f(z) = \int P\left(zv^{-1}\right) f(v) dm(v).$$

We shall show that if $f$ is continuous at a point $u \in \mathbb{T}$, then $H_f(z)$ tends to $f(u)$ when $z$ tends to $u$ remaining in the disc $D : |z| < 1$.

It is more convenient to replace $z$ by $zu$, to make $z$ tend to 1, and to start from the relation

$$(23.2) \quad H_f(zu) = \int P\left(zuv^{-1}\right) f(v)dm(v) = \int P\left(zw^{-1}\right) f(uw)dm(w).$$

To apply the method of Dirac sequences (n° 5) to the functions $u \mapsto P(zu)$ it is enough to show that they are positive, have total integral 1, and that, when $z \to 1$, the function $P\left(zw^{-1}\right)$ converges to 0 uniformly on every arc $J : |w - 1| > \delta$ of $\mathbb{T}$.

The function $P$ is clearly positive, by (22.14). To establish the relation

$$(23.3) \quad \int P\left(zw^{-1}\right) dm(w) = \int P(zw)dm(w) = 1 \quad \text{for } |z| < 1,$$

one observes that, for $|z| < 1$, the function $w \mapsto P\left(zw^{-1}\right)$ is an absolutely convergent Fourier series as (22.11') shows; the integral (3) is then (Chap. V, n° 5) equal to the term $n = 0$ of the series, obviously equal to 1.

It remains to verify uniform convergence on the arc $J$. Now

$$(23.4) \quad P\left(zw^{-1}\right) = \left(1 - |z|^2\right)/\left|1 - zw^{-1}\right|^2 = \left(1 - |z|^2\right)/|z - w|^2.$$

Since the uniform norm of $w \mapsto P\left(zw^{-1}\right)$ on $J$ is the product of $1 - |z|^2$, which tends to 0 and does not depend on $w$, and of the uniform norm of $w \mapsto 1/|z - w|^2$, it is enough to show that the latter is, for $z$ near to 1, majorised by a constant independent of $z$. But this is obvious since the relations $|w - 1| > \delta$ and $|z - 1| < \delta/2$ imply $|z - w| \geq \delta/2$ and thus $1/|z - w|^2 \leq 4/\delta^2$.

In sum, the conditions (D 1), (D 2) and (D 3) imposed on Dirac sequences in n° 5 do indeed hold. The fact that our functions depend on a complex parameter $z$ which tends to 1, rather than on an integer $n$ which increases indefinitely, clearly does not change the proofs. Since we may also make $z$ tend to 1 on the real axis, we finally obtain the following statement:

**Theorem 19.** *Let $f$ be a regulated function on $\mathbb{T}$. Then*

$$(23.5) \qquad f(u) = \lim_{\substack{z \to 1 \\ |z| < 1}} H_f(zu) = \lim_{\substack{r \to 1 \\ r < 1}} \sum_{\mathbb{Z}} r^{|n|} \hat{f}(n)u^n$$

*at every point $u \in \mathbb{T}$ where $f$ is continuous. If $f$ is continuous on an open arc $J$ of $\mathbb{T}$ then the limit (5) is uniform on every compact $K \subset J$ when $z$ or $r$ tends to 1.*

Translation into the language of periodic functions:

$$f(t) = \lim \sum r^{|n|} \hat{f}(n)\mathbf{e}_n(t)$$

at every point $t \in \mathbb{R}$ where $f$ is continuous, and uniform convergence on $\mathbb{R}$ if $f$ is continuous everywhere. We emphasis again that generally one cannot pass to the limit term-by-term in the series; if this were possible, as Poisson believed, every continuous function would be the sum of its Fourier series, which is not the case.

In this way we immediately recover Weierstrass' theorem: every continuous and periodic function $f$ is the uniform limit of trigonometric polynomials of the same period. The preceding theorem, with $J = \mathbb{T}$, shows in fact that $H_f(ru)$ converges uniformly on $\mathbb{T}$ to $f(u)$ as $r < 1$ tends to 1. But the series $H_f(ru) = \sum r^{|n|}\hat{f}(n)u^n$ is, for $r < 1$ given, normally convergent on $\mathbb{T}$, since $|\hat{f}(n)| \le \|f\|$. Its sum is therefore the uniform limit on $\mathbb{T}$ of its partial sums, which are trigonometric polynomials; in other words, one may approximate $f$ uniformly by functions that one may approximate uniformly by trigonometric polynomials. Qed.

We also recover the fact that, for a continuous function $f$ of period 1 such that

$$(23.6) \qquad\qquad \sum \left|\hat{f}(n)\right| < +\infty,$$

one has

$$(23.7) \qquad\qquad f(t) = \sum \hat{f}(n)\mathbf{e}_n(t) = \sum \hat{f}(n)e^{2\pi int}$$

for any $t$. The general term of the series $\sum r^{|n|}\hat{f}(n)u^n$ is actually majorised on the *closed* interval $0 \le r \le 1$ by $|\hat{f}(n)|$. Being a series of continuous functions of $r$ for $u$ given, this series is therefore normally convergent on this interval. Its sum is thus a continuous function of $r$ on $[0,1]$, so tends to its value $\sum \hat{f}(n)u^n$ for $r = 1$ when $r < 1$ tends to 1; but it also tends to $f(t)$ by Theorem 19, qed.

We leave it to the reader to extend Theorem 19 to the general case of a regulated function, i.e. to show that

$$(23.8) \qquad\qquad \lim H_f(zu) = \frac{1}{2}[f(u+) + f(u-)]$$

for any $u$.

One may also deduce the Parseval-Bessel equality from the preceding theorem, at the very least for the simple case where $f$ is continuous. Since $H_f(ru)$ converges uniformly to $f(u)$ it is clear that $|H_f(ru)|^2$ converges uniformly to $|f(u)|^2$, whence, integrating,

$$(23.9) \qquad\qquad \int |f(u)|^2\, dm(u) = \lim \int |H_f(ru)|^2\, dm(u).$$

But as the Fourier series $\sum \hat{f}(n)r^{|n|}u^n$ of $H_f(ru)$ is absolutely convergent for $r < 1$, Chap. V, n° 5 shows, "without knowing anything", that

$$(23.10) \qquad\qquad \int |H_f(ru)|^2\, dm(u) = \sum r^{|2n|}|\hat{f}(n)|^2.$$

When $r < 1$ tends to 1, the partial sums of the right hand side tend to those of the series $\sum |\hat{f}(n)|^2$; now they are majorised by the left hand side of

(10), which tends to the right hand side of (9). We conclude that the partial sums, and so the total sum, of the series $\sum |\hat{f}(n)|^2$ are majorised by the left hand side of (9), whence the Parseval-Bessel *inequality*. But then the right hand side of (10), considered as a series of continuous functions of $r$ on $[0,1]$, is dominated by the convergent series $\sum |\hat{f}(n)|^2$, so converges normally. We may therefore pass to the limit term-by-term (Chap. III, n° 8, Theorem 9 or n° 13, Theorem 17), whence the Parseval-Bessel *equality* using (9).

*Exercise* – For $f$ regulated we have

$$\lim \int |H_f(ru) - f(u)|^2 \, dm(u) = 0$$

(use Parseval-Bessel).

## 24 – Harmonic functions

The method of Fourier series applies to a class of functions closely linked to the holomorphic functions and which, historically, arose from mathematical physics (hydrodynamics, where d'Alembert had already written the Cauchy relations between the partial derivatives of a holomorphic function without having had the idea of going further, Newtonian potential, electrostatics, etc.) and transformed themselves in consequence, as always in such a case, into an occasion for the mathematicians to go very far beyond the needs of the users, and to generalise the situation. These functions are also linked to the $H_f$ that we have just studied.

Let $f(z) = P(x,y) + iQ(x,y)$ be a holomorphic function on an open set $U$. The Cauchy differential equation $f'_x = -if'_y$ can then be written, on separating the real and imaginary parts, in the form

$$(24.1) \qquad P'_x = Q'_y, \qquad P'_y = -Q'_x,$$

which, since $f' = f'_x = P'_x + iQ'_x$, shows in passing that

$$(24.2) \qquad f'(z) = P'_x - iP'_y = Q'_y + iQ'_x;$$

in other words, the knowledge of $P = \operatorname{Re}(f)$ or of $Q = \operatorname{Im}(f)$ determines $f'$ and so determines $f$ up to an additive constant. The function $f$, being analytic as a function of $z$ and *a fortiori* $C^\infty$ as a function of the real variables $x$ and $y$, so likewise are $P$ and $Q$, which allows us to differentiate the relations (1). A trivial calculation then shows that

$$(24.3) \qquad \Delta P = P''_{xx} + P''_{yy} = 0, \qquad \Delta Q = 0,$$

where

$$(24.4) \qquad \Delta = D_1^2 + D_2^2 = \partial^2/\partial x^2 + \partial^2/\partial y^2$$

is the *Laplace operator* which generalises in the obvious way to functions of
any number of variables. A function[43] $H(x,y)$ of class $C^2$ in an set open $U$
of $\mathbb{C}$ is said to be *harmonic* in $U$ if it satisfies the relation $\Delta H = 0$. One
may ask whether such a function, assuming it real valued, as we shall do in
all the rest of this §, is the real part of a holomorphic function. Though not
strictly correct, this conjecture is to a large measure true (but is of no help in
studying harmonic functions of more than two variables, which require very
different methods).

If, inspired by (2), we associate with $H$ the function

$$(24.5) \qquad g = H'_x - iH'_y = D_1 H - iD_2 H,$$

we see that *the Laplace equation means that $g$ is holomorphic.* If we could
find a function $f = P + iQ$ holomorphic in $C$ and such that $f' = g$, we would
have $H'_x = P'_x$, $H'_y = P'_y$, and so $H = P$ up to an additive constant. $H$ would
then be the real part of a holomorphic function in $U$ as hoped.

First assume, the simplest case, that $H$ is harmonic on a disc $D : |z| < R$.
The function $g$ is then a power series, so has a primitive $f(z) = \sum a_n z^n$ in
$D$ (Chap. II, n° 19), of which $H$ is, up to an additive constant, the real part,
as we have just seen. Putting $z = ru$ with $|u| = 1$, we then have

$$\begin{aligned}
2H(ru) &= \sum_{n\geq 0} a_n r^n u^n + \sum_{n\geq 0} \overline{a_n r^n u^n} = \\
&= \sum_{n\geq 0} a_n r^n u^n + \sum_{n\geq 0} \overline{a_n} r^n u^{-n} = \\
&= \sum_{\mathbb{Z}} c_n r^{|n|} u^n
\end{aligned}$$

with $c_n = a_n$ if $n \geq 0$, $c_n = \overline{a_{-n}} = \overline{c_{-n}}$ if $n < 0$ and $c_0 = 2\mathrm{Re}(a_0)$. Since
$r^{|n|}u^n$ is equal to $z^n$ for $n > 0$ and to $\bar{z}^{|n|}$ for $n < 0$, we finally obtain the
following result:

**Theorem 20.** *Every harmonic function $H$ on a disc $|z| < R$ has a series
expansion of the form*

$$(24.6) \quad H(z) = \sum c_n r^{|n|} u^n = c_0 + \sum_{n>0} [c_n(x+iy)^n + \overline{c_n}(x-iy)^n]$$

*with*

---

[43] The use of the letter $U$ is traditional in physics. The mathematicians more often
use $u$, which, in our case, might provoke confusion with the variable of integration
on the unit disc $\mathbb{T}$, while use of the letter $U$ would provoke confusion with open
sets, which we generally denote $U$. The use of the letter $H$ does not present these
risks, and, after all, is not absurd when treating harmonic functions ...

(24.7)
$$c_n r^{|n|} = \int H(ru)u^{-n}dm(u)$$

*for every $r < R$ and every $n \in \mathbb{Z}$.*

There is no greater problem of convergence for the series (6) than for the power series of $f$: they converge normally in every disc of radius $r < R$. The general term of the second series (6) is a homogeneous polynomial of degree $n$ in $x, y$, and clearly harmonic since it is the real part of $c_n z^n$.

**Corollary ("Theorem of the Mean").** *Let $H$ be a harmonic function in an open subset $U$ of $\mathbb{C}$. For every $a \in U$ and every $r > 0$ such that $U$ contains the closed disc $|z - a| \le r$, one has*[44]

(24.8)
$$H(a) = \int_{\mathbb{T}} H(a + ru)dm(u).$$

The argument is less easy – and the result less correct ... – in the case where $H$ is given in an annulus $C$. Consider the Laurent series $\sum b_n z^n$ of the function (5). After subtracting the term in $1/z$ it has, as we have seen at the end of n° 16, a pseudo primitive

(24.9)
$$f(z) = \sum a_n z^n$$

such that

(24.10)
$$g(z) = f'(z) + b_{-1}/z.$$

First we shall show that the residue $b_{-1}$ is *real*[45].

Now, by (5),

$$b_{-1} = \oint g(re(t))re(t)dt =$$
$$= \oint r\left[H'_x(re(t)) - iH'_y(re(t))\right](\cos 2\pi t + i \sin 2\pi t)dt,$$

whence

$$\mathrm{Im}\,(b_{-1}) = \oint \left[H'_x(re(t))r \sin 2\pi t - H'_y(re(t))r \cos 2\pi t\right]dt.$$

---

[44] One can show that the harmonic functions in an open subset $U$ of $\mathbb{C}$ are *characterised* by the fact that their value at the centre of any disc $D \subset U$ is equal to their mean value over the boundary of $D$. It is not even necessary to assume differentiability.

[45] The function $g = U'_x - iU'_y$ is not an arbitrary holomorphic function; it must be possible to put its real and imaginary parts $P$ and $Q$ in the form $P = U'_x$, $Q = U'_y$, which is not always the case.

Since $re(t)$ has coordinates $r\cos(2\pi t)$ and $r\sin(2\pi t)$, the chain rule shows that

$$\frac{d}{dt}H(re(t)) = -2\pi\left[H'_x(re(t))r\sin 2\pi t - H'_y(re(t))r\cos 2\pi t\right];$$

consequently, $\mathrm{Im}\,(b_{-1})$ is, up to a factor $-2\pi$, the variation between 0 and 1 of the function $t \mapsto H(re(t))$, zero by periodicity. The residue $b_{-1}$ is therefore real.

Now let us put $f = P + iQ$ with $P$ and $Q$ real. It follows that

$$H'_x - iH'_y = g(z) = P'_x - iP'_y + b_{-1}/(x + iy)$$

by (10), whence

$$\begin{array}{rcl} H'_x &=& P'_x + b_{-1}x/(x^2 + y^2), \\ H'_y &=& P'_y + b_{-1}y/(x^2 + y^2) \end{array}$$

since $b_{-1}$ is real. The function $R = H - P$ thus satisfies the relations

$$\begin{array}{rcl} R'_x &=& b_{-1}x/(x^2 + y^2), \\ R'_y &=& b_{-1}y/(x^2 + y^2). \end{array}$$

Now the function

$$L(x, y) = \log |z| = \log r = \frac{1}{2}\log\left(x^2 + y^2\right),$$

not to be confused with the pretend $\mathcal{L}$og of the complex number $z$, has partial derivatives

$$L'_x = x/(x^2 + y^2), \qquad L'_y = y/(x^2 + y^2).$$

The function $R - b_{-1}L$ thus has partial derivatives zero, so is constant, whence it follows that

(24.11) $$H(x, y) = P(x, y) + b_{-1}\log r + const.$$

Since $f(z) = P(x, y) + iQ(x, y) = \sum a_n z^n$ we find

$$\begin{array}{rcl} H(z) &=& b\log r + c + \frac{1}{2}\sum(a_n z^n + \overline{a_n z^n}) = \\ &=& b\log r + c + \frac{1}{2}\sum[a_n(x + iy)^n + \overline{a_n}(x - iy)^n] \end{array}$$

with real constants $b$ and $c$, summing over all nonzero $n \in \mathbb{Z}$. The expansion

(24.12) $$H(ru) = b\log r + c + \frac{1}{2}\sum_{n\neq 0}\left(a_n r^n + \overline{a_{-n}}r^{-n}\right)u^n$$

is deduced from this, and yields the general form of the Fourier coefficients of the function $H(re(t))$. One may put all this in the form

$$(24.13) \quad H(re(t)) = b.\log r + c + \sum_{n \geq 1} [b_n(r)\cos 2\pi nt + c_n(r)\sin 2\pi nt]$$

where the coefficients $b_n(r)$ and $c_n(r)$ are linear combinations with real coefficients of $r^n$ and $r^{-n}$.

*Exercise.* By using the equation $\Delta H = 0$, show directly that the Fourier series of $u \mapsto H(ru)$ has the form (13). (Argue as in n° 14).

The fact that $\log r$ and the negative powers of $r$ disappear when $H$ is harmonic on a disc is due to the continuity of $H$ on a neighbourhood of the origin: the Fourier coefficients of $H(re(t))$ must remain bounded when $r$ tends to 0.

One of the consequences of these calculations is that, in an annulus, a harmonic function is not always the real part of a holomorphic function. This is the case only if $b = 0$ in the expansion (13); direct calculation of the Fourier coefficients of $H(ru)$ shows that

$$(24.14) \qquad b\log r + c = \int H(ru)dm(u) = \oint H\left(re^{2\pi it}\right)dt,$$

the mean value of $H$ over the circle $|z| = r$. One may explain the appearance of the function $\log r$ by noting that it is the real part of the "function" $\mathcal{L}og\, z = \log r + i\arg z$, which would be holomorphic for $z \neq 0$ if one could forget the ambiguity inherent in the definition of the argument; this ambiguity being pure imaginary, the real part $\log r$ is, itself, a function in the strict sense – and it is harmonic. You can check this by calculating its Laplacian directly.

## 25 – Limits of harmonic functions

We have seen, in (24.8), that if a function $H$ is harmonic on an open disc of radius $R$, its value at the centre of the latter is equal to its mean value over every concentric circle of radius $r < R$. We deduce that *the maximum theorem* (Theorem 11 of n° 15) *and its corollary are valid for harmonic functions*; the proofs are precisely the same. In particular, if a function is continuous on the closure $K$ of a bounded domain $G$ and is harmonic in $G$, and is zero on the boundary $F$ of $G$, then it is identically zero since $\|H\|_G = \|H\|_F$.

If a function $H$ is harmonic on a disc $|z| < R$ of radius $R > 1$ and if one puts $f(u) = H(u)$ on $\mathbb{T}$, the series expansion

$$H(z) = \sum c_n r^{|n|} u^n \quad \text{with} \quad c_n r^{|n|} = \int H(ru)u^{-n}dm(u)$$

of Theorem 20, valid for $r < R$, holds in particular for $r = 1$ and shows that $c_n = \hat{f}(n)$. Hence

$$H(z) = H_f(z) \quad \text{for } |z| < 1.$$

In the case of an arbitrary radius $R$ one can, for $r < R$, apply this result to the function $z \mapsto H(rz)$, harmonic in the disc of radius $R/r > 1$. Whence

$$H(rz) = \int \frac{1 - |z|^2}{|z - u|^2} H(ru)dm(u) \quad (|z| < 1),$$

or, on replacing $z$ by $z/r$,

$$(25.1) \qquad H(z) = \int_{\mathbb{T}} \frac{r^2 - |z|^2}{|z - ru|^2} H(ru)dm(u) \quad \text{for } |z| < r.$$

This is the analogue for harmonic functions of Cauchy's integral formula of n° 1; the existence of such a formula is scarcely surprising, since, on a disc, a harmonic function is the real part of a holomorphic function.

Weierstrass' theorem on uniformly convergent sequences of holomorphic functions applies also to harmonic functions, but needs several preliminaries.

First, the formula (24.6), i.e.

$$(25.2) \qquad H(x, y) = c_0 + \sum_{n > 0} \left[ c_n (x + iy)^n + \overline{c_n}(x - iy)^n \right],$$

shows that *a harmonic function is of class $C^\infty$*; again not very surprising, since, locally, it is the real part of an analytic function. Further, if one differentiates the general term of the series (2) with respect to $x$ or $y$ one multiplies the coefficients of order $n$ by $n$ or $\pm in$; up to a factor of modulus 1 this is equivalent to replacing the two power series in $x + iy = z$ and $x - iy = \bar{z}$ that appear in (24.6) by their derived series; the resulting series, and more generally, those obtained by differentiating term-by-term *ad libitum* with respect to $x$ and $y$, converge under exactly the same conditions as (2). Theorem 20 of Chap. III, n° 17 would then show, if needed, that $H$ is indefinitely differentiable and that its partial derivatives of arbitrary order are obtained by differentiating the series (2) term-by-term with respect to $x$ and $y$. (The fact that we are dealing with functions of two variables is not important: the variable with respect to which one is not differentiating plays the rôle of a constant).

We deduce, after a small calculation, that the partial derivative

$$D_1^p D_2^q H = H^{(p,q)}$$

is given by

$$H^{(p,q)}(x, y) =$$
$$= \sum n(n-1) \dots (n - p - q + 1) \left[ i^q c_n (x + iy)^{n-p-q} + (-i)^q \overline{c_n}(x - iy)^{n-p-q} \right]$$

where one sums over $n \geq p + q$. In particular,

$$H^{(p,q)}(0,0) = (p+q)! \, [i^q c_{p+q} + i^q \overline{c_{p+q}}] = 2(p+q)! \mathrm{Re} \, (i^q c_{p+q}) \, .$$

Since $c_n r^n = \int H(ru) u^{-n} dm(u)$ for $n \geq 0$ we have

$$(25.3) \quad H^{(p,q)}(0,0) = (p+q)! r^{-p-q} \int H(ru) \left[ i^q u^{-p-q} + (-i)^q u^{p+q} \right] dm(u)$$

and consequently

$$(25.4) \qquad \left| H^{(p,q)}(0,0) \right| \leq 2 \frac{(p+q)!}{r^{p+q}} \sup_{|u|=1} |H(ru)|.$$

From this we have the analogue of Weierstrass' convergence theorem:

**Theorem 21.** *Let $G$ an open set in $\mathbb{C}$ and $(H_n)$ a sequence of harmonic functions in $G$ which converges uniformly on every compact $K \subset G$ to a limit function $H$. Then $H$ is harmonic, and, for any $p$ and $q$, the partial derivatives $H_n^{(p,q)}$ converge uniformly on every compact subset of $G$ to the partial derivative $H^{(p,q)}$ of $H$.*

We know thanks to Borel-Lebesgue (Chap. V, n° 6) that uniform convergence on every compact subset is a property of local character: to verify it for every compact $K \subset G$ it is enough to show that, for every $a \in G$, it holds on a closed disc of centre $a$.

So choose an $R > 0$ such that the disc $D : |z - a| \leq R$ is contained in $G$ and put $r = R/2$. For every $z$ such that $|z - a| \leq r$ the closed disc of centre $z$ and of radius $r$ is contained in $D$. By (4) we have, for every harmonic function $U$ in $G$,

$$U^{(p,q)}(z) \leq 2(p+q)! r^{-p-q} \sup_{|u|=1} |U(z+ru)|;$$

but for $|z - a| \leq r$, all the points $z + ru$ are in the large disc $D$, whence trivially

$$\sup_{|u|=1} |U(z+ru)| \leq \|U\|_D \, ,$$

and consequently

$$U^{(p,q)}(z) \leq 2(p+q)! r^{-p-q} \cdot \|U\|_D \quad \text{for } |z-a| \leq r.$$

Now we apply the general result to the functions $U = H_m - H_n$. Since the $H_n$ converge uniformly on every compact subset of $G$, and in particular on $D$, we have $\|H_m - H_n\|_D \leq \varepsilon$ for $m$ and $n$ large. The preceding inequality then shows that, for $m$ and $n$ large, we have

$$\left| H_m^{(p,q)}(z) - H_n^{(p,q)}(z) \right| \leq 2(p+q)! r^{-p-q} \varepsilon$$

at all the points of the disc $|z - a| \leq r$.

By Cauchy's criterion, the partial derivatives $H_n^{(p,q)}$ converge uniformly in this disc and more generally, since $a \in G$ is arbitrary, on every compact $K \subset G$. It follows that the function $H$ is $C^\infty$ like the $H_n$ and that the $H_n^{(p,q)}$ converge to $H^{(p,q)}$ for any $p$ and $q$. This allows us to pass to the limit in the Laplace equation $\Delta H_n = 0$, so that $H$ is again harmonic, qed.

If the domain $G$ is bounded and if the $H_n$ are continuous on the closure $K$ of $G$, the maximum theorem shows that, if the $H_n$ converge uniformly on the boundary $F = K - G$ of $G$, then they converge uniformly in $G$:

$$\| H_m - H_n \|_G = \| H_m - H_n \|_F .$$

The preceding theorem applies to this case (but do not believe that the partial derivatives converge uniformly on all of $G$: they converge uniformly only on every compact subset of $G$).

## 26 – The Dirichlet problem for a disc

As we saw in preceding n°, if a function $H$ is defined and harmonic on a disc of radius $R > 1$ and if one puts $f(u) = H(u)$ for $u \in \mathbb{T}$, then $H(z) = H_f(z)$ for $|z| < 1$. In this case Theorem 21 loses its interest: the series (24.6) converges normally in $|z| \leq r$ for every $r < R$, so for $r > 1$, so that the passage to the limit when $r < 1$ tends to 1 results from the continuity of $H$ in the closed disc $|z| \leq 1$, and even beyond.

The situation becomes more interesting if, given an arbitrary real regulated function $f$ on $\mathbb{T}$, one associates with it the function

$$(26.1) \qquad H_f(z) = \sum \hat{f}(n) r^{|n|} e_n(t) = \sum \hat{f}(n) r^{|n|} u^n =$$
$$= \int P\left(zu^{-1}\right) f(u) dm(u),$$

defined *a priori* for $|z| < 1$. Since $f$ is real we have $\hat{f}(-n) = \overline{\hat{f}(n)}$ and the function (1) is, up to the factor $\frac{1}{2}$, the real part of the power series $\sum_{n \geq 0} \hat{f}(n) z^n$ and so is harmonic; see also (22.15).

If $f$ is continuous, we know (Theorem 19) that $H_f(z)$ tends to $f(u)$ when $z$ converges (not necessarily along a radius) to a $u \in \mathbb{T}$ while remaining in the disc $|z| < 1$. This means that the function equal to $H_f(z)$ for $|z| < 1$, and to $f$ on $\mathbb{T}$, is continuous in the closed disc $|z| \leq 1$. This was proved using the fact that the functions $u \mapsto P(zu)$ have the properties of a Dirac sequence when $|z| < 1$ tends to 1. This result furnished us a second proof of Weierstrass' approximation theorem.

Granted this, one could give a simpler proof of Theorem 19. Since the function $u \mapsto P\left(zu^{-1}\right)$ is positive and has integral 1 on $\mathbb{T}$, the formula (1) shows that $|H_f(z)| \leq \|f\|$, whence the relation

(26.2)                                $\|H_f\|_D \leq \|f\|$

between the uniform norms of $f$ in $\mathbb{T}$ and of $H_f$ on the open disc $D : |z| < 1$; this is just the maximum principle for the harmonic function $H_f$, modulo the fact that we do not yet know (or we are pretending not to know yet) that $H_f$ is the restriction to the open disc of a continuous function on the closed disc. Now $f$ is the uniform limit on $\mathbb{T}$ of a sequence of trigonometric polynomials $f_n$, which one may assume real if $f$ is. For every trigonometric polynomial $g$ the series $H_g(z) = \sum \hat{g}(n) r^{|n|} u^n$ reduces to a finite sum, so is a continuous function of $z = ru$ on all $\mathbb{C}$. Denote by $H_n$ the harmonic function corresponding to $g = f_n$; by (2), we have $\|H_p - H_q\|_D \leq \|f_p - f_q\|$ for any $p$ and $q$; but since $H_p - H_q$ is defined and continuous on the *closed* disc $|z| \leq 1$ (and in fact on $\mathbb{C}$), we have

$$\|H_p - H_q\|_D = \|f_p - f_q\|$$

where $D$ is the closed disc $|z| \leq 1$. Consequently (Cauchy's criterion), the $H_n$ converge uniformly on $D$ and their limit is continuous there. Now they converge to $H_f$ in the *open* disc $|z| < 1$ since $\|H_f - H_n\|_D \leq \|f - f_n\|$ by (2), and to $f$ on $\mathbb{T}$. Whence the result:

**Theorem 22.** *Let $f$ be a continuous function on $\mathbb{T}$. Then the function equal to*

(26.3)                    $H_f(z) = \displaystyle\int_{\mathbb{T}} \frac{1 - |z|^2}{|z - u|^2} f(u) dm(u)$

*for $|z| < 1$ and to $f$ on $\mathbb{T}$ is continuous on the closed disc $|z| \leq 1$ and harmonic in the open disc $|z| < 1$. This is the only function possessing these properties.*

Uniqueness follows from the maximum theorem; see the beginning of the preceding n°.

We have resolved a very particular case of the *Dirichlet problem* which can be stated roughly as follows: given a bounded domain $G$ in $\mathbb{C}$ whose boundary is a not too savage curve, and, on the latter, a continuous function $f$, to construct a continuous function on the closure $\bar{G}$ of $G$, harmonic in $G$ and equal to $f$ on the boundary of $G$. Generalised to Euclidean spaces of arbitrary dimension, and to other differential operators than $\Delta$, this is one of the fundamental problems of the theory of partial differential equations. Let us make clear that, even in the classical case of the Laplacian in an open subset of $\mathbb{C}$, the case of the disc does not reflect the level of difficulty of the problem.

We remark moreover that a harmonic function in the open disc $|z| < 1$ in general has no reason to be continuable to a continuous function on the closed disc $|z| \leq 1$. The simplest counterexample is provided by the function $P(z) = \left(1 - |z|^2\right) / |z - 1|^2$ itself; it is harmonic in $\mathbb{C} - \{1\}$ but does not tend to any limit when $z$ tends to 1. A much more complicated case is obtained by starting from an arbitrary measure or even a distribution $\mu$ on $\mathbb{T}$ and considering the function

$$(26.4) \qquad H_\mu(z) = \int_{\mathbb{T}} \frac{1 - |z|^2}{|z - u|^2} \, d\mu(u) = \sum \hat{\mu}(n) r^{|n|} u^n;$$

its behaviour on a neighbourhood of the unit circle can be as strange as that of a holomorphic function. Again we do not obtain the most general harmonic functions $\sum c_n r^{|n|} u^n$ in this way, for one cannot have $c_n = \hat{\mu}(n)$ for a distribution $\mu$ unless the coefficients $c_n$ are of slow increase (n° 10, Theorem 6), which need not happen, even if the series converges for $r < 1$. Counterexample and exercise: $c_n = \exp\left(|n|^{1/2}\right)$ for every $n$.

# § 6. From Fourier series to integrals

In this §, the $\int$ sign denotes an integral extended over $\mathbb{R}$, while the sign $\oint$ denotes an integral extended over an interval of length 1. Recall the notation

$$\mathbf{e}(x) = e^{2\pi i x}, \qquad \mathbf{e}_y(x) = \mathbf{e}(xy)$$

for $y$ real.

## 27 – The Poisson summation formula

Recall also that given a regulated and absolutely integrable function $f$ on $\mathbb{R}$ one defines the Fourier transform of $f$ by the formula

$$(27.1) \qquad \hat{f}(y) = \int f(x) e^{-2\pi i x y} dx = \int \overline{\mathbf{e}(xy)} f(x) dx.$$

The integral converges since the exponential has modulus 1.

**Theorem 23.** *The Fourier transform of an absolutely integrable function is continuous and tends to 0 at infinity.*

Assume that $y$ remains in a compact subset $H$ of $\mathbb{R}$; the function $\mathbf{e}(xy)$ is continuous on $\mathbb{R} \times H$ and there exists a function $p(x)$ [namely 1] such that $|\mathbf{e}(xy)| \leq p(x)$ for every $y \in H$ and $\int p(x)|f(x)|dx < +\infty$. It then remains to apply Theorem 22 of Chap. V, n° 23, substituting $\mathbf{e}(xy)$ for $f(x, y)$ and $f$ for $\mu$. One could clearly argue directly: integrating over $[-N, N]$ instead of $\mathbb{R}$, one commits *whatever $y$ might be*, an error $\leq r$ if $N$ is large enough; so it suffices – uniform limits of continuous functions – to prove the continuity of the integral over $K = [-N, N]$. But since $(x, y) \mapsto \mathbf{e}(xy)$ is uniformly continuous on every compact subset of $\mathbb{R}^2$, the function $x \mapsto \mathbf{e}(xy)$ converges to $\mathbf{e}(xb)$ uniformly on $K$ when $y$ tends to a limit $b$; one may therefore pass to the limit in the integral over $K$.

It is clear that $f$ is bounded, with

$$(27.2) \qquad \|\hat{f}\| = \sup|\hat{f}(y)| \leq \int |f(x)|dx = \|f\|_1.$$

To show that $\hat{f}$ tends to 0 at infinity one proceeds from the simplest to the most general case.

(i) If $f$ is the characteristic function of a compact interval $[a, b]$,

$$\hat{f}(y) = \int_a^b e^{-2\pi i x y} dx = \frac{e^{-2\pi i x y}}{-2\pi i y} \Bigg|_a^b$$

for $y \neq 0$, whence the result in this case, hence also if $f$ is a step function vanishing outside a compact interval.

(ii) If $f$ is zero outside a compact interval $K$ and integrable on $K$, then for every $r > 0$ there is a step function $g$ zero outside $K$ such that $\int |f(x) - g(x)| dx \leq r$ as shown by the very definition of an integral (Chap. V, n° 2). Then $\left| \hat{f}(y) - \hat{g}(y) \right| \leq r$ for every $y$ by (2); since $|\hat{g}(y)| \leq r$ for $|y|$ large, we have $|\hat{f}(y)| \leq 2r$ for $|y|$ large, whence again the result.

(iii) In the general case, for every $r > 0$ there exists a compact interval $K$ such that the contribution of $\mathbb{R} - K$ to the total integral of $|f(x)|$ is $\leq r$; integrating over $K$ in (1), one commits an error $\leq r$ for every $y$, and since the integral over $K$ tends to 0, we again find $|\hat{f}(y)| \leq 2r$ for $|y|$ large, qed.

As we have already seen à *propos* the function cot or the elliptic functions, the "Eisenstein method", as Weil and Remmert call it, for constructing periodic functions on $\mathbb{R}$ consists of starting from non periodic functions $f(x)$ and considering the series

$$(27.3) \qquad F(x) = \sum f(x + n),$$

summing over $\mathbb{Z}$. If the series converges unconditionally, i.e. absolutely, the result is incontestably periodic since changing $x$ to $x + 1$ is equivalent to the permutation $n \mapsto n + 1$ in $\mathbb{Z}$. One may then try to expand the result as a Fourier series.

If one calculates formally, taking account of the periodicity of the exponentials,

$$
\begin{aligned}
(27.4) \quad \hat{F}(p) &= \oint \overline{\mathbf{e}_p(x)} dx \sum f(x + n) = \oint dx \sum f(x + n) \overline{\mathbf{e}_p(x + n)} \\
&= \sum \int_0^1 f(x + n) \overline{\mathbf{e}_p(x + n)} dx = \\
&= \sum \int_n^{n+1} f(x) \overline{\mathbf{e}_p(x)} dx = \hat{f}(p).
\end{aligned}
$$

where $\hat{f}$ is the Fourier transform of $f$. And since "every" periodic function is the sum of its Fourier series, we finally find the *Poisson summation formula* (though he never wrote it in this form)

$$(27.5) \qquad \sum f(x + n) = \sum \hat{f}(n) \mathbf{e}_n(x),$$

in particular, for $x = 0$,

$$(27.6) \qquad \sum f(n) = \sum \hat{f}(n).$$

All this is formal calculation. The first problem is to justify the permutation of the signs $\int$ and $\sum$ performed to obtain (4). It is simplest to assume first that $f$ is continuous and that the series $\sum f(x + n)$ converges normally on $[0, 1]$, in which case it is clear that it converges normally on every compact set, by periodicity; the presence of the factors $\mathbf{e}_p(x)$ does not change anything, since they are of modulus 1. If these conditions are satisfied then $F$ is continuous and the term-by-term integration in (4) is justified (Chap. V, n° 4, Theorem 4). Subject to these hypotheses, the function $f$ is moreover absolutely integrable on $\mathbb{R}$, for the integral of $|f(x)|$ over $(-n, n)$, the $n$th partial sum of the series $\sum \int |f(x+p)|dx$, where one integrates over $(0,1)$, is, for every $n$, less than the total sum of this series; the convergence of $\int |f(x)|dx$ follows from this (Chap. V, n° 22, Theorem 18). The formal calculation is therefore justified. It remains to justify the relation (5), which says that $F$ is everywhere equal to the sum of its Fourier series; to do this it is enough to assume that the latter is absolutely convergent, i.e. that $\sum |\hat{f}(p)| < +\infty$; convergence for any $x$ would suffice, by Fejér, but it is better, in this context, just to use a simple result:

**Theorem 24.** *Let $f$ be a function defined and continuous on $\mathbb{R}$ such that*

*(i) the series $\sum f(x + n)$ converges normally on every compact set,*
*(ii) $\sum |\hat{f}(n)| < +\infty$.*

*Then $f$ is absolutely integrable on $\mathbb{R}$ and*

$$(27.7) \qquad \sum f(x + n) = \sum \hat{f}(n)\mathbf{e}_n(x) \qquad \text{for every } x \in \mathbb{R}.$$

In practice, the convergence of the series $\sum f(x + n)$ is almost always obtained by estimating $f(x)$ for $|x|$ large. Assume for example

$$(27.8) \qquad f(x) = O\left(|x|^{-s}\right) \quad \text{at infinity, with } s > 1.$$

The continuous function $|x|^s f(x)$ being bounded for $|x|$ large, i.e. outside a compact set, is in fact bounded on $\mathbb{R}$, being bounded on every compact set; so likewise is $f$, so also $(1 + |x|^s) f(x)$, from which we have the estimate

$$|f(x)| \leq M/\left(1 + |x|^s\right) \qquad \text{for every } x,$$

with a constant $M > 0$. This shows that $f$ is absolutely integrable on $\mathbb{R}$ (Chap. V, n° 22). If $x$ remains in $[0,1]$, then $|x + n|$ varies between $|n|$ and $|n+1|$, so is $> |n|$ or $|n+1|$ according to the sign of $n$. The series $1/(1 + |n|^s)$ being convergent since $s > 1$, normal convergence of $\sum f(x + n)$ follows. As to the convergence of $\sum |\hat{f}(n)|$, this is assured, as we shall see later, if $f$ is sufficiently differentiable, as in the case of periodic functions.

The real problem, in the practical use of the Poisson formula, or more generally of the Fourier transform, is that we have to calculate the Fourier transforms explicitly. Sometimes this is easy, as we shall see, but the crude method – calculating a primitive of the integrand – is not any use in general, because the primitive does not reduce to "elementary" functions. We have therefore to find methods for calculating the integral over $\mathbb{R}$ (and not over an arbitrary interval) without knowing the primitive; it was the greatest success of Cauchy's residue calculus that it allowed this kind of calculation in cases unknown up till then. Nothing better has been found since then; we have many formulae for Fourier transforms in terms of special functions, Euler's $\Gamma$ function for example, but they are almost always obtained by Cauchy's *method*.

*Example 1.* Choose the function

$$f(t) = 1/(z+t)^s$$

where $z$ is a non real complex parameter and $s$ an integer $\geq 2$, with for example $\text{Im}(z) > 0$. The preceding considerations show that the series $\sum f(t+n)$ is normally convergent on every compact set, but it remains to calculate the Fourier transform

$$\hat{f}(n) = \int \exp(-2\pi int)\,(z+t)^{-s}\,dt$$

for $n$ real, not necessarily an integer. To seek a primitive, for example by integrating by parts, would lead, more complication, to integrals of the type $e^x x^n dx$ of Chap. V, n° 15, Example 2; they can be calculated immediately by hand for $n$ an integer $> 0$ but, for $n < 0$, and especially for $n = -2$, they resist every attempt at explicit calculation (and not only because we are working here at too elementary a level); Euler's gamma function would not have survived if one had been able to calculate a primitive of $e^{-x}x^s$. But, with his residue method, Cauchy succeeded in calculating in a general way the Fourier transform of a rational function $p/q$ having no real pole and decreasing sufficiently fast at infinity, i.e. such that $d(q) > d(p) + 1$. We may prove for example that, for $s$ integer $\geq 2$ (convergence!), we have

$$\int \exp(-2\pi iut)\,(z+t)^{-s}\,dt = \begin{cases} (-2\pi i)^s u^{s-1}\exp(2\pi iuz)/(s-1)! & \text{if } u > 0, \\[2mm] 0 & \text{if } u \leq 0 \end{cases}$$

(27.9)

on condition that $\text{Im}(z) > 0$. The summation formula $\sum f(n) = \sum \hat{f}(n)$ can be written, in this case, as

(27.10)    $$\sum_{\mathbb{Z}} \frac{1}{(z+n)^s} = \frac{(-2\pi i)^s}{(s-1)!} \sum_{n>0} n^{s-1} e^{2\pi inz} \quad \text{for } \text{Im}(z) > 0.$$

For $s = 2$, this is

(27.11)    $$\sum_{\mathbb{Z}} 1/(z-n)^2 = -4\pi^2 \sum_{\mathbb{N}} n e^{2\pi inz};$$

now we know (see (8.14) for example) that, for $z$ not an integer, the left hand side is equal to $\pi^2/\sin^2 \pi z$; for $\text{Im}(z) > 0$ we have $\left| e^{iz} \right| < 1$ and so

$$1/\sin^2 z = -4/\left(e^{iz} - e^{-iz}\right)^2 = -4e^{2iz}/\left(1 - e^{2iz}\right)^2 =$$
$$= -4e^{-2iz}\left(1 + 2e^{2iz} + 3e^{4iz} + \ldots\right)$$

using the power series of $1/(1 - x)^2$. In this way we see (11) as a consequence of the expansion of $1/\sin^2 z$ as a series of rational fractions, and vice-versa. Starting from (11) one might obtain the general case (10) by differentiating with respect to $z$: the right hand side of (11) is a series of holomorphic functions, so that, to legitimate the differentiations, it suffices, thanks to Weierstrass, to show that the right hand side of (11) converges normally on every compact subset of the half plane $\mathrm{Im}(z) > 0$; but on such a compact we have $\left|e^{2\pi inz}\right| = e^{-2\pi ny}$ where $y = \mathrm{Im}(z)$ remains larger than a strictly positive number $m$, for the distance from a *compact* set to the boundary of an *open* set containing it is always $> 0$; since $e^{-2\pi m} < 1$ normal convergence is then clear.

In fact, the residue calculus (Chap. VIII, n° 10, (ii)) allows one to extend the formula (9) and so (10) – replacing $(s - 1)!$ by $\Gamma(s)$ – to the case of a complex exponent $s$ satisfying only the condition $\mathrm{Re}(s) > 1$, so as to make the series (10) converge.

*Example 2.* Now choose

$$(27.12) \qquad\qquad f(x) = e^{-t|x|}$$

where $t$ is a parameter $> 0$, so that $f$ is integrable on $\mathbb{R}$. Then

$$\hat{f}(y) = \int \exp(-t|x| - 2\pi ixy)dx;$$

on each of the intervals $x < 0$ and $x > 0$ one has to integrate a function of the form $e^{cx}$, with $c$ complex, and since such a function has primitive $e^{cx}/c$ the calculation is immediate and yields the result:

$$(27.13) \qquad\qquad \hat{f}(y) = 2t/\left(t^2 + 4\pi^2 y^2\right).$$

Simple estimates show that Theorem 24 applies here, whence

$$\sum e^{-|n|t} = 2t \sum 1/\left(t^2 + 4\pi^2 n^2\right),$$

a formula strongly resembling the expansion of $\coth t$ as a series of rational fractions . . .

*Exercise.* Extend these calculations to the case where $t$ is complex, with $\mathrm{Re}(t) > 0$ (use Theorem 24 bis of Chap. V, n° 25).

## 28 – Jacobi's theta function

Another more spectacular application of Theorem 24 depends on the calculation of the Fourier transform of the function $f(x) = \exp(-\pi x^2)$. We have already met this in Chap. V, n° 25, Example 2, and we showed there, by differentiating under the $\int$ sign, that

$$\hat{f}(y) = cf(y) \quad \text{where } c = \hat{f}(0) = \int \exp(-\pi x^2)dx.$$

If one shows that Theorem 24 and in particular (27.6) applies to $f$, then one has $c\sum f(n) = \sum f(n)$ and so $c = 1$ since the $f(n)$ are all $> 0$.

Now the function $\exp(-\pi x^2)$ decreases at infinity faster than every negative power of $x$, so satisfies the condition (27.8). Since, for the same reason, the series $\sum \hat{f}(n)$ converges absolutely, Theorem 24 applies. Moreover, it yields the identity

$$(28.1) \qquad \sum \exp\left[-\pi(x+n)^2\right] = \sum \exp(-\pi n^2)\mathbf{e}_n(x),$$

valid for every $x \in \mathbb{R}$.

One may generalise, replacing the function $f(t) = \exp(-\pi t^2)$ by

$$(28.2) \qquad\qquad f(t, z) = e^{\pi i z t^2}$$

where $z = x + iy$ is a complex parameter. Then

$$|f(t, z)| = \exp\left(-\pi y t^2\right) = q^{t^2} \quad \text{where } q = e^{-\pi y};$$

this expression is $> 1$ if $y < 0$, whence $\sum |f(n, z)|$ has no chance of converging; if on the other hand, $y > 0$, then $q < 1$ so that, for $z$ given, $|f(t, z)|$ tends to 0 at infinity more rapidly than $t^{-N}$ for any $N > 0$ (Chap. IV, n° 5), whence normal convergence on every compact set of $\sum f(t+n, z)$. It remains to calculate the Fourier transform

$$(28.3) \qquad \hat{f}(u, z) = \int \exp\left(\pi i z t^2 - 2\pi i u t\right) dt = \int g(t, z)dt.$$

First assume $z = iy$ pure imaginary, so that $izt^2 = -yt^2$. The change of variable $t \mapsto y^{-1/2}t$ gives

$$\hat{f}(u, iy) = \int \exp\left(-\pi t^2 - 2\pi i u y^{-1/2} t\right) y^{-1/2} dt,$$

which leads us to the Fourier transform of $\exp(-\pi t^2)$ for the value $uy^{-1/2}$; thus

$$(28.4) \qquad \hat{f}(u, iy) = y^{-1/2} \exp\left(-\pi u^2/y\right) \quad \text{for } y > 0.$$

In the general case, since the function $g(t, z)$ under the $\int$ sign in (3) is, for given $t$, holomorphic in the half plane $U : \text{Im}(z) > 0$; $\hat{f}(u, z)$ is probably holomorphic too ($\hat{f} = f$ with a hat over it, see (28.3)). To confirm this, we very luckily find in Chap. V, n° 25, a Theorem 24 bis which presupposes the following hypotheses: (i) the integral (3) converges absolutely: obvious; (ii) the complex derivative $g'(t,z) = \pi i t^2 g(t,z)$ with respect to $z$ is a continuous function of $(t,z)$: obvious; (iii) for every compact $H \subset U$ there exists an integrable function $p_H(t)$ on $\mathbb{R}$ such that $|g'(t,z)| \leq p_H(t)$ for any $t \in \mathbb{R}$ and $z \in H$: this demands a proof. But since the compact set $H$ is contained in the open half plane $\mathrm{Im}(z) > 0$ there exists (see above) a number $m > 0$ such that $\mathrm{Im}(z) \geq m$ for every $z \in H$; then

$$|g'(t,z)| = \pi t^2 |g(t,z)| = \pi t^2 \exp\left(-\pi y t^2\right) \leq \pi t^2 \exp\left(-\pi m t^2\right) = p_H(t),$$

an integrable function on $\mathbb{R}$ because at infinity the function $\exp(-\pi m t^2)$ is $O(t^{-2N})$ for any $N > 0$; we would be happy with much less.

The function (3) is therefore holomorphic in the half plane $U : \mathrm{Im}(z) > 0$. Since we know how to calculate it for $z = iy$ pure imaginary we will obtain the general case by constructing on $\mathrm{Im}(z) > 0$ the one and only (principle of analytic continuation) holomorphic function which, on the imaginary axis, reduces to (4). Now

$$(28.5) \qquad \hat{f}(u,z) = (z/i)^{-1/2} \exp\left(-\pi i u^2/z\right)$$

for $z$ pure imaginary, agreeing that $(z/i)^{-1/2}$ is *positive* real for $z = iy$. The factor $\exp\left(-\pi i u^2/z\right)$ being holomorphic in $\mathbb{C}^*$, we have only to find a *holomorphic* function in the half plane $U$, which, for $z = iy$, reduces to $y^{-1/2}$; but, up to a few details, this is what we did at the end of n° 16. For $z \in U$, the ratio $z/i = \zeta$ indeed lies in the half plane $\mathrm{Re}(\zeta) > 0$ contained in $\mathbb{C} - \mathbb{R}_-$; in the latter one may define a uniform, i.e. holomorphic, branch of the "multiform function" $\zeta^{-1/2}$ by putting

$$\zeta^{-1/2} = |\zeta|^{-1/2} e^{-\frac{i}{2}\arg(\zeta)} \qquad \text{with } |\arg(\zeta)| < \pi.$$

Since the point $z = i$ corresponds to $\zeta = 1$ where $\arg(\zeta) = 0$, the holomorphic function we seek is therefore given by the formula

$$(28.6) \qquad (z/i)^{-1/2} = |z|^{-1/2} e^{-\frac{i}{2}\arg(z/i)} \qquad \text{with } |\arg(z/i)| < \pi$$

in the half plane $\mathrm{Im}(z) > 0$ in question (and even in $\mathbb{C}$ with the negative imaginary half-axis removed). This is equivalent to choosing

$$\arg(z/i) = \arg(z) - \pi/2 \qquad \text{with } 0 < \arg(z) < \pi,$$

a very natural choice: for $\arg(i) = \pi/2 + 2k\pi$ and the translation $-\pi/2$ moves the interval $]0, \pi[$ to the interval $]-\pi/2, \pi/2[$.

This point clarified, the Poisson summation formula gives us

$$(28.7) \qquad \sum \exp\left[\pi i z(t+n)^2\right] = (z/i)^{-1/2} \sum \exp\left(-\pi i n^2/z + 2\pi i n t\right).$$

Introducing the Jacobi function

(28.8)                    $\theta(z) = \sum \exp(\pi i n^2 z)$,     $\mathrm{Im}(z) > 0$

(or, for $z$ pure imaginary, the Poisson function), (7) reduces, for $t = 0$, to

(28.9)                    $\theta(-1/z) = (z/i)^{1/2}\theta(z)$;

note, a detail to remember, that

$$\mathrm{Im}(z) > 0 \implies \mathrm{Im}(-1/z) > 0.$$

These formulae are some of the "strange identities" of Chap. IV. On replacing $z$ by $-1/z$ we may rewrite (7) in the form

(28.10)   $\sum \exp\left(\pi i n^2 z + 2\pi i n t\right) = (z/i)^{1/2} \sum \exp\left[-\pi i (t + n)^2/z\right]$.

Now, putting $q = \exp(\pi i z)$, whence $|q| = \exp(-\pi y) < 1$, and[46] $x = \mathbf{e}(t)$, the left hand side is just the series

$$\sum q^{n^2} x^n$$

for which we wrote, in Chap. IV, eqn. (20.14), the curious expansion as an infinite product. With this notation, similarly

(28.11)   $\theta(z) = \sum q^{n^2} = 1 + 2q + 2q^4 + 2q^9 + \ldots$     $\left(q = e^{\pi i z}\right)$.

The series (8) is already, more or less, to be found in Fourier's *Théorie analytique de la chaleur*, though he set little importance on it. The relation (9) was published by Poisson in 1823. Jacobi and Abel studied series of the type (7) systematically, from 1825 on, by purely algebraic methods; Abel died too early to exploit them, but Jacobi drew such a mass of formulae and of results, mainly in theory of the elliptic functions, that his name has remained attached to them.

The connection to the theory of heat propagation is immediate. In an annulus the evolution of the temperature is controlled by the partial differential equation $f'_t = f''_{xx}$ with a numerical coefficient $> 0$ which depends on the physical constants; $t$ is the time and $x$ the polar angle. Fourier's idea was to seek solutions of the form $f(x, t) = g(x)h(t)$, whence $g(x)h'(t) = g''(x)h(t)$ and consequently $h'(t) = \lambda h(t)$, $g''(x) = \lambda g(x)$ where $\lambda$ is a constant. But $g$ must be of period $2\pi$, whence $g(x) = a \cos nx + b \sin nx$ and $\lambda = -n^2$, so that $h(t) = c.\exp\left(-n^2 t\right)$, where $a$, $b$ and $c$ are constants (Fourier eliminated the functions $\exp(+n^2 t)$ for obvious physical reasons). Fourier then postulated that the general solution of his equation is a sum

(28.12)        $f(x, t) = \sum \exp\left(-n^2 t\right)\left(a_n \cos nx + b_n \sin nx\right)$

---

[46] This $x$ is not the real part of $z$; the notation here has been chosen to fit with that of Chap. IV, n° 20.

of "decomposable" functions of this type, a method applicable to all sorts of other problems, mainly in classical or quantum physics. One may then calculate $f(x,t)$ if one can expand the initial state $f(x,0)$ of the system in a series of the form

$$f(x,0) = \sum a_n \cos nx + b_n \sin nx;$$

it was this problem which led Fourier to expand every periodic function as a trigonometric series.

The function

$$(28.13) \qquad \theta(x,t) = \sum \exp\left(-\pi n^2 t + 2\pi i n x\right) =$$
$$= 1 + 2\sum \exp\left(-\pi n^2 t\right) \cos 2\pi n x$$

satisfies the equation

$$\theta''_{xx} = 4\pi^2 \theta'_t$$

and so enters into the framework studied by Fourier; in fact, we now know that it dominates the problem, for if one writes (12) in the form

$$f(x,t) = \sum c_n \exp\left(-\pi n^2 t\right) \mathbf{e}_n(x),$$

summing over $\mathbb{Z}$, an immediate[47] calculation shows that

$$(28.14) \qquad f(x,t) = \oint \theta(x - y, t) f(y) dy \quad \text{for } t > 0,$$

where $f(y) = f(y,0)$ is the temperature distribution at the initial instant. For the Jacobi function the initial data

$$\theta(x,0) = \sum \exp(2\pi i n x) = \sum \mathbf{e}_n(x)$$

is not a true function; it is the Dirac measure at $x = 0$. Physically, the initial temperature is $+\infty$ at $x = 0$ and 0 elsewhere. This might not have made Dirac recoil, but Fourier did not go so far as to envisage this version of the *Big Bang* corresponding to what would happen if one set fire to an artillery piece whose barrel, curved, was a perfect torus.

## 29 – Fundamental formulae for the Fourier transform

The Poisson summation formula allows us to pass very rapidly from the theory of Fourier series to that of Fourier integrals. The proofs which follow are taken from N. Vilenkin, *Special functions and representations of groups*

---

[47] Use the general formula to calculate the Fourier coefficients of a convolution product.

(Moscow, 1965), and attributed to I. M. Gel'fand, 1960, the Soviet mathematician who has invented many more original ideas since the 1930s than one can give him credit for. Igor Sakharov relates in his *Mémoires* that, during 1950s, Gel'fand headed a team of mathematicians at the university of Moscow responsible for the calculations needed for the Soviet thermonuclear programme. Long forbidden to travel outside the national territory, he is now professor at Rutgers University, New Jersey, and travels often ... . Many other proofs of Fourier and Cauchy are now known, but there is little likelihood that this will ever be improved because of the total absence of explicit calculations[48]; this is the great difference from all the classical proofs.

One starts from a function $f$ satisfying the following hypotheses:

(H 1)  *$f$ is continuous,*
(H 2)  *the series $\sum f(x+n)$ converges normally on every compact set;*

it follows, as we have seen, that $f$ is bounded and absolutely integrable on $\mathbb{R}$. Let us put

$$(29.1) \qquad f_y(x) = f(x)\mathbf{e}(yx) = f(x)\mathbf{e}_y(x).$$

The exponential factors being of modulus 1, the series $\sum f_y(x+n)$ converges normally on every compact set for every $y$; on the other hand, the Fourier transform of the function $f_y$ is

$$(29.2) \qquad \widehat{f_y}(t) = \int f(x)\mathbf{e}(yx)\mathbf{e}(-tx)dx = \hat{f}(t-y).$$

If we now assume that
$$(29.3) \qquad \sum |\hat{f}(n-y)| < +\infty$$

the Poisson summation formula applies to $f_y$ and shows that

$$(29.4) \qquad \sum f(x+n)\mathbf{e}_y(x+n) = \sum \hat{f}(n-y)\mathbf{e}(nx).$$

Now $\sum f(x+n)\mathbf{e}_y(x+n) = \mathbf{e}_y(x) \sum f(x+n)\mathbf{e}_y(n)$; since, generally, $\mathbf{e}_y(x) = \mathbf{e}_x(y)$, (4) then leads to

$$(29.5) \qquad \sum f(x+n)\mathbf{e}_n(y) = \sum \hat{f}(n-y)\mathbf{e}_x(n-y).$$

For $x$ given, let us write $F_x(y)$ for the common value of the two sides. By (H 2), the left hand side of (5) is an absolutely convergent Fourier series in $y$.

---

[48] In fact, this is part of the theory of topological groups: one has a locally compact commutative group $G = \mathbb{R}$, a discrete subgroup $\Gamma = \mathbb{Z}$ such that the quotient group $G/\Gamma = K = \mathbb{T}$ is compact, and is concerned to pass from harmonic analysis on $K$ (Fourier series) to harmonic analysis on $G$ (Fourier integrals). This is what André Weil did in 1940, in the general case, in a book we have already cited and which may have inspired Gel'fand, who, at the same time, invented the subject in Moscow with D. A. Raïkov, using functional analytic methods not imposing any hypothesis on the structure of $G$.

The coefficients $f(x+n)$ are therefore found by integration (Chap. V, n° 5). Whence, for $n = 0$,

$$f(x) = \int_0^1 F_x(y)dy = \int_0^1 dy \sum \hat{f}(n-y)\mathbf{e}_x(n-y).$$

Let us now strengthen the hypothesis (3) and suppose

(H 3) *the series* $\sum \hat{f}(y+n)$ *converges normally on every compact set*;

then so does the series to be integrated, whence

$$f(x) \quad = \quad \sum \oint f(n-y)\mathbf{e}_x(n-y)dy =$$

$$= \quad \sum \oint f(n+y)\mathbf{e}_x(n+y)dy,$$

which is simply the *Fourier inversion formula*

$$(29.6) \qquad f(x) = \int \hat{f}(y)\mathbf{e}(xy)dy = \int \hat{f}(y)e^{2\pi ixy}dy$$

where one integrates over $\mathbb{R}$. We may write

$$(29.6') \qquad\qquad \widehat{\hat{f}}(x) = f(-x).$$

Let us now apply the Parseval-Bessel equality to (5), considered as a Fourier series in $y$. We find the relation

$$(29.7) \qquad \sum |f(x+n)|^2 \quad = \quad \oint dy \left| \sum \hat{f}(n-y)e^{2\pi ix(n-y)} \right|^2$$

$$= \quad \oint dy \left| \sum \hat{f}(n-y)e^{2\pi inx} \right|^2$$

since $e^{2\pi ixy}$, of modulus 1, is a factor of the series on the right hand side; the series on the left hand side is convergent by Parseval-Bessel, but since $f$ is bounded,

$$(29.8) \qquad\qquad |f(x+n)|^2 \leq \|f\|.|f(x+n)|;$$

so the series (7) converges normally on every compact set, by (H 2), and its sum is continuous.

Now let us integrate with respect to $x$ over $(0,1)$; we find $\int |f(x)|^2\, dx$, a convergent integral by (8) and the fact that $\int |f(x)|dx$ converges. On the right hand side the function $\sum \hat{f}(n-y)\mathbf{e}_n(x)$ is continuous in $(y,x)$, for if a convergent series $\sum v(n)$ dominates the series $\sum |\hat{f}(n-y)|$ on $I = [0,1]$, then it dominates the series in question on $I \times \mathbb{R}$. We may therefore interchange the order of integration (Chap. V, n° 9, Theorem 10), to obtain the double integral

$$\int_0^1 dy \int_0^1 dx \left| \sum \hat{f}(n-y) e_n(x) \right|^2 .$$

But the function to be integrated is, for $y$ given, the square of an absolutely convergent Fourier series in $x$. Its integral may therefore be calculated using Parseval-Bessel (Chap. V, n° 5 suffices), i.e. is equal to $\sum |\hat{f}(n-y)|^2$. On integrating with respect to $y$ and comparing with the preceding result we finally obtain the *Plancherel formula*

$$(29.9) \qquad \int |f(x)|^2 dx = \int |\hat{f}(y)|^2 dy;$$

this is the analogue of Parseval-Bessel for Fourier integrals. In conclusion:

**Theorem 25.** *Let $f$ be a continuous and absolutely integrable function on $\mathbb{R}$ such that the series $\sum f(x+n)$ and $\sum \hat{f}(y+n)$ converge normally on every compact set. Then*

$$f(x) = \int \hat{f}(y) e(xy) dy, \qquad \int |f(x)|^2 dx = \int |\hat{f}(y)|^2 dy,$$

*the three integrals over $\mathbb{R}$ being absolutely convergent.*

More generally, if two functions $f$ and $g$ satisfy the hypotheses of the theorem, then

$$(29.10) \qquad \int f(x)\overline{g(x)} dx = \int \hat{f}(y)\overline{\hat{g}(y)} dy;$$

one passes from the case $f = g$ to the general case as we did for Fourier series, i.e. by applying the Plancherel formula to the functions $f+g$, $f-g$, $f+ig$ and $f-ig$.

*Example.* In view of Example 2 of n° 27 we have

$$\int \frac{2t e^{2\pi i x y}}{t^2 + 4\pi^2 y^2} dy = e^{-t|x|} \qquad \text{for every } t > 0.$$

This formula essentially says no more than

$$\int \frac{e^{ixy} dy}{x^2 + 1} = \pi e^{-|x|}.$$

Cauchy's residue calculus would give this formula directly. To try to establish it "without knowing anything" is hopeless, except of course for $x = 0$, the only case where the primitive can be calculated.

*Exercise* (another proof of the inversion formula). Let $f$ be a continuous function on $\mathbb{R}$ satisfying

$$f(x) = O\left(1/|x|^a\right), \qquad \hat{f}(y) = O\left(1/|y|^b\right)$$

at infinity, with constants $a, b > 1$, so that $f$ and $\hat{f}$ are absolutely integrable. (i) Show that the Poisson summation formula applies to $f$ [use Theorem 2 of n° 6]. (ii) Show that, for every $T > 0$,

$$\sum f(x + nT) = \frac{1}{T} \sum \hat{f}(n/T) e^{2\pi i n x/T}.$$

(iii) Show that, when $T \to +\infty$, the left hand side tends to $f(x)$ and the other to $\int \hat{f}(y) e(xy) dy$.

## 30 – Extensions of the inversion formula

One may also write (29.10) in the often convenient form

$$(30.1) \qquad \int f(x)\hat{g}(x)dx = \int \hat{f}(y)g(y)dy;$$

it suffices, in (29.10), to replace $g(x)$ by $\overline{\hat{g}(x)}$ and to note that then $\overline{\hat{g}(y)}$ is replaced by $g(y)$. The relation (1) is in fact directly obvious if one calculates formally:

$$(30.2) \quad \int f(x)\hat{g}(x)dx =$$

$$= \int f(x)dx \int g(y)\overline{e(xy)}dy = \iint f(x)g(y)\overline{e(xy)}dxdy =$$

$$= \int g(y)dy \int f(x)\overline{e(xy)}dx = \int \hat{f}(y)g(y)dy.$$

But one has to justify the interchange of the integrations. In the Lebesgue theory it is enough to assume that $h(x, y) = f(x)g(y)\overline{e(xy)}$ is integrable on $\mathbb{R}^2$, i.e. that $f$ and $g$ are integrable, since $e(xy)$ is continuous and bounded; one then applies Fubini's theorem (the real one ...) to the function obtained.

In the Riemann theory, there is a more restricted, but nevertheless useful result:

**Lemma.** *Let $f$ and $g$ be two absolutely integrable regulated functions; then $\int f(x)\hat{g}(x)dx = \int \hat{f}(y)g(y)dy$.*

First assume $f$ and $g$ are zero outside compact intervals $K$ and $H$, and consider on $K$ and $H$ the measures $d\mu(x) = f(x)dx$, $d\nu(y) = g(y)dy$ (Chap. V, n° 30, Example 1). Since the function $e(xy)$ is continuous on $K \times H$ we have

$$\int d\mu(x) \int \overline{e(xy)}d\nu(y) = \int d\nu(y) \int \overline{e(xy)}d\mu(x)$$

(Chap. V, n° 30, Theorem 30). By definition of $\mu$ and $\nu$, this relation justifies the formal calculation (2). The lemma is therefore true under the hypotheses just formulated.

In the general case, let us write $f_n$ and $g_n$ for the functions equal to $f$ and $g$ on $[-n, n]$ and zero elsewhere, whence

$$(30.3) \qquad \int f_n(x)\widehat{g_n}(x)dx = \int \widehat{f_n}(y)g_n(y)dy.$$

It all reduces to showing that one may pass to the limit under the integration sign. Let us do this for the left hand side. It is enough to show that $\|f_n\widehat{g_n} - f\hat{g}\|_1$ tends to 0, since, generally, $|\int f| \leq \int |f|$. Now omitting the variable $x$,

$$\begin{aligned} |f_n\widehat{g_n} - f\hat{g}| &\leq |f_n - f| \cdot |\widehat{g_n}| + |f| \cdot |\widehat{g_n} - \hat{g}| \\ &\leq |f_n - f| \cdot \|\widehat{g_n}\| + |f| \cdot \|\widehat{g_n} - \hat{g}\| \\ &\leq |f_n - f| \cdot \|g_n\|_1 + |f| \cdot \|g_n - g\|_1 \end{aligned}$$

since

$$\|\hat{g}\| = \sup |\hat{g}(x)| = \sup \left| \int \overline{e(xy)}g(y)dy \right| \leq \|g\|_1$$

for every absolutely integrable function on $\mathbb{R}$. Whence, integrating over $\mathbb{R}$,

$$(30.4) \qquad \|f_n\widehat{g_n} - f\hat{g}\|_1 \leq \|f_n - f\|_1 \cdot \|g_n\|_1 + \|f\|_1 \cdot \|g_n - g\|_1$$

But

$$\|f_n - f\|_1 = \int_{|x|>n} |f(x)|dx$$

tends to 0 since $f$ is absolutely integrable, likewise $\|g_n - g\|_1$; the factor $\|f\|_1$ is independent of $n$ and the factor $\|g_n\|_1$ tends to $\|g\|_1$. The right hand side of (4) therefore tends to 0, qed.

For example let us choose for $g$ the function $e^{-t|x|}$ and for $f$ an absolutely integrable regulated function. By Example 2 of n° 27 we find

$$(30.5) \qquad \int \frac{2t}{t^2 + 4\pi^2 x^2} f(x)dx = \int e^{-t|y|}\hat{f}(y)dy \qquad \text{every } t > 0.$$

If we put

$$(30.6) \qquad u(x) = 2/(1 + 4\pi^2 x^2), \qquad u_n(x) = nu(nx),$$

the relation (5) can be written, for $t = 1/n$, in the form

$$(30.7) \qquad \int nu(nx)f(x)dx = \int e^{-|y|/n}\hat{f}(y)dy.$$

The function $u$ is continuous (and even $C^\infty$), positive, and its total integral is equal to 1, as an elementary calculation shows. Dirac's lemma of Chap. V, n° 27, the version of Example 1, then shows that, if $f$ is continuous at the origin and bounded on $\mathbb{R}$, the left hand side of (7) tends to $f(0)$. On the right hand side the exponential converges to 1 uniformly on every compact set while remaining $< 1$; if the function $\hat{f}$ is absolutely integrable, the right hand side of (7) then tends to the integral of the latter (dominated convergence), whence, in the limit,

$$(30.8) \qquad f(0) = \int \hat{f}(y)dy,$$

i.e. Fourier's inversion formula for $x = 0$.

In fact, it is not necessary to assume $f$ bounded. In Chap. V, n° 27, this hypothesis was used only to show that, for every $\delta > 0$, the integral $\int f(x)u_n(x)dx$ extended over the set $|x| > \delta$, tends to 0. Now it is clear that here

$$|x| > \delta > 0 \Longrightarrow |u_n(x)| < 1/2n\delta^2 x^2,$$

so that the function $x \mapsto u_n(x)$ converges uniformly to 0 on $|x| > \delta$; since, here, $f$ is assumed absolutely integrable, we have

$$\lim \int_{|x|>\delta} f(x)u_n(x)dx = 0$$

even if $f$ is not bounded.

To obtain the inversion formula at an arbitrary point $a \in \mathbb{R}$ one replaces $x \mapsto f(x)$ by $x \mapsto f(x + a)$. The Fourier transform becomes

$$\int f(x+a)\overline{\mathbf{e}(xy)}dx = \int f(x)\overline{\mathbf{e}(xy - ay)}dx = \hat{f}(y)\mathbf{e}(ay)$$

and by applying (8) to the new function one obviously obtains Fourier's inversion formula at the point $a$ if $f$ is continuous at this point. Consequently:

**Theorem 26.** *Let $f$ be a continuous absolutely integrable function on $\mathbb{R}$. Suppose that $\hat{f}$ is absolutely integrable. Then*

$$(30.9) \qquad f(x) = \int \hat{f}(y)\mathbf{e}(xy)dy \qquad \text{for every } x \in \mathbb{R}.$$

Note that the proof uses only the following facts: (i) formula (2), which we established using a "poor man's Fubini" *without* using Theorem 25, (ii) the perfectly elementary calculation of the Fourier transform of $e^{-t|x|}$, whence (5) directly, (iii) the fact, also totally elementary, that the functions $x \mapsto 2t/\left(t^2 + 4\pi^2 x^2\right)$ form a Dirac sequence when $t$ tends to 0.

The reader will doubtless observe that the functions of Theorem 25 satisfy the hypotheses of Theorem 26. So why state a useless Theorem 25 when Theorem 26 provides us the inversion formula under more general hypotheses and without passing through Theorem 25? The reason is simple: besides that Theorem 25 also gives us the Plancherel formula, its proof does not use, as we have said, *any* explicit calculation.

When $f$ is not continuous at 0 the relation $u_n(x) = u_n(-x)$ allows one to argue as we did *à propos* Dirichlet's theorem on Fourier series: the limit is $\frac{1}{2}[f(0+) + f(0-)]$. The formula we obtained resembles Dirichlet's, so we may conjecture that it is valid under more general hypotheses than integrability of $f$. This is an interesting *exercise*, though its usefulness to us is very small.

If one is inspired by the case of Fourier series, one replaces, in the inversion formula for $x = 0$, the "total" integral, not absolutely convergent, substituting for $\hat{f}$ its "partial" integrals

$$(30.10) \qquad s_N(0) = \int_{-N}^{N} \hat{f}(y)dy = \int_{-N}^{N} dy \int f(t)\mathbf{e}(yt)dt$$

and one interchanges the integration signs; the lemma established above authorises us to do this if $f$ is regulated and absolutely integrable: take for $g(y)$ the characteristic function of the interval $(-N, N)$. Then

$$(30.11) \qquad s_N(0) = \int f(t)\frac{\mathbf{e}(Nt) - \mathbf{e}(-Nt)}{2\pi it}\, dt = \int f(t)K_N(t)dt.$$

The function $K_N(t) = \sin(2\pi Nt)/\pi t$ is not absolutely integrable, but its integral over $\mathbb{R}$ is convergent since $1/t$ is monotone and tends to 0 at infinity (Chap. V, n° 24, Theorem 23; there is no problem at $t = 0$ since the function is continuous there). Let us put

$$(30.12) \qquad \int K_N(t)dt = 2c;$$

it will emerge that $c = \frac{1}{2}$, but we do not know this *a priori*. Since the function $K_N$ is even, we find, as in the case of Fourier series, that

$$s_N(0) - c[f(0+) + f(0-)] = \int_0^{+\infty} \frac{f(t) - f(0+)}{\pi t}\sin(2\pi Nt)dt +$$
$$+ \int_{-\infty}^0 \frac{f(t) - f(0-)}{\pi t}\sin(2\pi Nt)dt.$$

Assume now that the function $f$ has right and left derivatives at $t = 0$ and put

$$(30.13) \qquad g(t) = \begin{cases} [f(t) - f(0+)]/\pi t & \text{for} \quad t > 0, \\ ? & \text{for} \quad t = 0, \\ [f(t) - f(0-)]/\pi t & \text{for} \quad t < 0, \end{cases}$$

the sign ? indicating that the value attributed to $g$ at 0 is unimportant. We obtain a regulated function in $\mathbb{R}$ and

(30.14)    $s_N(0) - c[f(0+) + f(0-)] =$

$$= \int g(t) \sin(2\pi Nt)dt = [\hat{g}(N) - \hat{g}(-N)]/2i.$$

It all amounts to showing that $\hat{g}(y)$ tends to 0 as $|y|$ increases indefinitely.

This would be obvious if $g$ were absolutely integrable (n° 27, Theorem 23), but we are not in this case. Theorem 23 of Chap. V, n° 24 relative to integrals of the form $\int f(x)\sin(xy)dx$, where $f$ is monotone and tends to 0 at infinity, will resolve the problem.

We may decompose the integral in (14) into three parts relative to the intervals $(-\infty, -1)$, $(-1, 1)$ and $(1, +\infty)$. The integral extended over $(-1, 1)$ is the Fourier transform of a regulated function of compact support, so tends to 0 at infinity (Theorem 23). The integral extended over $(1, +\infty)$ can be written

$$\int_1^{+\infty} \frac{f(t)}{t} \sin(2\pi Nt)dt - f(0+) \int_1^{+\infty} \sin(2\pi Nt)dt/t;$$

this calculation is legitimate because $f(t)$ and *a fortiori* $f(t)/t$ are absolutely integrable, while the second integral converges; in fact, Theorem 23 of Chap. V, n° 24 even shows that it tends to 0. Likewise for the first, as the Fourier transform of an absolutely integrable function. One argues similarly for the interval $(-\infty, 1)$.

The integral in (14) thus tends to 0 and one obtains the following result:

**Theorem 27.** *Let $f$ be an absolutely integrable regulated function. Then*

(30.15)    $$\lim_{N\to\infty} \int_{-N}^{N} \hat{f}(y)e(xy)dy = \frac{1}{2}[f(x+) + f(x-)]$$

*at every point where $f$ has right and left derivatives.*

And why has the unknown constant $c$ transformed itself surreptitiously into $\frac{1}{2}$ ? Because, if one applies the formula to a sufficiently accommodating function, one already knows (Theorem 25) that the right hand side of (15) is equal to $f(x)$. So the constant $c$ has no choice ...

This small auxiliary result can be rewritten as

$$\int \sin(2\pi Nt)dt/t = \pi$$

or, by an obvious change of variable,

(30.16)    $$\int_{-\infty}^{\infty} \sin(t)dt/t = \pi,$$

a famous formula of Dirichlet's. You are advised not to try to establish this by looking for a primitive of $\sin(t)/t$.

## 31 – The Fourier transform and differentiation

As we have seen in Chap. V, n° 25, Example 1, if a regulated function $f$ on $\mathbb{R}$ satisfies[49]

$$(31.1) \qquad \int |f(x)|dx < +\infty, \qquad \int |xf(x)|dx < +\infty,$$

its Fourier transform is differentiable and

$$(31.2) \qquad \hat{f}'(y) = -2\pi i \int xf(x)\overline{e(xy)}dx,$$

the Fourier transform of $-2\pi i x f(x)$.

To formulate this result in a concentrated way, it is helpful to introduce "operators" transforming the (or certain) functions on $\mathbb{R}$ into other functions on $\mathbb{R}$:

(i)   the operator $M : f \mapsto Mf$ of multiplication by $-2\pi i x$;
(ii)  the differentiation operator $D : f \mapsto Df$;
(iii) the Fourier transform operator $F : f \mapsto Ff = \hat{f}$.

Then formula (2) can be written as

$$(31.2') \qquad DFf = FMf \qquad \text{or} \quad D \circ F = F \circ M,$$

the symbol $\circ$ as always denoting the composition of maps. One must remain aware of the fact that (2') assumes $Mf$ absolutely integrable.

One may iterate the argument so long as the functions $M^k f$ are integrable. Since the Fourier transform $F$ exchanges $M$ and $D$, one clearly finds the formula

$$(31.2'') \qquad D^k Ff = FM^k f$$

if $M^k f(x) = (-2\pi i x)^k f(x)$ is absolutely integrable. Whence a first result:

**Lemma 1.** *Let $f$ be a regulated function such that $\int |x^p f(x)|\, dx < +\infty$. Then $\hat{f}$ is of class $C^p$ and*

$$(31.3) \qquad D^k \hat{f}(y) = \int (-2\pi i x)^k f(x)e(xy)dx \quad \text{for every } k \leq p.$$

Limit case:

$$\hat{f} \text{ is } C^\infty \text{ if } \int |x^p f(x)|dx < +\infty \text{ for every } p.$$

---

[49] The second condition implies the first since $f$ is integrable on every compact set and $|f(x)| < |xf(x)|$ for $|x|$ large.

Now suppose that $f$ is of class $C^1$ (or, more generally, a primitive of a regulated function) and that $\int |f'(x)|dx < \infty$. Integrating by parts, one has, for $y \neq 0$,

$$(31.4) \quad \int_{-T}^{T} f(x)e^{-2\pi ixy}dx = f(x)\frac{e^{-2\pi ixy}}{-2\pi iy}\bigg|_{-T}^{T} + \frac{1}{2\pi iy}\int_{-T}^{T} f'(x)e^{-2\pi ixy}dx.$$

Since $f'$ is integrable on $\mathbb{R}$ the function

$$f(x) = f(0) + \int_0^x f'(t)dt$$

tends to a limit as $x$ tends to $+\infty$ or $-\infty$. This limit is zero since otherwise the integral $\int |f(t)|dt$ would be clearly divergent. So we see that in (4), the integrated part tends to 0 when $T \to +\infty$, and there remains

$$(31.5) \quad \widehat{f'}(y) = 2\pi iy\hat{f}(y),$$

which we may write in the form

$$(31.5') \quad FDf = -MFf \quad \text{or} \quad F \circ D = -M \circ F.$$

If $f$ is of class $C^p$, and if all its derivatives are absolutely integrable, we may apply the calculation $p$ times to obtain

$$(31.6) \quad \widehat{f^{(p)}}(y) = (2\pi iy)^p\hat{f}(y), \quad \text{i.e.} \quad FD^pf = (-1)^pM^pFf.$$

Now the left hand side tends to 0 at infinity (Theorem 23); consequently:

**Lemma 2.** *If $f$ is of class $C^p$ and if all its derivatives are absolutely integrable, then*

$$(31.7) \quad \hat{f}(y) = o\left(y^{-p}\right) \quad \text{when} \quad |y| \longrightarrow +\infty.$$

In other words: *the Fourier transform decreases at least as rapidly as the function $f$ has integrable derivatives.*

The ideal case is that where $f$ is indefinitely differentiable, with derivatives satisfying

$$(31.8) \quad f^{(p)}(x) = O\left(x^{-q}\right) \quad \text{for any } p \text{ and } q;$$

one then says (L. Schwartz) that $f$ is *indefinitely differentiable with rapid decrease*; the set of these functions is denoted $\mathcal{S}(\mathbb{R})$ or simply $\mathcal{S}$. We must not forget that the condition of decrease at infinity applies not only to $f$, but to all its derivatives. If $f$ is in $\mathcal{S}$, so also is the function $x^p f^{(q)}(x)$ for any $p$ and $q$, for on multiplying a derivative of arbitrary order of $x^p f^{(q)}(x)$ by a power of $x$ we obtain a linear combination of a finite number of functions of the form $x^k f^{(h)}(x)$, which are $O\left(x^{-N}\right)$ for any $N$ by (8) for $p = h$, $q = k+N$.

**Theorem 28.** *The Fourier transform maps $\mathcal{S}$ bijectively onto $\mathcal{S}$.*

It is clear that if $f \in \mathcal{S}$ one may apply Lemma 1 to it for any $p$, so that $\hat{f}$ is $C^\infty$. Since $x^p f(x)$ is also in $\mathcal{S}$ one may apply Lemma 2 for any $p$; consequently, $\hat{f}(y) = O\left(y^{-N}\right)$ for any $N$. But the derivatives of $\hat{f}$ are, up to constants factors, the Fourier transforms of the functions $x^p f(x)$, which are again in $\mathcal{S}$. They too are $O\left(y^{-N}\right)$ at infinity for any $N$.

Consequently, $f \in \mathcal{S}$ implies $\hat{f} \in \mathcal{S}$. But since $\hat{\hat{f}}(x) = f(-x)$, the condition $\hat{f} \in \mathcal{S}$ implies conversely that $f \in \mathcal{S}$. The map is therefore bijective, qed.

An immediate corollary is that *the Poisson summation formula, Fourier's inversion formula and Plancherel's formula apply to every $f \in \mathcal{S}$.*

Another important, and easy to establish, result in $\mathcal{S}$ is the formula

$$(31.9) \qquad \widehat{f \star g} = \hat{f}\hat{g}$$

which gives the Fourier transform of a convolution product

$$(31.10) \qquad f \star g(x) = g \star f(x) = \int f(x-y)g(y)dy,$$

an analogue to the formula (4.10) for periodic functions. Calculating formally:

$$
\begin{aligned}
\hat{f}(z)\hat{g}(z) &= \int f(x)\overline{\mathbf{e}(xz)}dx \int g(y)\overline{\mathbf{e}(yz)}dy = \iint \overline{\mathbf{e}_z(x+y)}f(x)g(y)dxdy = \\
&= \int g(y)dy \int \overline{\mathbf{e}_z(x+y)}f(x)dx = \int g(y)dy \int \overline{\mathbf{e}_z(x)}f(x-y)dx = \\
&= \int \overline{\mathbf{e}_z(x)}dx \int g(y)f(x-y)dy = \int f \star g(x)\overline{\mathbf{e}_z(x)}dx,
\end{aligned}
$$

whence the result. The interchange of the repeated integrals is justified by Theorem 25 of Chap. V, n° 26 since the exponential is of modulus 1 and the functions $f$ and $g$ are absolutely integrable and bounded in $\mathbb{R}$, so that the function $\overline{\mathbf{e}_z(x+y)}f(x)g(y)$ is, up to constants, dominated either by $|f(x)|$, or by $|g(y)|$.

The relation (9) is in fact valid under much wider hypotheses – it would be enough for $f$ and $g$ to regulated and absolutely integrable, the case where $f$ and $g$ are continuous of compact support being particularly obvious –, but since Lebesgue's integration theory yields it very easily in an even more general case, it is better to wait to deal with this.

Since the ordinary product of two functions of $\mathcal{S}$ is again in $\mathcal{S}$ (obvious!), (9) and Theorem 28 show that

$$(f \in \mathcal{S}) \ \& \ (g \in \mathcal{S}) \Longrightarrow f \star g \in \mathcal{S}.$$

*Exercise*: prove this directly, starting from (10).

Since we have already given three different proofs of the inversion formula (Theorem 25, Exercise of n° 29 and Theorem 26), we may as well give a fourth, based on the idea, dear to the physicists, that a "continuous spectrum" is the limit case of a "discrete spectrum" whose "lines" approach each other more and more closely, as Cavalieri, with his "indivisibles", would have had no trouble understanding. The method rests on a simple formal calculation, but one has to justify it, which is less easy.

One starts with a regulated function $f$ defined on $\mathbb{R}$ and, for every $T > 0$, considers the function $f_T$ of period $T$ satisfying

$$(31.11) \qquad f_T(x) = f(x) \quad \text{for} \quad -T/2 < x \le T/2.$$

"Clearly" we have a Fourier series expansion

$$(31.12) \qquad f_T(x) = \sum a_n(T)\mathbf{e}_n(x/T)$$

with

$$(31.13) \quad a_n(T) = \frac{1}{T} \int_{-T/2}^{T/2} f_T(x)\mathbf{e}_n(-x/T)dx = \frac{1}{T} \int_{-T/2}^{T/2} f(x)\mathbf{e}(-nx/T)dx.$$

For $T$ large the last integral is "almost" equal to the integral extended over all $\mathbb{R}$, i.e. to $\hat{f}(n/T)$, whence "manifestly", for $x = 0$ let us say, the formula

$$(31.14) \qquad f(0) \approx \frac{1}{T} \sum \hat{f}(n/T).$$

The right hand side is "obviously" the Riemann sum one would obtain in calculating $\int \hat{f}(y)dy$ using the subdivision of $\mathbb{R}$ by the abscissae $n/T$. Whence, in the limit, $f(0) = \int \hat{f}(y)dy$ and, by translation, the inversion formula at an arbitrary point. The same calculation also yields the Plancherel formula. The Parseval-Bessel theorem applied to the Fourier series of $f_T$ shows that

$$(31.15) \qquad \frac{1}{T} \int_{-T/2}^{T/2} |f_T(x)|^2 dx = \sum |a_n(T)|^2 \approx \frac{1}{T^2} \sum |\hat{f}(n/T)|^2,$$

the sign $\approx$ signifying that the right hand side is "almost" equal to the third. In the first term one can replace $f_T$ by $f$, whence an integral which tends to $\int |f(x)|^2 dx$; if one multiplies the third term by $T$ to eliminate the factor $1/T$ from the first, one finds again a Riemann sum which "clearly" tends to $\int |\hat{f}(y)|^2 dy$, "cqfd".

This is all very well, but there are several gaps to fill in, which explains why some textbooks for "users" confine themselves to the formal calculation and to the traditional mathematical variant of the argument from authority, accompanied by recourse to physical intuition.

If one restricts oneself to examining what happens for $x = 0$, which is no restriction in generality, the relation (12) assumes that $f_T$, i.e. $f$, is continuous at this point since otherwise the inversion formula has little chance of being correct. It also assumes, more seriously, that the Fourier series converges. Simplest is to assume $f_T$ of class $C^1$ on $\mathbb{R}$, so imposing the same hypothesis on $f$, but even in this case definition (11) shows that $f_T$ has every chance of being discontinuous at the points $\pm T/2$. A convenient procedure for eliminating the difficulty is to assume $f$ of *compact support* since, for $T$ large enough, $f$ is then zero on a neighbourhood of the end-points of the interval (11). If this is the case, the second integral is in fact extended over all $\mathbb{R}$ for $T$ large, whence $a_n(T) = \hat{f}(n/T)/T$ directly and the formula (14) follows.

One then has to pass from the series $\sum \hat{f}(n/T)/T$ to the integral of $\hat{f}$. This assumes at least that the latter converges. Since $f$ is of compact support, Lemma 2 above shows that this is the case if $f$ is $C^2$, since then $\hat{f}(y) = O(1/y^2)$ at infinity. This done, one may consider the series $\sum \hat{f}(n/T)/T$ as the integral over $\mathbb{R}$ of the function $\varphi_T$ equal to $\hat{f}(n/T)$ between $(n-1)/T$ and $n/T$; as $T$ increases $\varphi_T$ converges simply (and even uniformly on every compact set) to $\hat{f}$ since $\hat{f}$ is continuous. Since we have the global estimate $|\hat{f}(y)| \leq M/(1+y^2) = p(y)$, the same estimate applies to $\varphi_T$, and since the positive function $p$ is integrable on $\mathbb{R}$, the dominated convergence theorem shows that the integral of $\varphi_T$ tends to that of $\hat{f}$; one may therefore, in (14), replace the series by $\int \hat{f}(y)dy$, whence the inversion formula. The calculation leading to the Plancherel formula is justified by analogous arguments.

The necessary arguments become noticeably more difficult if one abandons the hypothesis that $f(x)$ is zero for $|x|$ large. Even under the much too strong hypothesis that $f \in \mathcal{S}$ the difficulty due to the fact that $f_T$ may have isolated discontinuities does not disappear: the Fourier series of $f_T$ does not converge *absolutely* and if one wants to pass to (14) one has to evaluate precisely the difference between $a_n(T)$ and $\hat{f}(n/T)/T$, which is easy since $\hat{f} \in \mathcal{S}$, and then pass to the limit term-by-term in the series (12).

## 32 – Tempered distributions

When Schwartz invented his theory of distributions (Chap. V, n° 34) he immediately asked the following question: can one define the Fourier transform of a distribution $T$ on $\mathbb{R}$ as on $\mathbb{T}$? Now a distribution is a linear form on the space $\mathcal{D} = \mathcal{D}(\mathbb{R})$ of the $C^\infty$ functions of compact support, satisfying certain conditions of continuity; since the exponentials are not of compact support, the standard formula makes no sense unless $T$ is a *bounded* Radon measure $\mu$ (Chap. V, n° 31, Example 1) on $\mathbb{R}$; one may, in this case, integrate every *bounded* continuous function with respect to $\mu$ (same method as for the measure $dx$: Chap. V, n° 22) and then define

$$\hat{\mu}(y) = \int \overline{\mathbf{e}(xy)}d\mu(x).$$

In the general case, the problem would have an immediate answer if we knew that $f \mapsto \hat{f}$ maps $\mathcal{D}$ bijectively onto $\mathcal{D}$: we would then define $\hat{T}$ so as to obtain the distribution $\hat{f}(y)dy$ if $dT(x) = f(x)dx$, i.e. by the formula

(32.1)        $$\int \varphi(y)d\hat{T}(y) = \int \hat{\varphi}(x)dT(x), \quad \text{i.e.} \quad \hat{T}(\varphi) = T(\hat{\varphi}),$$

directly inspired by (30.1).

Alas, *the Fourier transform of a function of compact support is never of compact support*. In this case

$$\hat{f}(y) = \int f(x)\exp(-2\pi ixy)dx = \sum(-2\pi iy)^{[n]} \int x^n f(x)dx$$

since we are integrating over a compact set $K$ a series that is normally convergent on $K$. Here we may even assume $y$ complex, so that $\hat{f}$ *is the restriction to $\mathbb{R}$ of an analytic function on $\mathbb{C}$*, i.e. of an entire function. The principle of analytic continuation then shows that, for $f$ regulated and of compact support, $\hat{f}$ cannot be of compact support (or zero on a nonempty open interval) unless $\hat{f} = 0$, which, for $f \in \mathcal{D}$ (and even for $f$ continuous: Theorem 26), implies $f = 0$. The situation is not what we met *à propos* Fourier series.

To cut through this dilemma, Schwartz had to introduce a particular class of distributions and, to do this, to substitute for $\mathcal{D}$ the space $\mathcal{S}$ endowed with a suitable topology. If one wants to define the Fourier transform of a distribution $T$ by formula (1) for $\varphi \in \mathcal{D}$, one has to be able to define the value of $T$ on the Fourier transforms of the $\varphi \in \mathcal{D}$, i.e. on functions which are in $\mathcal{S}$ but not in $\mathcal{D}$; supposing this point achieved, one again has to verify that the linear form $\varphi \mapsto T(\hat{\varphi})$ so obtained is continuous. The solution is then (i) to endow $\mathcal{S}$ with a topology making the map $f \mapsto \hat{f}$ of $\mathcal{S}$ into $\mathcal{S}$ *continuous*, (ii) to restrict oneself to the distributions $T : \mathcal{D} \to \mathbb{C}$ which can be extended to *continuous* linear forms $\mathcal{S} \to \mathbb{C}$.

Consider now the first problem. For every $f \in \mathcal{S}$ the numbers

(32.2)        $$N_{p,q}(f) = \sup \left| x^p f^{(q)}(x) \right|,$$

are finite by definition. Clearly

$$N_{p,q}(f + g) \leq N_{p,q}(f) + N_{p,q}(g)$$

and $N_{p,q}(cf) = |c|N_{p,q}(f)$ for every constant $c$; furthermore, it is clear that $N_{p,q}(f) = 0$ only if $f = 0$; each function $N_{p,q}$ is therefore a *norm* on the vector space $\mathcal{S}$. For every $r \in \mathbb{N}$ the function

(32.3)        $$N_r(f) = \sum_{p,q \leq r} N_{p,q}(f)$$

(no connection with the norms $N_p$ of integration theory) has again the same properties and $N_r \leq N_{r+1}$. One now defines a topology on $\mathcal{S}$ by calling every set defined by an inequality

$$N_r(f - g) < \rho,$$

where $\rho > 0$ and $r \in \mathbb{N}$ are chosen arbitrarily, a "ball of centre $f$", and declaring that a subset $U$ of $S$ is "open" if for every $f \in U$ the set $U$ contains a ball of centre $f$ (see[50] the Appendix to Chap. III, n° 8). Convergence in $S$ can then be translated into the condition

(32.4)     $$\lim N_r(f - f_n) = 0 \quad \text{for every } r;$$

equivalently, one demands that, for any $p$ and $q$,

(32.4')     $$\lim x^p \left[ f^{(q)}(x) - f_n^{(q)}(x) \right] = 0 \quad \text{uniformly on } \mathbb{R}.$$

This allows one to speak of continuous functions on $S$, for example of continuous maps from $S$ into $S$. If $U$ is such a map, denoted $f \mapsto U(f)$ or $Uf$ according to the case and to the author, one has, to express the continuity of $U$ at a "point" $f_0$ of $S$, to write that for every ball $B$ of centre $g_0 = Uf_0$ there exists a ball $B'$ of centre $f_0$ such that $U$ maps $B'$ into $B$; in other words that, for any $r \in \mathbb{N}$ and $\varepsilon > 0$, there exists an $r' \in \mathbb{N}$ and an $\varepsilon' > 0$ such that

$$N_{r'}(f - f_0) < \varepsilon' \implies N_r(Uf - Uf_0) < \varepsilon.$$

If $U$ is linear, the most frequent case, it clearly suffices to express continuity for $f_0 = 0$. If $U$ takes its values in $\mathbb{C}$, one replaces the inequalities $N_r(Uf - Uf_0) < \varepsilon$ by the single condition $|Uf - Uf_0| < \varepsilon$.

*Exercise* – Show that $f \mapsto f^2$ is a continuous map of $S$ into $S$.

With these definitions, one sees immediately that differentiation $D : f \mapsto f'$ is a continuous map of $S$ into $S$; indeed

$$N_{p,q}(f') = \sup \left| x^p f^{(q+1)}(x) \right| = N_{p,q+1}(f),$$

whence the inequality

(32.5)     $$N_r(f') \le N_{r+1}(f)$$

which yields the result.

Similarly, the operator $M$, multiplication by the function $-2\pi i x$, maps $S$ linearly into $S$ and is continuous. When one replaces $f(x)$ by $xf(x)$ the function $f^{(q)}(x)$ is replaced by $xf^{(q)}(x) + qxf^{(q-1)}(x)$, whence

$$
\begin{aligned}
N_{p,q}(Mf) &= 2\pi . \sup \left| x^{p+1} f^{(q)}(x) + qx^p f^{(q-1)}(x) \right| \le \\
&\le 2\pi N_{p+1,q}(f) + 2\pi q N_{p,q-1}(f),
\end{aligned}
$$

---

[50] One might also consider the sets $N_{p,q}(f - g) < \rho$ without changing the topology; using the $N_r$ is technically a little easier. On the other hand, note that the family of norms $N_r$ or $N_{p,q}$ is countable, so one could define the topology of $S$ using a single distance (Appendix to Chap. III, n° 8), so in fact $S$ is a metric space, and moreover complete (exercise!); but it is not a Banach space: the topology of $S$ cannot be defined by a single norm.

a finite result – whence $Mf \in \mathcal{S}$ – and implying the estimate

$$(32.6) \qquad N_r(Mf) \le c_r N_{r+1}(f)$$

with a constant $c_r$ whose exact value is of little importance, because (6) is enough to establish the continuity of $M$.

As a map of $\mathcal{S}$ into $\mathcal{S}$, the Fourier transform $F$ is also *continuous* in each sense. To see this without much calculating, first remark that

$$N_{p,q}(f) = N_0\left(M^p D^q f\right) = \|M^p D^q f\|_{\mathbb{R}}$$

and then $N_{p,q}(\hat{f}) = N_0\left(M^p D^q F f\right) = N_0\left(M^p F M^q f\right) = N_0\left(F D^p M^q f\right)$ by the "commutation formulae" (31.2") and (31.6). Now, in general,

$$N_0(Ff) = \sup\left|\int f(x)\overline{\mathbf{e}(xy)}dx\right| \le \int |f(x)|dx = \|f\|_1;$$

since the function $(x^2+1)f(x)$ is bounded by $N_2(f)$, by (3), one finds

$$N_0(Ff) \le N_2(f)\int \left(x^2+1\right)^{-1}dx,$$

with a convergent integral whose exact value, $c = \pi$, is not important. It follows that

$$N_{p,q}(\hat{f}) = N_0\left(FD^p M^q f\right) \le cN_2\left(D^p M^q f\right);$$

on applying (5) $p$ times to the function $M^q f$ one finds a result $\le N_{p+2}\left(M^q f\right)$ up to a constant factor, and by applying (6) $q$ times to $f$ one obtains a relation of the form $N_{p,q}(\hat{f}) \le N_{p+q+2}(f)$ up to a constant factor. Remembering the definition (3) of $N_r$, we finally have

$$(32.7) \qquad N_r(\hat{f}) \le c_r' N_{r+2}(f)$$

where $c_r'$ is a new constant. This proves the continuity of the Fourier transform. Since it is bijective and quasi identical to its inverse map by virtue of the relation $\widehat{\hat{f}}(x) = f(-x)$, we conclude that the Fourier transform is a bijective and bicontinuous map of $\mathcal{S}$ onto $\mathcal{S}$, in other words what in topology one calls a *homeomorphism* (linear too) of $\mathcal{S}$ onto $\mathcal{S}$.

With their systematic recourse to the operators $D$, $M$ and $F$, these calculations can appear a little abstract. But to write explicitly the integrals and derivatives which they mask would be even less enticing.

We can now return to distribution theory. Following Schwartz, we will call any *continuous* linear form $T : \mathcal{S} \to \mathbb{C}$ a *tempered distribution* on $\mathbb{R}$. The inequality $|T(f)| < \varepsilon$ has to hold for every $f \in \mathcal{S}$ "close enough" to 0; this means that there exists an $r \in \mathbb{N}$ and a $\delta > 0$ such that

$$N_r(f) < \delta \Longrightarrow |T(f)| < \varepsilon.$$

Continuity is expressed as follows: *there exist an* $r \in \mathbb{N}$ *and a constant* $M(T) \geq 0$ *such that*

$$(32.8) \qquad |T(f)| \leq M(T).N_r(f) \qquad \text{for every } f \in \mathcal{S};$$

the argument is the same as in the normed vector spaces of the Appendix to Chap. III, n° 6.

To justify the terminology, we have to show how $T$ defines a distribution in the sense of Chap. V, n° 34. Since $\mathcal{S}$ contains $\mathcal{D}$ it is clear that $T$ defines a linear form on $\mathcal{D}$, but again one has to prove continuity. If one works in the subspace $\mathcal{D}(K)$ of the $\varphi \in \mathcal{D}$ vanishing outside a compact subset $K$ of $\mathbb{R}$ one has

$$N_{p,q}(\varphi) = \sup \left| x^p \varphi^{(q)}(x) \right| < c(K)^p . \left\| \varphi^{(q)} \right\|$$

where $c(K)$ is the upper bound of $|x|$ on $K$. One deduces that

$$N_r(\varphi) \leq c_r(K) \left( \|\varphi\| + \ldots + \left\| \varphi^{(r)} \right\| \right) = c_r(K) \left\| \varphi \right\|^{(r)}$$

in the notation of Chap. V, (34.3), with again another constant $c_r(K)$ depending only on $K$ and on $r$. The inequality (8) then shows that the restriction of $T$ to the subspace $\mathcal{D}(K)$ satisfies the continuity condition $|T(\varphi)| \leq M_K(T). \|\varphi\|^{(r)}$ demanded of a distribution in Chap. V, (34.6).

It is equally necessary to show that two tempered distributions cannot define the same distribution on $\mathcal{D}$ unless they are identical[51]. By difference, it is enough to show that if $T(\varphi) = 0$ for every $\varphi \in \mathcal{D}$, then also $T(f) = 0$ for every $f \in \mathcal{S}$. Since $T$ is a *continuous* linear form on $\mathcal{S}$ it is enough to exhibit a sequence $f_n \in \mathcal{D}$ which converges to $f$ in $\mathcal{S}$, i.e. to show that $\mathcal{D}$ is *"everywhere dense"* in $\mathcal{S}$, like $\mathbb{Q}$ in $\mathbb{R}$, like the trigonometric polynomials in the space of continuous functions on $\mathbb{T}$, like the usual polynomials in the space of continuous functions on a compact interval, etc.

So let us start from a function $\varphi \in \mathcal{D}$ equal to 1 for $|x| < 1$, for example the function employed in Chap. V, n° 29, to prove the existence of $C^\infty$ functions having arbitrarily given derivatives at a point. Let us put $\varphi_n(x) = \varphi(x/n)$, a function equal to 1 for $|x| < n$. We shall see that, for every $f \in \mathcal{S}$, the $f_n(x) = \varphi_n(x)f(x)$, which are clearly in $\mathcal{D}$, answer the need, in other words that

$$(32.9) \qquad \lim N_r(f - f\varphi_n) = 0 \quad \text{for every } r \in \mathbb{N}.$$

This is equivalent to saying that all the functions $M^p D^q(f - f\varphi_n)$ converge to 0 *uniformly on* $\mathbb{R}$. Now, by Leibniz,

---

[51] The reader may accept the result, which is not of serious importance in what follows.

$$(32.10) \quad M^p D^q (f - f\varphi_n) \; = \; M^p \left[ D^q f - (D^q f.\varphi_n + \ldots + f.D^q\varphi_n) \right]$$
$$= \; M^p \left( 1 - \varphi_n \right) D^q f - M^p(\ldots)$$

where the terms inside the sign $(\ldots)$ contain derivatives of $\varphi_n$, i.e. functions of the form $n^{-k}\varphi^{(k)}(x/n)$ with $1 \le k \le q$. Such a function is everywhere bounded in modulus by $n^{-k} \left\| D^k\varphi \right\|$, so that the sum of the terms considered is, for every $x$, bounded in modulus by

$$\sum_{1\le k\le q} ?n^{-k} \left\| D^k\varphi \right\| . \left| D^{q-k} f(x) \right| ;$$

the signs ? denote binomial coefficients of no importance. If one applies the operator $M^p$ of multiplication by $(-2\pi i x)^p$ to these terms one obtains a function

$$\sum ?n^{-k} \left\| D^k\varphi \right\| . \left| x^p D^{q-k} f(x) \right|$$

with other coefficients ? independent of $f$ and of $n$. In passing to the sup for $x \in \mathbb{R}$ one finds a result less than

$$\sum ?n^{-k} \left\| D^k\varphi \right\| . N_r(f)$$

where $r = p+q$. Since the sum is over the $k \in [1,q]$ and since $n^{-k} \le 1/n$, the final result, up to a constant factor independent of $\varphi$ and of $r$, is bounded by $N_r(f)/n$. For $f$ given, it is therefore $O(1/n)$.

It remains to examine the term $M^p \left( 1 - \varphi_n \right) D^q f$ in (10). Since $\varphi_n(x) = 1$ for $|x| < n$, this term vanishes for $|x| < n$. Ignoring the factors $-2\pi i$, its uniform norm on $\mathbb{R}$ is then in fact equal to

$$\sup_{|x|>n} |1 - \varphi_n(x)| . |x^p D^q f(x)| .$$

Now $|1 - \varphi_n(x)| \le 1 + \|\varphi\| = c$. Since $f \in \mathcal{S}$ the function $\left| x^{p+1} D^q f(x) \right|$ tends to 0 at infinity, so is bounded on $\mathbb{R}$; we deduce estimates of the form

$$|x^p D^q f(x)| \le c_{pq}/|x|$$

valid for every $x \in \mathbb{R}$. The sup for $|x| > n$ is thus, also, $O(1/n)$.

Combining these two results, we see that

$$\left\| M^p D^q \left( f - f\varphi_n \right) \right\| = O(1/n)$$

for any $p$ and $q$, which proves that $f = \lim f\varphi_n$ in the topology of $\mathcal{S}$, qed.

This done, it is immediate to define the *Fourier transform* $\hat{T} = FT$ of a tempered distribution $T$: one puts, in Leibniz' notation,

$$(32.11) \qquad \int f(y) d\hat{T}(y) = \int \hat{f}(x) dT(x)$$

384     VII – Harmonic Analysis and Functions Holomorphic

or, in that of the inventor,

(32.11')     $\hat{T}(f) = T(\hat{f})$     for every $f \in \mathcal{S}$.

Since the map $f \mapsto \hat{f}$ of $\mathcal{S}$ in $\mathcal{S}$ is continuous, so likewise is $f \mapsto T(\hat{f})$, and so is $f \mapsto \hat{T}(f)$, whence one obtains a tempered distribution.

One may also, as in Chap. V, n° 35, define the *derivative* – again tempered – of $T$ by

(32.12)     $T'(f) = -T(f')$     for every $f \in \mathcal{S}$,

and iterate the operation. To calculate the derivative $D\hat{T}$ of $\hat{T}$ one has to write

$$D\hat{T}(f) = -\hat{T}(Df) = -T(FDf)$$

where $F$ is the Fourier transform in $\mathcal{S}$; but (31.5') shows that $FDf = -MFf$; thus

(32.13)     $$D\hat{T}(f) = T(MFf).$$

Putting $D\hat{T} = S$, this can be written

$$\int f(x)dS(x) = \int (-2\pi iy)\hat{f}(y)dT(y) = -\int \hat{f}(y)2\pi iy dT(y).$$

Thus we see the distribution "of density $2\pi iy$ with respect to $dT(y)$" appear; if $T$ were of the form $p(y)dy$ with a reasonable function $p$ we would thus obtain the distribution $-2\pi iyp(y)dy$. So it is natural to write $MT$ for the distribution $-2\pi iydT(y)$, the ordinary product of $T$ by the function $-2\pi iy$; it is again given by[52]

(32.14)     $MT(f) = T(Mf)$     for every $f \in \mathcal{S}$.

This done, (13) can be written

(32.15)     $DFT(f) = MT(Ff) = FMT(f)$

by definition of the Fourier transform $FMT$ of $MT$. In other words, the formula $DF = FM$ remains valid for tempered distributions. One can show similarly that $MFT = -FDT$: the Fourier transform exchanges the operators of differentiation and of multiplication by $-2\pi ix$ in the context of functions or of distributions.

---

[52] The formula has a meaning only because multiplication by $2\pi iy$ maps $\mathcal{S}$ continuously into $\mathcal{S}$. One may define $p(y)dT(y)$ for every function $p$ which is $C^\infty$ and such that $f \mapsto pf$ maps $\mathcal{S}$ into $\mathcal{S}$. This assumes that $p$ *and* its successive derivatives do not increase more rapidly at infinity than powers of $x$ ("tempered functions"): the product of a function "of slow increase" by a function "of rapid decrease" is again of rapid decrease.

If for example $f$ is a regulated function which is $O(|x|^N)$ at infinity for an integer $N > 0$, then the formula

$$T_f(\varphi) = \int \varphi(x)f(x)dx$$

has a meaning for every $\varphi \in \mathcal{S}$ and defines a tempered distribution; its Fourier transform is, by definition, the Fourier transform of $f$; it goes without saying that it is not a function in general.

In particular let us take $f(x) = x^p$ with $p \in \mathbb{N}$. Then

$$\widehat{T_f}(\varphi) = T_f(\hat{\varphi}) = \int \hat{\varphi}(y)y^p dy \quad \text{for } \varphi \in \mathcal{S};$$

multiplying by $(-2\pi i)^p$, one makes the function $M^p F\varphi = (-1)^p F D^p \varphi$ appear in the integral. Then

$$(2\pi i)^p \widehat{T_f}(\varphi) = \int F D^p \varphi(y)dy.$$

But since $D^p\varphi$ is in $\mathcal{S}$ we may apply the Fourier inversion formula to it, whence

$$(2\pi i)^p \widehat{T_f}(\varphi) = D^p\varphi(0) = \delta\left(D^p\varphi\right),$$

where $\delta$ is the Dirac measure at the origin, clearly a tempered distribution. In view of definition (12) of the derivative of a distribution, the result can be written

$$(2\pi i)^p \widehat{T_f} = \delta^{(p)},$$

the derivative of order $p$ of the distribution $\delta$. For $p = 0$, we see that the Fourier transform of the function 1 is the Dirac measure at the origin: this is exactly what formula $f(0) = \int \hat{f}(y)dy$, valid for $f \in \mathcal{S}$, means.

In conclusion we remark that all this generalises to functions of several variables. See for example the excellent Chap. 3 of Michael E. Taylor, *Partial Differential Equations. Basic Theory* (Springer, 1996) or the ultracondensed exposition of Lars Hörmander, *The Analysis of Linear Partial Differential Equations*, Vol. 1 (Springer, 1983).

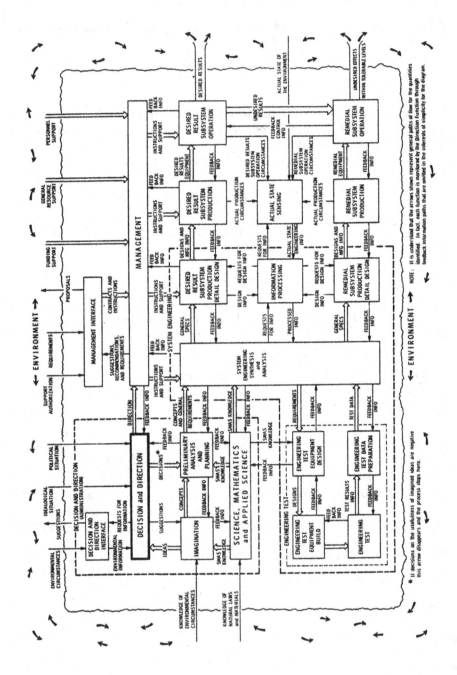

Extrait de C. Stark Draper, "Critical Systems and Technologies for the Future", in *International Cooperation in Space Operations and Exploration*, vol. 27, Science and Technology, 1971 (American Astronautical Society).

# Postface

# Science, technology, arms

The text below is only a part of the postface that was announced in the preface to volume I; the full text, with references to sources, is substantially longer than the French version, already 90 pages long. It will be available to interested readers on the Internet, at the following address:

www.springer.online.com/de/3-540-20921-2

Readers who wish to understand why a mathematics textbook includes the text below will find explanations in the preface to volume I.

I have tried to be as pedagogical as possible, but since this postface deals with many topics far removed from mathematics, it will, of course, require some work and good will from the reader to understand it.

Many sources have been used, and all of them will be found in the internet version. A few have been mentioned in the printed text.

Italics have been used for verbatim quotations in the main text.

## § 1. How to fool young innocents

The H-bomb was born in September 1941 at Columbia University in New York during a conversation between Enrico Fermi and Edward Teller. The explosion of an atomic bomb based on the fission of U-235 or Pu-239 nuclei could generate the tens or hundreds of million degrees necessary for the fusion of hydrogen nuclei, which in turn would generate amounts of energy hundreds of times greater than that of the atomic bomb itself. This was nothing more than a very rough idea, but Teller and others already knew (or believed) by 1942 that, if a 30 kg mass of U-235

> is used to detonate a surrounding mass of 400 kg of liquid deuterium, the destructiveness should be equivalent to that of more than 10,000,000 tons of TNT [the standard military explosive]. This should devastate an area of more than 100 square miles.

Yet the development of the A-bombs which destroyed Hiroshima (U-235) and Nagasaki (Pu) was top priority during the war, so that nothing much happened for several years although a few people around Teller continued their theoretical studies of the problem; after a team of physicists reviewed the issues in the spring of 1946, even Teller went back to theoretical physics at

Chicago. Calculations were very difficult to carry out, they neglected physical effects that opposed the fusion reaction, the choice of which isotopes of hydrogen to fuse was not easy, and the geometrical configurations they were drawing up could not work or, if they did, could not lead to the ultimate weapon, namely something with theoretically unlimited power. Last but not least, experimental verification of the computations was impossible short of exploding an actual weapon. In addition, many influential physicists were against the development of a weapon which they viewed as being far too powerful and which would most likely be imitated by the Soviet Union sooner or later.

The situation changed dramatically after the announcement in September 1949 by President Harry Truman of a first secret (but detected) Soviet atomic test. The General Advisory Committee (GAC) of the Atomic Energy Commission (AEC, now part of the Department of Energy, DoE) was convened to deal with the new situation at the end of October. Basically for ethical reasons, the GAC members (scientists J. Robert Oppenheimer, Arthur Compton, James Conant, Enrico Fermi, Lee A. DuBridge, Isidor I. Rabi, Cyril Stanley Smith, as well as the Bell Labs president, Oliver E. Buckley, and Hartley Rowe, an engineer) were unanimous in their opposition to the development and production of the H-bomb, though they were not against further theoretical studies; they recommended the production of more fission bombs – new types under development, up to 500 kilotons (KT), were deemed powerful enough to deter the Soviets -, including "tactical" ones (for use in Europe...), and they recommended providing by example *some limitation on the totality of war*; when briefed by Oppenheimer, the tough Secretary of State, Dean Acheson, a friend and admirer, replied: *How can you persuade a paranoid adversary to disarm "by example"* ? Other scientists, like Teller and Ernest Lawrence who were not GAC members, were also strongly in favor and briefed the president of the Congressional Committee on Atomic Energy and top men in the Air Force, who began to call for it. Three of the AEC administrators (including the AEC President) were against it, and the other two for it, including Lewis Strauss, a most influential and conservative Wall Street tycoon who, like Teller, was as "paranoid" as Joe Stalin, and did not hesitate to go straight to Truman. The H-bomb supporters rejected the idea that America might come out second in the H-bomb race; and in an America again made fiercely anti- Communist by the Soviet domination of Eastern Europe, by the 1948 attempt to blockade Berlin, and by the "loss" of China to the Communists in 1949, the overwhelming majority of people also wanted supremacy over, not parity with, the Soviet Union. Furthermore, the near total American demobilization in 1945 and the rejection of Universal Military Training meant that reliance on atomic weapons was America's only means of deterring, slowing down, or resisting the onrush of a Red Army which, after demobilizing, still retained about three million men and compulsory military training, even though the said onrush was considered by people in the know to be quite improbable within five years.

As President Truman said at the time, *there was actually no decision to make on the H-bomb*: he shared these arguments and, at the end of January 1950, after three months of totally secret discussions involving about one hundred people, he publicly announced that the development of the H-bomb would continue; he also forbade people connected with the AEC, including GAC members, to discuss the subject in public.

In early February, thanks to the partial decrypting of Soviet wartime telegrams, the unfolding of the Fuchs affair in Britain a few days earlier proved that the bright ex-German Communist physicist sent by the British to Los Alamos in 1943 had transmitted to the Soviets not only essential data on the A-bomb, but also most probably what was known on the future H-bomb up to April 1946: he had even taken a patent out on it, in common with von Neumann! The Soviets thus knew America was on the H-bomb trail, and America knew that the Soviets might also be working on it, as Teller had claimed – rightly, but without proof – long before. In March 1950, on the advice of the military, who did not need this new piece of information to make up their minds, Truman, this time secretly, made H-bomb production a top priority.

The correct physical principles were not even known. Numerical calculations carried out by mathematicians John von Neumann at Princeton and Stanislas Ulam at Los Alamos, and performed partly on the new but insufficiently powerful electronic machines, confirmed that Teller's ideas could not lead to the weapon he had been dreaming of since 1942; Teller's optimistic calculations still relied on incorrect hypotheses or missing data. One (theoretical) version of the weapon under consideration in 1950, which would develop a power of the order of 1,000 megatons, *was some 30 feet long, and a stunning 162 feet wide; the fission trigger alone weighed 30,000 pounds.* Technical follies, as Freeman Dyson would later say.

Anyway, developing the weapon had to be done at the Los Alamos laboratory where the A-bomb had been developed and where a reduced team had remained or had been recruited since Hiroshima. Although the outbreak of the Korean war led many top physicists to join the project, many members were on Oppenheimer's side as Teller knew full well, and he believed they were not enthusiastic enough to succeed. Supported again by Ernest Lawrence, the Air Force, and key Congressmen, Teller asked for the creation of a rival laboratory in 1950, but his request was denied by the Atomic Energy Commission. Teller was desperate at the end of 1950 and no longer sure a true H-bomb, with arbitrary large power, could be made.

But in January 1951, Ulam devised a new geometric configuration: to separate completely the atomic triggerfrom the material to be fused. It was seized by Teller who found an entirely new way to make the fusion work before the bomb blew up: the near-solid wave of neutrons flowing from the atomic explosion was too slow; instead, his idea was to use the X-ray burst from it to generate the necessary temperature and pressure. During a meeting at the Princeton Institute for Advanced Research (which Oppenheimer now headed) in June 1951, everyone enthusiastically agreed that this was the solution, and Teller got the laboratory he had asked for in September 1952. In November 1952, a test of the principles, using liquid deuterium and a good sized refrigeration installation, produced the 10 megatons (MT) predicted in 1942; it also vaporized a small island in the Pacific ocean. In April 1954, several tests of near-operational weapons using lithium deuteride, an easily stored white powder, produced between 10 and 15 MT – two or three times more than predicted, because one of the reaction phases had been overlooked. Operational weapons (10-15 MT) went aboard giant B-36 bombers from the end of 1954 to 1957; later ones never exceeded 5 MT and most were in the hundreds of KT range. All of these successes, and the great majority of later achievements too, were the work of Los Alamos people "lacking enthusiasm". The first true Soviet H-bomb was tested in November 1955 and produced about 1.6 megatons.

Set up at Livermore, not far from Berkeley, Teller's laboratory is now called the Lawrence Livermore Laboratory (LLL) and has been managed, at least officially, by the University of California since 1952, as Los Alamos has been since 1943. All American nuclear weapons were invented at these two places; while this still remains Los Alamos' basic activity, Livermore later concentrated a large part of its work on much more innovative scientific-military projects, as will be seen below. Lawrence won a Nobel prize for his invention at Berkeley in the 1930s of the first particle accelerators (cyclotrons). To a large extent, this was made possible by philanthropists attracted by the potential medical uses of radiation or artificial radio-elements available much more cheaply and abundantly than radium. During the war, Lawrence initiated and headed a massive electromagnetic isotope-separation process inspired by his cyclotrons; you can gauge Lawrence's influence from the fact that the Treasury Department lent him over thirteen thousand tons of silver to wire his "calutrons", despite an endless series of unexpected technical problems which brought operations to a complete halt as soon as the war ended. They nevertheless performed the final enrichment, at 80% of U-235, of much of the partly enriched uranium obtained from another massive factory, where uranium hexafluoride – a very nasty gas – was blown through thousands of porous metallic "barriers"; the very primitive Hiroshima bomb used some 60 kg of the final product. Together with Oppenheimer, Fermi, Arthur Compton and Conant, as well as the Secretaries of War and State, Lawrence had participated in the June 1945 top- level discussions concerning the use of the first available bombs. They had also recommended a well-financed research program in nuclear physics, military and civilian applications, as well as weapons production.

It was to this most influential operator, whose Berkeley Rad Lab had strong connections with Los Alamos, that the Atomic Energy Commission entrusted in 1952 the task of setting up a new development center for the H-bomb. Livermore needed a director, and Lawrence chose one of his assistants, Herbert York, then 30 years old.

After Sputnik (1957), York took charge for a while of all American military research and development. Health problems forced him to cut down on his activities, and he "retired" to a California university, while still participating in negotiations and meetings on arms control. From 1970 on, he wrote articles and books about the arms race, the absurdity and danger of which he could now clearly see.

In 1976 he wrote a short book, *The Advisors*, recounting the development of the thermonuclear project and, in particular, the discussions which had taken place at the end of 1949 on the opportunity to launch a H-bomb development program. His book reproduces in full the recently declassified report in which the AEC's General Advisory Committee explains the practical and ethical reasons against it.

With a rare frankness, York discloses the reasons which led him to participate in the project after the start of the Korean War (which led some opponents of the H-bomb, like Fermi and Bethe, to change their minds). There was first *the growing seriousness of the cold war, much influenced by my very close student-teacher relationship with Lawrence*, a fierce anti-Communist like Teller, Ulam, and von Neumann. There was also *the scientific and technical challenge of the experiment itself*: it's not every day you get the opportunity to explode the equivalent of ten million tons of TNT for the first time in history (it was actually done by Los Alamos). There was also, and perhaps most importantly as every young scientist can understand,

> my discovery that Teller, Bethe, Fermi, von Neumann, Wheeler, Gamow, and others like them were at Los Alamos and involved in this project. They were among the greatest men of contemporary science, they were the legendary yet living heroes of young physicists like myself, and I was greatly attracted by the opportunity of working with them and coming to know them personally.

Moreover,

> I was not cleared to see GAC documents or deliberations, and so I knew nothing about the arguments opposing the superbomb, except for what I learned second hand from Teller and Lawrence who, of course, regarded these arguments as wrong and foolish. (I saw the GAC report for the first time in 1974, a quarter of a century later!)

In less than one page, you have something similar to the corruption of a minor taking place in the scientific milieu: you are told that the enemy is threatening your country, the scientific problem is fascinating, great men you admire set the example, other great men you don't know personally are opposed to the project but their arguments are top secret, those great men who are luring you carefully refrain from honestly telling you what these arguments are, and, anyway, you'll be able to read the official documents in 25 or 30 years if you are American, in 50 or 60 if you are French or British, and, at the earliest, after the fall of the regime if you are a Soviet citizen. If you are still alive, your delayed comments will have no impact whatsoever because the project in which you participated was completed decades before, and its justifications have perhaps changed radically in the meantime.

This had already been seen in the A-bomb project: physicists were told (or claimed) in 1941 that the A-bomb was needed before the Nazis got one, it was discovered in May 1945, if not before, that they were years behind, but the bombs were still dropped: over a thoroughly defeated Japan. Quite a number of participants felt they had been fooled, even though they did not know, as we now do, that three weeks after Hiroshima, the Air Force sent General Groves, head of the Manhattan Project, a list of two dozen Soviet cities and asked him to provide the weapons (which was not done until 1948), while Stalin was giving absolute priority to his own atomic project. And nobody then – except perhaps Groves – imagined that tens of thousands of bombs would eventually be produced.

> Main references: Herbert York, *The Advisors. Oppenheimer, Teller, and the Superbomb* (Freeman, 1976), Stanislas Ulam, *Adventures of a Mathematician* (Scribner's, 1976), Richard Rhodes, *The Making of the Atomic Bomb* (Simon & Schuster, 1988) and *Dark Sun. The Making of the Hydrogen Bomb* (Simon & Schuster, 1995), Gregg Herken, *Brotherhood of the Bomb. The Tangled Lives and Loyalties of Robert Oppenheimer, Ernest Lawrence, and Edward Teller* (Henry Holt, 2002), Peter Goodchild, *Edward Teller. The Real Dr. Strangelove* (Harvard UP, 2004), David C. Cassidy, *Oppenheimer and the American Century* (PI Press, 2005).

York may not have been alone in this kind of situation; as Gordon Dean, AEC president 1950-1954, said at the Oppenheimer security hearing in 1954:

> We were recruiting men for that laboratory [Livermore], I would say practically all of whom came immediately out of school. They were young Ph.D.'s and some not Ph.D.'s (...) Under Lawrence's administration, with Teller as the idea man, with York as the man who would pick up the ideas and a whole raft of young imaginative fellows you had a laboratory working entirely – entirely – on thermonuclear work.

Livermore's then two divisions (thermonuclear and fission) were headed by Harold Brown, then 24, and John Foster, then 29; they both were later to head Livermore, then all military R&D, and even the Department of Defense (DoD). I don't know whether, once past the age of innocence, some of these "young imaginative fellows" reflected on their past as York did.

I do know of other similar cases though. Theodore B. Taylor (1925-2004), on hearing about Hiroshima, vowed never to have anything to do with atomic weapons, but he studied physics. In 1948, believing he was working for peace, he joined Los Alamos where he developed a fascination and a gift for improving atomic weapons. He invented the best A-weapons of the time, including a 500 KT fission weapon which, in May 1951, succeeded in fusing a few grams of deuterium; he also became an expert in predicting the effects of nuclear weapons. He left in 1956 for General Atomic (founded by one of Teller's colleagues) and the design of nuclear reactors, then headed the development of a spaceship propelled by multiple small atomic explosions and able to send people to Mars and beyond – the Nuclear Test Ban treaty prohibiting atmospheric tests killed that project in 1963; in 1964 he was put in charge of the maintenance of nuclear weapons, in 1966 he resigned and worked for a while with the international Vienna agency (AIEA) responsible for controlling the civilian nuclear energy business. His initial taste for weapons turned into its very opposite, notably after a visit to Moscow when, looking at the crowd in Red Square, he remembered he had helped the Air Force select the weapons best adapted to targets around the city, the Kremlin being most probably number one on the list. He spent the rest of his life advocating the abolition of nuclear weapons and nuclear energy which, he believed, would lead to an uncontrollable proliferation of weapons and even to their use by terrorists, a prospect he predicted in 1970 by emphasizing that the World Trade Center building could easily be felled by a small atomic explosion on its ground floor.

Recruiting young imaginative fellows at Livermore and other places is still going on, of course. William Broad, a *New York Times* science journalist who spent a week there in 1984 with a very special "O group" of young physicists twenty to thirty years old, explains in *Star Warriors* the role of the Hertz Foundation, founded shortly after Sputnik by Hertz Rent-a-Car's patriotic owner in order to maintain US technological preponderance (and to show his gratitude to a country which turned a poor immigrant into a very rich man). Every year the Foundation allocates about twenty five fellowships, valid for five years, to outstanding students; some of these are invited to spend a summer (or several years) at Livermore while preparing for their Ph.D. elsewhere. Those Broad met were asked to put their energies into problems at the cutting edge of technology with a not so obvious military interest: to build an optical computer using laser lines instead of electrical connections, to design from scratch and to miniaturize a supercomputer, to devise an X-ray laser, to elaborate a credible model of an atomic bomb using only published literature, etc. The group leader, Lowell Wood (who still sits on the Foundation Board together with several other Livermore or Los Alamos people), explained that:

The best graduate students tend to do very marvelous work because it's a win-or-die situation for them. There is no graceful second place.

If somebody else publishes the definitive results in the area, they go back to zero and start over (...) They don't realize how extremely challenging these problems are. So they are not dismayed or demoralized at first. By the time they begin to sense how difficult the problems are, they've got their teeth into them and made sufficient progress so that they tend to keep going. Most of them win. They occasionally lose, which is very sad to see (p. 31).

One of them, Peter Hagelstein, remembers his arrival in Teller's kingdom in 1975 when he was 20 years old:

The lab itself made quite an impression, especially the guards and barbed wire. When I got to the personnel department it dawned on me [!] that they worked on weapons here, and that's about the first I knew about it. I came pretty close to leaving. I didn't want to have anything to do with it [and his girlfriend was militantly opposed to it, which eventually destroyed their relationship]. Anyway, I met nice people, so I stayed. The people were extremely interesting. And I really didn't have anywhere else to go.

Hagelstein was asked to study the X-ray laser. He first spent four years, at the rate of 80-100 hours a week, learning the physics and doing computations with a very powerful program of his own. A senior Livermore physicist, George Chapline, had been trying for years to find a solution by using a nuclear explosion to get the energy needed to "pump" the laser (it is proportional to the cube of the frequency, which for X-rays is about 1000 times that of visible light). A first underground test in September 1978 was a failure because of a leak in a vacuum line. On Thanksgiving Day 1978, some senior physicists – including Wood, Chapline and an unwilling Hagelstein – were summoned to Teller's home to discuss the problem; Hagelstein was ordered by Teller to review the calculations done for Chapline – nothing more, but nothing less – and he had no choice but to comply. By the next day, he had to tell Wood there was a flaw in Chapline's theory, which put him in direct competition with Chapline. He found new ideas which he once dropped at a meeting in 1979, too tired after a 20-hour working day to realize what he was doing. They were seized upon at once and, he told Broad, he *had* [his] *arm twisted to do a detailed calculation* , under *political pressures like you wouldn't believe* . To his despair and with some prodding from Wood and Teller, his calculations and new ideas proved more and more promising, and in 1980 an underground test of his and Chapline's new designs proved Hagelstein's method was by far the better. He then had access to Livermore's gigantic laser lines, and his laser, though still virtual, got a name: Excalibur.

Hagelstein tells us of political pressures; no wonder. On the political side, for several years before Reagan's election, some very influential people – the Committee on the Present Danger – had been claiming that the Soviets were spending far more on defense than even the CIA said, and were re-arming to full capacity. As a matter of fact, since 1975 they had been deploying a few hundred new strategic missiles with multiple independent warheads (MIRV) (deployment of 840 American Minuteman-III MIRV missiles, and of 640 Poseidon submarine-launched similar missiles, had started in 1970 and 1971, respectively). They were also deploying very accurate middle-range SS-20s aimed at strategic targets in Western Europe and China. Their output of basic industrial goods (steel, coal, cement, etc.) was 50 to 100% higher than America's (but the American economy was converting to an "information society" far more efficient than Stalin's successors' taste for steel). They were discovering huge fields of oil and natural gas from which they got plenty of foreign currency, allowing them to buy (mostly American) grain and, much worse, advanced foreign machinery in spite of the US embargo on high-tech goods. "Marxists" were seizing power in several African states; unrest in Poland was repressed by the Polish army to avoid a Soviet intervention; the Red Army had intervened (unwillingly at first) in Afghanistan to defend the new Communist regime against its enemies, which many interpreted as a first Soviet step towards the proverbial Persian Gulf "warm waters" the Tsars had never managed to seize. The American deployment in Europe of equally dangerous American Pershing ballistic missiles and Tomahawk cruise missiles in answer to the SS-20s was opposed by strong "peace movements" that were suspected of being infiltrated by Soviet agents since, of course, ordinary German citizens were deemed too stupid to worry for themselves about these displays of atomic fire power. In short, the world had entered what became known as the *New Cold War* .

Thirty two members of the Committee on the Present Danger, including Reagan, occupied high administration offices after he came to the White House in January 1981. He immediately started to re-arm – the DoD budget, mostly financed by foreign capital attracted by high interest rates, went up from 181 BD in 1978 to 270 in 1984 in constant dollars -, and he continued to taunt the Soviets in speeches that culminated in his famous "Evil Empire" statement in 1983. However, many people in Washington, including Reagan himself, believed that in spite of its apparent strength, the USSR was under tremendous economic pressure with a grossly inflated military sector and a grossly underdeveloped civilian sector. They thought that a new round in the arms race would bankrupt the Soviets, or force them to agree to significant cuts in strategic armaments, or both.

There were already people in America trying to sell untested and wild anti-missile schemes, e.g. chemical lasers, 24 of which could supposedly destroy an entire fleet of Soviet missiles, or thousands of interceptors launched from hundreds of space stations. This led another bunch of conservative businessmen who had nothing to do with nuclear weapons, but were close to Reagan, to found a *High Frontier* committee, including Teller who wanted to sell his X-ray laser right away; they wanted to reach the White House without going through the Pentagon bureaucracy, where hard technical questions would be asked, of course. In this way, Teller was able to recommend a Los Alamos friend as Scientific Advisor, George Keyworth, who in turn appointed him to the new White House Science Council. A High Frontier report was sent to Reagan, and they got a fifteen-minute audience in January 1982; Teller apparently did not attend. They claimed that the Russians were well ahead in technology (as Teller had claimed to promote his H-bomb project), that they were close to deploying directed-energy weapons in space, thus altering the world balance of power. They recommended that America launch a major program to counter the Soviet threat in order to substitute *assured survival* for *assured destruction* , which suited Reagan quite well. Hagelstein's X-ray laser was the key to success and would be available within four years, followed by even more powerful versions. All of this rested on the secret results of a single test performed in a totally artificial underground environment.

Reagan, however, asked Keyworth to gather a team of experts from his Science Council in order to review the project before the end of 1982, if only to get an idea of the price tag. During this year, peace movements in America and Europe drew hundreds of thousands of people (and many scientists) demonstrating against the new arms race; many American Congressmen agreed. In June, a group of Livermore scientists who had the responsibility of continuing work after the first test, reported that the project would require ten more tests, six years, and 150-200 million a year to establish reliably that this laser was scientifically possible; it would then have to be transformed into an operational space weapon, which would require still more engineering, money, and years. This made Teller furious and all the more convinced that, as had been the case with the H-bomb, the project needed a lot of hype to take off. After complaining on TV that he had not yet met with President Reagan, he got an audience in September; some of those present interjected so many questions that Teller (and Keyworth) felt the meeting had been a disaster. In December, the House rejected funds for the production of a new and widely criticized generation of missiles, the MX, which could be randomly moved underground among many silos, most of them empty, in order to fool the Russian MIRV missiles. In January 1983, Teller got an audience with the Chief of Naval Operations; he was convinced by Teller's views and converted the Joint Chiefs of Staff; to them, it was at least *a way of convincing Moscow of the sheer financial power and technical superiority of the US* , as well as a new way to inflate the Defense budget since MX was becoming far too controversial. A meeting with Reagan in February 1984 ended in agreement; the military believed this would lead to an orderly development project, but Reagan did not wait. In March, to everyone's astonishment, he publicly announced his Strategic Defense Initiative (SDI, or Star Wars) project designed to protect the American people from Soviet missiles – a popular statement if ever there was one, which nevertheless did not placate the opposition.

In the meantime, in February 1983, a most famous nuclear physicist, Hans
Bethe, had gone to Livermore, reviewed Hagelstein's project and found it extremely clever physics, which did not mean clever weaponry. After Reagan
launched his Star Wars project, Hagelstein's laser became the most publi-
cized – and controversial – part of it though it was still, at best, years away
from any kind of operational status; a second test in March was actually
inconclusive due to a recording failure. The media explained that, propelled
into space by a single missile, individually oriented towards enemy missiles,
and "pumped" by a nuclear explosion, fifty X-ray lasers were expected to
destroy as many targets. Many physicists, foremost among them Hans Bethe
and Richard Garwin, were opposed to this new exotic hardware display and
said so publicly, because the chances of success were poor for many reasons –
the need for fantastically fast computers and communications (laser weapons
would be launched from submarines after the Soviet attack was detected,
they would have to spot missiles moving at a speed of four miles per second
and then orient the laser rods before firing, etc.) -, because nobody knew
whether the project would cost 150 or 3,000 billion dollars (BD) if successful,
and because it would only lead to one more spiral in the arms race and/or
could be easily defeated (as the Soviets at once remarked). Another official
panel reviewing the project came to rather pessimistic conclusions, relegat-
ing Reagan's dream to the year 2000 or so, and calling for a less ambitious
goal, while at the same time recommending one billion and top priority for
the laser, and 26 billion over seven years for the various other projects: SDI
had already acquired an immense political power by this time. During a pro-
paganda tour of Europe in 1985(?), SDI chief, General James Abrahamson
used plenty of sexy slides to explain it all at the Paris Ecole polytechnique (I
attended); this was a major contribution to the students' scientific education:
they (and I) did not know a thing about X-ray lasers, but you can trust them
to have "understood" everything within a week.

A few days after a successful test in December 1983, Teller sent an
overly optimistic report to Keyworth, without notifying anyone, not even Roy
Woodruff, a senior Livermore physicist who was deputy director for weapons
design and thus oversaw the X-ray laser group; Woodruff was furious and
wrote a corrective letter, which was blocked by Livermore's director. In the
Spring of 1984, other objections arose. According to Los Alamos scientists,
beryllium mirrors that were sending a fraction of the beam to recording instruments contained oxygen which, excited by the beam, possibly increased
the recorded brightness. The dispersion of the laser beam in space, the number of space stations and the power of the explosions needed were also publicly
criticized by independent scientists. But in Washington others noticed that
Soviet negotiators – who had been working for years on arms reduction –
were very concerned about this militarization of space and therefore might
be more accommodating, others again thought SDI would be a good oppor-

tunity to wreck the 1972 ABM treaty which seriously limited the deployment of anti-ballistic missiles.

By 1984, Hagelstein had lost his initial dislike for weapons:

> My view of weapons has changed. Until 1980 or so I didn't want to have anything to do with nuclear anything. Back in those days I thought there was something fundamentally evil with weapons. Now I see it as an interesting physics problem.

He did not have any illusions:

> I'm more or less convinced that one of these days we'll have World War III or whatever. It'll be pretty ugly. A lot of cities will get busted up.

In October 1984, Hagelstein and a team of forty people realized at long last the first "laboratory" X-ray laser, using a 150-meter long laser line pumped by capacitors discharging ten billion watts; this success was still very far from the operational weapon Teller was promising Reagan.

During this time, the Livermore group had devised the theoretical means to increase the laser power by several orders of magnitude, so that now "Super-Excalibur" lasers could be placed on a stationary orbit and still be able to kill missiles 20,000 miles away! At the end of 1984, Teller wrote through Wood to Paul Nitze, since 1950 the top expert in arms-control negotiations:

> a single X-ray laser module the size of an executive desk which applied this technology could potentially shoot down the entire Soviet land-based missile force, if it were to be launched into the module's field of view.

Woodruff was again by-passed but learned of the letter; he again tried to send a corrective one, which again was blocked. However, in February 1985, he was allowed a two- hour meeting with Nitze, who said that *it's always good to get a bright skeptical mind on a problem* . The initial results of a new and very elaborate test seemed so good in March that Teller's constant lobbying did pay off: hundreds of millions were released.

That same month, Mikhail Gorbachev came to power in the USSR, with a quasi-revolutionary program to transform the Soviet Union into a near-democracy and to terminate the arms race and the Cold War, which Reagan wanted too (but by other means). Although his scientists told him that SDI could be neutralized for 10% of the price to America, he decided to focus the US-Soviet arms-control talks on removing SDI in exchange for heavy cuts in missiles. Reagan met him in Geneva in November 1985 and, although the meeting was rather friendly, Gorbachev told him he should not count on bankrupting the Soviet Union or achieving military predominance, and that SDI would render impossible the expected 50% reduction in missiles. Reagan replied by extolling the virtues of defense, as usual. They continued to correspond for months in the hope of getting some kind of agreement; many of Reagan's aides and top military (not to mention Europeans, including France's president Mitterand) were appalled by Reagan's apparent willingness to dump American missiles provided he could keep SDI.

In October, at the annual conference of Los Alamos and Livermore people on nuclear weapons, Los Alamos scientists reiterated in detail their skepticism over the test results or even the existence of the X-ray laser; this allowed most members of the X-ray laser group to understand for the first time that these objections were serious. And Los Alamos people accused Livermore managers of abdicating their prerogatives to Teller and Wood, who, of course, claimed Los Alamos were trying to sabotage their project for political reasons or out of rivalry. This was enough for Woodruff, who resigned from his position. Teller's predictions, however, became somewhat more careful, and he emphasized that defense would be efficient even if it were only 20% effective because enough US missiles would survive to deter the Soviets attacking in the first place.

In November, a new and very expensive test (30 MD) resulted mostly in failure. Some Livermore scientists, who were already exasperated by Wood's authoritarian and sarcastic manner and by Teller's constant meddling in their work, left the project; as one of them said in 1989,

> To lie to the public, because we know that the public doesn't under-
> stand all this technical stuff, brings us down to the level of hawkers
> of snake oil, miracle cleaners and Veg-O-Matics.

Although he dismissed Los Alamos objections, Hagelstein too was disgusted by Teller's and Wood's extravagant public claims and by the bad faith the main protagonists displayed; as he told Goodschild in 2000, *I could not believe people behaved in that way* . However, it is easy to understand why they did. These people with plenty of willpower had for decades been in charge of designing the awesome weapons on which US security was supposed to rest. They were under enormous political pressure, and billions of dollars had already been spent on or budgeted for their pet project. Their reputations and the laboratory's were at stake.

Hagelstein quit Livermore for the MIT Research Laboratory in Electronics, which had been conducting military research since 1945, and worked in quantum electronics and, later, "cold fusion". This is a very controversial and to this day unproven method of generating energy at room temperature by means of fusion reactions among metallic compounds of hydrogen and deuterium. His scientific reputation suffered greatly as a result. As to Woodruff, he was exiled to a tiny office ("Gorky West") and his salary cut for several years, a good illustration of the contradicting ethics governing open and classified research; he joined Los Alamos in 1990.

The conflict between the two laboratories surfaced in the newspapers, triggering another public but inconclusive discussion, since the relevant technical data were top secret. In 1986 several thousand scientists publicly pledged not to participate in SDI in spite of the promise of exciting problems to solve and plenty of money for their laboratories. Politically, Teller won; he was supported by the military, by influential Congressmen, and by Reagan who understood nothing but trusted the "father of the H-bomb"; Teller hesitated neither to rely on Reagan's faith, nor to use his own scientific self-confidence, reputation and authority to ruthlessly counter opponents.

As for the Star Wars project, it survived until Bill Clinton's election in 1992. In January 1986, Gorbachev proposed to get rid of all Euromissiles on both sides, and to eliminate all nuclear weapons by 2000, provided America gave up developing, testing and deploying space weapons. Reagan proposed instead to reduce strategic warheads to 6,000 on each side (this was achieved four years later under George Bush) and to redress existing conventional imbalances. In July, Reagan proposed scrapping all ballistic missiles within ten years while continuing research on SDI which, when operational, would be made available to all (!). They had a second meeting in Reykjavik in October 1986 during which extraordinary proposals were made on both sides with a view to eliminating nuclear weapons entirely and reducing conventional forces. Once more, SDI killed the agreement at the last moment. Gorbachev's advisors (who were as bewildered as their American counterparts by these proposals) told him that Congress would kill SDI for him anyway. He did not follow their advice, but they were right: Congress cut the SDI budget by one third and prohibited tests in space in December 1987. In the meantime, a Livermore friend of Teller's had found a new miracle weapon, *Brilliant Pebbles* : space stations firing thousands of sophisticated projectiles, full of electronics, which would collide with Soviet warheads. A third Reagan-Gorbachev meeting in Washington a few weeks later led to the end of Euromissiles.

The Cold War died in 1990 and with it the Soviet Union and SDI; a few years later, the French Riviera was invaded by a new brand of Bolsheviks: oligarchs. The life expectancy of ordinary Russians began to decline. The European Union eastern boundaries (and with it those of NATO, a clever way of assuaging nationalist feelings in Russia) are now the pre-1939 boundaries of the former USSR. Last but not least, it has been "proved" that socialism is a dead end (especially if confronted with savage aggression followed by a ruinous fifty-year arms race led by a far more powerful opponent).

When asked why SDI did not work, Teller recently replied with a shrug: because the technology was not ready. The X-ray laser had cost 2.2 BD, and Star Wars a total of 30 BD. America is now spending a mere ten billion a year to develop anti-missile weapons against lesser threats than the Soviet arsenal, while Livermore (as well as the French Atomic Energy Commission) is trying to achieve, among other projects, controlled nuclear fusion of hydrogen isotopes by means of convergent laser beams in the hope, going back to 1950, of transforming nuclear fusion into an inexhaustible source of energy, as was done much earlier with nuclear fission. This also allows weapons designers to gain a deeper knowledge of fusion processes so as to improve their computer programs.

Main references: William J. Broad, *Star Warriors. The Weaponry of Space: Reagan's Young Scientists* (Simon and Schuster, 1985 or Faber and Faber, 1986), Goodchild, *Edward Teller* , Martin Walker, *The Cold War* (Vintage, 1994), John Prados, *The Soviet Estimate. US Intelligence Analysis and Soviet Strategic Forces* (Princeton UP, 1986), Stephen I. Schwartz, ed., *Atomic Audit. The Costs and Consequences of US Nuclear Weapons Since 1940* (Brookings, 1998).

Before having a look at Ken Alibek's Soviet career in biological weapons (BWs) from 1975 to the fall of the Soviet Union, let me sketch their previous development. After Pasteur, Koch, Metchnikoff and others had founded microbiology, it became possible to produce large amounts of vaccines. It also became obvious that, if required, similar techniques could be used to cultivate pathogens. That it was not pure theory was shown when the 1925 Geneva Convention prohibited it. The USA did not sign it, but the USSR and Japan did; it seems that USSR began to develop a typhus weapon in 1928, while Japan installed a very successful secret laboratory and production unit in Manchuria in the 1930s. Britain started to study vaccines after 1936 and, after the Nazis had advertized their brand of ethics at Warsaw and Rotterdam, thought it advisable to develop BWs as a hedge against similar German ones (they were not studied seriously until 1943 and came to almost nothing). British scientists worked mainly with anthrax, a bacterium which is easy to cultivate and store by transforming into spores that stay virulent for decades. Conclusive experiments on sheep were done at Gruinard Island, off Scotland; it was still contaminated and off limits fifty years later. They made anthrax cakes in sufficient quantities to be able to kill a lot of German cattle (and some people as well).

In America, studies on BWs began in 1940, and a National Academy of Sciences (NAS) committee was set up a month before Pearl Harbor. Although its February 1942 report was inconclusive in the absence of practical tests, it recommended studying all possibilities (for defense, of course) including anthrax, botulin toxin, and cholera. The program involved the Chemical Warfare Service, the Department of Agriculture for anti-crops weapons, and 28 universities. Although behind Britain until Pearl Harbor, American industry quickly developed a far bigger military potential than Britain, which, in this domain as in others (atomic bomb, radar, jet engines, etc.), contributed experts and knowledge, including penicillin which was industrialized in America during the war.

A research center was set up at Camp Detrick and, in Vigo, Indiana, a factory equipped with twelve 5,000-gallon fermenters could in principle produce 500,000 four-pound anthrax bombs a month, or 250,000 filled with botulin toxin (lethal dose: one milligram). The Americans also investigated brucellosis, a *more humane* weapon which kills few people, but is highly contagious and makes its victims ill for weeks or months, thus overwhelming the enemy's health system. Weapons for use against Japanese rice crops were also developed. But Roosevelt was not very interested in these matters about which he was very ill informed, and he never made his position clear one way or the other.

In any case, peace came before this program became operational, and Vigo was leased to a private manufacturer of penicillin. In 1945 BWs were considered potentially at least as efficient as, and much cheaper than, the atomic bomb; and since they don't destroy real estate, you don't have to compensate the enemy and allies after the victory. But atomic weapons were viewed as a sufficient deterrent, performing realistic tests of BWs was impossible, and the new German neurotoxic gases (tabun, sarin, soman) killed much faster – in a few minutes – than BWs. So, at first, work on BWs was limited to laboratory studies. During the Korean War, the Americans were accused of having experimented with BWs; it is now generally believed they had not, but the war accelerated the arms race in all domains, including BWs. In both the US and the USSR, all kinds of bacteria – anthrax, plague, tularemia, yellow fever -, and later viruses, were studied and mass produced. From 1947, the Soviets worked on smallpox which, by now eradicated, was still killing some 15 million people a year in the world in the 1960s. They built huge research centers and production units, some in cities, such as Sverdlovsk. The CIA had reason to suspect the worst as U-2 and satellite observations showed installations looking very much like the American ones, for instance a test range on an island in the Aral Sea.

During the 1950s, scientists in both countries discovered that instead of storing or spreading bacteria as liquid cultures, it was far better to dry and deep-freeze them (lyophilization); this kept them dormant for long periods, even at room temperature. The result was then milled into an ultra-fine powder which, after being carried by the wind over possibly tens of miles, became virulent again in people's lungs. This process worked particularly well with anthrax, the pulmonary form of which is normally rare and difficult to diagnose and kills 90% of its victims unless they are administered massive doses of penicillin very early.

The "top secret" American programs were actually known to plenty of people and, like the use of chemicals to destroy jungles in Vietnam, met with opposition from journalists, students, and biologists like Harvard's Matthew Meselson and Joshua Lederberg; the latter, who won a 1958 Nobel prize for his discovery of how bacteria can exchange genes in a natural setting, was in a good position to know that fast progressing molecular biology *can be bent to genocide* , as he wrote in the *Washington Post* in 1968. During the Vietnam war, opponents, particularly students, organized public demonstrations against Fort Detrick, as well as protests against military- university contracts and the National Academy of Sciences' involvement in recruiting young scientists for Fort Detrick. For their part, the military were not yet convinced of the usefulness of these weapons; proliferation was too easy and too cheap, and terrorist attacks were already being mentioned. Eventually, President Nixon unilaterally announced in November 1969 that America would limit herself to purely defensive work, and he ordered the destruction of stocks and the demilitarization of Fort Detrick, Pine Bluff and other centers; I remember a *Science* headline: *Is Fort Detrick really de-tricked*? In 1972, an international treaty between the US, the USSR and Britain, later approved by many other countries, prohibited the production and possession of biological weapons, but not defensive laboratory work; it did not provide for inspections either.

Before 1972, and although "weaponizing" pathogens required solving difficult technical problems, only natural bacteria and viruses were used. In 1972-1973, American biologists succeeded in systematically moving a gene from an organism to a bacterium in such a way that the modified bacterium would replicate itself as usual; their first experiment yielded a variant of the normally harmless Escherichia Coli that was resistant to penicillin. Thus genetic engineering was born and, with it, the possibility of discovering, by chance or on purpose, new pathogens from which no protection was known. But in the USSR, molecular biology and Mendelian genetics had been almost destroyed by Lysenko in the 1930s, and Soviet scientists were increasingly frustrated at the thought of being left behind. According to Alibek, the situation changed when a vice president of the Akademia Nauk, Yuri Ovchinnikov, explained to the Ministry of Defense and to President Brezhnev that bioengineering could lead to new weapons.

This led to the founding in 1973 of an officially civilian pharmaceutical organization, Biopreparat, under the Ministry of Health. Biopreparat's open mission was to develop and produce standard vaccines and antibiotics, but it enclosed a supersecret "Enzyme" project whose purpose was to develop and produce for intercontinental war *genetically altered pathogens, resistant to antibiotics and vaccines* , an outright violation of the 1972 treaty. It also led, as Ovchinnikov hoped, to a reversal of the taboo against genetics and molecular biology, and to new laboratories depending on the Moscow Academy since "purely scientific" work was paramount for "defense" against biological weapons. The timing was perfect: gene splicing had just been discovered, and its practical importance would soon be proved in the USA by using engineered bacteria to produce large amounts of insulin, hormones, etc. Enzyme, which was led by military scientists and administrators with KGB men everywhere, came to employ 32,000 workers, including many of the best biologists, epidemiologists, and biochemists, in addition to thousands of people working in Army labs.

Let us now go back to Alibek. Hoping to become a military physician who could save soldiers on the battlefields, he studied medicine at a military school and became interested in research. In 1973, he was ordered by one of his teachers to investigate a very unusual outbreak of tularemia which occurred around Stalingrad in 1942 among German troops before spreading to the Soviet army. After reading old documents, Alibek reported that this incident looked as though it had been *caused intentionally* . He was at once cut short by his teacher who told him he was only supposed to *describe how we handled the outbreak* , not what had caused it, and strongly advised *never to mention to anyone else what you just told me. Believe me, you'll be doing yourself a favor* . The lessons he drew from this episode are worth quoting:

> The moral argument for using any available weapon against an enemy threatening us with certain annihilation seemed to me irrefutable. I came away from this assignment fascinated by the notion that disease could be used as an instrument of war. I began to read everything I could find about epidemiology and the biological sciences.

In 1975, a mysterious and well tailored visitor came to interview him and other students; he said he was working for a no less mysterious *organization attached to the Council of Ministers* which *has something to do with biological defense* , a prospect which excited Alibek. He was handed a questionnaire and told: *Don't tell your friends or teachers about this conversation. Not even your parents* . A few weeks later, he learned he was assigned to the Council of Ministers of the Soviet Union together with four other students. He was overjoyed by the prospect of working in Moscow, but he was actually sent to a "post office box" hundreds of miles from Moscow. Like Hagelstein, he was impressed by the concrete wall and barbed wire surrounding the place and by the armed guards at the entrance. The huge Omutninsk Base where he arrived already employed some 10,000 people; it was part of the Enzyme project.

On arrival at Omutninsk, Alibek and his friends were not given any information about their research program. A KGB instructor however informed them that although an international treaty banning biological weapons had been signed in 1972, it was obviously *one more American hoax* , which they were quite prepared to believe; the Soviet Union therefore had to be ready to reply.

When Alibek began to discover Omutninsk's true mission – mass production of pathogens and not merely laboratory research -, he tried to get another job but was told he could not be spared. He thus remained and, after this classic early conscience crisis, adapted to the situation with enough success and enthusiasm to become Biopreparat's deputy director fifteen years later. The science and technique were fascinating and the career very rewarding provided you were bright, which he was, and made no big mistakes (such as inoculating yourself or being too talkative...).

The new recruits were trained in the culture of bacteria, the techniques being the same *whether they are intended for industrial applications, weaponization, or vaccination* . This is a difficult art which is first learned on harmless bacteria; one then has to learn how to infect lab animals with mildly pathogenic agents and conduct autopsies, until one may perhaps be allowed to work in "hot zones" with infected animals and where wearing the equivalent of a space suit is compulsory: half a dozen Ebola viruses will kill you in a month by destroying your blood vessels. A very competent colleague of Alibek once made a false move while inoculating an animal; after his death, they noticed the viruses in his body were particularly virulent, and therefore they weaponized this "Ustinov strain". One also has to learn industrial production processes.

Smallpox was modified to render all known vaccines useless. Diphtheria was grafted on plague. Sergei Popov, a bright colleague, improved Legionnella with fragments of myelin DNA to trigger metabolic reactions that devastate the brain and nervous system. The invention of a form of tularemia resistant to three of the main antibiotics, as well as studies on Ebola-like viruses took years of work. All in all, little produced by the genetic engineering programs was turned into weapons before the Soviet Union collapsed, according to Popov who has been living in the USA since 1992; Alibek also remains somewhat skeptical, though more pessimistic.

Incidents happened during this period. In April 1979, about sixty people died within a few weeks in the city of Sverdlovsk, an extremely unusual event. There was a Biopreparat branch located in the city, working round the clock on anthrax. A Russian magazine in West Germany broke the news of the outbreak in November, from which US intelligence agents again drew conclusions, despite claims that the deaths were due to contaminated meat. It is now known that a clogged air filter had been removed but not replaced for several hours...

In October 1989, Vladimir Pasechnik, a very bright scientist at the head of a civilian institute in Leningrad, went to France at the invitation of a pharmaceutical equipment manufacturer, and never came back. Since his institute had worked very efficiently for Biopreparat, he knew quite a lot. He was brought to Britain and debriefed.

Pasechnik's defection had serious consequences. In a memo to Gorbachev, KGB chairman Vladimir Kryuchkov recommended *the liquidation of our biological weapons production lines* , a stunning move which Alibek approved since, after all, *so long as we had the strains in our vaults, we were only three to four months away from full capacity* . Although many powerful people disapproved of Kryuchkov's initiative, Gorbachev issued a few weeks later a secret decree, prepared by Alibek and another fellow, ordering Biopreparat to *cease to function as an offensive warfare agency* ; but in transmitting Alibek's text to the Kremlin, his chief added a paragraph instructing the organization *to keep all of its facilities prepared for further manufacture and development* , which resurrected Biopreparat as a war organization, as Alibek says. He was furious but this, at any rate, allowed him to order an end to military development at some of the most important installations.

A second consequence was an agreement between the USA, UK and the USSR to organize inspections of suspected BW facilities. The first inspection of a few Biopreparat installations took place in January 1991; Alibek and the Russian side were very successful in showing as little as possible, but the visitors, who were aware of Pasechnik's disclosures, were not fooled.

In December 1991, during the week the Soviet Union collapsed, a visit to four American installations chosen by the Russians took place; they were known to anybody who had read *Science* magazine around 1970 (as I did). The Russian team included Alibek who could verify that these installations were in a dilapidated condition that precluded military work, or had been converted to medical research – work on the rejection of organ transplants fascinated the Russians -, or, in one case, had never done any military research. The Soviet delegation nevertheless reported to the contrary, and this convinced Alibek that official justifications for his work had been a KGB hoax rather than an American one.

He resigned from the Army, then from Biopreparat, got a job at once in a bank – *I had no aptitude for finance, but I was soon making deals like everyone else* -, and went on business trips abroad. His telephone was tapped, police watched him around Moscow, and some associates warned him that he had better not leave Russia for good and that in any case his family would never get permission to leave. In the meantime, a Yeltsin decree banned all offensive research and cut defense funding.

Alibek then went back to his native Kazakhstan, a newly independent country where a huge Biopreparat production center had been built years before. Local officials asked him to head a "medical-biological directorate" obviously intended for weapons research. He flatly rejected the offer, thus burning his bridges to both Russia and Kazakhstan, he tells us. Since he could still travel abroad for business, he was able to get in touch with Americans who were highly interested in his past and, with the help of a few Russians, managed to get himself and his family out in circumstances he obviously does not disclose.

While being debriefed in Washington, Alibek struck a friendship with his American counterpart, Bill Patrick, who had been at Fort Detrick for forty years and was then its chief scientist. Comparing the nature and timing of American and Soviet programs since the war, they came to the conclusion that at least one disciple of Klaus Fuchs must have been near the top of the US organization. After being kept under wraps for several years, Alibek went public and told his story in *Biohazard* (Delta Books, 1999). He is now the president of a new company, Advanced Biosystems, working on defense against biological weapons and employing, among other people, ex-Soviet scientists, e.g. Popov. And a good deal of cooperation with the US is helping former weaponeers in Russia to convert to peaceful research and to survive the rise in Lenin's country of the Robber Barons' variant of American capitalism.

Pyromaniacs, let us hope, are thus being transformed into firemen; a classic process. Nevertheless, the work is going on everywhere now, not only for "defensive" purposes in military laboratories, but also and mainly in perfectly harmless civilian labs by scientists who publish their findings in standard journals. Although many biologists have tried for decades to devise "ethical rules", knowledge is spreading, the techniques are becoming increasingly easier to learn, and weapons of mass destruction are now threatening their initiators in this domain, as atomic and chemical weapons did long ago.

References: Ken Alibek with Stephen Handelman, *Biohazard* (Delta Books edition, 2000), Judith Miller, Stephen Engelberg, and William Broad, *Germs. The Ultimate Weapon* (Simon & Schuster, 2001), Robert Harris and Jeremy Paxman, *A Higher Form of Killing. The Secret Story of Gas and Germ Warfare* (Granada Publishing edition, 1983). The potential of some of these weapons can be judged from Richard Preston's (real life) thriller, *The Hot Zone* (Random House, 1994, or Anchor, 1995).

The adventures of these weapons designers are, of course, extreme cases; I relate them here because extreme cases are extremely clear. In normal practice, a scientist and particularly a mathematician can only bring a small contribution to a complex weapons system. This does not raise such enormous and visible ethical problems as the development of H-bombs or biological weapons. But it only makes it easier for confusionists, mystifiers or corruptors to neutralize your objections.

More simply, one may be asked to solve a limited problem without being told of its military end. Although headed by the Department of Defense (DoD)*Advanced Research Projects Agency* (ARPA or DARPA), the Internet project – more accurately Arpanet, its predecessor – was to a large extent developed in a few university centers by many graduate students who were fascinated by it; many innovations are due to them. Contract holders ("Principal Investigators") had, of course, to provide ARPA with (sometimes vague or long term) military justifications, and some of the top people went from ARPA to universities or back. But, as Janet Abbate tells us in *Inventing the Internet* ,

> although Principal Investigators at universities acted as buffers between their graduate students and the Department of Defense, thus allowing students to focus on the research without necessarily having to confront its military implications, this only disguised and did not negate the fact that military imperatives drove the research (...) During the period during which the Arpanet was built, computer scientists *perceived* ARPA as able to provide research funding with few strings attached, and this perception made them more willing to participate in ARPA projects. The ARPA managers' skill at constructing an acceptable image of the ARPANET and similar projects for Congress ensured a continuation of liberal funding for the project and minimized outside scrutiny.

Military secrecy can only lead to similar situations.

That said, not everyone was fooled or seduced, as the case of Pierre Cartier shows. While a student at the Ecole normale supérieure in Paris around 1950, he was attracted by both mathematics and physics without at first being able to choose. He once told Yves Rocard – a physicist with strong industrial and military connections, who headed the physics lab at the school – that he wanted to work for a doctorate. Rocard then handed him a thick bundle of photographs; Cartier understood at once that these were a series of very close steps in an atomic explosion. Rocard proposed that he find a way of computing its power from these pictures, for instance from the propagation of the shock wave, or something similar. Cartier did not like the idea, still less Rocard's conditions: Rocard would help Cartier to get a good university position, but his thesis would remain secret, and he would have to sever his relations with his Communist friends, as well as with Rocard's son Michel, who was embarking on a political career (he became a Socialist Prime Minister thirty years later) and, at the time, had rather leftist opinions which were out of phase with Rocard's.

This decided Cartier to choose mathematics. He soon became a Bourbaki member and one of the best French mathematicians of his generation, still with a taste for mathematical physics, though not Rocard's brand. Of course, one can explain Cartier's reaction by the fact that, beside having strong religious beliefs, he was exposed to a much wider spectrum of political and philosophical opinions at the Ecole normale – where there are as many students in humanities as in science, all living together – than at Livermore or at a Soviet military school of medicine. Still, not everyone reacted the way he did. Thousands of scientists (and many more engineers) worked, and are still working, on military projects with no qualms.

## § 2. The evolution of R&D funding in America

All scientists of my generation know, if only vaguely and without proclaiming it too loudly, that WW II and the Cold War *did wonderful things* (I.I. Rabi) for science and technology; Rabi spent his whole career at Columbia University from 1928 to his death, was already a physics star by WW II, later a Nobel Prize winner, and a top government advisor for decades. I have sometimes been told by colleagues that a statement as "obvious" as Rabi's requires no proof, cafeteria gossip presumably being enough. If this is the case, then professional historians of science and technology might as well retire.

In this section, I'll first summarize the evolution of R&D in the USA since the war, since this country has clearly been the leader and even the model for half a century; Britain and France, as well as the Soviet Union, have always tried to follow America and to adopt its priorities, more or less, with differing results. R&D, for "Research and Development", means basic research (without any practical purpose in sight), applied research (with a more or less well defined practical purpose), and development, during which scientific results are used to design prototypes ready for production. These distinctions are not always very definite, and development usually requires solving many engineering problems, sometimes unexpected scientific ones, as well as extensive (and expensive) tests. Roughly speaking, basic and applied research cost 10 to 15% of R&D budgets each and development requires some 70% of it, but the proportions very much depend on the field.

The roughest measure of a country's R&D activities consists in comparing their total cost to the Gross National Product (GNP). In the USA, the proportion increased from 0.2% in 1930 and 0.3% in 1940 to 0.7% in 1945, 1.0% in 1950, 1.6% in 1954, 2.4% in 1958, and to a peak of 3.0% in 1964; at that time, US funds represented about 60% of all that was spent on R&D in OECD countries (North America, Western Europe, Japan, etc). As many articles, reports on "technological gaps", and books attested at the time, all other countries, and especially de Gaulle's France, looked at this 3% figure with an awe bordering on the mystical; someone joked that the optimal rate might be 3.14159...%. Since, moreover, the US GNP had climbed, in constant currency, from 100 BD in 1940 to about 300 MD in 1964, you can see that in this decisive quarter of a century, R&D expenses multiplied by ten in proportion with the GNP and by thirty in constant dollars! Such a miraculous growth rate could not, of course, be sustained: the R&D/GNP ratio began to fall as soon as it reached 3%, went down to 2.2% in 1978 and wavered between 2.6 and 2.8% between 1983 and 2000. The current and very optimistic goal of the European Community is to reach 3% by 2010.

In America as everywhere else, the two main sources of R&D funds are the Federal Government and private industry. Universities and not-for-profit private organizations also contribute, but on a much smaller scale, though their contributions to basic research may be important in some sectors. For instance, after having made a huge fortune at Hollywood, on the TWA airline, in buying hotels and casinos in Las Vegas and in selling planes to the Pentagon, Howard Hughes, like John D. Rockefeller long before him, set up a foundation whose trustees manage his little hoard, by now worth some 11 billion; the dividends support selected projects in medical research, by far the most popular field in America for a long time.

The relative importance of these two main sources of R&D funding has changed considerably since 1940. This is basically due to the nearly linear or weakly exponential growth of private industrial funds, while the fluctuations in federal funding were much larger, as will be shown.

In 1940, the figures (in current MD) for national total and for federal and industrial contributions were 345, 67 and 234, respectively. In 1945, they were 1520, 1070 and 430, respectively. In 1950, they were 2870, 1610 and 1180. Although data for these years are not entirely reliable, the trend is clear.

For each year between 1953 and 2000, data in *constant* (1996) MD are available in *Science and Engineering Indicators 2002* , an NSF publication easily available at nsf.gov/srs. It provides some significant figures:

|  | Total | Federal | Industry | Universities | Nonprofit |
|---|---|---|---|---|---|
| 1953 | 26805 | 14455 | 11670 | 190 | 286 |
| 1958 | 50439 | 32228 | 17130 | 256 | 492 |
| 1966 | 90236 | 57910 | 29971 | 673 | 1028 |
| 1975 | 89112 | 46289 | 39531 | 1078 | 1335 |
| 1982 | 122034 | 56200 | 61422 | 1821 | 1653 |
| 1987 | 162798 | 75468 | 80660 | 2916 | 2383 |
| 1994 | 176246 | 63316 | 103326 | 4100 | 3816 |
| 2000 | 247519 | 65127 | 169339 | 5583 | 5415 |

From less than 20% in 1940, federal contributions to the *total* R&D reached almost 62% in 1966, stayed over 50% until 1975, remained at 46% during the Reagan years (1980-1988) in spite of a sharp increase in federal (actually, military) funds, then decreased to 26% in 2000. It is only since 1980 that industry has been spending more than Washington. To a large extent, the proverbial "innovative capacity" of US private enterprise has been propelled by federal dollars for almost 40 years, and mainly by defense as shown below.

All federal agencies contribute to the funding of R&D. The Department of Defense (DoD) has been the most significant since 1941, followed by the Department of Energy (DoE, founded at the beginning of the 1970s, dealing with all kinds of energy, including the former Atomic Energy Commission, AEC, founded in 1946), NASA (or NACA, aeronautics, until 1958), the National Institutes of Health (NIH), and the National Science Foundation (NSF). Other federal departments together account for no more than 6% of the federal total, although their role, here too, is substantial in some fields. NSF annual statistics (*Federal Funds for Research and Development* ) provide a good, if probably not 100% accurate, view of their evolution.

In 1940, the government allocated 26 MD (current money) to defense R&D, 29 to agriculture and some to geology and mining; there was also a National Bureau of Standards which had been created in 1901 on the model of a German laboratory where much important research was conducted to determine accurate values for physical constants, weights, measures, etc. During WW I, the Washington Academy had created a National Research Council which did a lot of military research and was officially recognized after the war, but it got most of its small budget from private sources and spent it mostly on fellowships for young scientists. Otherwise, practically nothing went to research proper except for the creation in 1937 of a National Cancer Institute.

The picture had changed by 1945. Out of the 1590 MD in federal funds for R&D, agriculture still got 34, defense (atomic excluded) 513, the Manhattan Project (atomic) 859, and 114 went to the Office of Scientific Research and

Development (OSRD) created during the war to organize military research in all sectors. Not surprisingly, defense justified 90% of the total. During the war, industry spent less of its own funds on R&D than in 1940 in constant dollars, but, of course, received a flood of military contracts. Many universities received undreamed of amounts of money for military research: MIT 117 MD, CalTech 83 MD, Harvard 31 MD, Columbia 28 MD, to name but a few; new off-campus installations had to be set up for the most expensive projects. In 1950, out of 1083 MD in federal funds, agriculture got 53, DoD 652, AEC (essentially military at the time) 221, and NACA (similarly) 54 instead of 2 in 1940. Although Truman had considerably "restricted" the total defense budget after 1945 (13-14 BD until 1950, as against one in 1940), it remained large enough to finance a few large-scale technological projects, such as the development of the big jet bombers (B-47 and B-52) and supersonic jet fighters, progress in rockets and missiles, and the beginning of the development of nuclear submarines. The contributions of the main agencies are as follows for selected subsequent years, in *current* money:

|      | Total | DoD   | AEC/DoE | NASA | NIH   | NSF   |
|------|-------|-------|---------|------|-------|-------|
| 1953 | 1851  | 1275  | 278     | 84   | 59    | 0.151 |
| 1958 | 4774  | 3480  | 828     | 97   | 218   | 41    |
| 1966 | 16178 | 7099  | 1441    | 5327 | 1142  | 323   |
| 1975 | 19859 | 9179  | 2439    | 3207 | 2436  | 618   |
| 1982 | 37822 | 16786 | 5896    | 3708 | 3950  | 976   |
| 1987 | 57099 | 35708 | 5529    | 4096 | 6643  | 1531  |
| 1994 | 69450 | 34818 | 6959    | 8811 | 11141 | 2212  |
| 2000 | 77356 | 33215 | 6873    | 9754 | 18645 | 2942  |

These figures show the relative importance of the main federal sources of R&D money. DoD's contribution has always been, by far, the most important one, but to gauge the real size of defense-related funds, one should also take into account the AEC/DoE budget. In 1968, for example, out of a total of about 1600 MD, AEC's R&D budget included 400 for research proper (48 for weapons, 265 for physics, 86 for biology and medicine); 425 went to the development of weapons, 491 to the development of nuclear reactors, much of it for the Navy and Space, and 224 to construction work. It may also be assumed that NASA's R&D was not totally disconnected from defense even though the DoD itself spent between 500 and 1100 MD yearly on R&D for military astronautics between 1961 and 1965, and between 2 and 3 billion for the development of missiles. It may also be assumed that the CIA and the National Security Agency (NSA, cryptology, reconnaissance satellites, etc.), whose contributions are not reported, had sizable amounts to spend on R&D. And although much R&D for military industrial projects was to a large extent financed by the government even prior to any production, still some of it was private money.

On the other hand, the prospect of a federal budget surplus under Clinton prompted Congress to adopt a bill in 1998 to double the non-defense part of the federal R&D budget over ten years. This target was reached for the NIH by 2003, at least in current dollars, to the displeasure of specialists in other domains left behind.

The above table shows a substantial decrease of DoD funds after the Reagan years, but the trend was later reversed, courtesy of Mr Ben Laden. According to a recent analysis by the American Association for the Advancement of Science (www.aaas.org/spp/rd), out of the projected federal budget for R&D in the year 2005, the defense-related part, including 4.5 BD from the DoE, should amount to well over 74 billion, and the non-defense portion to over 57, of which NIH will get almost 30, Space over 10 and NSF 3.8. A new domain, antiterrorism R&D, will absorb 3 BD, of which 1.7 will go to NIH to fight bioterrorism, e.g. anthrax pocket weapons which are seen as a serious threat. Although the 2004 budget is the biggest ever since 1945, even in constant dollars, and far bigger than any other country's, America is able to afford it by devoting less than 4% of her GNP to total defense, as against at least 12% at the height of the Cold War. This is because GNP has grown at least five times in constant dollars since 1945.

The tables above make it possible to estimate the percentage of Defense money over total R&D, by converting current dollars into 1996 dollars. In 1958, defense-related federal funds for R&D accounted for 82% of all federal funds and 53.1% of national R&D expenses, hence more than industry's own contribution. In 1987 defense still accounted for almost one third of total R&D and 68% of federal R&D; it later decreased to a low of 13.6% in 2000 because of the growth of industry's own funding; Microsoft for instance is currently spending about 5 BD a year on R&D and presumably does not use the Pentagon's money to develop Windows, which may explain its quality... R&D is mostly development, but the importance of development in Defense is particularly striking: 2.9 BD out of 3.5 BD in 1958 and 28 BD out of 33 BD in 2000, with similar proportions in the interval. Industrial firms always get at least 60% of the DoD funds for R&D, while about 30% of the money is spent in DoD's own technical centers. According to the AAAS, only 5.18 BD should go to basic and applied research in 2005.

Some federal funds go to so-called Federally Funded Research and Development Centers (FFRDC). These were organized during or after WW II and are administered by industrial firms, universities, or nonprofit institutions. The first category includes huge centers such as Idaho, Oak Ridge, Sandia and Savanna River producing nuclear material or weapons, though on a very reduced scale now. The second includes the MIT Lincoln Lab (electronics, radar, SAGE, anti- missiles, etc.), the Jet Propulsion Lab (Cal Tech), Argonne (Chicago U.), Brookhaven (several universities) and huge installations for particle physics at Berkeley, Princeton, Stanford, etc.; last but not least, it also includes Los Alamos and Livermore labs initially founded for the development of nuclear weapons and administered by UC Berkeley, which didn't always relish it although it earned money from it. In the third category, there is the Rand Corporation which was organized in 1946 by Douglas aeronautics and the Air Force and soon became a research center financed by the Pentagon; it became famous in the 1950s for its development of operational research, game theory and mathematical programming, and for its slightly pathological strategic studies, particularly when Herman Kahn, in *Thinking the Unthinkable* and other books, made them popular by explaining nuclear war "escalation" theory (up to what he called a "nuclear spasm" or, as some said, "orgasm") as if it were a very funny poker game.

These cold figures should be supplemented with some more concrete information. As mentioned above, academic research got very little from Washington before the war; it was financed by university funds, philanthropic organizations and, in many engineering departments, by industry, enough to increase significantly the number of scientists during the inter-war period. The Rockefeller Foundation, which up to 1932 spent 19 million on academic research, spent a lot more on medicine than Washington. It also financed physics during the 1920s: thanks to its fellowships, many scientists, including future American designers of atomic bombs, learned their trade in Europe; European physicists were invited to America, some permanently; and the Foundation financed new laboratories in Copenhagen and Göttingen as well as the Poincaré, Institute in Paris. By 1930, and like many social scientists, it was having doubts over the value of physical sciences and technology: gas warfare in WWI had been rather bad publicity, as had the disruption of the American way of life and traditional values by technological advances. It therefore decided in 1932 to concentrate on applications of physics and chemistry to biology, which made it a prime sponsor for many of the future creators of molecular biology. Ernest Lawrence, and he alone, succeeded in attracting big money for his Berkeley cyclotrons: as much as one million in 1940 – a staggering sum at the time for physics – from the Foundation which betted on the prospect of cheap artificial radio-elements to fight cancer; otherwise, almost all of his money came from other philanthropists and the university. America had a good number of first class physicists by the 1930s; three dozen generally small particle accelerators were built in universities (Germany had none in 1940, France had one). In these depression years, particularly 1932-1934, attempts to get federal money were unsuccessful – almost all the New Deal relief money went to jobless people. Although senior scientists were generally comfortable, many younger ones were badly paid, and some unpaid ones spent part of their time making money to survive while continuing laboratory work. It is remarkable that the production of PhD's between 1930 and 1939, namely 980 in mathematics and 1924 in physics, was almost triple that in the preceding decade; this was mainly due to the strong growth of higher education in all domains. Without federal help to speak of, America was thus already the new dominant country in physics.

There were Jewish refugees in all intellectual domains after Hitler's seizure of power; though they were generally much younger, less well known than Albert Einstein, and not always welcome as Jews at the time, many American scientists helped them. After having a hard time until the war, most of the refugee scientists – almost 200 in mathematics and physics – were to find permanent university positions after 1945, and several dozen became leading scientists, or even stars. This also contributed to America's standing in these two domains, as in many others.

MIT, where many top American industrialists and engineers had been educated since the 1880s, already had the biggest electrical engineering department in the world, thanks to industrial contracts, gifts from alumni, and tuition fees. Private industry spent about 250 MD on R&D in 1940, partly in laboratories created fifteen or thirty years earlier by big companies like General Electric, AT&T, Westinghouse, or DuPont; they started doing some basic research in the 1920s. In 1925, AT&T, the private telephone monopoly, founded its Bell Labs, which soon became the largest industrial research laboratory in the world, with a 20 MD budget and some 2,000 employees by 1940; a physicist there won a Nobel prize for experiments on electron diffraction which confirmed the dual nature of elementary particles. Another Nobel Prize went to General Electric's physical chemist Irving Langmuir (who had its first success in 1913 in discovering that filling incandescent lamps with nitrogen greatly increased their life). At DuPont, a basic research program on polymers began in 1927, with initial funding of 250,000 dollars (to be compared with Columbia University physics department's budget of 15,000 dollars in 1939); from there came nylon in 1938, for the development of nylon in 1938; it cost about 2 MD and generated a 600 MD business twenty years later. There was also much R&D in the petroleum industry, with projects costing from a few hundred thousand to 15 MD. This figure looked enormous at the time.

References: David Noble, *America by Design. Science, Technology and the Rise of Corporate Capitalism* (Knopf, 1977), Daniel J. Kevles, *The Physicists. The History of a Scientific Community in Modern America* (Vintage Books, 1979), L.S. Reich, *The Making of American Industrial Research: Science and Business at GE and Bell, 1876-1926* (Cambridge UP, 1985), Pap Ndiaye, *Du nylon et des bombes. DuPont de Nemours, le marché et l'Etat américain, 1900-1970*, (Paris, Belin, 2001), Thomas P. Hughes, *American Genesis. A Century of Invention and Technological Enthusiasm, 1870-1970* (Chicago UP, 2004).

As previously mentioned, the war changed the picture. At MIT, a Radiation Lab was founded in order to develop radar; scientists of all levels worked there, including Hans Bethe (until 1943), Isidor I. Rabi and Lee A. DuBridge who headed the lab; Louis Alvarez and other young collaborators of Lawrence brought the expertise in electronics and high frequencies they had acquired in Berkeley; many of these people became very influential science advisors to the government after the war. At MIT and elsewhere, the work on radar required many advances in all domains of electronics, e.g. in high frequencies, or in semi-conductors because glass valves could not detect centimetric radar waves. Methods for purifying germanium were found at Purdue and were crucial to the invention of transistors a few years later, while Bell Labs did the same, with less success, for silicon. The size of the radar business can be gauged from the fact that the Rad Lab employed up to 4,000 people, while the industrial production proper cost almost 3 BD – more than the atomic bomb project.

Headed by General Groves, the Manhattan Project – that most spectacular success story, though less useful for winning the war – employed hundreds of scientists in Los Alamos and elsewhere; these included Fermi, Bethe, James Franck, Harold Urey, Arthur Compton, Lawrence, von Neumann, Alvarez, and even Niels Bohr, all of them (except von Neumann) past or future Nobel prizewin ners. Oppenheimer, a former Rockefeller fellow and the best native theoreti cian, headed Los Alamos with fantastic brio; he understood everything and made the whole enterprise succeed. He was under permanent surveillance by the FBI who were well aware of his pre-war leftist leanings and connections; this did not prevent the bombs' blueprints from quietly leaving Los Alamos for Moscow in a Plymouth driven by Klaus Fuchs in the summer of 1945. The project cost two billion, 70% of which was spent on the production elsewhere of U-235 in a gigantic isotopic separation factory or in Lawrence's calutrons, and of Pu in huge atomic piles. Most of the basic techniques later used in civilian nuclear energy were invented between 1942 and 1945, and this allowed General Electric, Westinghouse, DuPont and other companies to learn them and to become world leaders after the war in using nuclear power for electricity production, and first of all for the propulsion of submarines or aircraft carriers. More about this in the Internet file.

In 1945-1946, nuclear physicists were rewarded with millions left over from the Manhattan Project, which allowed them, among other consequences, to build new particle accelerators whose cost eventually came to billions (not millions). Before 1940, this prospect would have been dismissed as utterly insane. The AEC/DoE has funded this domain in America from 1947 to this day, while the Rockefeller Foundation withdrew its support after 1945 since the government could provide far more; in addition, since 1941 Lawrence and others had been hinting at spreading radioactive waste over or in front of enemy troops in case of war, which was not quite as glamorous as fighting cancer.

In a famous 1945 report, *Science, the Endless Frontier* , the chief of military R&D (OSRD) during the war, Vannevar Bush, advocated the establishment of a National Science Foundation funded by the government and whose president and programs would be chosen by scientists; the project was rejected by the President. It came into being in 1950 as a federal agency funded and governed by the government and controlled by Congress like other agencies, with, of course, plenty of scientific advisory committees; but it got very little money before Sputnik, as the table above shows. In the bio-medical sector, where a first National Cancer Institute had been founded in 1937, new National Institutes of Health were established; with strong backing from Congress and voters, they continued to grow and multiply and are now by far the most important non- defense source of federal money. Meanwhile, the Office of Naval Research founded in 1946 spent some 20 MD per year to help research in all domains, mainly to keep in touch -"in case" – with scientists and research; mathematics got about 10%, but a threateningly increasing part of it (up to 80% in 1950) was – already! – funding the development at MIT of a futuristic Whirlwind computer working in real time; a riot ensued, and Whirlwind would have died but for the birth of a far better sponsor in 1950, namely the air-defense system of the American continent, as we shall see later.

Private universities, where government interference was anathema before 1940, reversed their principles: ONR was very liberal and people got used to this new kind of "tainted money"; after all, nobody had ever asked trustees or benefactors of the rich universities how they became so wealthy; but it sometimes took several years before federal money (and possibly classified military contracts) were accepted. CalTech was still a small university in 1945; with a board of trustees made up of very conservative bankers and industrialists who approved the policy of basic research presided over by physicist and Nobel Prize winner Robert Millikan, it was several years before it bowed to the inevitable; meanwhile, the off-campus Jet Propulsion Laboratory founded by von Kármán prospered on guided missiles and DoD money, as was the case at Johns Hopkins with the Applied Physics Laboratory founded during the war. Julius Stratton, a future president of MIT who during the war had close ties with the higher echelons of the Pentagon – he was one of the stars of the MIT Radiation Lab -, wrote in October 1944 to MIT president:

> Twenty-five years ago everyone talked about the end of war; today we talk about World War III, and the Navy and Air Force, at least, are making serious plans to prepare for it. Inevitably this national spirit will react upon the policies of our educational and research institutions. It always has, and we might just as well face it (...) We shall have to deal with the Army and Navy and make certain concessions in order to meet their needs.

This means that by 1950, 85% of the MIT total research budget came from the military and AEC, with a still higher proportion for physical sciences in other elite universities. John Terman, another star in electronics, wrote in 1947 to his university's president that

Government-sponsored research presents Stanford, and our School
of Engineering, with a wonderful opportunity if we are prepared to
exploit it,

which of course they were. The importance to the military of these university departments was due not only to their research work, but also to their
educating thousands of scientists and engineers for defense work in particular.

References: Everett Mendelsohn, Merritt Roe Smith and Peter Wein-
gart, eds., *Science, Technology and the Military* (Kluwer, 1988), Paul
Forman & J.M. Sanchez- Ron, eds, *National Establishments and
the Advancement of Science and Technology* (Kluwer, 1996), Stu-
art W. Leslie, *The Cold War and American Science: The Military-
Industrial-Academic Complex at MIT and Stanford* (Columbia UP,
1993).

Various more or less successful attempts were made after 1950 to bring scientific advice to the highest levels of government, particularly the DoD; it
was Sputnik which brought scientists to the White House. Meanwhile, the
Korean War was an opportunity to organize "summer studies" during which
scientists, engineers and military men would gather for several weeks in order to study such (classified, i.e. secret) defense problems as anti-submarine
warfare, tactical nuclear weapons, air defense, etc.

The size of American defense activities in the 1950s and 1960s can easily
be explained by political factors and by reactions to perceived Soviet threats
(or counter-threats to perceived American threats: bombs, bombers to deliver
them, and the "encirclement" of the USSR by US air bases). As we have seen,
the first Soviet atomic test launched America into the race for the H-bomb.
In the spring of 1950, the celebrated NSC-68 report of the White House
National Security Council, vastly exaggerating the Soviet military threat and
supposed plans for world domination, recommended (among other things, e.g.
much stronger West European forces) a huge increase of the Defense budget;
the figures which were known but remained unwritten, namely 40- 50 billion
instead of 13-14, were judged excessive even by the military, who did not
know how to spend so much money. Truman did not agree either, but the
"Socialist camp" forced it on by sparking the Korean War. In particular, the
production capacity for U-235 and Pu was increased in a staggering way: five
new piles for the production of Pu, one for the production of tritium, and two
more huge isotopic separation units, with sixteen times the capacity of the
1945 factory, which had already been enlarged; up to 85 tons of U-235 could
be produced per year, which needed 6,000 megawatts of electricity, or 12% of
total US production. Nuclear weapons of all types grew in America at a rate
of several thousand per year, to reach 32,000 in 1964, with powers ranging
from a few tens of tons up to several megatons in TNT equivalent. This
was about fifteen times the Soviet arsenal at the time and could be delivered
by 800 intercontinental ballistic missiles (ICBM), 200 submarine-launched

missiles (SLBM), a thousand fighter- bombers based in Europe, the Middle East, Japan or on aircraft carriers, and strategic bombers (about 2,000 B-47 and 700 B-52 were built before 1962).

A gigantic system to defend America against Soviet bombers was built, as we are now going to see. The first Soviet atomic bomb led people at MIT to take the first steps to protect the USA from future Soviet bombers in 1950 (this threat was dismissed by Curtis LeMay, the Strategic Air Command (SAC) chief during the 1950s: his personal strategy was to wipe out Soviet planes, copies of the US B-29 bombers of 1944 vintage, before they could take off, but bypassing the President was slightly illegal...). This originally small Project Lincoln based on the Whirlwind computer led to the founding of a Lincoln Lab at MIT, and to the gigantic SAGE system of continental defense – a precursor to SDI -, at a cost of 30 billion (or 200 billion in 1996 money, and much more if personnel and other costs are included). Thousands of Bell Labs Nike-Hercules missiles, each carrying a 2 to 30 KT atomic warhead, could destroy *entire fleets of incoming aircraft* , assuming the Soviets were clever enough and able to send such fleets over the North Pole in suicide raids since, in any case, they could not make the round trip until big jets – never more than 200 – began to appear in 1955. Bell Labs, which had developed anti-aircraft rockets since 1945, managed everything while hundreds of subcontractors in practically all domains of technology helped develop the hardware and software needed in SAGE. SAGE was obsolete as soon as it became operational in 1960- 1962: bombers were replaced by unstoppable missiles after 1962, which led to the first and useless anti-missile systems, including the highly controversial Nike-Zeus missiles with 60 to 400 KT warheads, based around big cities and never deployed. The USSR's program evolved in similar fashion, but was even more expensive since missiles and bombers could come from many directions.

The SAGE project however played a major role in all kinds of technical advances, particularly long-range "over- the-horizon" radars, guided anti-aircraft weapons, and computers. In this last field, it led to magnetic core memories, video displays, light pens, graphics, simulation, synchronous parallel logic, analog-to-digital conversion and transmission of radar data over telephone lines via the first transistorized modems made by Bell, multiprocessing, automatic data exchange between different computers, etc. With its hundreds of thousand lines of code and hundreds of computer screens, SAGE provided the first opportunity to train several thousand programmers (most of whom later went to industry); this was done by the SDC branch of the Rand Corporation, which was founded in 1957 to that effect. Among many other machines, SAGE needed fifty six IBM AN/FSQ-7 and -8 (or "Whirlwind II") computers; there were twenty-four SAGE main command centers connected to a pharaonic installation under the Colorado mountains, itself connected to the White House and Pentagon; each of the centers used two of these IBM computers working in tandem to increase reliability. Made to order at a cost, in current money, of 30 million a piece, each of these machines weighed 275 tons, had some 60,000 valves, used 32-bit words, had a magnetic core memory – one of the great innovations from Whirlwind – of about 270 kilobits, twelve magnetic drums each storing 12,288 words of program, and was connected to about one hundred screens displaying enemy planes' trajectories and enabling operators to vector fighter planes graphically. It needed 750 kw of electric power to run and a hurricane to evacuate the heat it generated. These performances may look puny by 2005, but there was nothing more powerful at the time and, of course, the new techniques were put to good use in IBM's future commercial computers. All of the latter were transistorized after 1960, the first large ones (series 7090) being delivered to the three gigantic radars of the Ballistic Missiles Early Warning System in Alaska, Greenland and Scotland.

References on SAGE: chap. 4 of *Atomic Audit* , Edwards' chap. 3, Kent C. Redmond and Thomas M. Smith, *From Whirlwind to Mitre. The R&D Story of the SAGE Air Defense Computer* (MIT Press, 2000), very weak on technology, and Thomas P. Hughes, *Rescuing Prometheus* (Random House, 1998).

Then came Sputnik in October 1957, which scientists used very successfully to clamor for increased research funds. NACA was transformed into NASA, with very soon a budget in billions of dollars, while the Defense budget proper decreased. A scientific committee (PSAC) was instituted at the White House. The Advanced Research Projects Agency, ARPA, was founded by the DoD in order to fund and organize the most sophisticated research projects with military implications. Americans reacted to the "missile gap" with wild and shifting predictions on the size of the Soviet arsenal (100 in 1959, 500 by 1960 and 1,000 by 1961-1962) from the CIA, the Air Force, journalists, and democrat politicians, including Kennedy and especially Johnson, wishing to destroy the 1952-1960 Eisenhower republican administration. But radars from Turkey and Iran had detected Soviet missile tests in 1953-1954, and, from 1955 on, absolute priority was given to similar American programs, Atlas and Titan, soon followed by the silo protected Minuteman series of ICBMs, the Polaris missiles for nuclear submarines, and the first satellites for reconnaissance, infrared detection of missile firings, meteorology, communications, etc (1959-1961). Extended flights over Soviet territory first by U-2 spy planes, then satellites, proved in 1960 that there was indeed a big "missile gap": perhaps four Soviet operational missiles, to dozens of American ones.

Like the Korean War, Sputnik and Khrushchev's boasts proved to be a self-defeating move and another wonderful opportunity for the American and Soviet "scientific-military- industrial complexes". The Soviet arsenal, vastly outnumbered by the American arsenal until the 1970s, was nevertheless big enough to make an American attack unlikely, and in any case America's top political rulers found the Air Force's apocalyptic war plans quite repellent, although they knew they might have to "push the button" as a last resort (see my vol. I, p. 122; in 1960, over 150 weapons were reserved for the Moscow area alone, and quite a number of them would have destroyed each other). To paraphrase a journalist writing in *Science* , September 27, 1974, these huge defense systems were *the cathedrals of a century that future historians will characterize by its extraordinary technical capacities and its permanent devotion to the mortuary arts.* And so on, with ups and downs, until the fall of the Soviet Union. The most exotic parts of Reagan's Star Wars project were terminated, but a less ambitious anti-missile program is still going on, at the rate of several BD per year, with a first deployment in Alaska of weapons guided on a collision course with enemy missiles (a fascinating problem in Control Theory) although no one can guess who would be foolish enough to launch them. America's military doctrine is now undergoing a "Revolution in Military Affairs" based on "Space Dominance", which aims at fully integrating every weapon and everyone – from the President and the Pentagon warlords down to the GI on the battlefield – through all kinds of satellites, drones, telecommunications, information networks, etc. You will find an impressive survey of it in *Introduction au siècle des menaces* (Paris, Odile Jacob, 2004), by Jacques Blamont, a French specialist in Space Sciences with long and strong ties to the Jet Propulsion Lab (and, more recently, Soviet astronautics), and a member of the US Academy of Sciences. Another "revolution" has been under way since the Strategic Computing Initiative of the 1980s: substituting all kinds of "intelligent" robots for weak mortals on the battlefields of 2030, according to the New York Times (02/16/2005). Contracts worth 127 BD have already been issued for this Future Combat Systems project, which will contribute to boosting weapons acquisition costs from 78 BD now to 118 by 2010. Those who believed the end of the Cold War would slow down the technical progress of armaments were badly mistaken...

The development of nuclear weapons, fighter planes, bombers, missiles, nuclear submarines, aircraft carriers, SAGE, satellites for C4 RI (command, control of operations, communications, computers, reconnaissance and intelligence), etc. relied on and greatly encouraged technical progress in dozens of less spectacular domains: electronic components (from glass valves to transistors to printed circuits to integrated circuits to VLSI...), computer hardware and software, navigation and guidance systems, infrared detection, fire control devices, radar and sonar, microwave propagation, space telecommunications, materials, etc. The list is endless.

The development of transistors and integrated circuits is a good example. Semi-conductors had been known for a long time and were the first detectors used in wireless in the 1900s. Systematic experimental studies in the 1930s and during WW II, as well as the development of a solid-state theory using quantum mechanics, had led to a good understanding of the phenomena by 1945, and, at Purdue university, to methods of obtaining highly purified germanium (so named by its German discoverer), from which rectifying diodes were mass produced for radar detection. The Bell Labs did the same with silicon with less success at the time. After 1945, they tried to discover solid-state amplifiers, and the first very primitive point-contact transistors were made there in 1947 by two physicists, John Bardeen and Walter Brattain, headed by William Shockley who a few years later found a way to make industrialization far easier; all three shared a Nobel Prize. Transistors, patented by Bell in 1948, were expected to replace electronic valves and electro-mechanical switches in a myriad of devices used by the AT&T telephone system. But there was nothing urgent here – the capital invested in standard equipment was far too high to be scrapped – and, anyway, replacement would require years of further development and industrialization. AT&T, however, was under an anti-trust suit at the time and the military watched the development of transistors with great interest. Bell therefore organized a first information meeting at the beginning of 1951 for military and government officials only, then a symposium in September for some three hundred American and European engineers to whom the characteristics of a dozen transistors were disclosed. In 1952, Bell decided to sell its patents to 36 companies and, in April, to divulge the know-how to licensed companies. A first production unit for military transistors was built by Western Electric, the manufacturing branch of AT&T. The anti-trust suit ended in 1956 and, among other clauses, AT&T was ordered to limit its production to its own needs and to the government market, for which many Bell innovations were made; this favored other manufacturers. The Army Signal Corps had already issued production contracts to twelve makers for use in the forthcoming strategic missiles, and demanded 3,000 units of thirty different types per month. Since at that time only 5 to 15% of the production was free of defects, this required much higher production capacities, with very high unit costs. But the rate of rejects, and hence prices, soon dropped, and sales to less demanding buyers went from 14 million in 1956 to 28 million in 1958. The military were interested in transistors because they were small and light, consumed very little power, and were much less sensitive to shocks, vibrations and wear than valves. First models of transistorized computers were built at Bell Labs and Lincoln Lab (MIT) in the 1950s, for the military, of course.

The first civilian commercial uses of transistors were for hearing aids (Raytheon, 1954) costing 150-200 dollars; transistor portable radios came a few years later. It took at least ten years before a large commercial market developed because classical valves were far cheaper – one dollar instead of eight around 1953 -, had much better characteristics than early transistors, were much easier to make, and were much more familiar to most electronics engineers; the main advantages of transistors were not needed in most applications, though they attracted the military. Between 1954 and 1956, the markets for transistors and valves were $55 and over 1000 million respectively. And though several established valve manufacturers (General Electric, RCA, etc.) had 31% of the market in 1957, new and much smaller firms (Texas

Instruments, formerly a geophysical services company, Transistron, Hughes, etc.) had 64%.

Integrated circuits were invented in 1958 by Texas Instruments without military funding (military projects for miniaturizing electronic circuits all failed or came too late in the 1950s), but their mass production was made possible by the invention of the so-called planar process for silicon transistors by a group of eight physicists and engineers who left a company the insufferable Shockley had founded in 1954. The Fairchild Company which, since the 1920s, made aerial cameras and later components of analog computers (all mostly for the military), set up for them the Fairchild Semiconductor Corporation in 1957. Since they had their eye on the commercial market – some of them founded Intel a few years later -, they rejected military R&D contracts to remain free of having to develop products which, although militarily important, would be of little commercial interest. They nevertheless decided to concentrate first on the improvement and manufacture of high performance silicon transistors for the military market. This was the time the military was beginning to replace analog computers with digital ones in avionics and missiles because only silicon – and neither very expensive germanium, nor electronic valves – could stand the high temperatures, shocks and vibrations prevalent in many military systems. Their first customer was IBM which bought one hundred Fairchild "mesa" transistors at 150 dollars a piece for use in the navigational computer for the prototype of the B-70 supersonic bomber they had already made the analog computers for the B-52s, a much bigger market). They had no competitor other than Bell Labs, their mesa transistors immediately found many other avionics uses, and their sales jumped from 65,000 dollars in September 1958 to 2.8 MD for the first eight months of 1959. Their most important customer was Autonetics, in charge of developing the digital computer guidance system for the Minuteman missile. Other early uses included an air-to-air missile, a torpedo, and the Apollo space station. Problems of reliability led to the "planar process" to make much better transistors; the rate of defect-free components was 5% at first, but they were under such pressure from Autonetics, which demanded one year without failure, not to mention the now growing competition in mesa transistors, that they persisted, then developed ultra-reliable planar diodes for computers and eventually integrated circuits. The planar process made it possible to fabricate many components on the same silicon wafer and to connect them, again with a very low initial proportion of defect-free circuits. All of this looks very simple, but required extraordinary standards of cleanliness, manufacturing skills, and *an unprecedented level of discipline on the workforce* , as one of my sources said.

Total sales of ICs amounted to 4 MD in 1962, 41 in 1964, 148 in 1966 and 312 in 1968, while the average unit price dropped from 50 dollars to 2.33; in those same years the military bought 100%, 85%, 53% and 37% of the total sales. More generally, the military part of the electronics industry's total sales, which was 24% in 1950, climbed to 53-60% during the years 1952-1968. The general pattern in electronics at the time was that the first customers, namely the military and their industrial contractors, bought the initial product at prices which included most of the R&D and at least part of the tooling; prices then went down to a level which civilian industry and business could afford for their own uses, which in time lowered the prices again until the general public could buy solid-state gadgets like radios, TV sets or PCs. With a huge civilian market after 1980, chip makers like Intel could continue to improve their products with little help from the military; Intel even refused to work on highly sophisticated very high speed circuits (VHSIC) with no civilian uses.

The military actually benefited from this civilian market as they too needed a lot of standard electronics that could be purchased off-the-shelf at low prices. For this reason and to help American industry against Japanese competition, they became interested in "dual" technologies with military and civilian uses. The DoD still spends about 25% of its R&D budget on electronics and communications, but for more sophisticated products than personal computers...

The early development of computers was still more influenced by the military. Explaining it here would take too much space; see the Internet file. I'll merely point out that the 35 computers made between 1945 and 1955 were entirely financed by the DoD, with the exception of two in universities which my source does not know, and of the von Neumann Princeton computer which was financed by the Army, Navy, AEC and RCA (but its five copies were financed by AEC or, at Rand, by the Air Force). Almost all of these machines were one of a kind; only three companies made several production units: UNIVAC, the company Eckert-Mauchly had founded in 1947 in order to make huge data-processing machines with the commercial market (banking, insurance, etc.) in mind, although it also had military customers; ERA, founded by a team of former cryptologists from the Navy who made very advanced computers for the National Security Agency; and IBM which, at the start of the Korean War, decided to make digital machines. They looked for customers and found seventeen, either military or in the military industry. Of course a huge civilian market developed later – mainly after 1960 -, but the influence of military research contracts and procurement always was extremely powerful, and still is.,

References: Herman H. Goldstine, *The Computer: From Pascal to von Neumann* (Princeton UP, 1972), Kenneth Flamm, *Targeting the Computer* (1987, Brookings Inst. Press) and *Creating the Computer: Government, Industry, and High Technology* (1988, Brookings), Arthur L. Norberg and Judy E. O'Neill, *Transforming Computer Technology. Information Processing for the Pentagon, 1962-1986* (1996, Johns Hopkins UP), Donald MacKenzie, *Knowing Machines* (MIT Press, 1998), Janet Abbate, *Inventing the Internet*

(1999, MIT Press). National Academy of Sciences, *Funding a Revolution. Government Support for Computing Research* (NAS Press, 1999), very explicit and thankful to the DoD, Alex Rolland & Philip Shiman,*Strategic Computing: DARPA and the Quest for Machine Intelligence* (MIT Press, 2002).

Below industry level, all domains of science, from mathematics and computer science to nuclear physics, electronics, optronics,..., oceanography, geology (used e.g. for monitoring underground nuclear tests) and even to some extent biology and medicine, expanded tremendously since much of their results and many experts were needed in all domains of high technology and defense.

# § 3. Applied mathematics in America

In the entertaining chapter of his autobiography, *Un mathématicien aux prises avec le siècle* (Paris, Odile Jacob, 1997, trad. Birkhaser), which he devotes to his teaching at the Ecole polytechnique, Laurent Schwartz accuses (p. 355) the French *pure mathematicians* , and especially the Bourbaki group, of having *ostracized* their applied colleagues. As a matter of fact, for at least ten years there was nearly nobody to be "ostracized" before the rise of Jacques-Louis Lions (1928-2001), a very bright student of Schwartz who first worked on distributions and partial differential equations (PDEs) in the modern way made possible by the development of functional analysis. He discovered applied mathematics and computers in America in 1956 in circumstances that will be explained below, and later founded the very brilliant French School of Applied Mathematics; he himself was appointed a professor at Nancy in 1954, in Paris in 1963, at the Polytechnique (1965-1986), and at the Collège de France in 1973.

From 1980 to 1984, he headed the French government National Institute for Research in Informatics and Automatics (INRIA) with which he had been connected for ten years, the French NASA (CNES) from 1984 to 1992, and he won some of the highest international prizes; quite a victim of our ostracism, and otherwise a great mathematician with some 50 doctoral students and hundreds of "descendants" in the world. See a substantial biography by Roger Temam, one of his principal students, at www.siam.org/siamnews/07-01/lions.htm.

Schwartz decrees that *every mathematician must concern himself with the applications of what he is doing* without, it seems, being aware of the fact that "to concern oneself with" may have quite a number of different meanings, whether in French or in English. He provides neither a justification for his categorical imperative nor the slightest account of the very diverse applications of mathematics. The fact that applied mathematics *were undergoing a powerful expansion in the United States and USSR among others* seems to justify everything, without it being necessary to explain this strange and very new development in the two countries which led the arms race until 1990. The development of applied mathematics in the USA which so inspired Schwartz is not too difficult to explain, even though much remains to be done since physics and technology, being far more spectacular, have almost monopolized historians until now. The Soviet situation, although less well known, was certainly no better.

Before the war, "pure" mathematics prevailed in universities everywhere (except in the USSR, since this "bourgeois" concept was anathema to Marxism); engineers and physicists almost always solved their mathematical problems by themselves, even when the new quantum mechanics obliged physicists everywhere to rediscover strange mathematics. By the 1930s, the situation began to change in a few places, partly due to the arrival of European Jewish refugees. Richard Courant, Kurt Friedrichs, Fritz John and Hans Lewy brought to New York university some of the Göttingen tradition founded by Felix Klein forty years before. They dealt less with applied mathematics as we know them – computers had yet to come – than with often "modern" mathematics such as found in Courant and Hilbert's celebrated *Methoden der Mathematischen Physik*. In 1937, the Army Ballistic Research Laboratory at Aberdeen set up a scientific committee including von Neumann and von Kŕmń besides other luminaries. Von Kármán, formerly a student and later a competitor of Ludwig Prandtl, the foremost German aerodynamicist in Göttingen, had been at CalTech since 1934 (and part-time since 1926), where he founded the future Jet Propulsion Laboratory. In 1945 he became the Air Force's main scientific advisor and, in this capacity, one of the first promoters of atomic missiles. Classical Calculus being often sufficient, the WW II military R&D organization did not at first enlist mathematicians. Mainly at the request of mathematicians themselves, an Applied Mathematics Panel was set up in 1942 with teams in several universities put at everyone's disposal; they were, so to speak, the coalers of the R&D Dreadnoughts of which the officers were physicists. Stanislas Ulam, who later became chief mathematician at Los Alamos, had to ask his friend von Neumann for his help in getting war work in 1943. Applied (or, as Saunders McLane said, applicable) mathematics, much of it boring, blossomed in all kinds of fields, and some people converted to it for life. Shock waves propagation, surface waves in water of variable depth, "hydrodynamics computations" for the Nagasaki bomb, gas dynamics, statistical optimization of air bombings and anti-aircraft defense, operational research, statistical quality control for the mass production of weapons, etc. For anti-aircraft defense, Norbert Wiener invented statistical prediction methods based on harmonic analysis and analytical functions, but they were too sophisticated: he had been lured into the mathematics of the problem. Transmitting orders or conversations in a secure way, that is to say unintelligible to non-authorized people, was very difficult, particularly communications between such high level persons as Roosevelt,

Churchill or Eisenhower. This was intensively studied at Bell Labs, where digitalization of continuous speech was apparently invented, while separate frequency bands were encoded by adding random numbers and reduction modulo 6 (it took quite a while for Bell's engineers to discover it, although they were familiar with mod 2 arithmetic); each encoding system was used only once, and recorded on two highly precise phonograph records, one of which was used at the sending end and the other sent in advance to the receiving end; this involved a lot of very complex electronics using kilowatts of power to transmit milliwatts of speech, and the help of some people with mathematical abilities which the electronics engineers lacked. One of them was Claude Shannon, until 1941 at MIT and Princeton where he had studied applications of "Boolean algebra", i.e. set theory, to the analysis of electronic circuits; he derived from his work at Bell Labs the Information Theory that made him famous after the war. If you understand electrical engineering, see *A History of Engineering and Science in the Bell System. National Service in War and Peace (1925-1975)* (Bell Telephone Laboratories, 1978), pp. 291-316.

Most postwar standard mathematical publications, written by mathematicians who are too busy or too discreet to consult sources, contain only rather abstract and summary generalities about the relevant mathematics. But luck may help those who read books that mathematicians generally do not open, or know of, since they don't deal with mathematics.

The 1945 bombings on Japanese cities (and earlier ones on Germany) led to a fascinating problem: to determine the right proportion of explosive and incendiary bombs for maximum damage. A Berkeley statistician, Jerzy Neymann, was then called to help; he used methods which, after the war, made him a celebrity. Mathematical details are not to be found in my source, and it is likely that Neymann's contribution was less useful than those of scientists, led by Harvard chemist Louis Fieser, who in 1942 invented napalm, among other incendiaries, though it was not widely used until the war in Korea. During a bombing raid, planes were supposed to drop bomb clusters at 50-foot intervals, which would open at 2,000 feet and disperse 38 smaller bombs, starting a dozen fires; thus a B-29 was able to set fire to a 350x2,000-foot area. Relying on statistical computations to get the best results would thus have been a good idea (or a bad one, depending on your point of view). But recent books suggest that the method was discovered experimentally.

On the other hand, the task of choosing targets, based at first on their contributions to Japanese armaments, and of evaluating the weight of bombs needed, was conferred on a Committee of Operations Analysts which relied on methods developed in Britain, mostly by physicists like P.M.S. Blackett, initially for anti-submarine warfare, then for bombing operations. These problems involved fairly simple mathematics but gave rise soon after the war (first of all at the Rand Corporation) to an extravagant amount of hype in favor of game theory, Operations Research, and linear or dynamic programming; it was claimed they were the truly "modern" mathematics that could be applied to "solve the problems of society" – logistics, bombers basing, optimizing a massive nuclear strike in case of war, dispatching packs of Coca-Cola to troops in the field or grocery stores, etc. No wonder these disciplines, which were still rather primitive mathematics assisted by the first computers, did not attract everyone after the war even if they found harmless applications later:

What are we to think of a civilization which has not been able to talk about killing almost everybody, except in prudential and game-theoretical terms,

a good question Oppenheimer asked on TV in February 1950 or perhaps in 1959 – my sources do not agree.

In the atomic sector, where the most difficult problems were to be found, the development of the implosion bomb (Nagasaki, plutonium) forced theoreticians, headed by Hans Bethe, to solve numerically the PDEs governing the propagation of the convergent shock wave produced by classical explosives surrounding a sub-critical ball of plutonium. At hundreds of thousands of bars of pressure, plutonium behaves like a viscous fluid which you have to keep perfectly spherical, whence a "hydrodynamics" problem as they called it. To get the needed spherical shock wave required an assembly of 32 pentagonal pyramids of fast explosives, with a half-sphere ("lens") of slow explosives in the middle of each one. Ready in the Spring of 1945 after thousands of tests, this device required solving countless problems by American and British experts in explosives, many of them academics. Von Neumann contributed significantly to this effort in recommending that much larger amounts of conventional explosives be used than was projected, as well as in the design of the explosive lenses; after having learned chemical engineering at Zürich Polytechnicum in his youth, he had participated at Aberdeen in the development of "shaped charges" for anti-tank projectiles. Hans Bethe, a nuclear physicist who knew a lot of mathematics, wrote a 500- page report on shock waves at Los Alamos.

To solve the two-dimensional PDE (three-dimensional computations were beyond them until the 1980s), they first used the same classical finite difference method as for one-dimensional problems. It turned out that small variations in the dimensions of time and space steps led to large variations in the results: instability. Richard Courant was then called to the rescue. He explained to Bethe the successive approximations method that Friedrichs, Lewy and himself had used (Math. Annalen, 1928) to prove the existence of solutions: it prescribes non-obvious restrictions on the relative dimensions of the time and space steps used. It is at Los Alamos, it seems, that the first opportunities to use the method arose. Thanks to that,

very soon problems involving fluid dynamics, neutron diffusion and
transport, radiation flow, thermonuclear reactions and the like were
being solved on various machines all over the United States

writes Bethe's first successor as chief of theoretical physics at Los Alamos, D.
Richtmyer, in a 1957 book explaining, among other things, advances made
after the war by von Neumann and Peter Lax concerning the convergence and
stability of approximations; Banach spaces could now be used indirectly to
understand what went on inside a bomb, for obviously this is what everybody
was interested in at the time in Los Alamos. Lax, who spent his summers at
Los Alamos during the 1950s, was one of applied mathematics' rising stars
and, later, a strong opponent of Bourbaki's mathematics. He once wrote of
Vietnam war opponents who wanted to enlist the AMS that most of them
*specialize in branches of mathematics that are abstract, often esoteric, and
completely unmotivated by problems of the real world*, thus implying that, had
they instead busied themselves with, say, the mathematics of shock waves,
they would have had no qualms over B-52s flattening Laos...

J-L. Lions, mentioned above, said much later in an interview (Le Monde,
May 8, 1991) that he discovered applied mathematics and computers in
America in 1956 thanks to Lax, who told him of von Neumann's ideas; after
mentioning a few current civilian applications, Lions treats us to an eulogy
of von Neumann,

the father of the discipline who, at the end of the 1940s, was so
able to guess all the benefits that would result from the use of the
first computers to describe such complex systems as meteorological
phenomena,

and that he himself only *added one chapter which von Neumann had not
entered: the industrial chapter* (with enough success to be a member of the
board of several big French industrial companies during his last years). Von
Neumann's (and the Air Force's) interest in meteorology is well known but,
as the reader already knows, he was interested in other uses of computers.
By 1956,

[his] combination of scientific ability and practicality gave him a cred-
ibility with military officers, engineers, industrialists, and scientists
that no one else could match. He was the clearly dominant figure in
nuclear missilery.

This other eulogy is from Herbert York *Race to Oblivion* (Simon & Schuster, 1970, p. 85); the was a member of the *Teapot Committee* which, chaired
by von Neumann, chose in 1954-1955 the characteristics of ATLAS, the first
intercontinental missile. Lions may not have been told in 1956 of von Neumann's taste for military projects, but in 1960, the year he started a seminar
on numerical analysis in Paris, his first "really applied" paper was on nuclear reactors. That he did not even hint in a 1991 interview at the huge

military influence on the development of his discipline may be explained by the Russian principle: *show the best, hide the rest* . One of his best students, Roland Glowinski, tells us on the web that the A (for Automatics, i.e. Control) of the IRIA Institute of Research in Informatics and Automatics that Lions headed had been suggested by Pierre Faurre. A bright Polytechni cien well known among applied mathematicians, Faurre published a book on the mathematics of inertial guidance (1971) in a collection directed by Lions. In America, this technique made Charles Stark Draper and his Instrumentation Laboratory famous (it was the focus of student riots at MIT in 1969) and was developed first for strategic bombers, later missiles, and still later commercial planes; Faurre soon became the general secretary of SAGEM, a well-known company he eventually headed and which was making (among other things, e.g. telecommunication hardware and fire-control systems) inertial guidance systems for planes and missiles, whether civilian or military. One should not forget the multi-volume and multi-author treatise of *Analyse mathématique et calcul numérique pour les sciences et les techniques* (English trad. Springer) which Lions edited together with Robert Dautray, a Polytechnicien who, from 1955 to 1998, followed a bright career at the French AEC (CEA) up to the highest position. Dautray was appointed scientific director of its Military Applications Division (DAM) in 1967 in order to help its engineers extricate themselves from the complexities of H-bomb design; it seems he did this by asking questions to a well-known British expert who told him they had found, but not recognized, the solution. To be sure, none of these connections proves that Lions did actual military work, and it may well be that he was mainly interested in applications to astronautics, meteorology, the environment, industrial processes, etc. Let me say simply that I have read too many biographies by scientists to trust them automatically to tell the whole truth.

Richtmyer mentions "machines". At Los Alamos in 1943, numerical computations were first carried out on mechanical desk computers – distant descendants of Pascal's and Leibniz's machines -, as everywhere else. The enormity of the task led physicists to order commercial IBM punch card machines, improved to perform multiplications (!) and not merely additions. For months, Richard Feynmann headed dozens of (human) computers who had to push millions of punch cards into the machines.

Von Neumann devoted two weeks to learning how to use them, which explains the shock that was his chance discovery, in 1944, of the Eckert-Mauchly team who, at the University of Pennsylvania, were designing the first electronic computing machine, ENIAC, to help the Aberdeen Proving Ground accelerate its firing-tables business; though not yet automatically programmable, ENIAC was far faster than IBM's primitive machines were; it was not fully operational before the Fall of 1945 and was at once used for the H-bomb program, as was von Neumann's own machine when operational in 1952. Drawing in part on Eckert-Mauchly's ideas, von Neumann formalized in 1945-1946 what is now called the "von Neumann architecture", thus creating true computers, and (slowly) built one at Princeton; Maurice Wilkes built one in Britain in 1948, Eckert-Mauchly delivered their first commercial UNIVAC in 1950, while another small company, ERA, delivered very advanced machines for cryptological work to the National Security Agency (NSA) also before 1950, as already said, all on von Neumann's architecture. The Los Alamos and Livermore laboratories were first served with almost all the new "scientific" computers available, from copies of von Neumann's machine to the present teraflop supercomputers, of which they were always the most demanding users and often the promoters.

And while we are celebrating WW II applied mathematics in the United States, we might as well inquire about a country that is so often "forgotten" by most apostles of applied mathematics: Germany, which in some scientific and technical domains was well ahead of her enemies. At Göttingen, Prandtl's lab had been transformed during WW I into an Aerodynamischen Versuchanstalt (AVA) which, in 1925, became associated with the newly founded and more theory- oriented Kaiser-Wilhelm Institut (KWI) fr Strömungsforschung. The arrival of the Nazis opened the way to the new Luftwaffe, which was good for aerodynamics, and AVA expanded. Prandtl, who was much more an innocent than a Nazi, congratulated them publicly for it while trying, without success, to protect valuable scientists who were not 100% Jewish. Now running under the Luftwaffe ministry and almost entirely devoted to the needs of the aeronautical industry, AVA was separated from the KWI in 1937. Work at KWI, under Prandtl, while more "fundamental" than at AVA, was nevertheless increasingly devoted to studies for the Luftwaffe (high speed aerodynamics), or von Braun (supersonic aerodynamics), or the Navy (cavitation studies for fast torpedoes), as well as for meteorology. A young mathematician, Harry Görtler, took charge of numerical computations and devised simple ways of programming them for KWI's biological "computers", young girls with a high school degree and desk machines.

Outside fluid mechanics and ballistics, military research did not really start before 1942, when the *Blitzkrieg* myth was dispelled; as in 1914, most scientists had been mobilized like everyone else in 1939. Furthermore, Nazi Germany, a conglomerate of administrative feodalities fighting each other for power, lacked the centralized coordination of R&D that America set up even before Pearl Harbor. Most Nazi leaders, Hitler to begin with, could hardly understand the importance of revolutionary weapons, except for their psychological impact. The development of jet fighters was delayed by two years (fortunately for Allied bombers) and von Braun's V-2s production, though not development, longer still. In 1943, they changed their mind and tried to develop "miracle weapons" in earnest; engineers had plenty of these on their drawing boards, but it was too late for most of them.

Student numbers enrolling in aerodynamics and the like grew from a mere 80 in 1933 to reach 700 by 1939, while the Nazi policy had the opposite effect on mathematics – student enrolment fell by 90% at Göttingen – and physics, not only as a result of dismissing Jewish scientists, but also because the official ideology favored more virile prospects. In physics, mentioning "Jewish" Relativity theory was anathema, but most atomic physicists were not foolish enough to fall into this trap. There was also a "Deutsche Mathematik" gang trying to discredit some parts of mathematics and the mathematicians connected with it. Jewish- made transfinite numbers were fortunately not really needed to compute rocket trajectories.

Often at their own request mathematicians were eventually mobilized for military research. In Germany as in Allied countries, it was thus possible to protect scientists from the *chances of a Turkish bullet* , a fate which had so incensed Ernest Rutherford when one of British physics' rising stars, Philip Moseley, was killed in the Dardanelles in 1915 – a fate that should obviously be reserved for scientifically uneducated people. Some mathematical work remained rather theoretical, like Wilhelm Magnus' first version of the Magnus and Oberhettinger book on special functions, Erich Kamke's on differential equations, or Lothar Collatz's on eigenvalue calculations. Other studies were more directly applied to supersonic aerodynamics of shells and missiles, wing flutter, pursuit curves for self guided projectiles, cryptology, etc. Some well known "pure" mathematicians, like Helmut Hasse, Helmut Wielandt, Hans Rohrbach, even converted to it temporarily. Alwin Walther, Courant's former assistant, who before the war had founded a Practical Mathematics Institute (IPM) at the Darmstadt Teknische Hochschule, already worked for von Braun in 1939, and IPM became the main computing center for military research during the war. Walther's first task after the war was to direct the writing for the Allies of five reports on mathematics; he pointed out the similarity of German and American areas of work, *miraculously bearing witness to the autonomous life and power of mathematical ideas across all borders* . Courant agreed and invited Walther to emigrate to the US; to this moving reunion – applied mathematicians of all countries, unite! – Walther, now a "pacifist", preferred working for the reconstruction of his country.

In Germany also, a remarkably clever engineer, Konrad Zuse, who had attended Hilbert's lectures in mathematical logic, started in 1936, without any government help and ahead of the Americans, to build three computing machines using telephone relays. The last one, Z3, became operational during the last months of the war and was used to control the shape of mass-produced rocket wings. All these machines were damaged during the war. Components of an electronic machine (which would have used 2,000 tubes instead of ENIAC's 18,000) were built by his friend Wilhelm Schreyer; this aroused even less interest, and Schreyer later emigrated to Brazil to teach. At the end of the war, Zuse went to the Zürich Poly where he built a Z4, much more reliable than the first electronic machines, then enjoyed a successful technical and business career in computers, later at Siemens. He also invented a Plan Kalkül in 1945, i.e. a logical architecture for computers; but he was not in a position to compete with von Neumann, if not in software, at any rate in prestige and support.

References: Amy Dahan-Delmedico, *L'essor des mathématiques appliquées aux Etats-Unis: l'impact de la seconde guerre mondiale* (Revue d'histoire des mathématiques, 2 (1996), pp. 149-213) and two papers by the same author and Peter Galison in Amy Dahan et Dominique Pestre, eds, *Les sciences pour la guerre, 1940- 1960* (Paris, EHESS, 2004), the first one dealing in detail with a Soviet team at Gorky. On Germany, see H. Mehrtens, "Mathematics and War: Germany, 1900-1945", in Forman, *National Military Establishments* , Sanford L. Segal, *Mathematicians under the Nazis* (Princeton UP, 2003), Konrad Zuse, *The Computer. My Life* (Springer, 1993).

Going back to America, a long report on applied mathematics stated in 1956:

Let it also be said at the outset that, with very few exceptions, their organization does not antedate World War II and their continued existence is due to the intervention of the Federal Government. *Without the demands resulting from considerations of national security, applied mathematics in this country might be as dead as a door nail*
.

According to the report, government administrations – i.e., in those times, military de jure or, like AEC or NACA, de facto – and connected industries were practically alone in employing professional applied mathematicians. A 1962 report claimed that in 1960, out of 9,249 "professional mathemati cians" employed in government or industry, about 2,000 were in federal military centers, 1,000 at the AEC, while aeronautics and electronics employed 1,961 and 1,226 respectively in the private sector. These two fields consistently got about 60% and 25% of the federal R&D money going to industry.

In 1968, another report – this one about mathematics in general – recommended that the so-called mission-oriented agencies, namely Defense, AEC, NASA and NIH in that order, should continue to fund research in those domains most useful to their missions, and to propose their problems to the mathematical community. This report was edited during the Vietnam War by Lipman Bers, one of the main opponents to the war among mathematicians. He explained in the 1976 *Notices* of the AMS that he had agreed to do it only after being assured that the war would end before the report's publication; it ended five years later. A 1970 report finds 876 mathematicians (166 with PhDs) at AT&T, 170 at Boeing, 239 at McDonnellDouglas, 147 at Raytheon, 68 at Sperry Rand, 287 at TRW, 137 at Westinghouse, etc. All of these high-tech companies had large military markets.

In 1971, the DoD employed 81% of all mathematicians and statisticians employed by the government, 67% of all engineers, 41% of all physicists (but there was also the AEC), and 10% of all biologists and physicians. Serious work needing e.g. harmonic analysis, stochastic processes, information theory, differential equations and PDEs, etc., was performed most of the time via university contracts. This is where historians should look to get a more precise idea of the importance of "higher" mathematics in military or industrial applications, a huge program.

Applied mathematics and numerical analysis have many civilian applications nowadays, but their degree of militarization always remained very high in the USA if we are to judge from the amount of federal funds attributed to them. The same is true a fortiori for what is now called computer science or informatics (logical architecture of machines, programming, networks structure, etc., hardware excluded). Here is a simplified table, taken from NSF statistics, on the main sources of federal funds (in current MD) for basic and applied research (no development) in mathematics and computer science attributed to all public or private organizations concerned with these fields: Since one 1958 dollar is worth about six 2001 dollars, this means that our

|         | 1958 | 1964 | 1968 | 1974 | 1980 | 1987 | 1994  | 2001  |
|---------|------|------|------|------|------|------|-------|-------|
| Total   | 40.4 | 98   | 119  | 127  | 241  | 759  | 1,242 | 2,810 |
| DoD     | 36.4 | 69   | 79   | 70   | 137  | 453  | 593   | 947   |
| NSF     | 1.4  | 11.4 | 18.6 | 24   | 53   | 124  | 238   | 569   |
| NASA    | 0    | 6.3  | 3.7  | 1.9  | 3.7  | 70   | 26    | 85    |
| AEC/DoE | 1.9  | 5.1  | 5.8  | 5.6  | 11.6 | 38   | 201.8 | 824   |

field got about twelve times as much money in 2001 as in 1958, while between 1945 and 1950, it got about two million per year from ONR, a large part of it going to the Whirlwind computer. Here too the change of scale is stunning. The more recent increase in DoE funding is largely due to the development of 3D simulation methods for nuclear weapons, as well as to controlled fusion experiments designed to check the computations: 751 MD were allocated to it in 2004. The DoD was planning in 1998 to spend some 2.5 BD over several years on simulation and modelization.

Separating mathematics and computer science yields interesting results. The funding of informatics was still comparatively low in 1958; in 1980, out of a total 241 MD, computer science got 128 MD and mathematics 90, the remainder being a mixture of both. In 2001, mathematics got 396 MD and computer science 2,022. The difference is, of course, still more striking in applied research, for which maths (resp. computer science) got 23.8 (resp. 82) MD in 1980, then 95 (resp. 566) in 1994, then 105 (resp. 1,438) in 2001. The same year, Defense ARPA's funding was 8.7 MD for mathematics and 424 for informatics. All of these figures are from the NSF statistical series. A striking feature of this growth since the 1970s is the fact that basic research in computer science has been increasingly financed by the NSF and decreasingly by DoD, in part a consequence of Mansfield's amendment (1970) prohibiting DoD from funding research without explicit military relevance, in part the result of an increasing number of relatively small standard contracts with many new computer science departments which were then on the rise, while ARPA limited its grants to a few "centers of excellence". It was also due to the financing of specialized and costly equipment in universities, e.g. supercomputing centers connected to other places. The end result was that in 2001, the NSF spent 119 MD for basic research in mathematics and 450 in computer science.

Obviously, not all funding goes to universi ties. The following table gives some idea of recent trends in the federal funding of research in universities (in current MD).

|               | 1976 | 1984 | 1992 | 1999 |
|---------------|------|------|------|------|
| Total         | 57   | 182  | 478  | 662  |
| Mathematics   | 30   | 76   | 150  | 131  |
| Computer Sci. | 26   | 74   | 320  | 506  |

These data concern basic and applied research and represent a large part of the total, which also includes a small portion involving both sectors. For instance, in 1994, according to another NSF report which does not quite agree with the above data, the federal government attributed 196 MD to mathematics and 453 to computer science, while total expenses – funds specifically attributed to research by all sources – were 278 and 659 MD; this means that federal funds accounted for about two thirds of university research support in mathematics and computer science, the remainder being universities' own funds and, presumably, industrial contracts at least in computer science. In 2000, out of the total federal funding of university research in mathematics (resp. computer science) of 211.5 (resp. 568) MD, these fields got 29.5 (resp. 209.8) from DoD, 8.9 (resp. 6.1) from DoE, 75.2 (resp. 0.5) from NIH (as against at most 12 MD before 2000), 0.7 (resp. 18.3) from NASA, and 99.6 (resp. 336.6) from NSF. This is no longer the 1958 situation, when nearly all federal funds were military, and over 80% of military funding now goes to computer science.

These statistics, mainly for the early years, do not accurately reflect the importance of activities specifically devoted to direct military work. Before the 1960s, when NSF hardly existed, military contracts went to many people who specialized in "useless" and "abstract" maths. These contracts allowed the universities to recruit more people, to help graduate and post-graduate students, to invite foreign colleagues, including perhaps the present author, and, last but not least, to secure America's *preponderance of power* in mathematics as in everything else. However, it is not the bystander's duty to prove that a military contract commits its beneficiary; it is up to the beneficiary who disputes it to prove that it does not.

And how are we to explain that the life sciences sector, on the other hand, never benefited from proportionally equivalent DoD favors? In 1968, federal funding of life sciences totalled 1,534 MD, of which 105 came from the DoD;

in 1994, 9.3 BD, of which 265 MD; and in 2001, 23.057 billion in federal money, of which 1.052 billion from the DoD. Life sciences have been financed for fifty years essentially by the NIH (and, to a much lesser extent, by the NSF), and very strongly encouraged by Congress and the voters. As for the drug industry, it devotes billions to R&D without ever having received more than a few percentage points from the federal government, less than 4% in 1993 for instance. In 2001, the industry spent a total of 12.2 BD, and since it belongs to the chemical industry sector, and the NSF tells us elsewhere that it got 150 MD in federal funds, an upper limit of 1.4% in federal funding for the drug industry follows. To be sure, drug companies indirectly benefit from their university contracts, but their main source of R&D money is obviously the countless products which are sold around the world to all who can afford them.

After students rioted against the Vietnam War and military work in universities, a Congressional Mansfield's amendment forbade the DoD from financing research without a clear military interest, as already said. It was somewhat softened later, but its spirit remained, and military support of "pure" mathematics nearly vanished, except in cryptology. The main threat to "pure" mathematics now comes from the enormous development of applied mathematics, even though their applications may be mostly civilian. As we shall see in the next section, this is the most striking difference between post-WW II applied mathematics and Jacobi's mathematics *pour l'honneur de l'esprit humain* (or for mathematicians' entertainment...) which, to a very large extent, were preponderant from the 1820s to the eve of WW II.

# Index

# Table of Contents of Volume I

# Universitext